MW00331784

Introduction to Optical Microscopy

Second Edition

Introduction to Optical Microscopy

Second Edition

JEROME MERTZ

Boston University

CAMBRIDGE
UNIVERSITY PRESS

CAMBRIDGE
UNIVERSITY PRESS

University Printing House, Cambridge CB2 8BS, United Kingdom

One Liberty Plaza, 20th Floor, New York, NY 10006, USA

477 Williamstown Road, Port Melbourne, VIC 3207, Australia

314–321, 3rd Floor, Plot 3, Splendor Forum, Jasola District Centre, New Delhi – 110025, India

79 Anson Road, #06–04/06, Singapore 079906

Cambridge University Press is part of the University of Cambridge.

It furthers the University's mission by disseminating knowledge in the pursuit of
education, learning, and research at the highest international levels of excellence.

www.cambridge.org
Information on this title: www.cambridge.org/Mertz
DOI: 10.1017/9781108552660

First published 2019
Reprinted 2019

Printed in the United Kingdom by TJ International Ltd., Padstow, Cornwall

A catalogue record for this publication is available from the British Library.

Library of Congress Cataloging-in-Publication Data
Names: Mertz, Jerome, author.
Title: Introduction to optical microscopy / Jerome Mertz (Boston University).
Description: Second edition. | Cambridge, United Kingdom ; New York, NY :
    Cambridge University Press, 2019.
Identifiers: LCCN 2019007596 | ISBN 9781108428309 (hardback : alk. paper) |
    ISBN 1108428309 (hardback : alk. paper)
Subjects: LCSH: Microscopy.
Classification: LCC QH205.2 .M47 2019 | DDC 570.28/2–dc23
    LC record available at https://lccn.loc.gov/2019007596

ISBN 978-1-108-42830-9 Hardback

Additional resources for this publication at https://www.cambridge.org/Mertz

To my wonderful daughters
Adele and Lorine

CONTENTS

PREFACE TO SECOND EDITION

The first edition of this book was published in 2009. Since then, much has happened, the most significant of which, as far as this book is concerned, is that the original publisher ceased to exist and the book went out of print. As anyone who has written a book can well imagine, the prospect of seeing all that work go to waste was a source of concern. My vague and ephemeral notions of writing a second edition at some time in the distant future suddenly took on a sense of urgency. I am grateful to Cambridge University Press, in particular Heather Brolly, for having come to my aid.

The goal of this second edition remains unchanged from the first, namely to provide a self-contained, comprehensive overview of the fundamentals of optical microscopy, targeted at graduate and upper-level undergraduate students, and to researchers in the field in general. This second edition is more rigorous than the first, and its foundations are better solidified. This actually posed a problem. What I originally thought would entail some small revisions and the occasional addition of new material here and there turned into a wholesale revamping of the text from cover to cover, in order to maintain a relatively uniform level of rigor. Indeed, only the skeleton of the first edition remains, along with two or three chapters that were left more or less unscathed. The rest is entirely rewritten.

Some key differences with the first edition are that imaging is treated more in three-dimensional space than it was before, making more allowances for sample defocus and volumetric imaging. But as a result, to avoid the proliferation of unwieldy convolution-type integrals, more of the math is performed in frequency space than in real space. While such math is undeniably simpler, it may be less intuitive to some readers, at least initially.

Other differences are that I have added a fair amount of new material. For example, Chapters 19 and 20 are completely new. The first deals with various techniques, arbitrarily called pump-probe techniques, including incoherent and coherent Raman, Brillouin, photothermal, and photoacoustic microscopies. The second deals with imaging in scattering media, which, to my mind, remains the greatest challenge in microscopy to date, and something that I have always regretted not having included in the first edition. Of course, only a single chapter on the subject can hardly do it justice. Other sundry topics, such as extended depth of field, light-sheet microscopy, image scanning microscopy, programmable array microscopy, temporal focusing, etc., have also been added.

Yet another difference with the first edition is in my use of notation. This was a recurring complaint I received regarding the first edition. I therefore made efforts to better align my notation with convention, while still trying to minimize the overloading of variables. For example, what was previously referred to as the coherent spread function (CSF) and coherent transfer function (CTF), are now referred to as the amplitude point spread function (H) and the coherent transfer function ($\mathcal{H}$). The free-space propagator, previously denoted by H, is now denoted by D. Radiative power, previously denoted by W, is now denoted by Φ, etc. Other notational realignments are interspersed throughout.

Finally, a resource web page is now provided at www.cambridge.org/Mertz, where the solution sets for each chapter are available (for instructors only) along with the book illustrations (in pdf format).

It should be emphasized that this book is about the physical aspects of optical microscopy. What is missing, glaringly, is the computational aspects. I hesitated to include an additional chapter on computational microscopy, and in the end decided against it because this topic strikes me as too rapidly evolving to be properly encapsulated. Perhaps if there is a third edition of this book (in the distant future), then this topic will be included, as it seems an inevitable trend.

I would like to thank three people in particular. The first is Anne Sentenac, who did a sabbatical in my lab for a year. It was Anne who showed me how woefully incomplete was my two-dimensional vision of microscopy, and who made me see the light, as it were, in the form of the Born approximation. The second is Colin Sheppard who, unbeknownst to him, and little by little, revealed microscopy to me through the immense body of his work. There is a running joke in my lab that any new idea must first be checked against Colin's work to make sure it is indeed new. The third is Timothy Weber, one of my graduate students, whom I entrusted with reading this manuscript in its entirety. Were it not for his utmost care and meticulousness, an embarrassing number of errors would have gone unnoticed. Tim made this a better book.

But there is a final person I wish to thank: my wife, Diane Ma. Were it not for her enduring support through thin and thick, and her ability to keep everything together while I was absent at the keyboard, this second edition would not have been possible.

PREFACE TO FIRST EDITION

For anyone who has taught a graduate or upper-level undergraduate course in optical microscopy, the problem often arises of which textbook to use. While there are many excellent textbooks that deal with various applications of optical microscopy, and still more that deal with optics in general, very few textbooks deal specifically with optical microscopy from a physics or engineering standpoint. A notable exception is, or rather was, Pluta's Advanced Light Microscopy series, but unfortunately this has become out of date and is currently out of print. My own solution to the problem has been to heavily rely on course notes; however, over the years, these have become steadily more scattered to the point of becoming unusable. And so this book was conceived largely out of necessity, my goal being to provide a self-contained overview of the foundations and techniques of optical microscopy suitable for teaching at the graduate or advanced undergraduate level.

This book can be divided into two parts. The first part (Chapters 1–8) deals with properties of light, light propagation, and light detection and is meant to provide the groundwork for the remainder of the book. Most of this first part is centered on Fourier optics and statistical optics and is intended as an overview of some of the fundamental concepts in these fields, ranging from imaging transfer functions to partial coherence theory. Much of this material can be found in, and indeed was drawn from, many excellent textbooks, in particular, Goodman's Introduction to Fourier Optics and Statistical Optics. I tailored this material as much as possible to optical microscopy and included chapters on 3D imaging and radiometry for completeness. The first of these chapters provides a synthesis of the generalized imaging functions involved in 3D microscopy. The second discusses concepts of étendue and throughput as applied to microscope layouts. These concepts seem to have fallen by the wayside in many optics textbooks, perhaps because of their limited connection to Fourier optics. Nevertheless, they remain highly useful for facilitating a "first pass" intuitive understanding of optical layouts.

The second part of this book (Chapters 9–18) is specifically dedicated to microscopy techniques, in particular, to contrast mechanisms and to the repertoire of tricks, old and new, that has been devised for revealing contrast. I made no attempt to present these techniques in chronological order or in their historical context. Instead, I chose to group them under common themes to emphasize their underlying relations to one another. Indeed, techniques that might have appeared dissimilar when treated separately turn out in many cases to be implementations of the same basic idea. The material in these chapters was more difficult to distill from the literature and, for this reason, is likely to constitute the main interest of this book.

Chapters 9–12 deal with nonfluorescence imaging techniques, particularly those that revolve around phase contrast and interferometry. Chapter 9 is an introduction to the physical mechanisms that actually generate nonfluorescence contrast, such as scattering and absorption. Chapters 10–12 then present various imaging techniques designed to reveal nonfluorescence contrast. These techniques become progressively more quantitative and, with optical coherence tomography (Chapter 12), move into the domain of bona fide 3D imaging.

Chapters 13–18 deal mostly with 3D fluorescence imaging techniques, beginning with an introduction to the photophysics of fluorescence generation and contrast modalities (Chapter 13), followed by a progression through confocal (Chapter 14) and two-photon (Chapter 15) microscopies. Chapter 16 does not deal with fluorescence per se since it discusses nonlinear microscopy techniques that are coherent. Nevertheless, it seemed natural for this chapter to follow coverage of two-photon microscopy. Finally, Chapters 17 and 18 discuss more recent techniques employing structured illumination and superresolution and reflect various new directions in optical microscopy.

There is too much material in this book to be covered in a single-semester course (I have tried, to the chagrin of my students!). The instructor is therefore encouraged to parse the chapters for the most relevant material. This done, each chapter can be reasonably covered in a single session, save perhaps some of the longer chapters that might be better served in two sessions. Problem sets associated with each chapter can be found at the publisher's Web site www.robertspublishers.com.[1] These will be periodically updated and solution sets will be made available to instructors only.

Finally, a word on notation. I strove to maintain consistent notation and to minimize the overloading of variables throughout this book. This has led me on occasion to stray from convention. In particular, I introduced the variables H and $\mathcal{H}$ to refer to free-space propagators in real and Fourier space, respectively (in general, Roman and calligraphic variables are Fourier transforms of one another). In turn, the variables CSF and CTF refer to the coherent spread function and coherent transfer function of an imaging device (often denoted by h and H in the literature), and the variables PSF and OTF refer to the incoherent point spread function and optical transfer function of an imaging device (often denoted by $|h|^2$ and $\mathcal{H}$). Finally, there seems to be no consistently accepted variable to represent spatial frequencies. I therefore introduced a new variable κ to maintain resemblance with the universally accepted variable k representing angular spatial frequencies (the relation between the two being $k = 2\pi\kappa$).

The completion of this book did not arrive without setbacks and delays. I am especially indebted to Profs. Joseph Goodman and Manuel Joffre for wading through this manuscript from cover to cover as it slowly took form. Their guidance and insightful suggestions have been invaluable. Moreover, perhaps without their awareness, they have served as models to me of scientific and pedagogical excellence. I also owe my gratitude to several colleagues and students, in particular, Sidney Cahn, Rainer Heintzmann, Neil Switz, Kengyeh Chu, and Daryl Lim, who graciously agreed (or were forced) to review various chapters of this work.

Finally, I would like to thank my publisher, Ben Roberts, for his patience and encouragement, Phil Allen for his help in providing images, and my many students who, throughout the years, have made my job as a professor a pleasure.

---

[1] The first edition is now out of print, and the publisher's website does not exist.

1 Introduction

Light can be described as a field $E(\vec{r}, t)$ that varies in space and time. These variations can be decomposed into spatial frequencies $\vec{\kappa}$ and temporal frequencies ν respectively. That is, we write

$$E(\vec{r}, t) = \iint d^3\vec{\kappa}\, d\nu\, \mathcal{E}(\vec{\kappa}, \nu) e^{i2\pi\left(\vec{\kappa}\cdot\vec{r} - \nu t\right)} \tag{1.1}$$

$$\mathcal{E}(\vec{\kappa}, \nu) = \iint d^3\vec{r}\, dt\, E(\vec{r}, t) e^{-i2\pi\left(\vec{\kappa}\cdot\vec{r} - \nu t\right)} \tag{1.2}$$

where throughout this book $\mathcal{E}(\vec{\kappa}, \nu)$ will be called a radiant field and $\vec{\kappa} = (\kappa_x, \kappa_y, \kappa_z)$ is the wavevector associated with the field. Note that a distinction will be made here between a wavevector $\vec{\kappa}$ and an angular wavevector $\vec{k} = 2\pi\vec{\kappa}$, the latter being commonly found in the literature. Because $\vec{\kappa}$ is the Fourier conjugate variable of the position vector $\vec{r} = (x, y, z)$, it can be thought of as associated with momentum.

To each temporal frequency ν is associated a wavenumber κ (as distinct from an angular wavenumber), defined by

$$\kappa = \frac{n}{c}\nu \tag{1.3}$$

where n is the index of refraction of the surrounding medium ($n = 1$ for free space) and c is the speed of light in free space ($c = 3.0 \times 10^8$ m/s). To each wavenumber is also associated a free-space wavelength, defined by

$$\lambda = \frac{n}{\kappa} \tag{1.4}$$

The use of the variables $\vec{\kappa}$ to denote wavevector and κ to denote wavenumber is not accidental. As we will see in Chapter 2, for propagating fields the two are linked by a fundamental law known as the energy–momentum relation, given by

$$\left|\vec{\kappa}\right| = \kappa \tag{1.5}$$

Written in this notation this relationship appears self-evident, but one must bear in mind that the left-hand side represents the integration variables in Eq. 1.1 (related to momentum) while the right-hand side comes from Eq. 1.3 (related to energy). In other words, the wavevector here can have arbitrary direction but its magnitude is constrained to lie on a spherical shell defined by the optical frequency ν. Off-shell components of $\vec{\kappa}$ are possible, but these do not propagate. Since we will only concern ourselves with propagating fields throughout this book, the notation in Eq. 1.5 will be maintained. Equation 1.5 is perhaps the most fundamental relation in imaging theory and stems directly from the wave equation, as will be seen in the next chapter.

1.1 COMPLEX FIELDS

Throughout this book, the field $E(\vec{r}, t)$ will be treated as complex. That is, $E(\vec{r}, t)$ should not be confused with an electric field, since, by definition, an electric field is a physically measurable quantity that must be real. Nevertheless, $E(t)$ will serve as a *representation* of an electric field $E_{\text{elec}}(\vec{r}, t)$, such that

$$E_{\text{elec}}(\vec{r}, t) \propto \text{Re}\left[E(\vec{r}, t)\right] \qquad (1.6)$$

While this relation places a constraint on $\text{Re}\left[E(\vec{r}, t)\right]$, it allows $\text{Im}\left[E(\vec{r}, t)\right]$ to be chosen arbitrarily. By general convention, this arbitrariness is removed when $\text{Im}\left[E(\vec{r}, t)\right]$ is chosen to fulfill a second condition, applied now to the radiant field and given by

$$\mathcal{E}(\vec{\kappa}, \nu) = 0 \qquad \text{when} \qquad \nu < 0 \qquad (1.7)$$

This second condition implies that $E(\vec{r}, t)$ is an analytic function of t. It imposes the constraint that $\text{Re}\left[E(\vec{r}, t)\right]$ and $\text{Im}\left[E(\vec{r}, t)\right]$ are related by what is known as a Hilbert transform (in t), thereby specifying $E(\vec{r}, t)$ completely. While the conditions imposed by Eqs. 1.6 and 1.7 will rarely be mentioned again throughout this book, they should nevertheless always be kept in mind. For more information on analytic functions and Hilbert transforms, the reader can consult [1] or [3].

The purpose of this introductory chapter is to motivate some general concepts in optical imaging theory. In the most basic imaging applications, light from a two-dimensional (2D) plane (called the object plane) is mapped onto another 2D plane (called the image plane) some distance away. The goal of imaging theory is to analyze this mapping process. Because 2D planes are the starting point of imaging theory, our coordinate systems will be tailored accordingly and we write $\vec{r} = (\vec{\rho}, z)$, where $\vec{\rho}$ lies in the 2D plane of interest. We also begin by considering light whose time dependence contains only a single harmonic frequency ν_0. Such light is called monochromatic. The field in a 2D plane is then written in a simplified form as $E(\vec{\rho})$, where the harmonic time dependence is implicit. Correspondingly, the general Fourier transform relations linking a 2D field and 2D radiant field reduce to

$$E(\vec{\rho}) = \int \mathrm{d}^2\vec{\kappa}_\perp \, \mathcal{E}(\vec{\kappa}_\perp) e^{i2\pi \vec{\rho} \cdot \vec{\kappa}_\perp} \qquad (1.8)$$

$$\mathcal{E}(\vec{\kappa}_\perp) = \int \mathrm{d}^2\vec{\rho} \, E(\vec{\rho}) e^{-i2\pi \vec{\kappa}_\perp \cdot \vec{\rho}} \qquad (1.9)$$

where $\vec{\kappa} = (\vec{\kappa}_\perp, \kappa_z)$.

Having established this basic formalism for complex fields, we turn now to some intuitive notions of beam directionality and ray optics.

1.2   INTENSITY AND RADIANCE

Typical optical frequencies are on the order of $\nu_0 \sim 10^{15}$ Hz, which is far too fast to be directly measurable with detectors based on current technology. Standard detectors do not directly measure the light field but rather the light intensity, defined by

$$I(\vec{\rho}) = \langle E(\vec{\rho})E^*(\vec{\rho}) \rangle \tag{1.10}$$

where the brackets $\langle ... \rangle$ denote a temporal average over many temporal oscillations (this time average will be better defined in Chapter 7). Because $I(\vec{\rho})$ is a physically measurable parameter, it must be real. Throughout this book, $I(\vec{\rho})$ will provide a unit of reference. In particular, $I(\vec{\rho})$ has units of power per area, or W/m^2. Correspondingly, $E(\vec{\rho})$, from the relationship defined above, must have units $\sqrt{W}/m$ and $\mathcal{E}(\vec{\kappa}_\perp)$ must have units $\sqrt{W}m$. Again, these units are not those associated with the physically measurable electric field $E_{\text{elec}}(\vec{\rho})$ (units: V/m), and should only be interpreted as convenient units to verify dimensional consistency.

Technically, as defined by Eq. 1.10, $I(\vec{\rho})$ should be referred to as an irradiance rather than an intensity; however, in recent times the latter term seems to have come into favor, particularly amongst experimentalists. This book will yield to the new jargon.

To gain an intuitive picture of light propagation from one 2D plane to another, it is often convenient to think of light as composed of rays that transport power, as depicted in Fig. 1.1. The rigorous connection between the notion of light rays and the description of light as complex fields remains a difficult problem. One might be tempted to think of the direction of a light ray

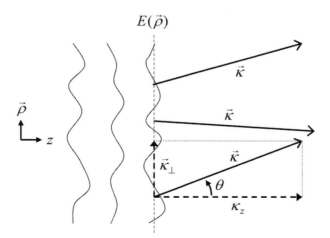

Figure 1.1. Connection between wave optics and ray optics.

as linked to a wavevector $\vec{\kappa}$. Following this line of reasoning, one might consider inferring this wavevector $\vec{\kappa}$ from a Fourier transform of $I(\vec{\rho}) = \langle E(\vec{\rho})E^*(\vec{\rho}) \rangle$. While this might have been a good starting point, it is clearly problematic because a Fourier transform of $\langle E(\vec{\rho})E^*(\vec{\rho}) \rangle$ is inherently non-local in space and does not fit our intuitive notion of light rays emanating from specific locations with specific directions. Instead, a commonly accepted connection between complex fields and ray optics was formulated by Walther [8] and Friberg [2], and is based on a parameter called the radiance [7], defined by

$$\mathcal{L}(\vec{\kappa}_{\perp}; \vec{\rho}) = \kappa^2 \int d^2\vec{\rho}' \, \langle E(\vec{\rho} + \tfrac{1}{2}\vec{\rho}')E^*(\vec{\rho} - \tfrac{1}{2}\vec{\rho}') \rangle \, e^{-i2\pi\vec{\kappa}_{\perp} \cdot \vec{\rho}'} \qquad (1.11)$$

By construction, the radiance is a *local* Fourier transform (often called a Wigner function), though not of the intensity but rather of the field autocorrelation function. The direction of the light ray emanating from point $\vec{\rho}$ is then prescribed by the value of $\vec{\kappa}_{\perp}$ about which $\mathcal{L}(\vec{\kappa}_{\perp}; \vec{\rho})$ is peaked. The local angle of propagation from point $\vec{\rho}$, accordingly, is defined by $(\theta_x, \theta_y) = \left(\sin^{-1}(\hat{\kappa}_x/\kappa), \sin^{-1}(\hat{\kappa}_y/\kappa) \right)$, where $(\hat{\kappa}_x, \hat{\kappa}_y)$ corresponds to this peak. This connection between radiance and ray direction is valid only for fields that are slowly spatially varying on the scale of a wavelength, meaning, in effect, it is valid only for small angles, a condition known as the paraxial limit, which will be invoked repeatedly. Indeed, the paraxial limit has been implicitly assumed in the definition of radiance provided above.

1.3 RAY OPTICS

The notion of light rays is very convenient in that it provides a simple and intuitive description of light propagation through basic elements of an optical imaging device. Since $\vec{\rho}$ and $\vec{\kappa}_{\perp}$ are Fourier conjugate coordinates, we may also think of (x, y) and $(n \sin\theta_x, n \sin\theta_y)$ as conjugate coordinates, or, in the paraxial limit, (x, y) and $(n\theta_x, n\theta_y)$. A light ray can then be described as a vector

$$\begin{pmatrix} x \\ n\theta_x \end{pmatrix} \qquad (1.12)$$

where, for purposes of discussion, we consider rays in the x–z plane only. This vector indicates the position and direction of the ray at a given optical plane (plane 0). The effect of an optical system, imaging or otherwise, is to transfer this ray to a new plane (plane 1) leading, in general, to a change both in position and direction of the ray. A linear optical system can then be characterized by a general transfer matrix $\mathbf{M}$, such that,

$$\begin{pmatrix} x_1 \\ n_1\theta_{x1} \end{pmatrix} = \mathbf{M} \cdot \begin{pmatrix} x_0 \\ n_0\theta_{x0} \end{pmatrix} \qquad (1.13)$$

as schematically depicted in Fig. 1.2, where allowances have been made for different indices of refraction on either side of the system. $\mathbf{M}$ is often referred to as an ABCD transfer matrix

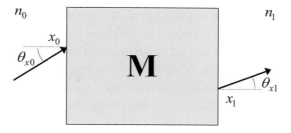

Figure 1.2. General ABCD matrix.

because it consists of four elements. Detailed discussions of ABCD transfer matrix formalism are provided in several optics textbooks, such as [4] and [6].

The following are a few basic results:

For propagation through an interface, then

$$\mathbf{M}_{\text{interface}} = \begin{pmatrix} 1 & 0 \\ 0 & 1 \end{pmatrix} \tag{1.14}$$

indicating that on either side of the interface we have $x_1 = x_0$ and $n_1\theta_1 = n_0\theta_0$, the latter being a statement of conservation of momentum along the x direction (also known as Snell's law).

For propagation an axial distance z through a medium of index n, then

$$\mathbf{M}_{\text{propagation}} = \begin{pmatrix} 1 & z/n \\ 0 & 1 \end{pmatrix} \tag{1.15}$$

For propagation through a thin lens of focal length f surrounded by a medium of index n, then

$$\mathbf{M}_{\text{lens}} = \begin{pmatrix} 1 & 0 \\ -n/f & 1 \end{pmatrix} \tag{1.16}$$

For a system that performs perfect imaging with magnification M, then

$$\mathbf{M}_{\text{image}} = \begin{pmatrix} M & 0 \\ 0 & 1/M \end{pmatrix} \tag{1.17}$$

Finally, for a system that exchanges the coordinates x and $n\theta_x$, then

$$\mathbf{M}_{\text{FT}} = \begin{pmatrix} 0 & L \\ -1/L & 0 \end{pmatrix} \tag{1.18}$$

where the length scale L is introduced for dimensional consistency. Since x and $n\theta_x$ can be thought of as Fourier conjugate coordinates, $\mathbf{M}_{\text{FT}}$ can be though of as performing a perfect Fourier transform where the coordinates are swapped, similar to Eqs. 1.8 and 1.9 (hence the subscript FT).

Two observations can be made. First, all of the above transfer matrices satisfy the condition $\det[\mathbf{M}] = 1$. This is in fact a general rule for lossless systems, resulting from the conservation of a fundamental quantity called the étendue of the light beam. Thus, for perfect imaging (Eq. 1.17), the magnification of the position by a factor M is necessarily accompanied by the de-magnification of the propagation angle by a factor $1/M$. A similar conclusion holds for a perfect Fourier transform (Eq. 1.18), though applied to the crossed terms. Much more will be said about étendue in Chapter 6.

A second observation relates to the effect of the index of refraction on propagation distances. A distinction can be made between physical distances, such as z and f, and their associated optical distances, such as nz and nf. For systems embedded in a medium, it is the latter that become important.

Optical imaging systems, in general, involve the transfer of light rays through a medium (or several media) and through lenses. We consider these transfers separately and then in combination. For simplicity, we begin with systems surrounded by free space ($n = 1$). A more general case involving unequal indices of refraction will be considered at the end of the chapter.

Propagation Through Free Space

In the simplest scenario, an arbitrary light ray starts at plane 0 and propagates an axial distance z, leading to

$$\begin{pmatrix} x_0 + z\theta_{x0} \\ \theta_{x0} \end{pmatrix} = \mathbf{M}_{\text{propagation}} \cdot \begin{pmatrix} x_0 \\ \theta_{x0} \end{pmatrix} \tag{1.19}$$

In the limit where z becomes very large, we can eventually neglect the starting position of the ray and write

$$\begin{pmatrix} x_1 \\ \theta_{x1} \end{pmatrix} \rightarrow \begin{pmatrix} \approx z\theta_{x0} \\ \theta_{x0} \end{pmatrix} \tag{1.20}$$

This limit is called the far-field or Fraunhofer limit. It is important because we find that in this limit the position of the ray scales directly with its initial starting angle, the scaling factor being simply z. In effect, far-field propagation can be thought of as performing a scaled Fourier transform of the initial field by exchanging x_0 with $z\theta_{x0}$. However, this Fourier transform is one-way since only x_0 is exchanged with θ_{x0} and not vice versa.

Propagation Through a Lens

Next, we examine the propagation of a light ray through a thin lens by taking the planes 0 and 1 to be on either side of the lens.

Referring to Eq. 1.16, we conclude that if a ray impinges upon the center of the lens at an arbitrary angle θ_{x0}, then it exits the center of the lens undeviated. That is,

$$\begin{pmatrix} 0 \\ \theta_{x0} \end{pmatrix} = \mathbf{M}_{\text{lens}} \cdot \begin{pmatrix} 0 \\ \theta_{x0} \end{pmatrix} \tag{1.21}$$

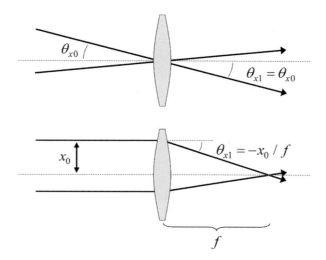

Figure 1.3. Propagation of rays through a lens of focal length f.

Similarly, if a ray travels parallel to the optical axis but impinges upon the lens at an arbitrary position, then it exits the lens at the same position, but deflected at an angle:

$$\begin{pmatrix} x_0 \\ -x_0/f \end{pmatrix} = \mathbf{M}_{\text{lens}} \cdot \begin{pmatrix} x_0 \\ 0 \end{pmatrix} \tag{1.22}$$

This deflection angle causes the outgoing ray to intersect the optical axis at a distance f from the lens, independently of the initial position of the ray at the lens. In other words, a lens causes *any* ray parallel to the optical axis to converge to the same focal point. Both the above properties are illustrated in Fig. 1.3.

1.4 BASIC TRANSFER PROPERTIES OF A LENS

Finally, we derive some basic transfer properties of a lens in free space by again considering planes 0 and 1 on either side of the lens, but this time generalizing the position of these planes to be arbitrary distances s_0 and s_1, respectively, from the lens. The transfer of a light ray from plane 0 to plane 1 is then governed by the composite matrix

$$\mathbf{M}_{\text{T}} = \mathbf{M}_{\text{free}}(s_1) \cdot \mathbf{M}_{\text{lens}}(f) \cdot \mathbf{M}_{\text{free}}(s_0) = \begin{pmatrix} 1 - \frac{s_1}{f} & s_0 + s_1 - \frac{s_0 s_1}{f} \\ -\frac{1}{f} & 1 - \frac{s_0}{f} \end{pmatrix} \tag{1.23}$$

Two specific cases are of particular interest:

1.4.1 Fourier Transform with a Lens

In the event that $s_0 = s_1 = f$, Eq. 1.23 simplifies to

$$\mathbf{M}_\mathrm{T} = \begin{pmatrix} 0 & f \\ -1/f & 0 \end{pmatrix} \tag{1.24}$$

and we recognize from Eq. 1.18 that in this specific configuration a lens performs a perfect Fourier transform accompanied by a scaling factor f. It should be emphasized that no Fraunhofer approximation was required to obtain this Fourier transform analogy, in contrast to the far-field propagation result derived above (the small angle approximation, however, remains valid). Moreover, the Fourier transform operation derived here is now two-way.

1.4.2 Imaging with a Lens

In the event that s_0 and s_1 satisfy a relation known as the thin-lens formula, given by

$$\frac{1}{s_0} + \frac{1}{s_1} = \frac{1}{f} \tag{1.25}$$

then Eq. 1.23 simplifies to

$$\mathbf{M}_\mathrm{T} = \begin{pmatrix} M & 0 \\ -1/f & 1/M \end{pmatrix} \tag{1.26}$$

where we have introduced the magnification factor

$$M = -\frac{s_1}{s_0} \tag{1.27}$$

A lens performs near-perfect imaging in this specific configuration, as illustrated in Fig. 1.4. The only defect in the imaging arises from the off-axis element $-1/f$ that imparts an extraneous dependence of the propagation angle at the image plane (plane 1) on the ray position at the object plane (plane 0), as is observed by carrying out the matrix multiplication

$$\begin{pmatrix} x_1 \\ \theta_{x1} \end{pmatrix} = \mathbf{M}_\mathrm{T} \cdot \begin{pmatrix} x_0 \\ \theta_{x0} \end{pmatrix} = \begin{pmatrix} Mx_0 \\ -x_0/f + \theta_{x0}/M \end{pmatrix} \tag{1.28}$$

This extraneous dependence corresponds to a coupling between the ray direction and the ray position at the image plane, which would be absent in the case of a perfect imaging system described by Eq. 1.17. One consequence of such coupling is apparent in Fig. 1.4. If s_1 is increased or decreased relative to its in-focus value, not only does the image become blurred (the rays do not converge properly), but the magnification increases or decreases as well. Such axially dependent magnification is often problematic in microscopy applications (more on this in Chapter 5). Imaging systems that do not exhibit this problem and where magnification remains a constant independent of defocus are called telecentric systems. Modern microscopes are almost always telecentric, and we will consider these in detail in subsequent chapters. In the

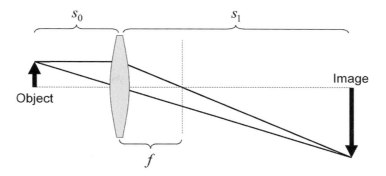

Figure 1.4. Imaging with a lens of focal length f.

meantime, we only note that the limit of telecentricity can be approached with a single lens when $|x_0| \ll |\theta_{x0} f/M|$, in which case Eq. 1.28 simplifies to

$$\begin{pmatrix} x_1 \\ \theta_{x1} \end{pmatrix} \rightarrow \begin{pmatrix} Mx_0 \\ \approx \theta_{x0}/M \end{pmatrix} \tag{1.29}$$

Optical planes connected by an imaging operation are called conjugate planes, whereas those connected by a Fourier transform operation are called Fourier planes. We will make use of this terminology throughout this book.

Axial Magnification

Though we have considered only transverse imaging magnifications so far, we can readily derive an axial imaging magnification from the thin lens formula. For small axial displacements, this axial magnification is defined by

$$M_z = \frac{ds_1}{ds_0} \tag{1.30}$$

Taking the derivative of both sides of Eq. 1.25 with respect to s_0, we find

$$M_z = -\left(\frac{s_1}{s_0}\right)^2 = -M^2 \tag{1.31}$$

1.4.3 Thick Lens

The transfer matrix $\mathbf{M}_{\text{lens}}$ (Eq. 1.16) is applicable to thin lenses only, so thin that a ray traversing the lens incurs only a negligible lateral displacement. In practice, however, a lens always possesses a finite thickness. In many cases it can even comprise several optical elements and be quite thick indeed, an example being a microscope objective. Nevertheless, even a thick lens, if properly designed, possesses a well-defined focal length. This focal length is defined not from the center of the lens but rather from what is known as a principal plane, as illustrated in Fig. 1.5.

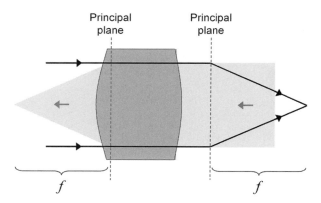

Figure 1.5. Principal planes of a thick lens of focal length f.

An important property of lenses in general is that they obey a principle of optical reciprocity wherein if a collimated bundle of rays parallel to the optical axis is focused to a point when incident from one side of the lens, then it is focused to another point when incident from the opposite side. In the case of a thin lens, these points are located a distance f from either side of the lens. In the case of a thick lens, however, they are located distances f from the associated principal planes of the lens, as illustrated in Fig. 1.5. The difference between thin and thick lenses is that for a thin lens the principal planes are merged into a single plane located exactly at the lens plane, whereas for a thick lens they are displaced from one another. This displacement can be quite large, often larger than the physical thickness of the lens itself. Moreover, the principal planes may be distributed asymmetrically relative to the physical lens center, and may even be on opposite sides of each other from what might be expected. In any case, regardless of the location of the principal planes the focal lengths of a thick lens in either direction are the same, provided only that the index of refraction of the media on either side of the lens is also the same.

It should be emphasized that this principle of reciprocity does not mean that a thick lens works just as well in either orientation. The angular acceptance of a thick lens can be quite different depending on its orientation, leading to very different fields of view in an imaging application. More will be said about this in Chapter 6.

Effect of Surrounding Media

So far, we have mostly considered systems surrounded by free space, and only made general allowances for the possibility of surrounding media other than free space. We close this chapter with a more detailed examination of the effects of different indices of refraction. For example, let us consider a thick lens, itself of refractive index n_f, surrounded on one side by refractive index n_0 and on the other by refractive index n_1, as shown in Fig. 1.6. The thin-lens formula in this case becomes more complicated (cf. [5]), and assumes the more general form given by

$$\frac{n_0}{s_0} + \frac{n_1}{s_1} = \frac{n_f - n_0}{R_0} - \frac{n_f - n_1}{R_1} \tag{1.32}$$

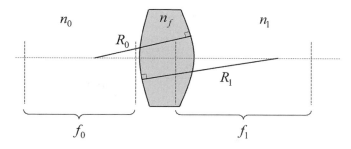

Figure 1.6. Media of different indices of refraction.

where R_0 and R_1 are the radii of curvature of the two lens faces (note: $R_0 > 0$ and $R_1 < 0$ for a bi-convex lens and for rays propagating from left to right). Equation 1.32 is known as the lens-maker's formula. To find the focal length of this lens, let us consider a collimated beam incident from the left, equivalent to setting s_0 to infinity. The focal length to the right of the lens is then given by s_1, and we thus have

$$\frac{1}{f_1} = \frac{1}{n_1}\left(\frac{n_f - n_0}{R_0} - \frac{n_f - n_1}{R_1}\right) \tag{1.33}$$

The subscript 1 is associated with this focal length because, as it happens, we find a different focal length when we instead consider a collimated beam incident from the right, in which case we have

$$\frac{1}{f_0} = \frac{1}{n_0}\left(\frac{n_f - n_1}{R_1} - \frac{n_f - n_0}{R_0}\right) \tag{1.34}$$

bearing in mind that both the ordering and the signs of R_0 and R_1 have switched (i.e. the terms in parentheses in Eqs. 1.33 and 1.34 are the same).

We conclude that two focal lengths are associated with this lens, depending on the direction of the light propagation. This result can be summarized by a generalized thin-lens equation given by

$$\frac{n_0}{s_0} + \frac{n_1}{s_1} = \frac{n_1}{f_1} = \frac{n_0}{f_0} \tag{1.35}$$

Again, f_1 and f_0 here correspond to the physical distances of the foci from their respective principal planes. The ABCD matrix for this lens (Eq. 1.16) thus contains $-n_1/f_1$ or $-n_0/f_0$ in the bottom left corner. What is apparent from these results is that the two focal lengths associated with a lens are not a fixed property of the lens itself, since they depend on the surrounding media.

Finally, to perform imaging with this lens, s_0 and s_1 must obey this new generalized thin-lens formula, and the resultant ABCD matrix for imaging becomes

$$\mathbf{M_T} = \begin{pmatrix} M & 0 \\ -\frac{n_0}{f_0} \text{ or } -\frac{n_1}{f_1} & 1/M \end{pmatrix} \tag{1.36}$$

where the magnification M is replaced by

$$M = -\frac{n_0}{n_1}\frac{s_1}{s_0} \tag{1.37}$$

We will revisit these results in Chapter 3.

1.5 PROBLEMS

Problem 1.1

Consider an "afocal" arrangement where two lenses in free space are separated by a distance $f_0 + f_1$.

(a) Calculate the ABCD transfer matrix between plane 0 located a distance s_0 in front of lens 0, and plane 1 located a distance s_1 behind lens 1.

(b) Calculate the ABCD transfer matrix when $s_0 = f_0$ and $s_1 = f_1$. This is called a 4f, or telecentric, imaging system. What is the resultant magnification?

(c) Calculate the ABCD transfer matrix when $s_0 = f_0$ and $s_1 = f_1 + \delta z$. What happens to the magnification for a beam parallel to the optical axis?

Problem 1.2

Consider a 4f imaging arrangement of the type described in Problem 1.1. That is, two lenses of focal lengths f_0 and f_1 are separated by distances $f_0 + f_1$. The object plane is located a distance f_0 in front of the lens 0. The corresponding image plane is located a distance f_1 behind lens 1. Consider a slight error such that lens 1 is displaced a distance ε from its nominal 4f position (where $\varepsilon \ll f_0 < f_1$).

(a) Derive the imaging transfer matrix for the case where the object plane remains at its initial position. What is the magnification? Why is this magnification not well defined?

(b) Where should the imaging plane be for the magnification to be well defined?

Problem 1.3

Consider a 4f imaging arrangement of the type described in Problem 1.1. That is, two lenses of focal lengths f_0 and f_1 are separated by distances $f_0 + f_1$. The object plane is located a distance f_0 in front of the lens 0. The corresponding image plane is located a distance f_1 behind lens 1. Insert an intermediate lens of focal length f a distance f_0 behind lens f_0.

(a) Where is the new image plane?

(b) What is the magnification at this new image plane?

Problem 1.4

Consider two free-space single-lens imaging systems with lenses f_0 and f_1 and magnifications M_0 and M_1 respectively. Place these two imaging systems in tandem (i.e. 3 conjugate planes).

(a) Calculate the ABCD transfer matrix from the first conjugate plane to the last conjugate plane. What is the net magnification? Is the imaging perfect (i.e. telecentric)?

(b) Now place a lens f exactly at the middle conjugate plane (this is called a field lens). Re-calculate the above ABCD matrix. Has the net magnification changed?

(c) At what value of f is there no crosstalk between ray position and angle?

(d) A field lens is also useful for increasing the field of view. That is, given that lenses have finite diameters, a field lens can allow the imaging of bigger objects. Can you explain why (qualitatively)?

Problem 1.5

In the free-space thin lens formula, distances s_0 and s_1 are measured relative to the lens itself. Consider instead measuring distances relative to the lens front and back focal planes. That is, write $z_0 = s_0 - f$, and $z_1 = s_1 - f$. Show that the thin lens formula can be expressed equivalently by the so-called Newtonian lens formula $z_0 z_1 = f^2$.

References

[1] Bracewell, R. N. *The Fourier Transform and its Applications*, McGraw-Hill (1965).

[2] Friberg, A. T. "Propagation of a generalized radiance in paraxial optical systems," *Appl. Opt.* 30, 2443–2446 (1991).

[3] Goodman, J. W. *Statistical Optics*, 2nd edn, Wiley (2015).

[4] Hecht, E. *Optics*, Addison Wesley (2002).

[5] Jenkins, F. A. and White, H. E. *Fundamentals of Optics*, McGraw-Hill (1957).

[6] Siegman, A. E. *Lasers*, University Science Books (1986).

[7] Walther, A. "Radiometry and coherence," *J. Opt. Soc. Am.* 58, 1256–1259 (1968).

[8] Walther, A. "Propagation of the generalized radiance through lenses," *J. Opt. Soc. Am.* 68, 1606–1610 (1978).

2

Monochromatic Wave Propagation

In the most general case, a light wave is characterized as a vectorial field

$$\vec{E}(\vec{r},t) = E_x(\vec{r},t)\hat{x} + E_y(\vec{r},t)\hat{y} + E_z(\vec{r},t)\hat{z} \tag{2.1}$$

where $\hat{x}$, $\hat{y}$ and $\hat{z}$ are unit vectors that define orthogonal polarization directions (conventionally, $\hat{x}$, $\hat{y}$ are transverse and $\hat{z}$ is axial). Each of the field components $E_x(\vec{r},t)$, $E_y(\vec{r},t)$, and $E_z(\vec{r},t)$ is complex. Moreover, in the event that the wave is traveling through an isotropic medium, such as free space, each of these components can be treated independently. We will make this assumption here and consider only a single transverse component of the field, say along $\hat{x}$, enabling us to considerably simplify our notation with the substitution

$$E_x(\vec{r},t)\hat{x} \rightarrow E(\vec{r},t) \tag{2.2}$$

The spatiotemporal evolution of this field component is governed by the well-known Maxwell equations. These equations are discussed in detail in all electricity and magnetism textbooks (e.g. [6, 8]) and only a brief review is presented here of some of their ramifications pertaining to optical imaging. We begin by considering fields that vary arbitrarily with time, and follow this by considering fields that oscillate in time at a given temporal frequency ν, called monochromatic fields. As a reminder, we use the notation $\vec{r} = (x,y,z) = (\vec{\rho},z)$ and $\vec{\kappa} = (\kappa_x, \kappa_y, \kappa_z) = (\vec{\kappa}_\perp, \kappa_z)$.

2.1   TIME-DEPENDENT WAVE EQUATION

Maxwell's equations, when applied to a single field component in a medium with isotropic index of refraction n, lead to the time-dependent wave equation [1, 9], given by

$$\left(\nabla^2 - \frac{n^2}{c^2}\frac{\partial^2}{\partial t^2}\right)E(\vec{r},t) = S(\vec{r},t) \tag{2.3}$$

where $S(\vec{r},t)$ corresponds to the physical source, such as atoms or molecules, from which the field component originates. The speed of the wave in the medium is, accordingly, c/n.

To solve for $E(\vec{r},t)$, we adopt a standard approach from linear systems. That is, we first derive the field produced by a point source located at position $\vec{r}_0$ at time t_0, and then invoke the principle of linear superposition to calculate the field resulting from the source distribution $S(\vec{r}_0, t_0)$. The point-source response is called the Green function, which is the solution to

$$\left(\nabla^2 - \frac{n^2}{c^2}\frac{\partial^2}{\partial t^2}\right) G(\vec{r}, t) = \delta^3(\vec{r})\delta(t) \tag{2.4}$$

Defining the four-dimensional (4D) Fourier transform of $G(\vec{r}, t)$ to be

$$\mathcal{G}(\vec{\kappa}, \nu) = \iint d^3\vec{r}\, dt\, G(\vec{r}, t)\, e^{-i2\pi\left(\vec{\kappa}\cdot\vec{r} - \nu t\right)} \tag{2.5}$$

and taking the Fourier transform of both sides of Eq. 2.4, we obtain

$$\mathcal{G}(\vec{\kappa}, \nu) = -\frac{1}{4\pi^2\left(\left|\vec{\kappa}\right|^2 - \frac{n^2}{c^2}\nu^2\right)} \tag{2.6}$$

This is the Fourier representation of the Green function for the time-dependent wave equation. We notice that it diverges when $\left|\vec{\kappa}\right| = \frac{n}{c}\nu$. However, it also makes allowances for the possibility $\left|\vec{\kappa}\right| \neq \frac{n}{c}\nu$, which seems to go against the energy–momentum relationship that was posited in Chapter 1 (Eqs. 1.3 and 1.5). This is because $\mathcal{G}(\vec{\kappa}, \nu)$ includes the possibility of non-propagating, or evanescent, waves. As noted in Chapter 1, we will not concern ourselves with these, and restrict ourselves only to the case $\left|\vec{\kappa}\right| = \frac{n}{c}\nu$. But the resulting divergence in $\mathcal{G}(\vec{\kappa}, \nu)$ must be treated carefully (see [12] for a detailed exposition). The conventional approach involves applying a boundary condition to $G(\vec{r}, t)$ to ensure it obeys causality (i.e. $G(\vec{r}, t) = 0$ when $t < 0$). Performing an inverse Fourier transform of Eq. 2.6 by contour integration, we then obtain

$$G(\vec{r}, t) = -\frac{c}{4\pi nr}\delta\left(r - \frac{c}{n}t\right) \qquad (t > 0) \tag{2.7}$$

This is the (causal) Green function for the time-dependent wave equation. Once again, it corresponds to the field resulting from an impulsive point source, located here at the origin. We observe that $G(\vec{r}, t)$ is zero everywhere except on a spherical shell that expands outwardly with speed c/n. It also diminishes as $1/r$ as the shell expands, in accordance with the principle of conservation of energy. The solution to the field given by Eq. 2.3 resulting from an arbitrary source becomes finally

$$E(\vec{r}_1, t_1) = \iint d^3\vec{r}_0\, dt_0\, G(\vec{r}_1 - \vec{r}_0, t_1 - t_0) S(\vec{r}_0, t_0)$$
$$= -\frac{1}{4\pi}\int d^3\vec{r}_0\, \frac{S\left(\vec{r}_0, t_1 - \frac{n}{c}\left|\vec{r}_1 - \vec{r}_0\right|\right)}{\left|\vec{r}_1 - \vec{r}_0\right|} \tag{2.8}$$

2.2 TIME-INDEPENDENT WAVE EQUATION

Equation 2.3 defines the field $E(\vec{r}, t)$ produced by a source that varies arbitrarily with time. It is often convenient to express both the source and resultant field in terms of their harmonic components. That is,

$$S(\vec{r}, t) = \int d\nu \, S(\vec{r}; \nu) e^{-i2\pi\nu t} \tag{2.9}$$

$$E(\vec{r}, t) = \int d\nu \, E(\vec{r}; \nu) e^{-i2\pi\nu t} \tag{2.10}$$

When substituted into Eq. 2.3, these lead directly to the inhomogeneous, time-independent wave equation

$$\left(\nabla^2 + 4\pi^2\kappa^2\right) E(\vec{r}; \nu) = S(\vec{r}; \nu) \tag{2.11}$$

also known as the Helmholtz equation, which will serve as the basis for our treatment of light propagation for monochromatic waves.

Just as a general solution to the time-dependent wave equation was obtained by first deriving its Green function, the same can be done with the time-independent wave equation. The most straightforward approach involves simply taking the Fourier transform of Eq. 2.7, leading to

$$G(\vec{r}; \nu) = -\frac{e^{i2\pi n\nu r/c}}{4\pi r} \tag{2.12}$$

or, suppressing the explicit temporal frequency dependence for ease of notation,

$$G(\vec{r}) = -\frac{e^{i2\pi\kappa r}}{4\pi r} \tag{2.13}$$

This is the outgoing Green function associated with the Helmholtz equation. Its interpretation is intuitive: a harmonically varying point source at the origin produces a harmonically varying spherical wave that expands outward from the origin. An alternative manner of writing this Green function that is useful when considering propagation into either half-space $z > 0$ or $z < 0$ comes from the Weyl representation of spherical waves, and is given by

$$G(\vec{r}) = \frac{1}{i4\pi} \int d^2\vec{\kappa}_\perp \frac{1}{\sqrt{\kappa^2 - \kappa_\perp^2}} e^{i2\pi\left(\vec{\kappa}_\perp \cdot \vec{\rho} + \sqrt{\kappa^2 - \kappa_\perp^2}\,|z|\right)} \tag{2.14}$$

Regardless of the representation of $G(\vec{r})$, the monochromatic field that results from an arbitrary distribution of harmonically varying sources is formally given by the convolution

$$E(\vec{r}_1) = \int d^3\vec{r}_0 \, G(\vec{r}_1 - \vec{r}_0) S(\vec{r}_0) \tag{2.15}$$

where, again, $E(\vec{r}_1)$ and $S(\vec{r}_0)$ are shorthand notation for $E(\vec{r}_1; \nu)$ and $S(\vec{r}_0; \nu)$.

Still another representation of $G(\vec{r}) = G(\vec{\rho}, z)$ is given by the partial Fourier transform of Eq. 2.14 over the coordinate $\vec{\rho}$, leading to

$$\mathcal{G}(\vec{\kappa}_\perp; z) = \frac{e^{i2\pi|z|\sqrt{\kappa^2 - \kappa_\perp^2}}}{i4\pi\sqrt{\kappa^2 - \kappa_\perp^2}} \tag{2.16}$$

Equations 2.13, 2.14, and 2.16 are equivalent representations of the *outgoing* Green function. That is, they correspond to a spherical wave propagating outward from the origin. Similarly,

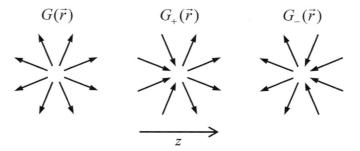

Figure 2.1. Outward-, forward-, and backward-propagating Green functions.

a solution to the Helmholtz equation can be defined for a point sink (as opposed to a point source), leading to the *incoming* Green function, $G^*(\vec{r})$ or $\mathcal{G}^*(-\vec{\kappa}_\perp; z)$, corresponding to a spherical wave that contracts inward toward the origin.

But, ultimately, we will find that the most useful Green functions to work with in the context of imaging applications are hybrid functions that are both incoming and outgoing (see Fig. 2.1). Accordingly, we construct the *forward-propagating* Green function

$$G_+\left(\vec{r}\right) = \begin{cases} G\left(\vec{r}\right) & (z > 0) \\ -G^*\left(\vec{r}\right) & (z < 0) \end{cases} \tag{2.17}$$

and the *backward-propagating* Green function

$$G_-\left(\vec{r}\right) = \begin{cases} -G^*\left(\vec{r}\right) & (z > 0) \\ G\left(\vec{r}\right) & (z < 0) \end{cases} \tag{2.18}$$

The Weyl representations for these functions are given by

$$\mathcal{G}_\pm(\vec{\kappa}_\perp; z) = \frac{e^{\pm i2\pi z\sqrt{\kappa^2 - \kappa_\perp^2}}}{i4\pi\sqrt{\kappa^2 - \kappa_\perp^2}} \tag{2.19}$$

Note the absence of the absolute value operator for z in the exponent. As before, the field produced by a collection of sources is given by a convolution

$$E(\vec{r}_1) = \int d^3\vec{r}_0\, G_\pm(\vec{r}_1 - \vec{r}_0)S\left(\vec{r}_0\right) \tag{2.20}$$

Whereas in Eq. 2.15 the field $E(\vec{r}_1)$ propagates forward or backward depending on the sign of $z_1 - z_0$, here the direction of propagation is independent of the sign of $z_1 - z_0$, and depends only on which Green function is used, either G_+ or G_-. This will be made clear below.

2.2.1 Propagating Versus Evanescent Waves

While it was stressed from the outset that we will concern ourselves only with propagating waves and not evanescent waves, it is worth taking a moment to clarify the distinction

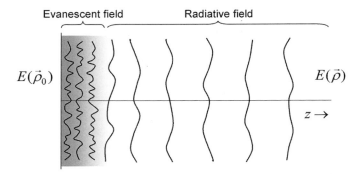

Figure 2.2. Radiative field propagation versus evanescent field decay.

between the two. This clarification can be made by way of Eq. 2.14, where it is apparent that the field resulting from a harmonically-varying point source (i.e. $G\left(\vec{r}\right)$) behaves in two qualitatively different manners depending on the magnitude of $\kappa_\perp$. In the case $\kappa_\perp < \kappa$, the quantity $\sqrt{\kappa^2 - \kappa_\perp^2}$ is real and the exponential in the integrand oscillates with increasing $|z|$. In the case $\kappa_\perp > \kappa$, the quantity $\sqrt{\kappa^2 - \kappa_\perp^2}$ is imaginary and the exponential instead decays with increasing $|z|$, to become vanishingly small. We can thus distinguish two components to the field $G\left(\vec{r}\right)$. The first is composed of transverse spatial frequencies $\vec{\kappa}_\perp$ such that $\kappa_\perp < \kappa$. This component is propagating, or radiative, since it transports radiant energy to arbitrarily large $|z|$. The second is composed of transverse spatial frequencies $\vec{\kappa}_\perp$ such that $\kappa_\perp > \kappa$. This component is non-propagating, or evanescent, since it cannot transport radiant energy away from the source. Instead the radiant energy is confined to the vicinity of the source (see Fig. 2.2). The larger the value of $\kappa_\perp$, the more restricted this confinement.

As it happens, the principle of causality imposes the condition that any field produced by any source must contain both propagating and evanescent components (see, for example, [10]). However, the fate of the various transverse spatial frequencies in these fields is different. Low frequencies ($\kappa_\perp < \kappa$) propagate to arbitrarily large distances, while higher frequencies ($\kappa_\perp > \kappa$) decay exponentially and become irretrievably lost upon propagation. Almost all optical microscopes, and indeed all microscopes considered in this book, are based on the detection of propagating fields, meaning that the information retrieved from these fields is inherently bandwidth-limited (though, as we will see in the next chapters, practical bandwidth limitations are usually much more severe than those imposed by propagation alone). Microscopes that extend beyond this bandwidth limitation by specifically implicating evanescent fields are referred to as near-field microscopes. Again, we will not consider these. For detailed discussions of these microscopes the reader is referred to specialized textbooks, such as [4, 10, 11].

Bearing in mind that we will consider only propagating fields, we close this section with yet another representation of the forward and backward-propagating Green functions, this time as a function of three-dimensional (3D) spatial frequencies. By applying a Fourier transform in z to Eq. 2.19, we arrive at

$$\mathcal{G}_{\pm}(\vec{\kappa}_{\perp}, \kappa_z) = \frac{1}{i4\pi\sqrt{\kappa^2 - \kappa_{\perp}^2}} \delta\left(\kappa_z \mp \sqrt{\kappa^2 - \kappa_{\perp}^2}\right) \tag{2.21}$$

where the distinction between forward and backward propagating is now explicit. Forward propagating corresponds to $\kappa_z > 0$; backward propagating corresponds to $\kappa_z < 0$.

2.3 RAYLEIGH–SOMMERFELD DIFFRACTION

Green functions, either time-dependent or time-independent, provide solutions for the fields produced by physical sources. Needless to say, these will become indispensable when we examine various types of microscopes. But before this, we will spend the next several chapters considering a different problem. Specifically, instead of asking the question what field is produced by a given source, we will ask the question what field in some region of space arises from the field in another region. For example, a common imaging geometry involves the propagation of light from one plane, say at $z = z_0$, to another plane at $z_1 > z_0$. From a knowledge of the light field at plane z_0, is it possible to infer the light field at arbitrary $z_1 > z_0$? That is, can $E(\vec{\rho}_1, z_1)$ be derived from $E(\vec{\rho}_0, z_0)$? Such a problem is called a boundary-value problem.

To simplify the problem, we make a few assumptions. For there to be a field at z_0 in the first place, sources must be present somewhere. Let us assume these vary harmonically with monochromatic frequency ν. Moreover, let us assume they are all located in the region $z < z_0$. In other words, the region between the planes z_0 and z_1 remains source-free (Fig. 2.3). Finally, let us adopt what are called the Kirchhoff boundary conditions [7], namely:

1. The field at z_0 is bounded by a finite aperture (i.e. the field vanishes outside the aperture).
2. The field inside the aperture is unaffected by the presence of the aperture.

With these assumptions, the derivation of $E(\vec{\rho}_1, z_1)$ from $E(\vec{\rho}_0, z_0)$ becomes tractable using Green's theorem (see Appendix B.1). Detailed accounts of this derivation for scalar fields are presented in [2] and [5]. Derivations involving a full vectorial treatment of light are presented in [3] and [6]. These accounts go beyond the scope of this book and only the most relevant results are cited here.

We begin with the answer:

$$E(\vec{\rho}_1, z_1) = \int \mathrm{d}^2\vec{\rho}_0 \, D_+(\vec{\rho}_1 - \vec{\rho}_0, z_1 - z_0) E(\vec{\rho}_0, z_0) \tag{2.22}$$

In brief, $D_+(\vec{\rho}, z)$ is a propagator that plays a similar role as a Green function (compare with Eq. 2.20), but it is not a Green function. Instead it is an auxiliary function that is constructed from the Green function using a method of images in order to satisfy what is known as

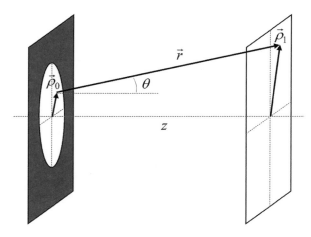

Figure 2.3. Geometry for free-space propagation with Kirchhoff boundary conditions.

the Dirichlet boundary condition (namely, the condition that $E(\vec{\rho}_0, z_0)$ is known, but not necessarily its derivative along z). With this boundary condition, we have

$$D_+ \left(\vec{\rho}_1 - \vec{\rho}_0, z_1 - z_0 \right) = -\frac{\partial}{\partial z_0} \left[G \left(\vec{\rho}_1 - \vec{\rho}_0, z_1 - z_0 \right) - G \left(\vec{\rho}_1 - \vec{\rho}_0, -z_1 - z_0 \right) \right]_{z_0 = 0}$$

(2.23)

where the point of the subscript $+$ will be made clear below.

Equations 2.22 and 2.23 tell us how a field propagates through free space (recall, there are no sources beyond z_0). Collectively, they are referred to as the Rayleigh–Sommerfeld diffraction formula, and will serve as one of the foundations of imaging theory.

Inserting Eq. 2.13 into Eq. 2.23, we find

$$D_+ \left(\vec{\rho}, z \right) = -i\kappa \frac{z}{r} \left(1 - \frac{1}{i2\pi\kappa r} \right) \frac{e^{i2\pi\kappa r}}{r}$$

(2.24)

where $r = \sqrt{\rho^2 + z^2}$.

Alternatively, inserting Eq. 2.14 into Eq. 2.23, we find (for $z > 0$)

$$D_+ \left(\vec{\rho}, z \right) = \int d^2\vec{\kappa}_\perp \, e^{i2\pi \left(\vec{\kappa}_\perp \cdot \vec{\rho} + \sqrt{\kappa^2 - \kappa_\perp^2} z \right)}$$

(2.25)

Both expressions for $D_+ \left(\vec{\rho}, z \right)$ are equivalent for $z > 0$.

As we will discover throughout the next few chapters, it is often easier to work not with a spatial representation of $D \left(\vec{\rho}, z \right)$, but rather a mixed representation involving both spatial and spatial-frequency coordinates. We define accordingly

$$\mathcal{E}(\vec{\kappa}_\perp; z) = \int d^2\vec{\rho} \, E(\vec{\rho}, z) e^{-i2\pi \vec{\kappa}_\perp \cdot \vec{\rho}}$$

(2.26)

$$\mathcal{D}_+(\vec{\kappa}_\perp; z) = \int d^2\vec{\rho}\, D_+(\vec{\rho}, z) e^{-i2\pi \vec{\kappa}_\perp \cdot \vec{\rho}} \tag{2.27}$$

where the former represents the radiant field (see Section 1.1), and the latter, from Eq. 2.25, takes on the remarkably simple form

$$\mathcal{D}_+(\vec{\kappa}_\perp; z) = e^{i2\pi z \sqrt{\kappa^2 - \kappa_\perp^2}} \tag{2.28}$$

Technically, our derivation so far has made the assumption $z > 0$. However, it is clear that if the sources producing the known field at $E(\vec{\rho}, z = 0)$ are displaced to arbitrarily large negative values of z, then Eq. 2.28 becomes valid for both $z > 0$ and $z < 0$, provided only that no sources are present in the zone of propagation. A fully 3D spatial-frequency representation of $\mathcal{D}_+(\vec{\kappa}_\perp; z)$ can be derived then from its Fourier transform along z, obtaining

$$\mathcal{D}_+(\vec{\kappa}) = \delta\left(\kappa_z - \sqrt{\kappa^2 - \kappa_\perp^2}\right) \tag{2.29}$$

Clearly, while $\mathcal{D}_+(\vec{\kappa}_\perp; z)$ can accommodate both positive and negative values of z, $\mathcal{D}_+(\vec{\kappa})$ is non-zero only for positive κ_z. In other words, the transfer function $\mathcal{D}_+(\vec{\kappa})$ is inherently restricted to forward-propagating waves, in the same manner as $\mathcal{G}_+(\vec{\kappa})$. Similarly, we can derive the transfer functions $\mathcal{D}_-(\vec{\kappa}_\perp; z)$ and $\mathcal{D}_-(\vec{\kappa})$ that are inherently restricted to backward-propagating waves, given by

$$\mathcal{D}_-(\vec{\kappa}_\perp; z) = e^{-i2\pi z \sqrt{\kappa^2 - \kappa_\perp^2}} \tag{2.30}$$

$$\mathcal{D}_-(\vec{\kappa}) = \delta\left(\kappa_z + \sqrt{\kappa^2 - \kappa_\perp^2}\right) \tag{2.31}$$

This is a good occasion to note that both forms of $\mathcal{D}_\pm(\vec{\kappa})$ enforce the condition $\kappa_\perp^2 + \kappa_z^2 = \kappa^2$, thereby providing a posteriori justification to the notation used in Eq. 1.5.

Finally, from the convolution theorem (see Eq. A.6), Eq. 2.22 can be recast as

$$\mathcal{E}(\vec{\kappa}_\perp; z_1) = \mathcal{D}_+(\vec{\kappa}_\perp; z_1 - z_0)\mathcal{E}(\vec{\kappa}_\perp; z_0) \tag{2.32}$$

or, equivalently, making use of the Fourier shift theorem (see Eq. A.3),

$$\mathcal{E}(\vec{\kappa}) = e^{-i2\pi \kappa_z z_0}\mathcal{D}_+(\vec{\kappa})\,\mathcal{E}(\vec{\kappa}_\perp; z_0) \tag{2.33}$$

A potential misconception should be clarified here. Equation 2.32 suggests that field propagation occurs from location z_0 to location z_1. This interpretation, while correct in the case $z_1 > z_0$, is misleading in general. In fact, the field propagation direction is defined not by the order of the spatial locations z_0 and z_1, but rather by our use of the propagator D_+, which enforces the condition $\kappa_z > 0$. For example, Eq. 2.32 is equally valid for $z_1 < z_0$, in which case the field propagates instead from location z_1 to location z_0. Equation 2.32 should be thought of as simply establishing the relation between different locations of the same field. In the case of Eq. 2.32, this field is forward propagating. If instead we wrote $\mathcal{E}(\vec{\kappa}_\perp; z_1) = \mathcal{D}_-(\vec{\kappa}_\perp; z_1 - z_0)\mathcal{E}(\vec{\kappa}_\perp; z_0)$, the field would be backward propagating.

2.3.1 Primary Versus Secondary Sources

A comparison of G_+ and D_+ is instructive, where we restrict ourselves here to forward-propagating fields (a similar comparison can be made for backward-propagating fields). The Green-function propagator G_+ determines the field resulting from a physical, or primary, source (see Eq. 2.20); the Rayleigh–Sommerfeld propagator D_+ determines the field at one location resulting from the field at another (see Eq. 2.22). In the latter case, the field $E(\vec{\rho}_0, z_0)$ itself may be thought of as a virtual, or secondary, source. In both cases, the sources may be thought of as comprised of distributions of elemental point sources from which the fields emanate.

While there are similarities between both propagators, there are also fundamental differences. The main differences become apparent when we recast Eq. 2.24 in the form

$$D_+ \left(\vec{r} \right) = i4\pi\kappa \cos\theta \left(1 - \frac{1}{i2\pi\kappa r} \right) G_+ \left(\vec{r} \right) \tag{2.34}$$

where θ denotes the off-axis tilt angle, and the term $\cos\theta = \frac{z}{r}$ is called an obliquity factor.

A further simplification comes from considering propagation distances much larger than the wavelength, in which case $\kappa r \gg 1$. This is known as the radiative approximation, which leads to

$$D_+ \left(\vec{r} \right) = i4\pi\kappa \cos\theta G_+ \left(\vec{r} \right) \tag{2.35}$$

or, equivalently,

$$\mathcal{D}_+(\vec{\kappa}_\perp; z) = i4\pi \sqrt{\kappa^2 - \kappa_\perp^2} \, \mathcal{G}_+(\vec{\kappa}_\perp; z) \tag{2.36}$$

Manifestly, $D_+ \left(\vec{r} \right)$ and $G_+ \left(\vec{r} \right)$ differ by an obliquity factor. But this difference is relatively minor, particularly when we consider highly forward-propagating light where $\cos\theta \approx 1$ or $\kappa_\perp \approx 0$. A much more fundamental difference is the presence of the additional prefactor i in $D_+ \left(\vec{r} \right)$. This prefactor indicates that, in the radiative limit, $D_+ \left(\vec{r} \right)$ is phase-shifted relative to $G_+ \left(\vec{r} \right)$ by $\frac{\pi}{2}$. This is, in fact, the defining difference between propagation from primary versus secondary sources (see [3]), and will play a crucial role in our analysis of microscope contrast mechanisms.

It should be noted that the concept of secondary sources may seem a bit peculiar given that the sources are virtual and appear to have been created out of thin air (not to mention vacuum!). Nevertheless, the concept is very powerful, as was recognized by Huygens as early as the 17th century. Indeed, the waves emanating from elemental secondary sources are often referred to as Huygens wavelets.

2.3.2 Propagator Properties

We begin with the Rayleigh–Sommerfeld propagator. We observe that $\mathcal{D}(\vec{\kappa}_\perp; z)$ is Hermitian in the variable z (see Appendix B). That is,

$$\mathcal{D}_\pm(\vec{\kappa}_\perp; -z) = \mathcal{D}_\pm^*(\vec{\kappa}_\perp; z) \tag{2.37}$$

Moreover, when two Rayleigh–Sommerfeld propagators are combined, we obtain

$$\mathcal{D}_\pm(\vec{\kappa}_\perp; z_2 - z_1)\mathcal{D}_\pm(\vec{\kappa}_\perp; z_1 - z_0) = \mathcal{D}_\pm(\vec{\kappa}_\perp; z_2 - z_0) \tag{2.38}$$

and hence

$$\mathcal{D}_\pm(\vec{\kappa}_\perp; z_0 - z_1)\mathcal{D}_\pm(\vec{\kappa}_\perp; z_1 - z_0) = 1 \tag{2.39}$$

While these last two properties are obvious in the spatial-frequency domain, they are less obvious when translated to real space, where they prescribe respectively

$$\int d^2\vec{\rho}_1\, D_\pm(\vec{\rho}_2 - \vec{\rho}_1, z_2 - z_1)D_\pm(\vec{\rho}_1 - \vec{\rho}_0, z_1 - z_0) = D_\pm(\vec{\rho}_2 - \vec{\rho}_0, z_2 - z_0) \tag{2.40}$$

and

$$\int d^2\vec{\rho}_1\, D_\pm(\vec{\rho}_0' - \vec{\rho}_1, z_0 - z_1)D_\pm(\vec{\rho}_1 - \vec{\rho}_0, z_1 - z_0) = \delta^2(\vec{\rho}_0' - \vec{\rho}_0) \tag{2.41}$$

The first of these (Eq. 2.38 or 2.40) is a statement that the application of two propagators in series is equivalent to the application of a single propagator over the combined propagation distance. The second (Eq. 2.39 or 2.41) is a statement that the round trip application of the propagator from one plane to another and then back again leaves the field unchanged.

When the Rayleigh–Sommerfeld propagator is combined with the Green-function propagator, we have

$$\mathcal{D}_\pm(\vec{\kappa}_\perp; z_2 - z_1)\mathcal{G}_\pm(\vec{\kappa}_\perp; z_1 - z_0) = \mathcal{G}_\pm(\vec{\kappa}_\perp; z_2 - z_0) \tag{2.42}$$

or, equivalently,

$$\int d^2\vec{\rho}_1\, D_\pm(\vec{\rho}_2 - \vec{\rho}_1, z_2 - z_1)G_\pm(\vec{\rho}_1 - \vec{\rho}_0, z_1 - z_0) = G_\pm(\vec{\rho}_2 - \vec{\rho}_0, z_2 - z_0) \tag{2.43}$$

These last relations will have important ramifications throughout this book, and embody one of the main advantages of using propagating Green functions (Eq. 2.19) rather than the outgoing Green function (Eq. 2.16). As an example, let us consider a cluster of primary sources, and calculate the forward-propagating (i.e. right-going) field produced by theses sources. Specifically, let us calculate this field at two different planes, one arbitrarily traversing the sources at location z_1, and another completely to the right of the sources at location z_2 (see Fig. 2.4). The latter field is straightforward to calculate, and is given by either

$$E(\vec{r}_2) = \int d^3\vec{r}_0\, G(\vec{r}_2 - \vec{r}_0)S(\vec{r}_0) = \int d^3\vec{r}_0\, G_+(\vec{r}_2 - \vec{r}_0)S(\vec{r}_0) \tag{2.44}$$

since in this case $G = G_+$.

In the absence of the sources, this same field can be evaluated at arbitrary locations using the Rayleigh–Sommerfeld propagator. In particular, it can be calculated at plane z_1, obtaining

$$E(\vec{r}_1) = \int d^2\vec{\rho}_2\, D_+(\vec{r}_1 - \vec{r}_2)E(\vec{r}_2) \tag{2.45}$$

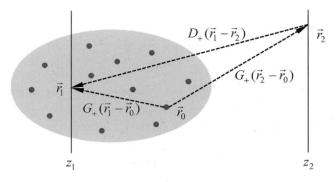

Figure 2.4. Calculation of forward-propagating fields from primary sources. Note that the arrows do not depict the direction of light propagation (which goes from left to right).

It is important to note that $E(\vec{r}_1)$ here corresponds to the forward-propagating field that results from *all* the primary sources in the cluster, including the sources to the right of z_1. This may appear counter-intuitive since, at first glance, it appears that the sources to the right of z_1 cannot produce forward-propagating fields at z_1. But we must remember that $E(\vec{r}_1)$ is not the actual field at plane z_1, but rather the forward-propagating component of this field. Moreover, since we are dealing with monochromatic fields, these essentially span all of space. In effect, $E(\vec{r}_1)$ should be thought of as a fictitious field at plane z_1 that, if left to propagate through free space to plane z_2 (or beyond), would result in the same field as actually produced by the cluster of primary sources.

The advantage of using G_+ rather than G becomes apparent when we observe that Eqs. 2.44 and 2.45 can be expressed much more succinctly as

$$E(\vec{r}_1) = \int d^3\vec{r}_0\, G_+(\vec{r}_1 - \vec{r}_0) S(\vec{r}_0) \qquad (2.46)$$

where we have made use of Eq. 2.43. In other words, the forward-propagating field produced by the cluster of sources can be calculated directly from G_+ at arbitrary locations to the right or left or even within the cluster, without having to worry about the sign of $z_1 - z_0$ (which would not be the case if we had tried to calculate $E(\vec{r}_1)$ using G). Of course, we must decide beforehand what field component we are interested in, whether forward or backward propagating. In microscopy applications, this is usually evident. For example, if a detector is placed to the right of the sources, then we need only concern ourselves with forward-propagating light. But we are getting ahead of ourselves, since we will only start considering such applications in Chapter 10.

2.4 APPROXIMATIONS

The Rayleigh–Sommerfeld propagator $D_+(\vec{\rho}, z)$ defined by Eq. 2.24 is often difficult to work with, and one must generally resort to approximations (the same applies to $D_-(\vec{\rho}, z)$, but henceforth, for simplicity, we consider only forward propagation).

We have already encountered one such approximation, namely the radiative approximation, where the propagation distance is assumed much larger than the wavelength ($\kappa r \gg 1$), leading to

$$D_+\left(\vec{\rho}, z\right) = -i\kappa \cos\theta \frac{e^{i2\pi\kappa r}}{r} \qquad \text{(radiative)} \qquad (2.47)$$

where, again, $r = \sqrt{\rho^2 + z^2}$ and $\cos\theta = \frac{z}{r}$.

In many applications, the light propagation can be assumed quasi-axial, either because of the geometry of the problem or because one is interested only in very forward propagating light directions. In these cases, a small-angle approximation can be adopted, known as the paraxial approximation, which amounts to setting $\cos\theta \to 1$ and $r \to z$ in the denominator of Eq. 2.47, leading to

$$D_+\left(\vec{\rho}, z\right) = -i\frac{\kappa}{z} e^{i2\pi\kappa r} \qquad \text{(paraxial)} \qquad (2.48)$$

We will return to both these approximations in later chapters.

2.4.1 Fresnel Diffraction

Despite the simplification afforded by the paraxial approximation, Eq. 2.48 remains difficult to work with because of the variable r in the exponential. This difficulty is alleviated when the paraxial approximation is restrained further by expanding r as a Taylor series in $(\rho/z)^2$, and keeping only the first-order term. That is, we approximate

$$r \to z + \frac{\rho^2}{2z} \qquad (2.49)$$

From Eq. 2.48, we arrive then at the Fresnel approximation for the Rayleigh–Sommerfeld propagator, given by

$$D_+(\vec{\rho}, z) = -i\frac{\kappa}{z} e^{i2\pi\kappa z} e^{i\pi\frac{\kappa}{z}\rho^2} \qquad \text{(Fresnel)} \qquad (2.50)$$

which is valid for $z^3 \gg \kappa\rho^4$. The Fresnel approximation for $D_+(\vec{\rho}, z)$ is extremely useful, and will be used repeatedly throughout this book. When inserted into the general free-space propagation integral given by Eq. 2.22, it leads to the well-known Fresnel diffraction integral

$$E(\vec{\rho}_1, z_1) = -i\frac{\kappa}{z} e^{i2\pi\kappa z} \int d^2\vec{\rho}_0\, E(\vec{\rho}_0, z_0) e^{i\pi\frac{\kappa}{z}|\vec{\rho}_1 - \vec{\rho}_0|^2} \qquad (2.51)$$

where $z = z_1 - z_0$. This can be rewritten in the form

$$E(\vec{\rho}_1, z_1) = -i\frac{\kappa}{z} e^{i2\pi\kappa z} e^{i\pi\frac{\kappa}{z}\rho_1^2} \int d^2\vec{\rho}_0\, E(\vec{\rho}_0, z_0) e^{-i2\pi\frac{\kappa}{z}\vec{\rho}_1 \cdot \vec{\rho}_0} e^{i\pi\frac{\kappa}{z}\rho_0^2} \qquad (2.52)$$

which is often more convenient to work with.

As before, $D_+(\vec{\rho}, z)$ can be cast in a mixed representation. Taking the Fourier transform of Eq. 2.50, or, alternatively, adopting the small propagation-angle approximation $\kappa_\perp^4 \ll \kappa^4$, which allows the substitution

$$\sqrt{\kappa^2 - \kappa_\perp^2} \rightarrow \kappa \left(1 - \frac{\kappa_\perp^2}{2\kappa^2}\right) \tag{2.53}$$

into Eq. 2.28, we obtain

$$\mathcal{D}_+(\vec{\kappa}_\perp; z) = e^{i2\pi\kappa z} e^{-i\pi \frac{z}{\kappa} \kappa_\perp^2} \qquad \text{(Fresnel)} \tag{2.54}$$

It can be noted that the identities for the combination of free-space propagators listed in Eqs. 2.38, 2.40, 2.39, and 2.41 remain valid in the Fresnel approximation.

Examples of Fresnel Diffraction

The following are examples of Fresnel diffraction from simple secondary source configurations that are of general utility.

Point Aperture

A field emanating from a point-like aperture at $z = 0$ can be written as

$$E(\vec{\rho}, 0) = \kappa^{-2} E_0 \, \delta^2(\vec{\rho}) \tag{2.55}$$

where the prefactor κ^{-2} has been introduced for dimensional consistency. However, κ^{-2} plays another important role. If a point source is truly infinitesimally small, then the field it produces must contain spatial frequencies $\vec{\kappa}_\perp$ that are infinitely large. However, as we saw in Section 2.2.1, propagating fields can only comprise spatial frequencies limited to the radiative bandwidth $\kappa_\perp < \kappa$. As far as radiative fields are concerned, then, the apparent area of a point source can appear no smaller than $\approx \kappa^{-2}$, thus justifying the introduction of this minimum apparent area in Eq. 2.55. More will be said about this in Chapter 6.

By inserting Eq. 2.55 into Eq. 2.52, Fresnel diffraction from an elemental secondary source reduces to

$$E(\vec{\rho}, z) = -\frac{i}{\kappa z} E_0 e^{i2\pi\kappa z} e^{i\pi \frac{\kappa}{z} \rho^2} \tag{2.56}$$

which corresponds to the Fresnel approximation of a Huygens wavelet. We observe that, in addition to exhibiting a $\frac{\pi}{2}$ phase shift, expected from an elemental secondary source, the field $E(\vec{\rho}, z)$ becomes attenuated along the z axis. Equation 2.56 should be used with caution as it suggests that $E(\vec{\rho}, z)$ is constant in amplitude along the transverse direction, which is clearly unphysical. Again, this equation is valid only within the confines of the Fresnel approximation (see above).

Plane Wave

A plane wave propagating through a very large aperture at $z = 0$ can be approximated as

$$E(\vec{\rho}, 0) = E_0 \tag{2.57}$$

where, without loss of generality, the arbitrary phase of the plane wave at the aperture has been set to zero. We note that this plane wave extends laterally to a infinite distance and hence technically violates the Kirchhoff boundary requirement of a finite aperture. Nevertheless, we will adopt the approximation that the aperture is so large in size that it can be effectively treated as infinite for calculation purposes.

From Eq. 2.52, Fresnel diffraction leads to

$$E(\vec{\rho}, z) = E_0 e^{i2\pi\kappa z} \tag{2.58}$$

meaning that the diffracted wave remains a plane wave, as expected. In contrast to the example of a point aperture, we observe here that the plane wave becomes neither attenuated along the z axis nor does it exhibit a $\frac{\pi}{2}$ phase shift, the latter having quite remarkably disappeared as a result of interference from multiple secondary sources. To obtain the above result, we have taken the limit of an effectively infinite Kirchhoff aperture and made use of the "useful" integral provided by Eq. B.2.

Gaussian Beam

A third example which we will return to frequently, particularly when dealing with laser beams, is that of a Gaussian field. A Gaussian field at $z = 0$ can be written as

$$E(\vec{\rho}, 0) = E_0 e^{-\rho^2/w_0^2} \tag{2.59}$$

where w_0 is called the beam waist. Again, we take the limit of a Kirchhoff aperture that is so large, in this case so much larger than w_0, that its role may be neglected.

Fresnel diffraction then leads to

$$E(\rho, z) = \frac{E_0}{1 + i\zeta} e^{i2\pi\kappa z} e^{-\rho^2/\left[w_0^2(1+i\zeta)\right]} \tag{2.60}$$

where the scaled axial distance $\zeta = \frac{z}{\pi w_0^2 \kappa}$ has been introduced. Upon free-space propagation the Gaussian beam is found to remain Gaussian in profile, but with a width that expands in proportion to z (for $\zeta^2 \gg 1$). Moreover, the beam incurs an extra $\frac{\pi}{2}$ phase shift as $z \to \infty$. This is called the Gouy phase shift and stems from the Huygens phase shift. Indeed, as $z \to \infty$ the Gaussian beam effectively appears like a field emanating from a secondary point source at $z = 0$, resembling Eq. 2.56. Gaussian beams will be treated in more detail in Section 5.1.1.

2.4.2 Fraunhofer Diffraction

In the previous section, we considered a small-angle approximation, which ultimately led to the Fresnel diffraction integral (Eq. 2.51). We turn our attention now to a different approximation, the small-aperture approximation, for which it will be convenient to change our notation slightly. Specifically, we take the origin of our coordinate system to lie somewhere in the aperture plane, as illustrated in Fig. 2.5. That is, the field $E(\vec{r}_0)$ within the aperture is defined

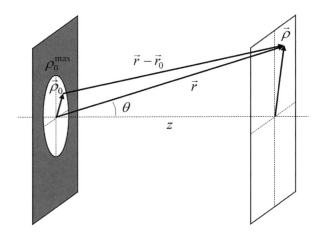

Figure 2.5. Geometry for Fraunhofer diffraction.

over the coordinates $\vec{r}_0 = (\vec{\rho}_0, 0)$. As before, our goal is to calculate a resultant field $E\left(\vec{r}\right)$ at a downstream location $\vec{r} = (\vec{\rho}, z)$, where we now suppress the subscript 1 for ease of notation.

In compliance with the Kirchhoff boundary requirements, the field $E(\vec{\rho}_0, 0)$ at the aperture plane must vanish for $\rho_0 > \rho_0^{\max}$, either because of the presence of a physical aperture or because the field inherently decays to zero off axis. If this field is taken to propagate over a distance so large that the relation $r \gg \kappa \left(\rho_0^{\max}\right)^2$ is satisfied, we can make the substitution

$$\left|\vec{r} - \vec{r}_0\right| \to r - \frac{\vec{r}_0 \cdot \vec{r}}{r} = r - \frac{\vec{\rho}_0 \cdot \vec{\rho}}{r} \tag{2.61}$$

This is variously known as the Fraunhofer or far-field approximation, and is tantamount to a small-aperture approximation. An application of the Fraunhofer approximation to the radiative Rayleigh–Sommerfeld diffraction integral (Eqs. 2.22 and 2.47), leads to

$$E(\vec{\rho}, z) = -i\frac{\kappa}{r}e^{i2\pi\kappa r}\cos\theta \int d^2\vec{\rho}_0\, E(\vec{\rho}_0, 0)e^{-i2\pi\frac{\kappa}{r}\vec{\rho}_0 \cdot \vec{\rho}} \tag{2.62}$$

where $\cos\theta = \frac{z}{r}$. This is the well-known Fraunhofer diffraction integral. In contrast to Fresnel diffraction, where $E(\vec{\rho}, z)$ and $E(\vec{\rho}_0, 0)$ are related by a convolution integral, in Fraunhofer diffraction they are related by a 2D Fourier transform (recall our discussion surrounding Eq. 1.20). For example, in the case of cylindrical symmetry, we obtain

$$E(\vec{\rho}, z) = -i\frac{\kappa}{r}e^{i2\pi\kappa r}\cos\theta\, \mathcal{E}(\kappa\sin\theta; z = 0) \tag{2.63}$$

where we have made the association $\kappa_\perp = \kappa\sin\theta$.

In cases where the small-aperture (Fraunhofer) and small-angle (paraxial) approximations are both valid, Eq. 2.62 can be further simplified to

$$E(\vec{\rho}, z) = -i\frac{\kappa}{z}e^{i2\pi\kappa z}e^{i\pi\frac{\kappa}{z}\rho^2}\int d^2\vec{\rho}_0\, E(\vec{\rho}_0, 0)e^{-i2\pi\frac{\kappa}{z}\vec{\rho}_0 \cdot \vec{\rho}} \tag{2.64}$$

referred to as the paraxial Fraunhofer diffraction integral. This is identical to the Fresnel diffraction integral given by Eq. 2.52 except that the quadratic phase factor $e^{i\pi\frac{\kappa}{z}\rho_0^2}$ has now vanished.

Examples of Fraunhofer Diffraction

The following are examples of Fraunhofer diffraction from simple secondary source configurations.

Circular Aperture

Throughout this book, we will often make use of a very common aperture configuration called a circular aperture. This is an unobstructed circular opening of radius a. A plane wave through such an aperture is given by

$$E(\vec{\rho}_0, 0) = \begin{cases} E_0 & \rho_0 \leq a \\ 0 & \rho_0 > a \end{cases} \tag{2.65}$$

$$\mathcal{E}(\vec{\kappa}_\perp; 0) = \pi a^2 E_0 \operatorname{jinc}(2\pi\kappa_\perp a) \tag{2.66}$$

where the jinc function is defined by Eq. A.16.

Such a field is cylindrically symmetric. From Eq. 2.63, Fraunhofer diffraction then leads to

$$E(\vec{\rho}, z) = -i\pi a^2 E_0 e^{i2\pi\kappa r}\frac{\kappa}{r}\cos\theta \operatorname{jinc}\left(2\pi a\kappa \sin\theta\right) \tag{2.67}$$

corresponding to the far field produced by a uniform secondary circular source. We note again the presence of the Huygens $\frac{\pi}{2}$ phase shift.

Point Aperture

As discussed above, an elemental secondary point source is effectively equivalent to an aperture of area $\pi a^2 \approx \kappa^{-2}$.

From Eq. 2.67, Fraunhofer diffraction then leads to

$$E(\vec{\rho}, z) = -\frac{i}{\kappa r}E_0 e^{i2\pi\kappa r}\cos\theta \operatorname{jinc}\left(2\sqrt{\pi}\sin\theta\right) \tag{2.68}$$

In the event only small angles are of interest, the paraxial Fraunhofer approximation applies and Eq. 2.68 further simplifies to

$$E(\vec{\rho}, z) = -\frac{i}{\kappa z}E_0 e^{i2\pi\kappa z(1+\frac{1}{2}\theta^2)} \tag{2.69}$$

where $\theta = \frac{\rho}{z}$. As expected, this is identical to the result based on the Fresnel approximation (Eq. 2.56), where it should be emphasized that in the paraxial approximation the tilt angle θ is confined to only the central lobe of the jinc function in Eq. 2.68.

We close this chapter with a schematic (Fig. 2.6) of the various free-space propagation approximations that are used throughout this book, along with their conditions of validity.

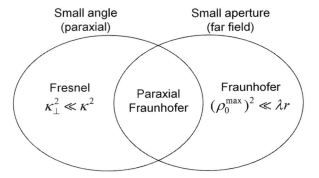

Figure 2.6. Approximations applied to the Rayleigh–Sommerfeld diffraction integral.

2.5   PROBLEMS

Problem 2.1
(a) Derive Eq. 2.24 from Eqs. 2.13 and 2.23.
(b) Derive Eq. 2.25 from Eqs. 2.14 and 2.23.

Problem 2.2
A paraxial wave propagating in the z direction may be written as

$$E(\vec{r}) = A(\vec{r})e^{i2\pi\kappa z}$$

where the envelope function $A(\vec{r})$ is slowly varying. The conditions for $A(\vec{r})$ to be slowly varying are

$$\lambda \frac{\partial A(\vec{r})}{\partial z} \ll A(\vec{r})$$

$$\lambda \frac{\partial^2 A(\vec{r})}{\partial z^2} \ll \frac{\partial A(\vec{r})}{\partial z}$$

(a) Show that in free space (no sources), the envelope function of a paraxial wave satisfies a simplified version of the Helmholtz equation given by

$$\left(\nabla_{\perp}^2 + i4\pi\kappa \frac{\partial}{\partial z}\right) A(\vec{r}) = 0$$

This equation is called the paraxial Helmholtz equation.

(b) The Fresnel free-space propagator may be written as a paraxial wave, such that

$$D_{+}(\vec{\rho}, z) = D_A(\vec{\rho}, z)e^{i2\pi\kappa z}$$

where $D_A(\vec{\rho}, z) = -i\frac{\kappa}{z}e^{i\pi\frac{\kappa}{z}\rho^2}$ is the associated envelope function. Show that $D_A(\vec{\rho}, z)$ satisfies the paraxial Helmholtz equation.

(c) The radiant field associated with a paraxial wave may be written as

$$\mathcal{E}(\vec{\kappa}_\perp; z) = \mathcal{A}(\vec{\kappa}_\perp; z)e^{i2\pi\kappa z}$$

Show that $\mathcal{A}(\vec{\kappa}_\perp; z)$ satisfies a mixed-representation version of the paraxial Helmholtz equation given by

$$\left(\pi\kappa_\perp^2 - i\kappa\frac{\partial}{\partial z}\right)\mathcal{A}(\vec{\kappa}_\perp; z) = 0 \tag{2.70}$$

(d) Finally, show that a field that satisfies the Fresnel diffraction integral also satisfies the paraxial Helmholtz equation. (Hint: this is much easier to demonstrate in the frequency domain.)

Problem 2.3

Let $A(\vec{r})$ be the envelope function of a paraxial wave, as defined in Problem 2.2. That is, $A(\vec{r})$ satisfies the paraxial Helmholtz equation. In general, $A(\vec{r})$ is complex and can be written as

$$A(\vec{r}) = \sqrt{I(\vec{r})}e^{i\phi(\vec{r})}$$

where $I(\vec{r})$ is the wave intensity and $\phi(\vec{r})$ is a phase, both real-valued.

Show that $I(\vec{r})$ and $\phi(\vec{r})$ satisfy the equation

$$2\pi\kappa\frac{\partial I(\vec{r})}{\partial z} = -\vec{\nabla}_\perp \cdot I(\vec{r})\,\vec{\nabla}_\perp\phi(\vec{r})$$

This is called the transport of intensity equation.

Problem 2.4

When a field is focused into a glass slab, the refraction at the slab interface produces aberrations in the field that cause the focus to distort. These aberrations are commonly characterized by their effect on the phase of the radiant field. Specifically, consider focusing a Gaussian field into a glass slab of index of refraction n, where the slab interface is located at $z = 0$ (see figure). When no slab is present ($n = 1$), the Gaussian field produces a focus of beam waist w_0 (see Eq. 2.59) at a distance z_0. When the slab is present, the focus is both distorted and axially displaced.

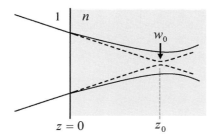

(a) Start by calculating the radiant field incident on the slab interface at $z = 0$, bearing in mind that, by symmetry, transverse momentum $\vec{\kappa}_\perp$ must be conserved. That is, $\vec{\kappa}_\perp$ must be the same on both sides of the interface. (Hint: use the Rayleigh–Sommerfeld transfer function.)

(b) Next, calculate the radiant field inside the slab. You should find that the phase of the radiant field is given by

$$\phi\left(\vec{\kappa}_\perp; z\right) = 2\pi \left(z\sqrt{n^2\kappa^2 - \kappa_\perp^2} - z_0\sqrt{\kappa^2 - \kappa_\perp^2}\right)$$

(c) While there are different ways to define the location of the new beam focus, one way is where $\phi\left(\vec{\kappa}_\perp; z\right)$ is as flat as possible about $\vec{\kappa}_\perp = 0$. Find the axial displacement of the new focus. (Hint: expand $\phi\left(\vec{\kappa}_\perp; z\right)$ in orders of $\kappa_\perp/\kappa$.)

Problem 2.5

Consider two point sources located on the x_0 axis at $x_0 = \frac{d}{2}$ and $x_0 = -\frac{d}{2}$. Use the Fresnel and Fraunhofer diffraction integrals to calculate the resultant fields $E_{\text{Fresnel}}(x, 0, z)$ and $E_{\text{Fraunhofer}}(x, 0, z)$ obtained after propagation a large distance z. Derive the corresponding intensities $I_{\text{Fresnel}}(x, 0, z)$ and $I_{\text{Fraunhofer}}(x, 0, z)$ (note: these are observed to form fringes).

(a) Derive the fringe envelope functions of $I_{\text{Fresnel}}(x, 0, z)$ and $I_{\text{Fraunhofer}}(x, 0, z)$. In particular, what is the ratio of these envelope functions at the location $x = z$?

(b) Derive the fringe periods of $I_{\text{Fresnel}}(x, 0, z)$ and $I_{\text{Fraunhofer}}(x, 0, z)$. In particular, what is the ratio of these periods at the location $x = z$? (note: the periods may vary *locally*)

(c) Which approximation, Fresnel or Fraunhofer, is better off axis?

References

[1] Arfken, G. B. and Weber, H. J. *Mathematical Methods for Physicists*, 4th edn, Academic Press (1995).
[2] Barrett, H. H. and Myers, K. J. *Foundations of Image Science*, Wiley-Interscience (2004).
[3] Born, M. and Wolf, E. *Principles of Optics*, 6th edn, Cambridge University Press (1980).
[4] Courjon, D. *Near-field Microscopy and Near-field Optics*, Imperial College Press (2003).
[5] Goodman, J. W. *Introduction to Fourier Optics*, 4th edn, W. H. Freeman (2017).
[6] Jackson, J. D. *Classical Electrodynamics*, 3rd edn, Wiley (1998).
[7] Kirchhoff, G. "Zur Theorie der Lichtstrahlen," *Weidemann Ann.* (2) 18, 663 (1883).
[8] Marion, J. B. and Heald, M. A. *Classical Electromagnetic Radiation*, 2nd edn, Academic Press (1980).
[9] Morse, P. M. and Feshbach, H. *Methods of Theoretical Physics*, McGraw-Hill (1953).
[10] Nieto-Vesperinas, M. *Scattering and Diffraction in Physical Optics*, 2nd edn, World Scientific (2006).
[11] Novotny, L. and Hecht, B. *Principles of Nano-Optics*, Cambridge University Press (2006).
[12] Sheppard, C. J. R., Kou, S. S. and Lin, J. "The Green-function transform and wave propagation," *Front. Phys.* 2, 1–10 (2014).

3 | Monochromatic Field Propagation Through a Lens

A lens is one of the most basic elements in any optical imaging device, and this chapter will consider the effect of lenses on light propagation. As motivated previously, a lens can be used to effect a Fourier transform of a field. Indeed, this application of a lens is the starting point of an entire discipline known as Fourier optics, the pioneers of which go as far back as Abbe [1] and Lord Rayleigh [9] in the late 19th century. Somewhat more recently, work by Duffieux provided the field a firm theoretical foundation [3].

The goal of this chapter is to derive some basic elements of Fourier optics that will be required throughout the remainder of this book. We will make use of the Fresnel approximation, as defined in Section 2.4.1. That is, we will consider lenses whose focal lengths cause only weak deviations of light rays. This will allow us to continue treating different polarization components of light as independent, meaning we will continue to regard a light wave as a scalar field.

Most of this chapter is derived from the classic textbook on Fourier optics by Goodman [4], though the reader is also referred to [5, 6, 10]. For more elaborate discussions that involve a fully vectorial treatment of light, the reader is referred to [2] and [8]. Finally, it should be noted that a complete description of lenses should also include a discussion of aberrations; however, these fall largely beyond the scope of this book. Detailed treatments of aberration theory can be found, for example, in [2] and [11].

3.1 LENS TRANSMISSION

For simplicity, we consider here only a thin lens, bearing in mind that a thick lens can be treated in a similar manner when the gap between principal planes is accounted for (see discussion in Section 1.4.3). A thin lens, by definition, and by virtue of its being thin, changes only the local phase of a field and not its amplitude. For complete generality, however, we will make an allowance for possible amplitude changes by defining the field transmission function of a thin lens to be

$$t(\vec{\xi}) = P(\vec{\xi})e^{-i\phi(\vec{\xi})} \tag{3.1}$$

where $\vec{\xi}$ denotes the transverse coordinates in the lens plane and $P(\vec{\xi})$ is a pupil function. For example, $P(\vec{\xi})$ can take into account the finite size of a lens by vanishing outside the lens area.

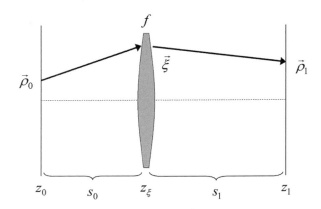

Figure 3.1. Geometry for optical propagation through a lens.

An ideal lens converts a plane wave incident from one side of the lens into a spherical wave that converges to a focal point a distance f from the other side. To obey this condition, the phase function $\phi(\vec{\xi})$ should equal $2\pi\kappa\sqrt{f^2 + \xi^2}$. In this chapter we will consider an approximation to this ideal condition, namely the Fresnel approximation, where $\phi(\vec{\xi})$ is expanded only to first order in $(\xi/f)^2$. That is, we write

$$t(\vec{\xi}) = P(\vec{\xi})e^{-i\pi\frac{\kappa}{f}\xi^2} \tag{3.2}$$

neglecting any phase shift that is constant across the lens area, since this plays no role in imaging.

We consider the imaging geometry shown in Fig. 3.1 where a field propagates from a plane z_0 to a plane z_1 by way of an intervening lens located at z_ξ. For simplicity, we assume that the index of refraction on either side of the lens is the same, relegating the more general case of unequal indices to the end of this chapter. Our goal will be to derive the field at z_1 by proceeding in steps. For brevity, we adopt the notational convention

$$E(\vec{\rho}_0, z_0) = E_0(\vec{\rho}_0) , \quad \text{etc.} \tag{3.3}$$

The first step is to derive the field incident on the lens at the lens plane. Since we have adopted the Fresnel approximation, this derivation is readily performed by making use of the Fresnel diffraction integral (Eq. 2.52), obtaining

$$E_\xi(\vec{\xi}) = -i\frac{\kappa}{s_0}e^{i2\pi\kappa s_0}e^{i\pi\frac{\kappa}{s_0}\xi^2}\int d^2\vec{\rho}_0\, e^{-i2\pi\frac{\kappa}{s_0}\vec{\rho}_0\cdot\vec{\xi}}e^{i\pi\frac{\kappa}{s_0}\rho_0^2}E_0(\vec{\rho}_0) \tag{3.4}$$

where $s_0 = z_\xi - z_0$.

The field emanating from the lens at the lens plane is then $E'_\xi(\vec{\xi}) = t(\vec{\xi})E_\xi(\vec{\xi})$, which, by our definition of the lens transmission function (Eq. 3.2) and assuming an infinitely large lens for now, leads to

$$E'_\xi(\vec{\xi}) = -i\frac{\kappa}{s_0}e^{i2\pi\kappa s_0}e^{i\pi\kappa\xi^2\left(\frac{1}{s_0}-\frac{1}{f}\right)}\int d^2\vec{\rho}_0\, e^{-i2\pi\frac{\kappa}{s_0}\vec{\rho}_0\cdot\vec{\xi}}\,e^{i\pi\frac{\kappa}{s_0}\rho_0^2}E_0(\vec{\rho}_0) \tag{3.5}$$

Finally, the field at z_1 is derived by again invoking the Fresnel diffraction integral, leading to

$$E_1(\vec{\rho}_1) = -\frac{\kappa^2}{s_0 s_1}e^{i2\pi\kappa(s_0+s_1)}e^{i\pi\frac{\kappa}{s_1}\rho_1^2}\iint d^2\vec{\xi}\,d^2\vec{\rho}_0\, e^{-i2\pi\kappa\vec{\xi}\cdot\left(\frac{\vec{\rho}_0}{s_0}+\frac{\vec{\rho}_1}{s_1}\right)}e^{i\pi\kappa\xi^2\left(\frac{1}{s_0}+\frac{1}{s_1}-\frac{1}{f}\right)}e^{i\pi\frac{\kappa}{s_0}\rho_0^2}E_0(\vec{\rho}_0) \tag{3.6}$$

where $s_1 = z_1 - z_\xi$. This cumbersome result applies for arbitrary distances s_0 and s_1 and is more general than we require. In particular, we turn to specific cases of interest where s_0 and s_1 are chosen judiciously.

3.2 FOURIER TRANSFORM WITH A LENS

To begin, we consider the case where $s_0 = f$. In this case, following an application of the "useful" integral (Eq. B.2), Eq. 3.6 simplifies to

$$E_1(\vec{\rho}_1) = -i\frac{\kappa}{f}e^{i2\pi\kappa(f+s_1)}\int d^2\vec{\rho}_0\, e^{-i2\pi\frac{\kappa}{f}\vec{\rho}_0\cdot\vec{\rho}_1}e^{i\pi\frac{\kappa}{f}\rho_0^2\left(1-\frac{s_1}{f}\right)}E_0(\vec{\rho}_0) \tag{3.7}$$

This does not constitute a remarkable simplification since the integrand still contains a quadratic phase factor. Nevertheless, Eq. 3.7 is included here since it will be useful in our discussion of holographic microscopy (Chapter 11). Integrals of the form above containing a quadratic phase factor within the integrand are called Fresnel transforms.

A much more significant simplification is occasioned when we consider $s_1 = f$. Again following an application of the "useful" integral (Eq. B.2), Eq. 3.6 becomes

$$E_1(\vec{\rho}_1) = -i\frac{\kappa}{f}e^{i2\pi\kappa(s_0+f)}e^{i\pi\frac{\kappa}{f}\rho_1^2\left(1-\frac{s_0}{f}\right)}\int d^2\vec{\rho}_0\, e^{-i2\pi\frac{\kappa}{f}\vec{\rho}_0\cdot\vec{\rho}_1}E_0(\vec{\rho}_0) \tag{3.8}$$

In this case the quadratic phase factor has been transferred from inside the integral to outside. That is, the integral has been reduced from a Fresnel transform to a simple Fourier transform. A comparison of Eq. 3.8 with Eq. 2.64 reveals that the lens configuration with $s_1 = f$ is equivalent to paraxial Fraunhofer propagation. In effect, the lens can be thought of as having displaced the propagation plane in Section 2.4.2 from a distant location in the far-field to a more proximal location a distance f from the lens, and this with the advantage that the previous constraints on the aperture size at plane 0 have now been relaxed.

Yet a further simplification to Eq. 3.8 is occasioned when $s_0 = s_1 = f$. This last geometry, illustrated in Fig. 3.2, is called a 2f configuration and serves as an important building block in optical imaging systems, which will make use of on several occasions. In a 2f configuration, Eq. 3.6 finally simplifies to

$$E_1(\vec{\rho}_1) = -i\frac{\kappa}{f}e^{i4\pi\kappa f}\int d^2\vec{\rho}_0\, e^{-i2\pi\frac{\kappa}{f}\vec{\rho}_0\cdot\vec{\rho}_1}E_0(\vec{\rho}_0) \tag{3.9}$$

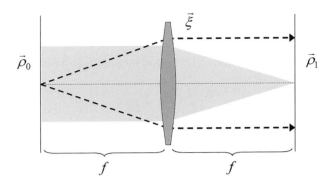

Figure 3.2. 2f configuration.

Equation 3.9 is quite remarkable. It indicates that a lens in a 2f configuration performs a perfect Fourier transform, aside from a scaling factor and an overall fixed axial phase. A Fourier reciprocal relationship can thus be established between planes z_0 and z_1, given by

$$E_1(\vec{\rho}_1) = -i\frac{\kappa}{f}e^{i4\pi\kappa f}\mathcal{E}_0\left(\frac{\kappa}{f}\vec{\rho}_1\right) \tag{3.10}$$

$$\mathcal{E}_1(\vec{\kappa}_\perp) = -i\frac{f}{\kappa}e^{i4\pi\kappa f}E_0\left(-\frac{f}{\kappa}\vec{\kappa}_\perp\right) \tag{3.11}$$

Planes in a 2f configuration relative to one another are referred to as Fourier planes. The capacity of a lens to perform a perfect Fourier transform, which was motivated in Section 1.4.1 in a hand-waving manner, has now been verified with Fresnel diffraction theory.

3.2.1 Gaussian Pupil

To arrive at Eq. 3.9 we made the implicit assumption that the lens diameter was infinite. This obviously cannot be the case in practice, raising the issue of how large a lens must be to perform a perfect (or near-perfect) Fourier transform.

Returning to the definition of lens transmission given by Eq. 3.2, the finite physical size of the lens can be incorporated into its pupil function $P(\vec{\xi})$. In most cases, this pupil function has a hard edge associated with the physical hard edge of the lens. However, in many cases it is mathematically convenient to approximate this hard edge with a Gaussian roll-off. Such a Gaussian pupil approximation leads to

$$P(\vec{\xi}) = e^{-\xi^2/w_\xi^2} \tag{3.12}$$

where w_ξ can be thought of as a "soft" lens radius.

Following the same procedure as outlined above, Eq. 3.6 becomes, in the 2f configuration,

$$E_1(\vec{\rho}_1) = -i\varsigma\frac{\kappa}{f}e^{i4\pi\kappa f}e^{i\pi\frac{\kappa}{f}\rho_1^2}\int d^2\vec{\rho}_0\, e^{-i\pi\varsigma\frac{\kappa}{f}\left|\vec{\rho}_0+\vec{\rho}_1\right|^2}e^{i\pi\frac{\kappa}{f}\rho_0^2}E_0(\vec{\rho}_0) \tag{3.13}$$

where we have introduced the variable

$$\varsigma = \left(1 + i\frac{f}{\pi w_\xi^2 \kappa}\right)^{-1} \tag{3.14}$$

If $\varsigma \to 1$ then we recover Eq. 3.9. We thus infer that the infinite-lens approximation assumed in the previous section is acceptable provided the lens radius obeys $w_\xi^2 \gg \lambda f$.

3.3 IMAGING WITH A LENS

Referring again to Eq. 3.6 describing a general field transfer through a lens, we turn to yet another configuration, namely that specified by the thin-lens formula:

$$\frac{1}{s_0} + \frac{1}{s_1} = \frac{1}{f} \tag{3.15}$$

as illustrated in Fig. 3.3.

We have encountered this formula already in Section 1.4.2, where it was found to lead to imaging, albeit imperfect. The insertion of Eq. 3.15 into Eq. 3.6 leads to another important simplification

$$E_1(\vec{\rho}_1) = -\frac{\kappa^2}{s_0 s_1} e^{i2\pi\kappa(s_0+s_1)} e^{i\pi\frac{\kappa}{s_1}\rho_1^2} \int d^2\vec{\rho}_0 \, \delta^2\left(\frac{\kappa}{s_1}\vec{\rho}_1 + \frac{\kappa}{s_0}\vec{\rho}_0\right) e^{i\pi\frac{\kappa}{s_0}\rho_0^2} E_0(\vec{\rho}_0) \tag{3.16}$$

where the definition of a 2D delta-function has been invoked from Appendix A, and the lens size has been taken to be effectively infinite. The above integration leads to

$$E_1(\vec{\rho}_1) = \frac{1}{M} e^{i2\pi\kappa(s_0+s_1)} e^{-i\pi\frac{\kappa}{fM}\rho_1^2} E_0\left(\frac{1}{M}\vec{\rho}_1\right) \tag{3.17}$$

where $M = -\frac{s_1}{s_0}$. Overlooking the quadratic phase factor for the moment, we conclude that the field E_1 is a perfect magnified replica of the field E_0, where the magnification factor is M.

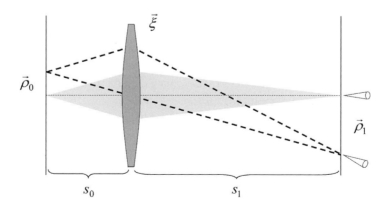

Figure 3.3. Imaging with a single lens.

In other words, the field E_1 is an image of the field E_0, and accordingly the planes z_0 and z_1 are referred to as object and image planes, respectively. The planes z_0 and z_1 are also called conjugate planes, a designation we will use repeatedly.

Unfortunately, perfect imaging cannot exist for practical and fundamental reasons:

1. A lens cannot have an infinite radius.
2. The derivation of Eq. 3.17 required the Fresnel approximation.

Both of these constraints impose limits on imaging resolution, as will be detailed below.

3.3.1 Arbitrary Pupil

To deal with the issue of a finite lens size, let us consider an arbitrary pupil function $P(\vec{\xi})$. This pupil function can characterize not only the lens size, but also any deviations from the quadratic phase transmission specified by Eq. 3.2. That is, $P(\vec{\xi})$ can be complex in general. The Fourier transform of the pupil function is defined by

$$\mathcal{P}(\vec{\kappa}_\perp) = \int d^2\vec{\xi}\, e^{-i2\pi \vec{\kappa}_\perp \cdot \vec{\xi}} P(\vec{\xi}) \tag{3.18}$$

Again following the procedure outlined above to derive Eq. 3.6, we conclude that the inclusion of an arbitrary lens pupil leads to a modification in Eq. 3.16, which now reads

$$E_1(\vec{\rho}_1) = -\frac{\kappa^2}{s_0 s_1} e^{i2\pi\kappa(s_0+s_1)} e^{i\pi \frac{\kappa}{s_1}\rho_1^2} \int d^2\vec{\rho}_0\, \mathcal{P}\left(\frac{\kappa}{s_1}\vec{\rho}_1 + \frac{\kappa}{s_0}\vec{\rho}_0\right) e^{i\pi \frac{\kappa}{s_0}\rho_0^2} E_0(\vec{\rho}_0) \tag{3.19}$$

The only difference here is that the 2D delta function δ^2 in Eq. 3.16 has been replaced by the pupil Fourier transform $\mathcal{P}$ defined by Eq. 3.18.

To proceed further, two routes can be taken. The first involves adopting yet another approximation. Specifically, if the imaging is assumed quasi-perfect (i.e. $\mathcal{P}$ is assumed to be well localized), then the integral in Eq. 3.19 becomes significant only when $\vec{\rho}_1 \approx M\vec{\rho}_0$, and the quadratic phase factor inside the integral can be transferred to the outside without significant error. That is,

$$E_1(\vec{\rho}_1) \approx -\frac{\kappa^2}{s_0 s_1} e^{i2\pi\kappa(s_0+s_1)} e^{-i\pi \frac{\kappa}{\bar{J}M}\rho_1^2} \int d^2\vec{\rho}_0\, \mathcal{P}\left(\frac{\kappa}{s_1}\vec{\rho}_1 + \frac{\kappa}{s_0}\vec{\rho}_0\right) E_0(\vec{\rho}_0) \tag{3.20}$$

In imaging applications where we are interested only in the detected intensity at the image plane, this quadratic phase factor outside the integral plays no role and can be discarded. In the event this quadratic phase factor is important for some reason, or if $\mathcal{P}$ is not as well localized as one would like, then we can restrict ourselves to conditions where the

Fraunhofer approximation applies. That is, we can invoke the limits $\frac{\kappa \rho_0^2}{s_0} \to 0$ and $\frac{\kappa \rho_1^2}{s_1} \to 0$, obtaining

$$E_1(\vec{\rho}_1) \approx -\frac{\kappa^2}{s_0 s_1} e^{i2\pi\kappa(s_0+s_1)} \int d^2\vec{\rho}_0 \, \mathcal{P}\left(\frac{\kappa}{s_1}\vec{\rho}_1 + \frac{\kappa}{s_0}\vec{\rho}_0\right) E_0(\vec{\rho}_0) \qquad (3.21)$$

It should be noted that it is precisely the quadratic phase factor in Eq. 3.20 (or Eq. 3.17) that leads to a coupling between the position variable $\vec{\rho}_1$ in the image plane and the effective direction of the light ray from this position, as depicted in Fig. 3.3. The coupling between position and direction was already encountered in our purely ray-optic description of imaging prescribed by the transfer matrix in Eq. 1.26, which was termed as non-telecentric because of the off-diagonal coupling element $-1/f$.

We turn now to an alternative route for proceeding beyond Eq. 3.19 that eliminates the presence of quadratic phases altogether.

3.3.2 4f Imaging

In the previous section, imaging was based on the use of a single lens. This configuration led to Eq. 3.19, which in turn imposed certain assumptions and approximations to arrive at Eq. 3.21. The presence of the quadratic phase factors in Eq. 3.19 is a drawback of imaging with a single lens. A simple and rigorous way to eliminate these phase factors is based on an alternative imaging geometry that involves not one but two lenses, distributed in what is known as a 4f configuration, illustrated in Fig. 3.4. This geometry is equivalent to two 2f configurations in tandem, meaning that it effectively performs two Fourier transforms in series. By applying Eq. 3.9 twice, we directly arrive at

$$E_1(\vec{\rho}_1) = -\frac{\kappa^2}{f_0 f_1} e^{i4\pi\kappa(f_0+f_1)} \int d^2\vec{\rho}_0 \, \mathcal{P}\left(\frac{\kappa}{f_1}\vec{\rho}_1 + \frac{\kappa}{f_0}\vec{\rho}_0\right) E_0(\vec{\rho}_0) \qquad (3.22)$$

where the pupil function $P(\vec{\xi})$ is now located at the intermediate plane between the two lenses. This intermediate plane is known as an aperture plane. It is a Fourier plane relative to both

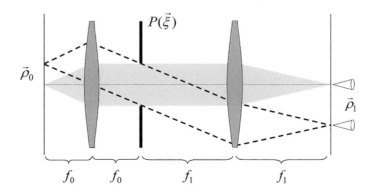

Figure 3.4. 4f configuration.

the object plane (z_0) and the image plane (z_1), which are themselves conjugate planes. Aside from the Fresnel approximation, no additional assumptions or approximations were required to arrive at Eq. 3.22.

As illustrated in Fig. 3.4, the coupling between position and direction at the image plane has been eliminated, and imaging is said to be telecentric.

3.3.3 Amplitude Spread and Transfer Functions

To gain a better understanding of the imaging relation given by Eq. 3.22, we introduce what we will refer to throughout this book as the amplitude point spread function H, defined by

$$H(\vec{\rho}) = \left(\frac{\kappa}{f_0}\right)^2 \int d^2\vec{\xi} \, e^{i2\pi \frac{\kappa}{f_0} \vec{\rho} \cdot \vec{\xi}} P(\vec{\xi}) \tag{3.23}$$

(also variously known as the amplitude spread function or impulse response, or coherent spread function).

As can be noted from a comparison with Eq. 3.18, H is simply a scaled version of $\mathcal{P}$. That is, we can write

$$H(\vec{\rho}) = \left(\frac{\kappa}{f_0}\right)^2 \mathcal{P}\left(-\frac{\kappa}{f_0}\vec{\rho}\right) \tag{3.24}$$

If we are dealing with single-lens imaging (Eq. 3.21) rather than 4f imaging, then f_0 must be replaced by s_0 in the above definitions, but otherwise these definitions remain unchanged.

The imaging relation given by Eq. 3.22 (or Eq. 3.21) can now be recast in the more revealing form

$$E_1(\vec{\rho}_1) = \frac{1}{M}e^{i4\pi\kappa(f_0+f_1)} \int d^2\vec{\rho}_0 \, H\left(\frac{1}{M}\vec{\rho}_1 - \vec{\rho}_0\right) E_0(\vec{\rho}_0) \tag{3.25}$$

where M is the magnification factor defined by $-\frac{f_1}{f_0}$ (or $-\frac{s_1}{s_0}$). The image and object fields $E_1(\vec{\rho}_1)$ and $E_0(\vec{\rho}_0)$ are now manifestly related by a convolution. That is, the function $H(\vec{\rho})$ acts as an imaging propagator in an entirely analogous way as $D(\vec{\rho})$ acted as a free-space propagator in Chapter 2. Note that the prefactor $e^{i4\pi\kappa(f_0+f_1)}$ corresponds to a fixed uniform phase in the image field that is generally discarded since it plays no role in imaging.

Continuing our analogy with free-space propagation, we recall from our discussion of Rayleigh-Sommerfeld diffraction in Chapter 2 that the radiant-field free-space transfer function $\mathcal{D}$ is defined as the Fourier transform of the field free-space propagator D. In an identical way, a radiant-field imaging transfer function can be defined as

$$\mathcal{H}(\vec{\kappa}_\perp) = \int d^2\vec{\rho} \, e^{-i2\pi\vec{\kappa}_\perp \cdot \vec{\rho}} H(\vec{\rho}) \tag{3.26}$$

which is referred to as the coherent transfer function. This is simply a scaled version of the pupil function. That is,

$$\mathcal{H}(\vec{\kappa}_\perp) = P\left(\frac{f_0}{\kappa}\vec{\kappa}_\perp\right) \tag{3.27}$$

The convolution relation given by Eq. 3.25 translates in the frequency domain to a simple product relation,

$$\mathcal{E}_1(\vec{\kappa}_\perp) = M e^{i4\pi\kappa(f_0+f_1)} \mathcal{H}(M\vec{\kappa}_\perp)\mathcal{E}_0(M\vec{\kappa}_\perp) \tag{3.28}$$

where again the prefactor $e^{i4\pi\kappa(f_0+f_1)}$ is generally discarded.

Equations 3.24, 3.25, 3.27, and 3.28 are fundamental imaging relations which, together with Eq. 3.10, serve as the foundation of the Fourier optics theory of imaging. The functions H and $\mathcal{H}$ are referred to as "coherent" because they apply specifically to imaging with monochromatic fields. We will consider imaging with non-monochromatic fields in the next chapter, but before turning to this we must derive some basic properties of H and $\mathcal{H}$.

3.3.4 Spatial Bandwidth and Resolution

To begin, let us consider the response of an imaging system to a point-like field in the object plane. In other words, let us write, as we did in Section 2.4.1, $E_0(\vec{\rho}_0) \propto \delta^2(\vec{\rho}_0)$, neglecting dimensions and prefactors. From Eq. 3.25, the resultant field produced at the image plane is $E_1(\vec{\rho}_1) \propto H(\frac{1}{M}\vec{\rho}_1)$. This image field is not a point, but rather a field whose profile is spread over a width defined both by the magnification M and by H. The *apparent* object field, that is the field projected back to the object plane by correcting for the image magnification M, has a spread defined by H only. The role of H thus becomes clear: it corresponds to the response of the imaging system to a point-like field in the object plane (see Fig. 3.5). The reader is reminded that we have already encountered a similar phenomenon of apparent spreading of a point-like field. Specifically, in 2.4.1 we concluded that radiative propagation alone was sufficient to cause an apparent spreading of a point-like field to an area no smaller than $\approx \kappa^{-2}$. As we will see below, the apparent spreading caused by imaging (i.e. by H) is more severe.

Before proceeding, let us turn our attention to the coherent transfer function $\mathcal{H}$ of an imaging system and introduce two fundamental concepts: spatial bandwidth and spatial resolution. The spatial bandwidth is denoted by $\Delta\kappa_\perp$, corresponding to the range of transverse wavevectors spanned (or supported) by $\mathcal{H}$. Associated with this spatial bandwidth $\Delta\kappa_\perp$ is the spatial resolution $\delta\rho$, defined by $1/\Delta\kappa_\perp$. Again, these definitions apply specifically to imaging with monochromatic fields.

As an example, let us consider a 4f imaging system where $\mathcal{H}$ is defined by an unobstructed circular pupil of radius a, with pupil function

$$P(\vec{\xi}) = \begin{cases} 1 & \xi \leq a \\ 0 & \xi > a \end{cases} \tag{3.29}$$

$$\mathcal{P}(\vec{\kappa}_\perp) = \pi a^2 \mathrm{jinc}(2\pi\kappa_\perp a) \tag{3.30}$$

where the jinc function is defined by Eq. A.16. Such a pupil function is routinely encountered in imaging applications and describes an imaging system with no aberrations.

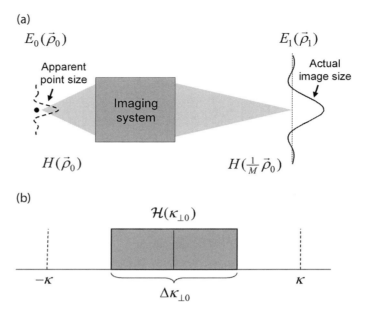

Figure 3.5. (a) Apparent spreading of a point-like object field. (b) Spatial bandwidth defined by $\mathcal{H}$.

From the definition of $\mathcal{H}$ provided by Eq. 3.27, the full span of $\vec{\kappa}_\perp$ is found to be

$$\Delta\kappa_{\perp 0} = 2\kappa\frac{a}{f_0} \tag{3.31}$$

where the range $\Delta\kappa_{\perp 0}$ is the spatial bandwidth of the imaging system (see Fig. 3.5), and the spatial resolution is accordingly

$$\delta\rho_0 = \frac{1}{\Delta\kappa_{\perp 0}} = \frac{f_0}{2\kappa a} \tag{3.32}$$

The subscript 0 has been explicitly introduced in our notation for $\Delta\kappa_{\perp 0}$ and $\delta\rho_0$ to emphasize that these correspond to the apparent spatial bandwidth and resolution of the imaging system as referred to the object plane. To make this point clear, let us return to the example of a point-like field in the object plane. Because of H, the resultant field distribution at the image plane is spread to a width no smaller than $\delta\rho_1$, meaning that the apparent width of the field when projected back to the object plane is no smaller than $\delta\rho_0 = \delta\rho_1/\,|M|$.

An equivalent argument holds true in the frequency domain and is depicted in Fig. 3.6. If the actual field distribution in the object plane contains a large spread of spatial frequencies, then only the subset of these frequencies supported by $\mathcal{H}$ can be projected to the image plane, yielding a maximum spatial bandwidth at the image plane defined by $\Delta\kappa_{\perp 1}$. When projected back to the object plane, this apparent maximum spatial bandwidth becomes $\Delta\kappa_{\perp 0}$. The relationships between actual resolution and bandwidth in the image plane versus apparent resolution and bandwidth in the object plane are summarized by

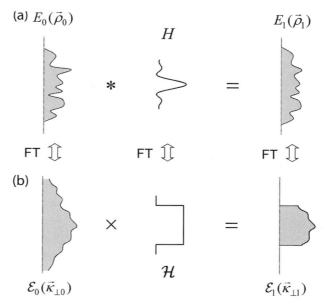

Figure 3.6. (a) Effect of H on an object field. (b) Effect of $\mathcal{H}$ on an object radiant field.

$$\frac{\delta\rho_1}{\delta\rho_0} = \frac{\Delta\kappa_{\perp 0}}{\Delta\kappa_{\perp 1}} = |M| \tag{3.33}$$

In imaging applications, it is conventional to characterize the angular width of the aperture pupil by its numerical aperture, defined by

$$\text{NA}_0 = n\sin\theta_0^{\text{max}} \tag{3.34}$$

where θ_0^{max} denotes the maximum half-angular width allowed by the pupil, as seen from the object plane, and n denotes index of refraction. Making the association $\kappa_{\perp 0} = \kappa\sin\theta_0$ and recalling that $\kappa = \frac{n}{\lambda}$, where λ is the free-space optical wavelength, Eq. 3.34 can be rewritten as

$$\text{NA}_0 = \frac{n\Delta\kappa_{\perp 0}}{2\kappa} = \frac{\lambda\Delta\kappa_{\perp 0}}{2} \tag{3.35}$$

and Eq. 3.32 can be rewritten as

$$\delta\rho_0 = \frac{\lambda}{2\text{NA}_0} \tag{3.36}$$

This is the well-known diffraction-limited resolution, which we will have ample occasion to revisit.

A potential source of confusion should be dispelled here, since, at first glance, it might appear that Eqs. 3.32 and 3.36 are incompatible. Indeed, we are tempted to associate $\frac{a}{f_0}$ with $\tan\theta_0^{\text{max}}$ rather than $\sin\theta_0^{\text{max}}$. There are two ways to resolve this apparent incompatibility. First, if we restrict ourselves to the Fresnel approximation, then θ_0^{max} is small and $\tan\theta_0^{\text{max}} \approx \sin\theta_0^{\text{max}}$. But we can actually go beyond the Fresnel approximation. Lenses in modern microscopes are

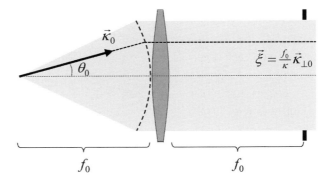

Figure 3.7. Abbe sine condition.

generally designed to obey what is known as the Abbe sine condition (see [7]). In effect, this condition establishes a linear correspondence between transverse wavevector components $\vec{\kappa}_{\perp 0}$, in the object (or image) plane, and spatial coordinates $\vec{\xi}$ in the pupil plane. This correspondence is depicted in Fig. 3.7. A consequence of the Abbe sine condition is that it establishes the relations

$$\sin \theta_0^{\max} = \frac{\Delta \kappa_{\perp 0}}{2\kappa} = \frac{a}{f_0} \tag{3.37}$$

(with similar relations for tilt angles and spatial frequencies viewed from the image plane). The source of confusion thus comes from our assumption of a flat principal plane, as associated with a thin lens. For a thick lens obeying the Abbe sine condition, a principal plane can be thought of instead as effectively spherical.

It should be emphasized that a linear correspondence between $\vec{\kappa}_{\perp 0}$ and $\vec{\xi}$ represents a mainstay of Fourier imaging theory, as summarized by Eqs. 3.24 or 3.27. But throughout this chapter, this correspondence was derived from the Fresnel approximation. The fact that the Abbe sine condition can extend this correspondence beyond the Fresnel approximation explains why it is routine to find Fourier imaging theory applied to microscopes with numerical apertures even as high as ≈ 0.7.

Techniques for Increasing Bandwidth

Based on a consideration of Eq. 3.31, a straightforward way to increase spatial bandwidth is to increase the size of the pupil aperture. In practice, however, this is usually difficult because it requires increasing the numerical aperture of lenses, which becomes technically problematic for large numerical apertures. In any event, even if it were possible to increase the aperture pupil to infinite size, the spatial bandwidth would still be limited by the constraints of radiative propagation (see Section 2.2.1), meaning that the bandwidth would remain fundamentally bounded by

$$\Delta \kappa_{\perp} \leq 2\kappa = 2\frac{n}{\lambda} \tag{3.38}$$

as depicted in Fig. 3.5. Such a bound corresponds to $NA_0 \leq n$. In practice, imaging systems rarely come close to approaching this limit.

A more fundamental technique for increasing spatial bandwidth is to increase κ by either reducing λ or increasing n. In the limit that $\Delta\kappa_\perp$ becomes very large, then even if the aperture pupil is finite in size we have $\mathcal{H}(\vec{\kappa}_\perp) \to 1$ and $H(\vec{\rho}) \to \delta^2(\vec{\rho})$. This limit leads to Eq. 3.17, corresponding to "perfect" imaging. As an example, matter waves generally exhibit much smaller wavelengths than optical waves, and for this reason electron microscopes typically provide much higher resolution that optical microscopes. More will be said about resolution in Chapter 19.

3.3.5 Properties of Amplitude Spread Functions

The following are useful properties and identities related to H's:

Useful Properties
- If $P(\vec{\xi}) = P^*(\vec{\xi})$ (i.e. pupil is real), then $H(\vec{\rho}) = H^*\left(-\vec{\rho}\right)$ (i.e. $H(\vec{\rho})$ is Hermitian).
- If $P(\vec{\xi}) = P(-\vec{\xi})$ (i.e. pupil is symmetric) then $H(\vec{\rho}) = H\left(-\vec{\rho}\right)$.
- If the pupil is real and symmetric, then H is real and symmetric.

Useful Identities
- $\int d^2\vec{\rho}_c\, H\left(\vec{\rho}_c + \tfrac{1}{2}\vec{\rho}_d\right) H^*\left(\vec{\rho}_c - \tfrac{1}{2}\vec{\rho}_d\right) = \left(\frac{\kappa}{f}\right)^2 \int d^2\vec{\xi}\, e^{i2\pi \frac{\kappa}{f}\vec{\rho}_d \cdot \vec{\xi}} P(\vec{\xi})P^*(\vec{\xi})$
- $\int d^2\vec{\rho}_d\, H\left(\vec{\rho}_c + \tfrac{1}{2}\vec{\rho}_d\right) H^*\left(\vec{\rho}_c - \tfrac{1}{2}\vec{\rho}_d\right) = \left(\frac{2\kappa}{f}\right)^2 \int d^2\vec{\xi}\, e^{i4\pi \frac{\kappa}{f}\vec{\rho}_c \cdot \vec{\xi}} P(\vec{\xi})P^*(-\vec{\xi})$
- $\iint d^2\vec{\rho}_c\, d^2\vec{\rho}_d\, H\left(\vec{\rho}_c + \tfrac{1}{2}\vec{\rho}_d\right) H^*\left(\vec{\rho}_c - \tfrac{1}{2}\vec{\rho}_d\right) = |P(0)|^2$
- If $\left|P(\vec{\xi})\right|^2 = P(\vec{\xi})$ (i.e. $P(\vec{\xi}) = 0$ or 1), then $\int d^2\vec{\rho}_c\, H\left(\vec{\rho}_c + \tfrac{1}{2}\vec{\rho}_d\right) H^*\left(\vec{\rho}_c - \tfrac{1}{2}\vec{\rho}_d\right)$
$= H\left(\vec{\rho}_d\right)$

Normalization
In a 4f imaging configuration, the wavevectors from the object plane that contribute to imaging span a solid angle Ω_0, which is defined by the pupil aperture. Using the identity

$$d^2\vec{\Omega}_\kappa = \frac{d^2\vec{\kappa}_\perp}{\kappa\,|\kappa_z|} \tag{3.39}$$

we can calculate this solid angle to be

$$\Omega_0 = \frac{1}{\kappa} \int \frac{d^2\vec{\kappa}_\perp}{|\kappa_z|} \left|P\left(\frac{f_0}{\kappa}\vec{\kappa}_\perp\right)\right|^2 \tag{3.40}$$

where the absolute-value square of the pupil function is taken rather than the pupil function itself, as a measure of the light power through the pupil (see Chapter 4).

In the case of a circular pupil defined by Eq. 3.29, we find, upon integration

$$\Omega_0 = 2\pi \left(1 - \sqrt{1 - \left(\frac{\Delta \kappa_{\perp 0}}{2\kappa} \right)^2} \right) \approx \pi \left(\frac{\Delta \kappa_{\perp 0}}{2\kappa} \right)^2 \tag{3.41}$$

where the last relation comes from the paraxial approximation. The solid angle Ω_0 can thus be interpreted as a normalized apparent 2D spatial bandwidth of the object (2D because it refers to an "area" in frequency space). Correspondingly, the normalized actual 2D spatial bandwidth of the field at the image plane is $\Omega_1 = \Omega_0 / M^2$.

In the case of an unobstructed circular pupil, H and $\mathcal{H}$ are normalized as follows:

$$H(0) = \int d^2 \vec{\kappa}_\perp \, \mathcal{H}(\vec{\kappa}_\perp) = \frac{\pi}{4} \Delta \kappa_{\perp 0}^2 \approx \kappa^2 \Omega_0 \tag{3.42}$$

$$\mathcal{H}(0) = \int d^2 \vec{\rho} \, H(\vec{\rho}) = 1 \tag{3.43}$$

In the more general case where $P(0)$ need not be equal to 1, then

$$\mathcal{H}(0) = \int d^2 \vec{\rho} \, H(\vec{\rho}) = P(0) \tag{3.44}$$

3.4 EFFECT OF SURROUNDING MEDIA

Throughout this chapter, we have made the assumption that the object and image, and the lenses themselves, are surrounded by a medium of uniform index of refraction n. In practice, one often encounters situations that are more complicated. For example, the object can be immersed in one medium (e.g. water) while the image is immersed in another (typically air). Figure 3.8 depicts a 4f imaging configuration that is more general, where an allowance is made for three different indices of refraction n_0, n_ξ, and n_1. Associated with these indices of refraction are three different wavenumbers κ_0, κ_ξ, and κ_1. But these are not the only differences to take into account. We must also bear in mind that the focal lengths on either side of each lens are different (see Section 1.4.3). As an example, when considering propagation from left to right in Fig. 3.8, the field transmission through the leftmost lens formerly defined by Eq. 3.2 should be modified to read

$$t(\vec{\xi}) = P(\vec{\xi}) e^{-i\pi \frac{\kappa_\xi}{f_0'} \xi^2} \tag{3.45}$$

A derivation of the results in this chapter taking into account these changes is straightforward, and only the final results are presented. Making use of the equalities from Section 1.4.3

$$\frac{\kappa_0}{f_0} = \frac{\kappa_\xi}{f_0'} \qquad \text{and} \qquad \frac{\kappa_\xi}{f_1'} = \frac{\kappa_1}{f_1} \tag{3.46}$$

these results may be summarized succinctly. Basically, two key modifications must be made. First, throughout the equations in this chapter, all instances of $\frac{\kappa}{f_0}$ and $\frac{\kappa}{f_1}$ should be replaced by

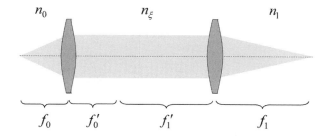

Figure 3.8. 4f geometry with different surrounding media.

$\frac{\kappa_0}{f_0}$ and $\frac{\kappa_1}{f_1}$, respectively (the same applies to instances of $\frac{\kappa}{s_0}$, etc.). For example, $\mathcal{H}$ (Eq. 3.27) now reads

$$\mathcal{H}(\vec{\kappa}_\perp) = P\left(\frac{f_0}{\kappa_0}\vec{\kappa}_\perp\right) \tag{3.47}$$

The second modification is in our definition of magnification, which should be replaced by $M = -\frac{n_0}{n_1}\frac{f_1}{f_0}$ (or $M = -\frac{n_0}{n_1}\frac{s_1}{s_0}$).

For completeness, a third modification should also be made, which is much less consequential. Prefactors such as $e^{i4\pi\kappa(f_0+f_1)}$ should everywhere be replaced by $e^{i2\pi\left(\kappa_0 f_0 + \kappa_\xi f_0' + \kappa_\xi f_1' + \kappa_1 f_1\right)}$ (the same for $e^{i4\pi\kappa(s_0+s_1)}$). These prefactors simply take into account the overall phase accumulation that comes from propagation from one plane to another, and their role is generally discarded.

3.5   PROBLEMS

Problem 3.1

Consider a 4f imaging system of unit magnification (i.e. both lenses of focal length f), with an unobstructed circular pupil of radius a.

(a) Derive $H(\rho)$ in the case where an obstructing disk of radius $b < a$ is inserted into the pupil.

(b) Derive $H(\rho)$ in the case where the disk is transmitting but produces a phase shift of $90°$.

(c) Derive $H(\rho)$ in the case where the disk is transmitting but produces a phase shift of $180°$.

(d) Consider imaging an on-axis point source of light with either of the above systems. Compared to the unobstructed pupil system, is it possible to obtain an increase in the image intensity on axis? If so, under what conditions? Is it possible to obtain a null in the image intensity on axis? If so, under what conditions?

Problem 3.2

Consider inserting a thin wedge into an otherwise unobstructed circular pupil of radius a of a 4f imaging system (both lenses of focal length f). The wedge induces a phase shift that varies linearly from 0 at the far left to 2ϕ at the far right of the pupil. Derive H for this imaging system. (Hint: use the Fourier shift theorem.)

Problem 3.3

Consider imparting a spiral phase onto an otherwise azimuthally symmetric pupil. That is, if the pupil coordinates are $\vec{\xi} = (\xi \cos \varphi, \xi \sin \varphi)$, then the pupil function is given by $P(\vec{\xi}) = P(\xi)\, e^{im\varphi}$, where m is a positive integer. Assume unit-magnification 4f imaging, with lenses of focal length f. Derive a general expression for the amplitude point spread function $H(\vec{\rho})$ associated with this spiral-phase pupil. (Hint: make use of the Fourier transform properties of separable functions in cylindrical coordinates found in Appendix A.3. Your result should be in the form of a simple integral containing J_m.)

Problem 3.4

Derive Eqs. 3.16 and 3.17 from Eq. 3.6.

Problem 3.5

(a) Show that if $P(\vec{\xi})$ is binary (i.e. $P(\vec{\xi}) = 0$ or 1), then

$$\int d^2\vec{\rho}_c\, H(\vec{\rho}_c + \tfrac{1}{2}\vec{\rho}_d)H^*(\vec{\rho}_c - \tfrac{1}{2}\vec{\rho}_d) = H(\vec{\rho}_d)$$

(b) What is the implication of the above relation? In particular, what does it say about the imaging properties of two identical, unit-magnification, binary pupil imaging systems arranged in series?

References

[1] Abbe, E. "Beiträge zur Theorie des Mikroskops und der Mikroskopischen Wahrnehmung," *Archiv. Mikroskopische Anat.* 9, 413–468 (1873).
[2] Born, M. and Wolf, E. *Principles of Optics*, 6th edn, Cambridge University Press (1980).
[3] Duffieux, P. M. *l'Intégrale de Fourier et ses Applications à l'Optique*, Faculté des Sciences, Besançon (1946).
[4] Goodman, J. W. *Introduction to Fourier Optics*, 4th edn, W. H. Freeman (2017).
[5] Hecht, E. *Optics*, 4th edn, Addison Wesley (2002).
[6] Klein, M. V. and Furtak, T. E. *Optics*, 2nd edn, Wiley (1986).
[7] Mansuripur, M. *Classical Optics and its Applications*, Cambridge University Press (2002).
[8] Novotny, L. and Hecht, B. *Principles of Nano-Optics*, Cambridge University Press (2006).
[9] Lord Rayleigh, "On the theory of optical images with special reference to the microscope," *Phil. Mag.* (5) 42, 167 (1896).
[10] Saleh, B. E. A. and Teich, M. C. *Fundamentals of Photonics*, 2nd edn, Wiley Interscience (2007).
[11] Smith, W. J. *Modern Optical Engineering*, 3rd edn, SPIE Press and McGraw Hill (2000).

4 Intensity Propagation

By definition, optical imaging involves the detection of light. Because standard detectors measure light intensities and not light fields, the characterization of an optical imaging system requires an understanding of how light fields from an object interact to form a final detected intensity at an image plane. At first glance, it might seem we have already solved this problem in the preceding chapter, since we managed to fully characterize the propagation of fields through an imaging system by the system amplitude point spread function H (or the coherent transfer function $\mathcal{H}$). That is, if the field at the image plane is known, then the intensity must be known as well. However, we made a fundamental assumption in the preceding chapter. We dealt only with monochromatic fields, meaning that we assumed the time dependence of the fields to be harmonic. While this assumption might be valid in certain cases (for example when imaging with laser light), it is not valid in general. For example, though the local amplitude of an object field might be well defined, its local phase might be randomly varying as a function of time or space. If such is the case, both the phase and amplitude of the resulting image field are also random, since the image field is constructed from a superposition of object fields. When such a situation arises, light fields cannot be treated deterministically, as in the preceding chapter, but instead must be treated statistically.

In particular, in Section 1.2, the intensity of a light field was defined to be

$$I(\vec{\rho}) = \left\langle \left| E(\vec{\rho}, t) \right|^2 \right\rangle \tag{4.1}$$

For generality, the fields are taken here to be explicitly time dependent. This time dependence might be harmonic, as for a monochromatic field, or some other function of time that deviates from harmonic. In the former case, the time dependence can be characterized by a single frequency; in the latter case it involves a range a frequencies, called a frequency bandwidth. As we will find in Chapter 7, the effects of a non-zero frequency bandwidth can be quite dramatic on short time scales. However, we will also find that these effects can basically be ignored when averaging over longer time scales. The duration that separates long from short time scales is called the coherence time.

Our goal in this chapter is to relax the constraint of monochromaticity and consider fields whose time dependence is not necessarily deterministic. That is, we will allow fields to exhibit extended frequency bandwidths. The conclusions established in this chapter will be valid for long time averages only, specifically *longer* than the coherence time (though this term has yet to be defined). Such time averaging, or temporal filtering, is denoted by brackets $\langle ... \rangle$. We further assume that, upon temporal filtering, the light intensity becomes fairly constant, allowing us to drop the explicit time dependence from our notation yet again, though for different reasons than in the preceding chapter. Most of this chapter is derived from the textbooks by Goodman [4] and [5], though other references such as [1, 3, 6, 7, 9, 10] are also recommended. While we will briefly touch upon the theory of coherence both here and in future chapters, the reader may wish to consult Mandel and Wolf [8] for an authoritative discussion. A highly readable and condensed introduction to the theory of coherence is also presented in [11].

Since the aim of this chapter is to provide basic insight into the behavior of non-deterministic fields, we avoid dealing with the subtleties arising from the $\cos \theta$ obliquity factors encountered in Chapter 2, and systematically restrict ourselves to the paraxial, or Fresnel, approximation. These subtleties will be revisited in Chapter 6.

4.1 FIELD CORRELATIONS

Let us consider a detected intensity at an image plane. As defined by Eq. 4.1, this intensity involves a product of two fields, namely a field and its complex conjugate. Though these fields are co-localized at the image plane, they presumably arise from any pairwise combination of fields that may or may not be co-localized at the object plane. As we will see, the degree to which these pairwise combinations contribute to the final image intensity depends on the correlations between these the object fields. Much of this chapter will be devoted to establishing a formalism to characterize these field correlations in space, and thus introduce the concept of spatial coherence. We begin with a few definitions:

Mutual Intensity

The mutual intensity characterizes the time-averaged correlation between two fields at two different points in space. For arbitrary points in 3D space, the mutual intensity is defined by

$$J(\vec{r}, \vec{r}') = \left\langle E(\vec{r})E^*(\vec{r}') \right\rangle \tag{4.2}$$

In the restricted case where the two points are located within the same 2D transverse plane, the mutual intensity reduces to

$$J(\vec{\rho}, \vec{\rho}') = \left\langle E(\vec{\rho})E^*(\vec{\rho}') \right\rangle \tag{4.3}$$

Throughout this chapter, we will be concerned mostly with field correlations in 2D, and thus limit ourselves to this last definition. In terms of mutual intensity, the intensity at location $\vec{\rho}$ becomes simply

$$I(\vec{\rho}) = J(\vec{\rho}, \vec{\rho}) \tag{4.4}$$

A variable change will often prove to be very convenient, and we introduce the new coordinates

$$\begin{aligned} \vec{\rho}_c &= \tfrac{1}{2}(\vec{\rho} + \vec{\rho}') \\ \vec{\rho}_d &= \vec{\rho} - \vec{\rho}' \end{aligned} \tag{4.5}$$

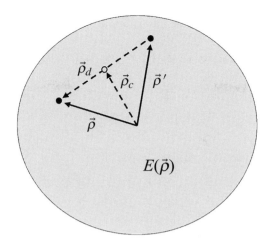

Figure 4.1. Coordinate systems for calculation of mutual intensity.

which denote respectively the mean (center) and difference vectors between the pair of points $\vec{\rho}$ and $\vec{\rho}'$, as illustrated in Fig. 4.1. Useful identities related to this coordinate transformation are

$$\mathrm{d}^2\vec{\rho}\, \mathrm{d}^2\vec{\rho}' = \mathrm{d}^2\vec{\rho}_c\, \mathrm{d}^2\vec{\rho}_d \tag{4.6}$$

$$\rho^2 - \rho'^2 = 2\vec{\rho}_c \cdot \vec{\rho}_d \tag{4.7}$$

The mutual intensity and intensity are thus re-defined as

$$J(\vec{\rho}_c, \vec{\rho}_d) = \left\langle E\left(\vec{\rho}_c + \tfrac{1}{2}\vec{\rho}_d\right) E^*\left(\vec{\rho}_c - \tfrac{1}{2}\vec{\rho}_d\right) \right\rangle \tag{4.8}$$

and

$$I(\vec{\rho}_c) = J(\vec{\rho}_c, 0) \tag{4.9}$$

Both definitions of mutual intensity will be variously adopted depending on which coordinate system, $(\vec{\rho}, \vec{\rho}')$ or $(\vec{\rho}_c, \vec{\rho}_d)$, is most appropriate.

Radiant Mutual Intensity

From Eq. 1.9, the radiant field associated with a field in a 2D axial plane is given by

$$\mathcal{E}(\vec{\kappa}_\perp) = \int \mathrm{d}^2\vec{\rho}\, E(\vec{\rho}) e^{-i2\pi\vec{\kappa}_\perp \cdot \vec{\rho}} \tag{4.10}$$

We can define the radiant mutual intensity in a similar manner as the mutual intensity, though in the frequency domain

$$\mathcal{J}(\vec{\kappa}_\perp, \vec{\kappa}_\perp') = \left\langle \mathcal{E}(\vec{\kappa}_\perp)\mathcal{E}^*(\vec{\kappa}_\perp') \right\rangle \tag{4.11}$$

Again, this radiant mutual intensity is a restricted version which is a function of transverse coordinates only. A more general version will be considered in the next chapter.

The mutual and radiant mutual intensities are thus related by a double Fourier transform

$$\mathcal{J}(\vec{\kappa}_\perp, \vec{\kappa}_\perp') = \iint d^2\vec{\rho}\, d^2\vec{\rho}'\, J(\vec{\rho}, \vec{\rho}') e^{-i2\pi(\vec{\kappa}_\perp \cdot \vec{\rho} - \vec{\kappa}_\perp' \cdot \vec{\rho}')} \tag{4.12}$$

Adopting the same variable change in the frequency domain, we write

$$\begin{aligned}
\vec{\kappa}_{\perp c} &= \tfrac{1}{2}(\vec{\kappa}_\perp + \vec{\kappa}_\perp') \\
\vec{\kappa}_{\perp d} &= \vec{\kappa}_\perp - \vec{\kappa}_\perp'
\end{aligned} \tag{4.13}$$

which leads to the useful identities

$$d^2\vec{\kappa}_\perp d^2\vec{\kappa}_\perp' = d^2\vec{\kappa}_{\perp c} d^2\vec{\kappa}_{\perp d} \tag{4.14}$$

$$\kappa_\perp^2 - \kappa_\perp'^2 = 2\vec{\kappa}_{\perp c} \cdot \vec{\kappa}_{\perp d} \tag{4.15}$$

and

$$\vec{\kappa}_\perp \cdot \vec{\rho} - \vec{\kappa}_\perp' \cdot \vec{\rho}' = \vec{\kappa}_{\perp d} \cdot \vec{\rho}_c + \vec{\kappa}_{\perp c} \cdot \vec{\rho}_d \tag{4.16}$$

In this new coordinate system, the radiant mutual intensity is redefined as

$$\mathcal{J}(\vec{\kappa}_{\perp c}, \vec{\kappa}_{\perp d}) = \left\langle \mathcal{E}(\vec{\kappa}_{\perp c} + \tfrac{1}{2}\vec{\kappa}_{\perp d}) \mathcal{E}^*(\vec{\kappa}_{\perp c} - \tfrac{1}{2}\vec{\kappa}_{\perp d}) \right\rangle \tag{4.17}$$

or

$$\mathcal{J}(\vec{\kappa}_{\perp c}, \vec{\kappa}_{\perp d}) = \iint d^2\vec{\rho}_c\, d^2\vec{\rho}_d\, J(\vec{\rho}_c, \vec{\rho}_d) e^{-i2\pi(\vec{\kappa}_{\perp c} \cdot \vec{\rho}_d + \vec{\kappa}_{\perp d} \cdot \vec{\rho}_c)} \tag{4.18}$$

Note that there has been a reversal in the associations of Fourier conjugate variables. While in Eq. 4.12 these associations were $\vec{\rho} \leftrightarrow \vec{\kappa}_\perp$ and $\vec{\rho}' \leftrightarrow \vec{\kappa}_\perp'$, here they are $\vec{\rho}_c \leftrightarrow \vec{\kappa}_{\perp d}$ and $\vec{\rho}_d \leftrightarrow \vec{\kappa}_{\perp c}$. This association reversal stems from Eq. 4.16 and is an outcome of our new coordinate system. As we will see below, this reversal will have significant ramifications.

Radiant Intensity

In correspondence with our definition of intensity (Eq. 4.1), the radiant intensity is defined to be

$$\mathcal{R}(\vec{\kappa}_\perp) = \kappa^2 \left\langle \left| \mathcal{E}(\vec{\kappa}_\perp) \right|^2 \right\rangle \tag{4.19}$$

where we note here the presence of an additional prefactor κ^2. This prefactor is valid in the paraxial approximation, and takes on a slightly more complicated form when the paraxial approximation is relaxed, as in Chapter 6.

In our alternative coordinate system, the radiant intensity is given by

$$\mathcal{R}(\vec{\kappa}_{\perp c}) = \kappa^2 \mathcal{J}(\vec{\kappa}_{\perp c}, 0) \tag{4.20}$$

In terms of the mutual intensity, it is given by

$$\mathcal{R}(\vec{\kappa}_{\perp c}) = \kappa^2 \iint d^2\vec{\rho}_c\, d^2\vec{\rho}_d\, J(\vec{\rho}_c, \vec{\rho}_d) e^{-i2\pi\vec{\rho}_d \cdot \vec{\kappa}_{\perp c}} \tag{4.21}$$

This definition of radiant intensity is slightly different from the conventional radiometric definition which has dimensions of power/steradian. In particular, as will be discussed in Chapter 6, it lacks an obliquity prefactor of $|\kappa_z|/\kappa$. In other words, $\mathcal{R}(\vec{\kappa}_\perp)$ here may be interpreted as equivalent to the conventional radiometric definition of radiant intensity, but only in the paraxial approximation where $|\kappa_z| \approx \kappa$.

Intensity Spectrum

Since microscopes are generally based on the recording of an intensity distribution $I(\vec{\rho})$ at a detection plane, a very useful quantity will be the Fourier transform of this intensity distribution, called the intensity spectrum. This is given by

$$\mathcal{I}(\vec{\kappa}_\perp) = \int d^2\vec{\rho}\, I(\vec{\rho}) e^{-i2\pi\vec{\rho}\cdot\vec{\kappa}_\perp} \tag{4.22}$$

In terms of the radiant mutual intensity, the intensity spectrum becomes

$$\mathcal{I}(\vec{\kappa}_{\perp d}) = \int d^2\vec{\kappa}_{\perp c}\, \mathcal{J}(\vec{\kappa}_{\perp c}, \vec{\kappa}_{\perp d}) \tag{4.23}$$

We will make ample use of this definition throughout this book. An important feature of the intensity spectrum is that it must be Hermitian. In other words, in must obey the property $\mathcal{I}^*(\vec{\kappa}_\perp) = \mathcal{I}(-\vec{\kappa}_\perp)$, which stems from the fact that $I(\vec{\rho})$ is real.

Power

Finally, the total power traversing an axial plane is defined by

$$\Phi = \int d^2\vec{\rho}\, I(\vec{\rho}) = \frac{1}{\kappa^2} \int d^2\vec{\kappa}_\perp\, \mathcal{R}(\vec{\kappa}_\perp) \tag{4.24}$$

where Parseval's theorem (see Eq. A.9) has been invoked. The first of these definitions is familiar and represents an integral of intensity over cross-sectional area. The second definition is less familiar. Bearing in mind that the differential solid angle associated the 3D wavevector $\vec{\kappa}$ is given by $d\Omega_\kappa = \frac{d^2\vec{\kappa}_\perp}{\kappa|\kappa_z|}$, we observe that this second definition may be interpreted as an integral of the radiant intensity over solid angle, though, again, restricted to the paraxial approximation where $|\kappa_z| \approx \kappa$.

4.2 FREE SPACE PROPAGATION

Equipped with the above definitions, we can now proceed to re-examine the issue of free-space propagation. In Chapter 3, we considered the propagation of monochromatic fields from one plane to another. Here, we generalize our discussion to fields that are not necessarily monochromatic, by treating them pairwise and averaging over long durations (i.e. longer than the coherence time). We will thus base our discussion on the propagation of mutual intensity or of radiant mutual intensity. Throughout the remainder of this chapter, we will confine ourselves to the Fresnel approximation for simplicity, meaning we will consider fields that are mostly axially directed.

From the definition of mutual intensity (Eq. 4.3) and the Fresnel diffraction integral (Eq. 2.51), the general relation describing the propagation of the mutual intensity from plane 0 to plane 1 (z being the free-space distance separating these planes) is given by

$$J_1(\vec{\rho}_1, \vec{\rho}_1') = \left(\frac{\kappa}{z}\right)^2 \iint d^2\vec{\rho}_0\, d^2\vec{\rho}_0'\, e^{i\pi\frac{\kappa}{z}\left(\left|\vec{\rho}_1 - \vec{\rho}_0\right|^2 - \left|\vec{\rho}_1' - \vec{\rho}_0'\right|^2\right)} J_0(\vec{\rho}_0, \vec{\rho}_0') \qquad (4.25)$$

This relation is sometimes called the Zernike propagation law.

Applying the coordinate transformation prescribed by Eq. 4.5, this can be rewritten as

$$J_1(\vec{\rho}_{1c}, \vec{\rho}_{1d}) = \left(\frac{\kappa}{z}\right)^2 \iint d^2\vec{\rho}_{0c}\, d^2\vec{\rho}_{0d}\, e^{i2\pi\frac{\kappa}{z}\left(\vec{\rho}_{1c} - \vec{\rho}_{0c}\right)\cdot\left(\vec{\rho}_{1d} - \vec{\rho}_{0d}\right)} J_0(\vec{\rho}_{0c}, \vec{\rho}_{0d}) \qquad (4.26)$$

The benefit of our coordinate transformation is immediately apparent, as it has removed the quadratic phase dependences present in Eq. 4.25. These expressions can be further simplified if we consider instead the propagation of radiant mutual intensity, in which case the Zernike propagation law becomes

$$\mathcal{J}_1(\vec{\kappa}_\perp, \vec{\kappa}_\perp') = e^{-i\pi\frac{z}{\kappa}\left(\kappa_\perp^2 - \kappa_\perp'^2\right)} \mathcal{J}_0(\vec{\kappa}_\perp, \vec{\kappa}_\perp') \qquad (4.27)$$

or

$$\mathcal{J}_1(\vec{\kappa}_{\perp c}, \vec{\kappa}_{\perp d}) = e^{-i2\pi\frac{z}{\kappa}\vec{\kappa}_{\perp c}\cdot\vec{\kappa}_{\perp d}} \mathcal{J}_0(\vec{\kappa}_{\perp c}, \vec{\kappa}_{\perp d}) \qquad (4.28)$$

where we have made use of Eqs. 2.32 and 2.54.

Equations 4.26 and 4.28 apply to arbitrary $J_0(\vec{\rho}_{0c}, \vec{\rho}_{0d})$ and $\mathcal{J}_0(\vec{\kappa}_{\perp c}, \vec{\kappa}_{\perp d})$ and lay the groundwork for the remainder of our discussion on free-space propagation. For starters, they provide some basic insight into the concept of radiant intensity. For example, from Eq. 4.28 and making use of Eq. 4.20, we conclude that $\mathcal{R}_1(\vec{\kappa}_\perp) = \mathcal{R}_0(\vec{\kappa}_\perp)$, meaning that radiant intensity is conserved upon propagation through free space. This principle is quite general, and applies beyond the Fresnel approximation invoked here (cf. [6]).

As another example, from Eq. 4.9 we can establish the relation between the intensity at plane 1 and the mutual intensity at plane 0:

$$I_1(\vec{\rho}_{1c}) = \left(\frac{\kappa}{z}\right)^2 \iint d^2\vec{\rho}_{0c}\, d^2\vec{\rho}_{0d}\, e^{-i2\pi\frac{\kappa}{z}\vec{\rho}_{0d}\cdot\left(\vec{\rho}_{1c} - \vec{\rho}_{0c}\right)} J_0(\vec{\rho}_{0c}, \vec{\rho}_{0d}) \qquad (4.29)$$

If the propagation distance z is so large, or the extent of the source field so small, that the condition $\left(\vec{\rho}_{0d} \cdot \vec{\rho}_{0c}\right)_{max} \ll \lambda z$ applies, then from Eq. 4.21 we find that Eq. 4.29 simplifies to

$$I_1(\vec{\rho}_{1c}) \approx \frac{1}{z^2} \mathcal{R}_0\left(\frac{\kappa}{z}\vec{\rho}_{1c}\right) \qquad (4.30)$$

The above condition is the same as the Fraunhofer condition (see Section 2.4.2). In other words, we find that for free-space propagation the light intensity at a distant plane 1 is simply given by the scaled radiant intensity at plane 0 from which it originates. One must be a little

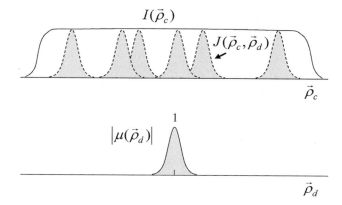

Figure 4.2. Intensity and coherence functions of a quasi-homogeneous beam.

careful with this derivation since it too relied on the Fresnel approximation, meaning that, strictly speaking, it is valid only within the paraxial Fraunhofer approximation. However, as shown in [8], Eq. 4.30 also applies beyond the Fresnel approximation, and this relation between far-field intensity and radiant intensity is quite general.

4.2.1 Quasi-Homogeneous Beams

For an important class of fields, the mutual intensity can be factored into independent functions of $\vec{\rho}_c$ and $\vec{\rho}_d$, enabling us to write

$$J(\vec{\rho}_c, \vec{\rho}_d) = I(\vec{\rho}_c)\mu(\vec{\rho}_d) \tag{4.31}$$

where $I(\vec{\rho}_c)$ is the intensity function, which we have seen before, and $\mu(\vec{\rho}_d)$ is referred to as a coherence function. This latter function, illustrated in Fig. 4.2, plays an important role in defining the spatial coherence of a field. The concept of spatial coherence, only briefly alluded to so far, will be elaborated on both here and in the next two chapters. By definition, fields that allow factoring as expressed in Eq. 4.31 are called quasi-homogeneous [2]. We recall from Eq. 4.8 that $J(\vec{\rho}_c, \vec{\rho}_d)$ is defined for long time averages only (i.e. longer than the coherence time), meaning that the factorization condition prescribed in Eq. 4.31 need only occur post averaging. As such, the functions $I(\vec{\rho}_c)$ and $\mu(\vec{\rho}_d)$ are also defined post averaging.

Correspondingly, the radiant mutual intensity of a quasi-homogeneous field is defined by

$$\mathcal{J}(\vec{\kappa}_{\perp c}, \vec{\kappa}_{\perp d}) = \mathcal{I}(\vec{\kappa}_{\perp d})\hat{\mu}(\vec{\kappa}_{\perp c}) \tag{4.32}$$

where the intensity spectrum $\mathcal{I}(\vec{\kappa}_{\perp d})$ is defined by Eq. 4.22 and $\hat{\mu}(\vec{\kappa}_{\perp})$ is the Fourier transform of $\mu(\vec{\rho}_d)$. That is

$$\hat{\mu}(\vec{\kappa}_{\perp c}) = \int d^2\vec{\rho}_d\, e^{-i2\pi\vec{\kappa}_{\perp c}\cdot\vec{\rho}_d}\mu(\vec{\rho}_d) \tag{4.33}$$

where we again note the reversal in the associations of Fourier conjugate variables, stemming directly from the reversal noted in our derivation of Eq. 4.18.

By construction, the functions $\mu(\vec{\rho})$ and $\hat{\mu}(\vec{\kappa}_\perp)$ obey the normalization properties

$$\mu(0) = 1 \tag{4.34}$$

$$\int d^2\vec{\kappa}_\perp \, \hat{\mu}(\vec{\kappa}_\perp) = 1 \tag{4.35}$$

4.2.2 Reciprocity Between Coherence and Intensity

By definition of a quasi-homogeneous beam, the coherence function $\mu(\vec{\rho}_d)$ is independent of $\vec{\rho}_c$. For this to be possible, the coherence function must generally span an area much smaller than the total beam area, meaning that $\mu(\vec{\rho}_d)$ is generally peaked and localized relative to $I(\vec{\rho}_c)$. The condition $\left(\vec{\rho}_{0d} \cdot \vec{\rho}_{0c}\right)_{\max} \ll \lambda z$ required for the derivation of Eq. 4.30 can now be applied without fully conforming to the Fraunhofer condition, since now the spans of $\vec{\rho}_{0d}$ and $\vec{\rho}_{0c}$ are decoupled. That is, the condition $\left(\vec{\rho}_{0d} \cdot \vec{\rho}_{0c}\right)_{\max} \ll \lambda z$ can be satisfied even when the span of $\vec{\rho}_{0c}$ (defined by $I(\vec{\rho}_{0c})$) is fairly broad or when the propagation distance z is modest, provided only that the span of $\vec{\rho}_{0d}$ (defined by $\mu(\vec{\rho}_{0d})$) is narrow enough. Assuming the above condition is indeed satisfied, then Eq. 4.26 reduces to

$$J_1(\vec{\rho}_{1c}, \vec{\rho}_{1d}) = \left(\frac{\kappa}{z}\right)^2 e^{i2\pi \frac{\kappa}{z} \vec{\rho}_{1c} \cdot \vec{\rho}_{1d}} \iint d^2\vec{\rho}_{0c} \, d^2\vec{\rho}_{0d} \, e^{-i2\pi \frac{\kappa}{z}(\vec{\rho}_{0c} \cdot \vec{\rho}_{1d} + \vec{\rho}_{1c} \cdot \vec{\rho}_{0d})} I_0(\vec{\rho}_{0c}) \mu_0(\vec{\rho}_{0d}) \tag{4.36}$$

for a quasi-homogeneous beam.

While $J_1(\vec{\rho}_{1c}, \vec{\rho}_{1d})$ is not exactly separable as prescribed by Eq. 4.31, it can be recast in the form

$$J_1(\vec{\rho}_{1c}, \vec{\rho}_{1d}) = e^{i2\pi \frac{\kappa}{z} \vec{\rho}_{1c} \cdot \vec{\rho}_{1d}} I_1(\vec{\rho}_{1c}) \mu_1(\vec{\rho}_{1d}) \tag{4.37}$$

where approximate separability can be achieved when the condition $\left(\vec{\rho}_{1d} \cdot \vec{\rho}_{1c}\right)_{\max} \ll \lambda z$ is met, allowing us to discard the exponential prefactor. This is the same condition as discussed above, but now applied to the field in plane 1 rather than plane 0.

The intensity and coherence functions in Eq. 4.37 are accordingly defined as

$$I_1(\vec{\rho}_{1c}) = \Phi \left(\frac{\kappa}{z}\right)^2 \int d^2\vec{\rho}_{0d} \, e^{-i2\pi \frac{\kappa}{z} \vec{\rho}_{1c} \cdot \vec{\rho}_{0d}} \mu_0(\vec{\rho}_{0d}) \tag{4.38}$$

and

$$\mu_1(\vec{\rho}_{1d}) = \frac{1}{\Phi} \int d^2\vec{\rho}_{0c} \, e^{-i2\pi \frac{\kappa}{z} \vec{\rho}_{1d} \cdot \vec{\rho}_{0c}} I_0(\vec{\rho}_{0c}) \tag{4.39}$$

where the total power Φ has been introduced in the prefactors to preserve the normalization condition specified by Eq. 4.34. Equation 4.38 we have seen before, and is a restatement of Eq. 4.30 applied to a quasi-homogeneous field. Equation 4.39 is new.

Equations 4.38 and 4.39 are fundamental. They indicate that a quasi-homogeneous beam remains quasi-homogeneous upon free space propagation, provided the conditions $\left(\vec{\rho}_{0d} \cdot \vec{\rho}_{0c}\right)_{\max} \ll \lambda z$ and $\left(\vec{\rho}_{1d} \cdot \vec{\rho}_{1c}\right)_{\max} \ll \lambda z$ are met. Moreover, they indicate that $I_1(\vec{\rho}_{1c})$

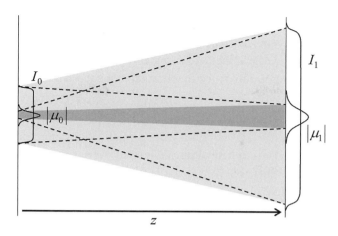

Figure 4.3. Free-space propagation of intensity and coherence functions. Dashed lines denote scaled Fourier transforms.

is derived from $\mu_0(\vec{\rho}_{0d})$ by a scaled Fourier transform. Similarly $\mu_1(\vec{\rho}_{1d})$ is derived from $I_0(\vec{\rho}_{0c})$, also by a scaled Fourier transform. This reciprocity relation between the intensity and coherence functions is depicted in Fig. 4.3, and will play a major role in our treatment of radiometry presented in Chapter 6.

Following the usual properties of Fourier transforms, we find that the smaller the width of $\mu_0(\vec{\rho}_{0d})$, the larger the width of $I_1(\vec{\rho}_{1c})$, and vice versa. Similarly, the larger the width of $I_0(\vec{\rho}_{0c})$ the smaller the width of $\mu_1(\vec{\rho}_{1d})$, and vice versa. It should be noted that the shaded regions in Fig. 4.3 are by no means intended to suggest a one-to-one correspondence between $\mu_1(\vec{\rho}_{1d})$ and $\mu_0(\vec{\rho}_{0d})$ or between $I_1(\vec{\rho}_{1d})$ and $I_0(\vec{\rho}_{0d})$, since neither would be correct (the correspondences are prescribed instead by the dashed lines). The shading is meant to illustrate that, in accord with their definitions in Eqs. 4.38 and 4.39, both the intensity and coherence functions expand with free-space propagation in proportion to the normalized propagation distance $\frac{z}{\kappa}$. This scaling with $\frac{z}{\kappa}$ is made explicit if we recast Eqs. 4.38 and 4.39 in the forms

$$I_1(\vec{\rho}_{1c}) = \Phi\left(\frac{\kappa}{z}\right)^2 \hat{\mu}_0\left(\frac{\kappa}{z}\vec{\rho}_{1c}\right) \tag{4.40}$$

$$\mu_1(\vec{\rho}_{1d}) = \frac{1}{\Phi}\mathcal{I}_0\left(\frac{\kappa}{z}\vec{\rho}_{1d}\right) \tag{4.41}$$

Again, the first of these is a restatement of Eq. 4.30, applied here to quasi-homogeneous beams. More remarkable is the second of these, which indicates that a quasi-homogeneous beam gains coherence simply as a result of free-space propagation. This conclusion applies independently of the beam's initial state of coherence, as we will see below.

4.2.3 Incoherent Beam Propagation

By definition, the narrower the coherence function $\mu\left(\vec{\rho}_d\right)$ of a field, the less the field is spatially coherent. However, as discussed in Sections 2.2.1 and 3.3.4, if a field is to be radiative then its spatial bandwidth must be limited by the wavenumber κ. Accordingly, the spatial bandwidth of the mutual intensity must also be limited, and as a result, the coherence function $\mu\left(\vec{\rho}_d\right)$ cannot be infinitely narrow even if the field is spatially incoherent. Nevertheless, it is often convenient to formally express the coherence function of an incoherent field by

$$\mu\left(\vec{\rho}_d\right) \rightarrow \kappa^{-2}\delta^2(\vec{\rho}_d) \tag{4.42}$$

This expression clearly does not satisfy the normalization condition imposed by Eq. 4.34 and as such is only valid when applied within an integral. As before, the prefactor κ^{-2} is introduced for dimensional consistency, but also to indicate that the minimum area spanned by the coherence function of a radiative incoherent beam can be no smaller than $\approx \kappa^{-2}$. More will be said about this minimum area in Chapter 6.

As prescribed by Eq. 4.42, the mutual intensity of an incoherent beam is then formally written as

$$J_0(\vec{\rho}_{0c}, \vec{\rho}_{0d}) \rightarrow \kappa^{-2}I_0(\vec{\rho}_{0c})\delta^2(\vec{\rho}_{0d}) \tag{4.43}$$

which is manifestly separable in the coordinates $\vec{\rho}_{0c}$ and $\vec{\rho}_{0d}$. An incoherent beam is therefore quasi-homogeneous, by definition, meaning that the free-space propagation results derived above for quasi-homogeneous beams also apply to incoherent beams. Thus, when Eq. 4.43 is inserted into Eq. 4.26, we readily obtain

$$J_1(\vec{\rho}_{1c}, \vec{\rho}_{1d}) = \frac{1}{z^2}e^{i2\pi\frac{\kappa}{z}\vec{\rho}_{1c}\cdot\vec{\rho}_{1d}} \int d^2\vec{\rho}_{0c}\, e^{-i2\pi\frac{\kappa}{z}\vec{\rho}_{1d}\cdot\vec{\rho}_{0c}} I_0(\vec{\rho}_{0c}) \tag{4.44}$$

This is the well-known van Cittert–Zernike theorem. Unlike the more general propagation law (Eq. 4.26) which is valid for arbitrary states of coherence, the van Cittert–Zernike theorem is a special case that is valid only when the field at plane 0 is incoherent. Nevertheless, it remains one of the most important theorems in imaging theory.

Several conclusions can be drawn from the van Cittert–Zernike theorem. The first is that, aside from the phase factor outside the integral in Eq. 4.44, the mutual intensity at plane 1 is simply a scaled Fourier transform of the intensity at plane 0. As before, this mutual intensity takes the form of Eq. 4.37, where the coherence function $\mu_1(\vec{\rho}_{1d})$ is given by Eq. 4.39. Thus, the width of $\mu_1(\vec{\rho}_{1d})$ continues to grow with propagation distance in proportion to $\frac{z}{\kappa}$, meaning that even an incoherent beam acquires partial coherence provided the propagation distance is sufficiently large compared to the size of the beam. In certain circumstances, it can even acquire full coherence, as will be explained in Chapter 6.

The reader is reminded that the conditions required for the derivation of the van Cittert–Zernike theorem, as expressed by Eq. 4.44, were twofold. First, the Fresnel approximation was applied, meaning that the angular deviations of the light beam from the optical axis are assumed to be weak. This condition goes against Eq. 4.38, which suggests that an incoherent beam must spread quite rapidly since $\mu_0(\vec{\rho}_{0d})$ is narrow. However, the application of the Fresnel condition to an incoherent beam is reconciled when only the portion of the propagating beam close to the axis is considered.

The second condition required for the derivation of Eq. 4.44 was more implicit. Equation 4.26, which applies for arbitrary states of coherence, contains the phase term $e^{i2\pi\frac{\kappa}{z}\vec{\rho}_{0c}\cdot\vec{\rho}_{0d}}$ inside the integral. This phase term disappeared from Eq. 4.44 as a result of our use of the expression 4.43 for the mutual intensity of an incoherent beam. However, again, this expression was only a formal approximation since the coherence function cannot be infinitely narrow even for an incoherent beam. More rigorously, it was found that the phase term can be neglected when the condition $\left(\vec{\rho}_{0d}\cdot\vec{\rho}_{0c}\right)_{\max} \ll \lambda z$ is met. Since for an incoherent beam the width of $\mu(\vec{\rho}_{0d})$ is of order a wavelength, this condition translates to the requirement $z \gg (\rho_{0c})_{\max}$. In other words, Eq. 4.44 is valid for propagation distances much larger than the size of the incoherent beam itself. We note that, in this limit, all the phase terms inversely dependent on z as well as the prefactor in Eq. 4.44 are well behaved.

In the opposite limit, when one is considering propagation over distances smaller than the size of the beam, it is more convenient to treat incoherent beams in terms of their radiant mutual intensities rather than their mutual intensities. From Eq. 4.43, we write in this case for an incoherent beam,

$$\mathcal{J}_0(\vec{\kappa}_{\perp c}, \vec{\kappa}_{\perp d}) = \kappa^{-2}\mathcal{I}_0\left(\vec{\kappa}_{\perp d}\right) \tag{4.45}$$

The propagation of this radiant mutual intensity a distance z then leads to, from Eq. 4.28,

$$\mathcal{J}_1(\vec{\kappa}_{\perp c}, \vec{\kappa}_{\perp d}) = \kappa^{-2}e^{-i2\pi\frac{z}{\kappa}\vec{\kappa}_{\perp c}\cdot\vec{\kappa}_{\perp d}}\mathcal{I}_0\left(\vec{\kappa}_{\perp d}\right) \tag{4.46}$$

Equation 4.46 is a Fourier representation of the van Cittert–Zernike theorem appropriate for small propagation distances, where the phase term in the exponential is also well behaved. This remarkably simple representation will be useful in later chapters.

Finally, as an added note concerning incoherent beams, we observe that by inserting Eq. 4.43 into Eq. 4.21 (or alternatively Eq. 4.45 into Eq. 4.20), the radiant intensity of an incoherent beam reduces simply to $\mathcal{R}(\vec{\kappa}_{\perp}) = \Phi$. In other words, the radiant intensity is found to be independent of $\vec{\kappa}_{\perp}$, meaning that the radiation is found to be formally isotropic. Again, because we have adopted the Fresnel approximation, this conclusion is valid only when applied to the roughly forward propagation direction. In practice, the radiant intensity of an incoherent beam is found to more generally obey a $|\kappa_z|/\kappa$ dependence on wavevector direction, leading to the Lambert cosine law. We will revisit this law in Chapter 6.

4.3 PROPAGATION THROUGH A 2f SYSTEM

We now turn to the propagation of mutual intensity through a lens of focal length f in a 2f configuration. In Section 3.2, we found that the effect of a 2f configuration on a monochromatic field is essentially the same as that of far-field free-space propagation, although without the constraints imposed by a Fraunhofer approximation. We will see here that the same holds true for mutual intensity.

To demonstrate this, we refer to Eqs. 3.9 and 4.8 to obtain for the 2f configuration

$$J_1(\vec{\rho}_{1c}, \vec{\rho}_{1d}) = \left(\frac{\kappa}{f}\right)^2 \iint d^2\vec{\rho}_{0c}\, d^2\vec{\rho}_{0d}\, e^{-i2\pi\frac{\kappa}{f}\left(\vec{\rho}_{1c}\cdot\vec{\rho}_{0d}+\vec{\rho}_{1d}\cdot\vec{\rho}_{0c}\right)} J_0(\vec{\rho}_{0c}, \vec{\rho}_{0d}) \qquad (4.47)$$

leading to the reciprocity relations connecting mutual intensities and radiant mutual intensities:

$$J_1(\vec{\rho}_{1c}, \vec{\rho}_{1d}) = \left(\frac{\kappa}{f}\right)^2 \mathcal{J}_0\left(\frac{\kappa}{f}\vec{\rho}_{1c}, \frac{\kappa}{f}\vec{\rho}_{1d}\right) \qquad (4.48)$$

$$\mathcal{J}_1(\vec{\kappa}_{\perp c}, \vec{\kappa}_{\perp d}) = \left(\frac{f}{\kappa}\right)^2 J_0\left(-\frac{f}{\kappa}\vec{\kappa}_{\perp c}, -\frac{f}{\kappa}\vec{\kappa}_{\perp d}\right) \qquad (4.49)$$

These equations are valid for arbitrary $J_0(\vec{\rho}_c, \vec{\rho}_d)$ or $\mathcal{J}_0(\vec{\kappa}_{\perp c}, \vec{\kappa}_{\perp d})$. In the event one is interested only in the light intensity at plane 1, Eqs. 4.48 and 4.20 lead directly to

$$I_1(\vec{\rho}_{1c}) = \frac{1}{f^2}\mathcal{R}_0\left(\frac{\kappa}{f}\vec{\rho}_{1c}\right) \qquad (4.50)$$

This is identical to our result obtained for free space propagation (4.30) except that it is now exact, and the previously required constraints imposed by the Fraunhofer approximation have been dispensed with. In other words, 2f propagation through a lens has effectively translated the "far-field" from being located a distance very far away to a much more manageable distance f, as discussed in Section 3.2. Moreover, the link between intensity and radiant intensity provided by Eq. 4.50 is entirely analogous to the link between field and radiant field provided by Eq. 3.10.

Quasi-homogeneous Field

When dealing with a quasi-homogeneous field, $J_0(\vec{\rho}_{0c}\cdot\vec{\rho}_{0d})$ can be separated into independent intensity and coherence functions, as prescribed by Eq. 4.31. By inserting these into Eq. 4.47, we find that the mutual coherence at plane 1 is also separable, and can be written as

$$J_1(\vec{\rho}_{1c}, \vec{\rho}_{1d}) = I_1(\vec{\rho}_{1c})\mu_1(\vec{\rho}_{1d}) \qquad (4.51)$$

with the definitions

$$I_1(\vec{\rho}_{1c}) = \Phi\left(\frac{\kappa}{f}\right)^2 \int d^2\vec{\rho}_{0d}\, e^{-i2\pi\frac{\kappa}{f}\vec{\rho}_{1c}\cdot\vec{\rho}_{0d}}\mu_0(\vec{\rho}_{0d}) \qquad (4.52)$$

and

$$\mu_1(\vec{\rho}_{1d}) = \frac{1}{\Phi}\int d^2\vec{\rho}_{0c}\, e^{-i2\pi\frac{\kappa}{f}\vec{\rho}_{1d}\cdot\vec{\rho}_{0c}}I_0(\vec{\rho}_{0c}) \qquad (4.53)$$

These results are also identical to the results obtained for free-space propagation (Eqs. 4.38 and 4.39), with the difference that the separation of $J_1(\vec{\rho}_{1c}, \vec{\rho}_{1d})$ into intensity and coherence functions is now exact. Specifically, Eq. 4.51 no longer contains the spurious phase prefactor occasioned in Eq. 4.37.

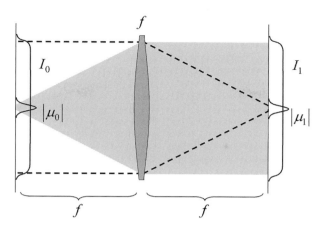

Figure 4.4. Propagation of intensity and coherence functions in a 2f configuration.

Expressed in separable forms, Eqs. 4.48 and 4.49 become

$$J_1(\vec{\rho}_{1c}, \vec{\rho}_{1d}) = \left(\frac{\kappa}{f}\right)^2 \hat{\mu}_0\left(\frac{\kappa}{f}\vec{\rho}_{1c}\right) \mathcal{I}_0\left(\frac{\kappa}{f}\vec{\rho}_{1d}\right) \tag{4.54}$$

$$\mathcal{J}_1(\vec{\kappa}_{\perp c}, \vec{\kappa}_{\perp d}) = \left(\frac{f}{\kappa}\right)^2 I_0\left(-\frac{f}{\kappa}\vec{\kappa}_{\perp c}\right) \mu_0\left(-\frac{f}{\kappa}\vec{\kappa}_{\perp d}\right) \tag{4.55}$$

A schematic of the propagation of the intensity and coherence functions through a 2f configuration is illustrated in Fig. 4.4.

Let us consider the example of a quasi-homogeneous field at plane 0 that is spatially incoherent (i.e. $\mu_0(\vec{\rho}_{0d}) \to \kappa^{-2}\delta^2(\vec{\rho}_{0d})$). This example is particularly relevant in microscopy applications since microscopes are often based on 2f illumination configurations (called Köhler illumination configurations) where a sample at plane 1 is illuminated by an incoherent source at plane 0 through a lens (called a condenser). The benefit of this configuration is that, as prescribed by Eq. 4.52, the intensity distribution incident on the sample plane is a uniform $I_1 = \Phi/f^2$, independent of the distribution of $I_0(\vec{\rho}_0)$ at the source.

In the more specific example where a circular pupil of radius a delimits the distribution of an otherwise constant source intensity I_0, we find, from Eq. 4.53,

$$\mu_1(\vec{\rho}_{1d}) = \text{jinc}\left(2\pi\kappa\frac{a}{f}\rho_{1d}\right) \tag{4.56}$$

where the jinc function is defined by Eq. A.16. Following the same line of reasoning as in Section 3.3.4, we conclude that the width of $\mu_1(\vec{\rho}_{1d})$ is given by

$$\delta\rho_{1d} = \frac{\lambda}{2\text{NA}} \tag{4.57}$$

where we have retrieved the definition of numerical aperture given by $\text{NA} = n\frac{a}{f}$ (n being the index of refraction at plane 1). The coherence function at plane 1, therefore, has the same width

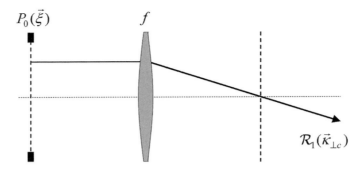

Figure 4.5. Connection between pupil function and radiant intensity in 2f configuration.

as a diffraction-limited spot associated with a pupil of radius a and a lens of focal length f. This result is a basic property of Köhler illumination, which will be discussed in more detail in Section 6.5.2.

More generally, in the case of an arbitrary pupil $P_0\left(\vec{\rho}_0\right)$, Köhler illumination with an incoherent source leads to

$$\mathcal{J}_1\left(\vec{\kappa}_{\perp c}, \vec{\kappa}_{\perp d}\right) = \kappa^{-2} I_0 \left| P_0 \left(-\frac{f}{\kappa}\vec{\kappa}_{\perp c}\right) \right|^2 \delta^2\left(\vec{\kappa}_{\perp d}\right) \tag{4.58}$$

From Eqs. 4.55 and 4.20, this last equation can be expressed equivalently as $\mathcal{J}_1\left(\vec{\kappa}_{\perp c}, \vec{\kappa}_{\perp d}\right) = (\kappa f)^{-2} \mathcal{R}_1(\vec{\kappa}_{\perp c})\delta^2\left(\vec{\kappa}_{\perp d}\right)$, meaning that the pupil function $\left| P_0\left(-\frac{f}{\kappa}\vec{\kappa}_{\perp c}\right)\right|^2$ at plane 0 can be directly associated with the radiant intensity at plane 1. An illustration of this association is provided in Fig. 4.5. Different source locations $\vec{\xi} = \frac{f}{\kappa}\vec{\kappa}_{\perp c}$ in plane 0 produce different wavevectors $-\vec{\kappa}_{\perp c}$ at plane 1 that propagate to the far field. In turn, the 2D delta function $\delta^2\left(\vec{\kappa}_{\perp d}\right)$ ensures that these wavevectors are uncorrelated.

3D Coherence Function
Though this chapter deals primarily with fields and radiant fields confined to 2D transverse planes, we make an exception here. In particular, it will be useful to examine the general 3D mutual intensity $J_1\left(\vec{r}_1, \vec{r}_1'\right)$ obtained by Köhler illumination with an incoherent source, which, with our familiar coordinate transformation, can be recast as

$$J_1\left(\vec{r}_{1c}, \vec{r}_{1d}\right) = \left\langle E_1\left(\vec{r}_{1c} + \tfrac{1}{2}\vec{r}_{1d}\right) E_1^*\left(\vec{r}_{1c} - \tfrac{1}{2}\vec{r}_{1d}\right)\right\rangle \tag{4.59}$$

From the previous chapter (Eq. 3.7), the relation between $E_1(\vec{r}_1)$ and $E_0(\vec{\rho}_0)$ was found to be

$$E_1(\vec{r}_1) = -i\frac{\kappa}{f}e^{i2\pi\kappa(f+z_1)} \int \mathrm{d}^2\vec{\rho}_0\, e^{-i2\pi\frac{\kappa}{f}\vec{\rho}_0\cdot\vec{\rho}_1} e^{i\pi\frac{\kappa}{f}\rho_0^2\left(1-\frac{z_1}{f}\right)} E_0(\vec{\rho}_0) \tag{4.60}$$

where z_1 (formerly s_1) corresponds to axial distance from the lens plane (a similar result is obtained for $E_1^*(\vec{r}_1')$).

A lengthy but straightforward calculation making use of Eq. 4.43 leads to

$$J_1\left(\vec{r}_{1c}, \vec{r}_{1d}\right) = \frac{1}{f^2} e^{i2\pi\kappa z_{1d}} \int d^2\vec{\rho}_{0c}\, e^{-i2\pi\frac{\kappa}{f}\vec{\rho}_{0c}\cdot\vec{\rho}_{1d}} e^{-i\pi\frac{\kappa}{f^2}z_{1d}\rho_{0c}^2} I_0\left(\vec{\rho}_{0c}\right) \tag{4.61}$$

In turn, we can define a 3D coherence function by

$$\mu\left(\vec{r}_d\right) = \frac{J\left(\vec{r}_c, \vec{r}_d\right)}{J\left(\vec{r}_c, 0\right)} = \frac{J\left(\vec{r}_c, \vec{r}_d\right)}{I\left(\vec{r}_c\right)} \tag{4.62}$$

leading finally to

$$\mu_1\left(\vec{r}_{1d}\right) = \frac{1}{\Phi_0} e^{i2\pi\kappa z_{1d}} \int d^2\vec{\rho}_{0c}\, e^{-i2\pi\frac{\kappa}{f}\vec{\rho}_{0c}\cdot\vec{\rho}_{1d}} e^{-i\pi\frac{\kappa}{f^2}z_{1d}\rho_{0c}^2} I_0\left(\vec{\rho}_{0c}\right) \tag{4.63}$$

where $\Phi_0 = \int d^2\vec{\rho}_{0c}\, I_0\left(\vec{\rho}_{0c}\right)$ is the total illumination power. As expected, $\mu_1\left(\vec{r}_{1d}\right)$ reduces to Eq. 4.53 when $z_{1d} \to 0$.

We note that $\mu_1\left(\vec{r}_{1d}\right)$ is independent of $\vec{r}_{1c}$. In other words, the statistical properties of Köhler illumination remain homogeneous throughout a *volume* about the focal plane. This is true regardless of the shape of $I_0\left(\vec{\rho}_{0c}\right)$, provided only that $I_0\left(\vec{\rho}_{0c}\right)$ is spatially incoherent. We will revisit this remarkable feature of Köhler illumination on several occasions.

4.4 PROPAGATION THROUGH A 4f SYSTEM: IMAGING

Finally, we consider the propagation of mutual intensity through a 4f configuration, as described in Section 3.3.2, which is a basic imaging configuration. Making use of Eqs. 4.8 and 4.17, as well as Eq. 3.24, we arrive at the general equations

$$J_1(M\vec{\rho}_{1c}, M\vec{\rho}_{1d}) =$$
$$\frac{1}{M^2} \iint d^2\vec{\rho}_{0c}\, d^2\vec{\rho}_{0d}\, H(\vec{\rho}_{1c} + \tfrac{1}{2}\vec{\rho}_{1d} - \vec{\rho}_{0c} - \tfrac{1}{2}\vec{\rho}_{0d}) H^*(\vec{\rho}_{1c} - \tfrac{1}{2}\vec{\rho}_{1d} - \vec{\rho}_{0c} + \tfrac{1}{2}\vec{\rho}_{0d}) J_0(\vec{\rho}_{0c}, \vec{\rho}_{0d}) \tag{4.64}$$

and

$$\mathcal{J}_1\left(\frac{1}{M}\vec{\kappa}_{\perp c}, \frac{1}{M}\vec{\kappa}_{\perp d}\right) = M^2\, \mathcal{H}(\vec{\kappa}_{\perp c} + \tfrac{1}{2}\vec{\kappa}_{\perp d}) \mathcal{H}^*(\vec{\kappa}_{\perp c} - \tfrac{1}{2}\vec{\kappa}_{\perp d}) \mathcal{J}_0(\vec{\kappa}_{\perp c}, \vec{\kappa}_{\perp d}) \tag{4.65}$$

where M is the imaging magnification. In a 4f configuration, M is defined by $-f_1/f_0$ (see Fig. 3.4).

The above equations apply for mutual intensities and radiant mutual intensities that are arbitrary. These equations are more general than will be required throughout most of this book, and we therefore focus on specific cases.

In particular, in most imaging applications the parameter of importance is the light intensity at the image plane (plane 1), since this is the parameter that is directly measurable by a standard detector. With the definition of intensity given by Eq. 4.9, Eq. 4.64 reduces to

$$I_1(M\vec{\rho}_{1c}) = \frac{1}{M^2} \iint d^2\vec{\rho}_{0c}\, d^2\vec{\rho}_{0d}\, H(\vec{\rho}_{1c} - \vec{\rho}_{0c} - \tfrac{1}{2}\vec{\rho}_{0d}) H^*(\vec{\rho}_{1c} - \vec{\rho}_{0c} + \tfrac{1}{2}\vec{\rho}_{0d}) J_0(\vec{\rho}_{0c}, \vec{\rho}_{0d}) \tag{4.66}$$

Similarly, with the definition of intensity spectrum given by Eq. 4.23, Eq. 4.65 reduces to

$$\mathcal{I}_1\left(\frac{1}{M}\vec{\kappa}_{\perp d}\right) = \int \mathrm{d}^2\kappa_{\perp c}\, \mathcal{H}(\vec{\kappa}_{\perp c} + \tfrac{1}{2}\vec{\kappa}_{\perp d})\mathcal{H}^*(\vec{\kappa}_{\perp c} - \tfrac{1}{2}\vec{\kappa}_{\perp d})\mathcal{J}_0(\vec{\kappa}_{\perp c}, \vec{\kappa}_{\perp d}) \tag{4.67}$$

In the more specific case where the field at the object plane (plane 0) is known to be incoherent, we can invoke the formal limit provided by Eq. 4.43 to write

$$I_1(M\vec{\rho}_{1c}) = \frac{1}{M^2\kappa^2}\int \mathrm{d}^2\vec{\rho}_{0c}\, \left|H(\vec{\rho}_{1c} - \vec{\rho}_{0c})\right|^2 I_0(\vec{\rho}_{0c}) \tag{4.68}$$

and

$$\mathcal{I}_1\left(\frac{1}{M}\vec{\kappa}_{\perp d}\right) = \frac{1}{\kappa^2}\mathcal{I}_0(\vec{\kappa}_{\perp d})\int \mathrm{d}^2\vec{\kappa}_{\perp c}\, \mathcal{H}(\vec{\kappa}_{\perp c} + \tfrac{1}{2}\vec{\kappa}_{\perp d})\mathcal{H}^*(\vec{\kappa}_{\perp c} - \tfrac{1}{2}\vec{\kappa}_{\perp d}) \tag{4.69}$$

These equations may be further simplified by introducing two of the most important functions in incoherent imaging theory, namely the point spread function (PSF), defined by

$$\mathrm{PSF}(\vec{\rho}) = \frac{1}{\kappa^2}\left|H(\vec{\rho})\right|^2 \tag{4.70}$$

and the optical transfer function (OTF), defined by

$$\mathrm{OTF}(\vec{\kappa}_{\perp d}) = \frac{1}{\kappa^2}\int \mathrm{d}^2\vec{\kappa}_{\perp c}\, \mathcal{H}(\vec{\kappa}_{\perp c} + \tfrac{1}{2}\vec{\kappa}_{\perp d})\mathcal{H}^*(\vec{\kappa}_{\perp c} - \tfrac{1}{2}\vec{\kappa}_{\perp d}) \tag{4.71}$$

(It should be noted these definitions of PSF and OTF are unnormalized, and thus differ from alternative definitions found elsewhere – e.g. see [4] and the previous edition of this book.)

Just as D and $\mathcal{D}$ from Chapter 2, and H and $\mathcal{H}$ from Chapter 3 are Fourier transform pairs, so too are PSF and OTF. That is, we have

$$\mathrm{OTF}(\vec{\kappa}_\perp) = \int \mathrm{d}^2\vec{\rho}\, \mathrm{PSF}(\vec{\rho})e^{-i2\pi\vec{\rho}\cdot\vec{\kappa}_\perp} \tag{4.72}$$

Finally, from these definitions we arrive at the fundamental equations for incoherent imaging:

$$I_1(M\vec{\rho}_1) = \frac{1}{M^2}\int \mathrm{d}^2\vec{\rho}_0\, \mathrm{PSF}(\vec{\rho}_1 - \vec{\rho}_0)I_0(\vec{\rho}_0) \tag{4.73}$$

$$\mathcal{I}_1\left(\frac{1}{M}\vec{\kappa}_\perp\right) = \mathrm{OTF}(\vec{\kappa}_\perp)\mathcal{I}_0(\vec{\kappa}_\perp) \tag{4.74}$$

In effect, the PSF characterizes imaging with an incoherent intensity in an analogous way as H characterizes imaging with a monochromatic field. Similarly, the OTF characterizes imaging with an incoherent intensity spectrum in an analogous way as $\mathcal{H}$ characterizes imaging with a monochromatic radiant field. These relations are of paramount importance and will serve as mainstays of our discussion on microscopy techniques.

Properties of the PSF and OTF

The following are useful identities and normalization properties related to the PSF and OTF:

- $\mathrm{PSF}(\vec{\rho})$ is real and positive.
- $\mathrm{OTF}(-\vec{\kappa}_\perp) = \mathrm{OTF}^*(\vec{\kappa}_\perp)$.
- If $\mathrm{PSF}(\vec{\rho})$ is symmetric, then $\mathrm{OTF}(\vec{\kappa}_\perp)$ is real and symmetric.

By construction, the PSF and OTF are normalized such that

$$\mathrm{PSF}(0) = \int d^2\vec{\kappa}_\perp \, \mathrm{OTF}(\vec{\kappa}_\perp) = \frac{1}{\kappa^2} \left| \int d^2\vec{\kappa}_\perp \, \mathcal{H}(\vec{\kappa}_\perp) \right|^2 = \frac{\kappa^2}{f_0^4} \left| \int d^2\vec{\xi} \, P(\vec{\xi}) \right|^2 \tag{4.75}$$

$$\mathrm{OTF}(0) = \int d^2\vec{\rho} \, \mathrm{PSF}(\vec{\rho}) = \frac{1}{\kappa^2} \int d^2\vec{\kappa}_\perp \left| \mathcal{H}(\vec{\kappa}_\perp) \right|^2 = \frac{1}{f_0^2} \int d^2\vec{\xi} \left| P(\vec{\xi}) \right|^2 \tag{4.76}$$

4.4.1 Circular Pupil

To close this chapter, we return to the example previously discussed in Section 3.3.4 and routinely found in practice, namely that of 4f imaging with an unobstructed circular aperture pupil.

In particular, a circular aperture pupil of radius a is defined by

$$P(\vec{\xi}) = \begin{cases} 1 & \xi \le a \\ 0 & \xi > a \end{cases} \tag{4.77}$$

from which $\mathcal{H}$ is derived immediately from Eq. 3.27, obtaining

$$\mathcal{H}(\vec{\kappa}_\perp) = \begin{cases} 1 & \kappa_\perp \le \frac{1}{2}\Delta\kappa_{\perp 0} \\ 0 & \kappa_\perp > \frac{1}{2}\Delta\kappa_{\perp 0} \end{cases} \tag{4.78}$$

In deriving this result, we have made use of the apparent object bandwidth defined in Section 3.3.4, given by

$$\Delta\kappa_{\perp 0} = 2\kappa\frac{a}{f_0} \tag{4.79}$$

The corresponding H associated with the imaging system then follows from a simple Fourier transform of $\mathcal{H}$ (Eq. 3.26), leading to

$$H(\rho) = \frac{\pi}{4}\Delta\kappa_{\perp 0}^2 \,\mathrm{jinc}\left(\pi\Delta\kappa_{\perp 0}\rho\right) \tag{4.80}$$

In the case of incoherent intensity imaging, the associated PSF is defined by Eq. 4.70, obtaining

$$\frac{\mathrm{PSF}(\rho)}{\mathrm{PSF}(0)} = \mathrm{jinc}^2\left(\pi\Delta\kappa_{\perp 0}\rho\right) \tag{4.81}$$

where $\mathrm{PSF}(0) = \kappa^2 \left(\frac{\pi\Delta\kappa_{\perp 0}^2}{4\kappa^2}\right)^2$. Equation 4.81 is called an Airy pattern.

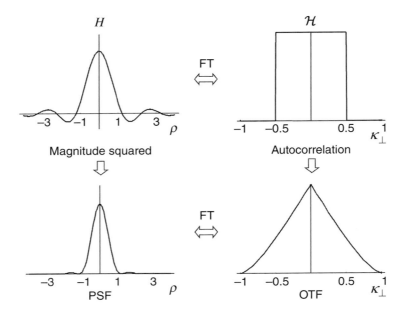

Figure 4.6. Basic imaging functions for a circular aperture pupil.

Similarly, the associated OTF is defined by Eq. 4.71, obtaining

$$\frac{\text{OTF}(\kappa_\perp)}{\text{OTF}(0)} = \begin{cases} \frac{2}{\pi}\left[\cos^{-1}\left(\frac{\kappa_\perp}{\Delta\kappa_{\perp 0}}\right) - \frac{\kappa_\perp}{\Delta\kappa_{\perp 0}}\sqrt{1 - \left(\frac{\kappa_\perp}{\Delta\kappa_{\perp 0}}\right)^2}\right] & \kappa_\perp \leq \Delta\kappa_{\perp 0} \\ 0 & \kappa_\perp > \Delta\kappa_{\perp 0} \end{cases} \tag{4.82}$$

where $\text{OTF}(0) = \frac{\pi\Delta\kappa_{\perp 0}^2}{4\kappa^2}$.

These four functions and their relationships are plotted in Fig. 4.6. We observe that the OTF extends to a bandwidth twice as large as $\mathcal{H}$. However, the OTF response is not uniform over this larger bandwidth: lower spatial frequencies are more efficiently transferred to the image plane than higher spatial frequencies.

The width of the PSF is conventionally defined to be

$$\delta\rho_0 = \frac{\lambda}{2\text{NA}_0} \tag{4.83}$$

where, recalling from Section 3.3.4, the numerical aperture is defined by $\text{NA}_0 = n\sin\theta_0$ (n being the index of refraction at the object and θ_0 the half-angle spanned by the aperture, as seen by the object). That is, we have

$$\text{OTF}(0) = \int d^2\vec{\rho}\,\text{PSF}(\vec{\rho}) = \frac{\pi}{4}\left(\frac{\Delta\kappa_{\perp 0}}{\kappa}\right)^2 = \frac{\pi}{n^2}\text{NA}_0^2 = \Omega_0 \tag{4.84}$$

where Ω_0 is the solid angle spanned by the pupil as seen from the object plane and, again, we have restricted ourselves to the paraxial approximation (see Eq. 3.41).

As before, Eq. 4.83 represents the diffraction-limited resolution of the imaging system, although this time applied to incoherent intensity imaging. This resolution is the same as that derived for a monochromatic field in Section 3.3.4 (see Eq. 3.36). Note that the two resolutions refer to different physical quantities and thus should not be directly compared. Equation 4.83 refers to an intensity resolution whereas Eq. 3.36 refers to a field resolution. The reason the two resolutions appear to be the same here, again, stems largely from convention.

4.5   PROBLEMS

Problem 4.1
Derive the variable change identity given by Eq. 4.6. (Hint: use a Jacobian.)

Problem 4.2
For a circular pupil imaging system, an alternative definition of resolution is given by what is known as the Rayleigh criterion. This criterion states that two point objects are resolvable if they are separated by a minimum distance $\delta\rho_{\text{Rayleigh}}$ such that the maximum of the PSF(ρ) of one point lies at the first zero of the PSF(ρ) of the other point. That is, $\delta\rho_{\text{Rayleigh}}$ is defined as the minimum distance such that PSF$(\delta\rho_{\text{Rayleigh}}) = 0$.

(a) Derive $\delta\rho_{\text{Rayleigh}}$ in terms of λ and NA (you will have to do this numerically).
(b) Consider a circular pupil imaging system where the pupil is partially obstructed by a circular opaque disk (centered) whose radius is η times smaller than the pupil radius ($\eta < 1$). Derive the PSF for this annular pupil system. What is the ratio PSF$_{\text{annular}}(0)/\text{PSF}_{\text{circular}}(0)$?
(c) Provide a numerical plot of PSF$_{\text{annular}}(\Delta\kappa_\perp\rho)$ and PSF$_{\text{circular}}(\Delta\kappa_\perp\rho)$ for $\eta = 0.9$ (normalize both plots to unit maximum). What does the Rayleigh resolution criterion say about the resolution of the annular pupil system compared to that of the circular pupil system? Would you say the annular system has better or worse resolution?

Problem 4.3
The 3D coherence function of Köhler illumination is given by Eq. 4.63. Sketch the range of 3D spatial frequencies $\{\vec{\kappa}_\perp, \kappa_s\}$ spanned by this coherence function, assuming that $I_0(\vec{\rho}_{0c})$ is a uniform disk of radius a. This is called the frequency support associated with Köhler illumination.

Problem 4.4
Consider the specific example where the intensity distribution of the incoherent source is given by

$$I_0(\vec{\rho}_0) = \tfrac{1}{2}I_0\left[1 + \cos(2\pi\rho_0^2/a^2)\right]$$

as illustrated in the figure.

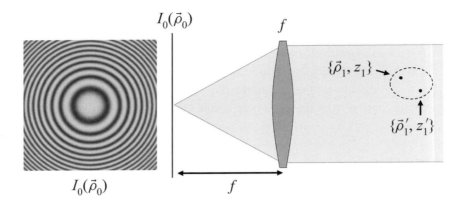

(a) You will find that $\mu_1(\vec{\rho}_{1d}, z_{1d})$ (Eq. 4.63) is peaked when $\{\rho_{1d}, |z_{1d}|\} \rightarrow \{0, 0\}$, as expected; but it is also peaked for another value of $\{\rho_{1d}, |z_{1d}|\}$. What is this value?

(b) Draw a sketch for what happens to the coherence peaks when the pattern in $I\left(\vec{\rho}_0\right)$ is shifted upward.

Problem 4.5

Consider a pinhole camera, as shown in the figure. A 2D incoherent intensity distribution $I_0\left(\vec{\rho}_0\right)$ is projected through a pinhole of pupil function $P(\vec{\xi})$ and creates an image $I_1\left(\vec{\rho}_1\right)$. The object and image planes are distances s_0 and s_1, respectively, from the pinhole plane. Show that, under the Fresnel approximation,

$$\mathcal{I}_1\left(\frac{1}{M}\vec{\kappa}_\perp\right) = \frac{1}{s_0^2}\mathcal{I}_0\left(\vec{\kappa}_\perp\right)\int d^2\vec{\xi}_c\, P\left(\vec{\xi}_c + \frac{s_0}{2\kappa}\vec{\kappa}_\perp\right) P^*\left(\vec{\xi}_c - \frac{s_0}{2\kappa}\vec{\kappa}_\perp\right) e^{i2\pi(1-1/M)\vec{\xi}_c\cdot\vec{\kappa}_\perp}$$

with magnification $M = -\frac{s_1}{s_0}$. Note that the OTF here is not simply the pupil autocorrelation function, as it is for a 4f system. (Hint: one can proceed by making use of Eq. 2.50 to propagate $E_0\left(\vec{\rho}_0\right)$ to $E_1\left(\vec{\rho}_1\right)$, and then calculate $I_1\left(\vec{\rho}_1\right)$. The replacement $\left\langle E_0\left(\vec{\rho}_0\right)E_0^*\left(\vec{\rho}_0'\right)\right\rangle \rightarrow \kappa^{-2}I_0\left(\vec{\rho}_0\right)\delta^2\left(\vec{\rho}_0 - \vec{\rho}_0'\right)$ from Eq. 4.43 then leads to the above result.)

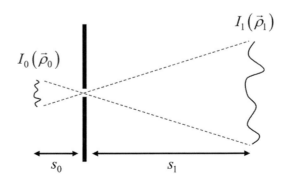

References

[1] Barrett, H. H. and Myers, K. J. *Foundations of Image Science*, Wiley-Interscience (2004).

[2] Carter, W. H. and Wolf, E. "Coherence and radiometry with quasi-homogeneous planar sources," *J. Opt. Soc. Am.* 67, 785–796 (1977).

[3] Françon, M. *Optical Image Formation and Processing*, Academic Press (1979).

[4] Goodman, J. W. *Introduction to Fourier Optics*, 4th edn, W. H. Freeman (2017).

[5] Goodman, J. W. *Statistical Optics*, 2nd edn, Wiley (2015).

[6] Ishimaru, A. *Wave Propagation and Scattering in Random Media*, IEEE and Oxford University Press (1997).

[7] Klein, M. V. and Furtak, T. E. *Optics*, 2nd edn, Wiley (1986).

[8] Mandel, L. and Wolf, E. *Optical Coherence and Quantum Optics*, Cambridge University Press (1995).

[9] Saleh, B. E. A. and Teich, M. C. *Fundamentals of Photonics*, 2nd edn, Wiley Interscience (2007).

[10] Thompson, B. J. "Image formation with partially coherent light," *Progress in Optics Vol. VII*, Ed. E. Wolf, North Holland (1969).

[11] Wolf, E. *Introduction to the Theory of Coherence and Polarization of Light*, Cambridge University Press (2007).

5 | 3D Imaging

So far, we have considered the propagation of fields or intensities from one transverse plane to another, whether through free space or through lens configurations. As such, we have considered only 2D descriptions of propagation. For example, in Section 3.1, the propagation of a monochromatic field from an object to image plane via a simple lens was found to be described by Eq. 3.6. This equation is technically three-dimensional in the sense that it is valid for arbitrary axial locations of the planes, but we refrained from considering such arbitrary axial locations because Eq. 3.6 is quite cumbersome. Instead we restricted ourselves to the simplifying cases where the object and image planes were fixed either by the thin-lens formula (Eq. 3.15) in a single-lens imaging geometry, or by the focal lengths f_0 and f_1 in a 4f imaging geometry. When these restrictions are applied, the imaging is said to be in focus.

Our goal in this chapter will be to relax the restrictions of in-focus imaging, and to re-examine the possibility that the object and image planes might be out of focus relative to one another. As it turns out, we have already developed the building blocks to formulate this. Indeed, out-of-focus imaging can be thought of as a combination of two operations: in-focus imaging followed by additional defocus. The functions describing both of these operations have already been established. In particular, the operational functions for in-focus imaging are given by H or $\mathcal{H}$ for monochromatic fields (Chapter 3), or by the PSF or OTF for incoherent intensities (Chapter 4). Since defocus corresponds to additional free-space propagation from an in-focus to an out-of-focus plane, the operational functions for this too have been derived in Chapter 2, and are given by D or $\mathcal{D}$.

Our aim will be to consolidate the operations of in-focus imaging and defocus into a combined single operation, which we refer to as 3D imaging. We begin this chapter by deriving operational functions for 3D imaging in mixed and fully spatial representations, and we end with fully 3D spatial-frequency representations, which are useful in defining the concept of frequency support. Detailed derivations of the operational functions for 3D imaging can be found in many optics textbooks such as [1, 3, 10], and a overview of techniques extending beyond the Fresnel approximation is presented in [7]. Most of the discussion presented here stems from the seminal work by [6, 8, 11, 13, 15], and the reader is referred to these original sources for more detail.

5.1   DEFOCUS

Defining $z = 0$ to be the position of an in-focus plane, then $z \neq 0$ corresponds to the position of an out-of-focus plane. That is, the operation of defocus may be functionally regarded as free-space propagation from $z = 0$ to $z \neq 0$. Such free-space propagation was described in Section 2.3 as a simple 2D convolution operation over the spatial coordinates $\vec{\rho}$ (Eq. 2.22),

$$E(\vec{\rho}, z) = \left[D_+(\vec{\rho}, z) * E(\vec{\rho}, 0) \right]_{\vec{\rho}} \tag{5.1}$$

where D_+ is the forward-propagating Rayleigh–Sommerfeld (Eq. 2.24) or Fresnel (Eq. 2.50) free-space propagator, depending on the degree of approximation. The associated free-space radiant field propagation is then (Eq. 2.32)

$$\mathcal{E}(\vec{\kappa}_\perp; z) = \mathcal{D}_+(\vec{\kappa}_\perp; z)\mathcal{E}(\vec{\kappa}_\perp; 0) \tag{5.2}$$

where $\mathcal{D}_+$ is the corresponding mixed-representation transfer function in Fourier space (Eq. 2.28 or 2.54). That is, $E(\vec{\rho}, z)$ can be calculated in two ways, either directly from Eq. 5.1, or indirectly via a sequence of a Fourier transformations:

$$E(\vec{\rho}, z) = \text{FT}^{-1} \left\{ \mathcal{D}_+(\vec{\kappa}_\perp; z) \, \text{FT} \left\{ E(\vec{\rho}, 0) \right\} \right\} \tag{5.3}$$

This is schematically depicted in Fig. 5.1.

For example, using this second approach with the Fresnel approximation, we arrive at

$$E(\vec{\rho}, z) = e^{i2\pi\kappa z} \int d^2\vec{\kappa}_\perp \, \mathcal{E}(\vec{\kappa}_\perp; 0) e^{i2\pi\vec{\rho}\cdot\vec{\kappa}_\perp - i\pi\frac{z}{\kappa}\kappa_\perp^2} \tag{5.4}$$

Equation 5.4 is useful in 3D imaging and will be utilized on several occasions. It is mathematically equivalent to Eq. 2.52 and as such it is a modified version of the Fresnel

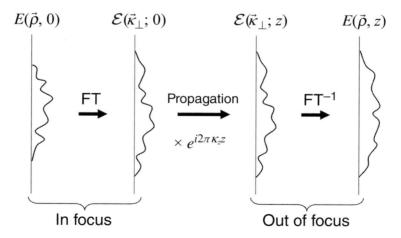

Figure 5.1. Field propagation from an in-focus to an out-of-focus plane. Note that $\kappa_z = \pm\sqrt{\kappa^2 - \kappa_\perp^2}$.

diffraction integral. The question of which Fresnel diffraction integral is preferable depends on the problem at hand. A notable advantage of Eq. 5.4 over the usual Fresnel diffraction integral (Eq. 2.52) is that it contains no $1/z$ factor either outside the integral or inside the exponent. The absence of this factor makes Eq. 5.4 particularly well suited for treating small defocus values that approach zero. However, for large defocus values, the scaling with z of the exponent in Eq. 5.4 becomes computationally problematic and can lead to aliasing. For large defocus values, then, the usual diffraction integral (Eq. 2.52) becomes preferable as the exponent in this integral becomes well behaved. This same issue came up, for example, when we considered the van Cittert–Zernike theorem in Section 4.2.3.

Throughout this chapter we will concentrate on small defocus values.

5.1.1 Gaussian Beam

Although we have already briefly considered the example of a Gaussian beam in Section 2.4.1, we will consider it again here since it serves as a quintessential application of the concept of defocus.

If at an in-focus plane a field exhibits a Gaussian amplitude profile of waist w_0 and a constant phase (set to 0), such that,

$$E(\rho, 0) = E_0 e^{-\rho^2/w_0^2} \tag{5.5}$$

then the radiant field at this focal plane is

$$\mathcal{E}(\kappa_\perp; 0) = E_0 \pi w_0^2 e^{-\pi^2 w_0^2 \kappa_\perp^2} \tag{5.6}$$

From the prescription outlined in Fig. 5.1, and abiding by the Fresnel approximation (i.e. Eq. 5.4), the field away from focus is then readily calculated to be

$$E(\vec{\rho}, z) = \frac{E_0}{1 + i\frac{z}{z_R}} e^{i2\pi\kappa z} e^{-\rho^2/\left[w_0^2\left(1+i\frac{z}{z_R}\right)\right]} \tag{5.7}$$

where the distance $z_R = \pi w_0^2 \kappa$ is known as the Rayleigh length. This is identical to our result in Eq. 2.60. By introducing $w(z) = w_0\sqrt{1 + z^2/z_R^2}$ the above equation can be recast in the more convenient form

$$E(\vec{\rho}, z) = E_0 \frac{w_0}{w(z)} e^{i2\pi\kappa z} e^{-\rho^2/w(z)^2} e^{i\pi\kappa\rho^2/R(z)} e^{-i\eta(z)} \tag{5.8}$$

where the wavefront radius of the Gaussian field $R(z) = z\left(1 + z_R^2/z^2\right)$ and the Gouy phase shift $\eta(z) = \tan^{-1}(z/z_R)$ have been explicitly identified. Note that for a defocus z less than z_R, the Gaussian field remains relatively unchanged and the phase front remains roughly planar. For a defocus z greater than z_R, the Gaussian field begins to diverge with a full angle $\Delta\theta = 2w_0/z_R = 2/\pi\kappa w_0$, as shown in Fig. 5.2. For more details on Gaussian beams, the reader is referred to [9, 12].

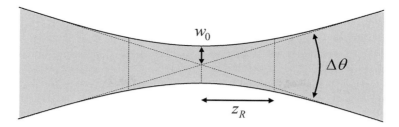

Figure 5.2. Gaussian beam profile.

5.2 3D IMAGING FUNCTIONS

Our goal now will be to combine in-focus imaging with defocus, thereby deriving generalized, arbitrary-focus imaging operators applicable to both monochromatic and incoherent fields. Throughout this chapter we assume unit magnification by setting $M = 1$. We choose this for notational simplicity, bearing in mind that a 4f system with lenses of equal magnification in reality has magnification $M = -1$. In other words, our choice of $M = 1$ is tantamount to flipping the transverse coordinate $\vec{\rho}$ in either object or image space. Moreover, we will assume a homogeneous index of refraction. An allowance for different indices of refraction, if needed, can be made following the recipe prescribed in Section 3.4.

5.2.1 3D Coherent Transfer Function

We begin by considering the case where the object plane is in focus but the image plane is out of focus. The radiant field at the image plane, in the case of forward-propagating light ($\kappa_z > 0$), is then readily expressed by

$$\mathcal{E}_1(\vec{\kappa}_\perp; z) = \mathcal{D}_+(\vec{\kappa}_\perp; z)\mathcal{H}(\vec{\kappa}_\perp)\mathcal{E}_0(\vec{\kappa}_\perp; 0) \tag{5.9}$$

where $\mathcal{H}(\vec{\kappa}_\perp)$ effects the in-focus imaging operation and $\mathcal{D}_+(\vec{\kappa}_\perp; z)$ effects the supplementary defocus operation (recall that even though we are considering forward-propagating light, the sign of z can be arbitrary – see Section 2.3.2). The field at the image plane is obtained from the radiant field $\mathcal{E}_1(\vec{\kappa}_\perp; z)$ by a simple inverse Fourier transform.

Alternatively, a situation that is more routinely found in practice is where the object plane is out of focus but the image plane is in focus. In this case, we have

$$\mathcal{E}_1(\vec{\kappa}_\perp; 0) = \mathcal{H}(\vec{\kappa}_\perp)\mathcal{D}_+(\vec{\kappa}_\perp; -z)\mathcal{E}_0(\vec{\kappa}_\perp; z) \tag{5.10}$$

where now the defocus operation is performed prior to the imaging operation, leading to a change in sign in z in the defocus operation.

Henceforth, we will use the convention that $z > 0$ when the image or object plane moves *away* from the imaging lens (or lenses). In this manner, we can define a forward-propagating 3D coherent transfer function that is common to both scenarios, given by

$$\mathcal{H}_+(\vec{\kappa}_\perp; z) = \mathcal{H}(\vec{\kappa}_\perp)\mathcal{D}_+(\vec{\kappa}_\perp; z) \tag{5.11}$$

Similarly, in the case of backward-propagating light ($\kappa_z < 0$) we can define a backward-propagating 3D coherent transfer function given by

$$\mathcal{H}_-(\vec{\kappa}_\perp; z) = \mathcal{H}(\vec{\kappa}_\perp)\mathcal{D}_-(\vec{\kappa}_\perp; z) \tag{5.12}$$

Depending on the approximation in use, these become

$$\mathcal{H}_\pm(\vec{\kappa}_\perp; z) = \begin{cases} \mathcal{H}(\vec{\kappa}_\perp)e^{\pm i2\pi z\sqrt{\kappa^2 - \kappa_\perp^2}} & \text{(Rayleigh–Sommerfeld)} \\ \mathcal{H}(\vec{\kappa}_\perp)e^{\pm i2\pi\kappa z}e^{\mp i\pi\frac{z}{\kappa}\kappa_\perp^2} & \text{(Fresnel)} \end{cases} \tag{5.13}$$

Inasmuch as our previous (in-focus) transfer function $\mathcal{H}(\vec{\kappa}_\perp)$ was defined by the pupil of the imaging device (Eq. 3.27), these new transfer functions $\mathcal{H}_\pm(\vec{\kappa}_\perp; z)$ can be thought of as defined by a generalized 3D pupil [11].

Useful Properties of 3D Coherent Transfer Functions

The following properties of $\mathcal{H}_\pm(\vec{\kappa}_\perp; z)$ are listed for reference:

- Normalization: $\mathcal{H}_\pm(0; z) = e^{\pm i2\pi\kappa z}P(0)$.
- $\mathcal{H}_+(\vec{\kappa}_\perp; z) = \mathcal{H}_-(\vec{\kappa}_\perp; -z)$.
- If the pupil is real (i.e. $P(\vec{\xi}) = P^*(\vec{\xi})$), then $\mathcal{H}_\pm(\vec{\kappa}_\perp; z) = \mathcal{H}_\pm^*(\vec{\kappa}_\perp; -z)$.
- If the pupil is symmetric (i.e. $P(\vec{\xi}) = P(-\vec{\xi})$), then $\mathcal{H}_\pm(\vec{\kappa}_\perp; z) = \mathcal{H}_\pm(-\vec{\kappa}_\perp; z)$.
- If the pupil is Hermitian (i.e. $P(\vec{\xi}) = P^*(-\vec{\xi})$), then $\mathcal{H}_\pm(\vec{\kappa}_\perp; z) = \mathcal{H}_\pm^*(-\vec{\kappa}_\perp; -z)$.
- If the pupil is binary (i.e. $P(\vec{\xi}) = 0$ or 1) then:

$$\mathcal{H}_\pm(\vec{\kappa}_\perp; z_1)\mathcal{H}_\pm(\vec{\kappa}_\perp; z_2) = \mathcal{H}_\pm(\vec{\kappa}_\perp; z_1 + z_2)$$
$$\mathcal{H}_\pm(\vec{\kappa}_\perp; z_1)\mathcal{H}_\pm^*(\vec{\kappa}_\perp; z_2) = \mathcal{H}_\pm(\vec{\kappa}_\perp; z_1 - z_2)$$
$$\mathcal{H}_\pm(\vec{\kappa}_\perp; z_1)\mathcal{H}_\mp(\vec{\kappa}_\perp; z_2) = \mathcal{H}_\pm(\vec{\kappa}_\perp; z_1 - z_2)$$
$$\mathcal{H}_\pm(\vec{\kappa}_\perp; z_1)\mathcal{H}_\mp^*(\vec{\kappa}_\perp; z_2) = \mathcal{H}_\pm(\vec{\kappa}_\perp; z_1 + z_2)$$

5.2.2 3D Amplitude Spread Function

Having derived a 3D coherent transfer function, the procedure for calculating the associated 3D amplitude point spread function is straightforward since the two are related by a Fourier transform. That is,

$$H_\pm(\vec{\rho}, z) = \text{FT}^{-1}\left\{\mathcal{H}_\pm(\vec{\kappa}_\perp; z)\right\} \tag{5.14}$$

From Eq. 5.11, we obtain

$$H_\pm(\vec{\rho}, z) = \int d^2\vec{\kappa}_\perp \, e^{i2\pi\vec{\rho}\cdot\vec{\kappa}_\perp}\mathcal{H}(\vec{\kappa}_\perp)\mathcal{D}_\pm(\vec{\kappa}_\perp; z) \tag{5.15}$$

For reference, an alternative version of Eq. 5.15 is included here, applicable in the case of cylindrical symmetry (see Appendix A), and given by

$$H_\pm(\rho, z) = 2\pi \int_0^\infty d\kappa_\perp \, \kappa_\perp J_0(2\pi\kappa_\perp\rho)\mathcal{H}(\kappa_\perp)\mathcal{D}_\pm(\vec{\kappa}_\perp; z) \tag{5.16}$$

where J_0 is the zeroth order cylindrical Bessel function (not to be confused with mutual intensity).

In either case, depending on whether the light is forward or backward propagating, the object and image fields are related by the convolution integral

$$E_1(\vec{\rho}_1, z_1) = \int d^2\vec{\rho}_0\, H_\pm(\vec{\rho}_1 - \vec{\rho}_0,\ z_1 - z_0)E_0(\vec{\rho}_0, z_0) \tag{5.17}$$

where z_0 and z_1 correspond to object and image defocus, respectively. Again, Eq. 5.17 is valid for unit magnification. When the Fresnel approximation is valid, non-unit magnification can be readily taken into account by making the replacement $E_1(\vec{\rho}_1, z_1) \rightarrow M E_1(M\vec{\rho}_1, M^2 z_1)$ on the left-hand side. The origin of the M^2 magnification along the axial coordinate was briefly motivated in Section 1.4, and is more thoroughly explained in [5].

As a reminder, there is no integration over z_0 in Eq. 5.17 indicating that $E_1(\vec{\rho}_1, z_1)$ remains implicitly a function of z_0. Thus we are still dealing with 2D imaging from one plane to another, though these planes may be located arbitrarily. Some important examples are presented below.

Circular Pupil

The unobstructed circular pupil defined by Eq. 4.77 leads to an associated coherent transfer function defined by Eq. 4.78. Because unit magnification is assumed ($f_0 = f_1 = f$) the field bandwidth is the same in both the object and image planes, and is given by

$$\Delta\kappa_\perp = 2\kappa\frac{a}{f} \tag{5.18}$$

From Eqs. 5.16 and 5.13, and using the Rayleigh-Sommerfeld approximation for $\mathcal{D}_\pm(\vec{\kappa}_\perp; z)$, we obtain then

$$H_\pm(\rho, 0) = \frac{\pi}{4}\Delta\kappa_\perp^2\, \text{jinc}\,(\pi\Delta\kappa_\perp\rho) \tag{5.19}$$

$$H_\pm(0, z) = \frac{1}{2\pi z^2}\left[e^{\pm i2\pi\kappa z}\left(1 \mp i2\pi\kappa z\right) - e^{\pm i2\pi z\sqrt{\kappa^2 - \Delta\kappa_\perp^2/4}}\left(1 \mp i2\pi z\sqrt{\kappa^2 - \Delta\kappa_\perp^2/4}\right)\right] \tag{5.20}$$

where the jinc function is defined by Eq. A.16.

The first of these equations we have seen before in Sections 3.3.4 and 4.4.1. The second is new. Note that both expressions for $H_\pm$ converge to the same value of $\frac{\pi}{4}\Delta\kappa_\perp^2$ as ρ or z converges to zero, in accord with our previously derived normalization condition (Eq. 3.42). Equation 5.20 can be simplified if we adopt the Fresnel approximation for $\mathcal{D}_\pm(\vec{\kappa}_\perp; z)$ in Eq. 5.16, obtaining

$$H_\pm(0, z) = 2\frac{\kappa}{z}e^{\pm i2\pi\kappa z\left(1 - (\Delta\kappa_\perp/4\kappa)^2\right)}\sin\left(2\pi\kappa z\,(\Delta\kappa_\perp/4\kappa)^2\right) \tag{5.21}$$

which closely tracks Eq. 5.20 even for moderate values of $\Delta\kappa_\perp/\kappa$. Equation 5.21 reveals some key properties of $H_\pm(0, z)$, namely that it oscillates rapidly, but with a carrier frequency given by $\kappa\left(1 - (\Delta\kappa_\perp/4\kappa)^2\right)$ rather than κ. Moreover, this oscillation is amplitude modulated by a lower frequency sinusoid that decays as $1/|z|$ with increasing defocus. We recall, from Eq. 3.37, that the recurring parameter $\Delta\kappa_\perp/4\kappa$ corresponds to $\frac{1}{2}\sin\theta_{\max}$, where $\theta_{\max}$ corresponds to the maximum half-angle of the field spreading due to defocus. In other words, the reduction in the carrier frequency may be loosely interpreted as arising from the averaged obliquity of the wavevectors traversing the focal plane.

Gaussian Pupil

Alternatively, in the case of a Gaussian aperture pupil defined by Eq. 3.12, we have

$$\mathcal{H}(\vec{\kappa}_\perp) = e^{-2\kappa_\perp^2/\Delta\kappa_\perp^2} \tag{5.22}$$

where the bandwidth is defined by $\Delta\kappa_\perp = \sqrt{2}\kappa\frac{w_\xi}{f}$. An application of Eq. 5.16 with the Fresnel approximation for $\mathcal{D}_\pm(\vec{\kappa}_\perp; z)$ then leads to

$$H_\pm(\rho, z) = \frac{\pi}{2}\frac{\Delta\kappa_\perp^2}{(1 \pm i\zeta)}e^{\pm i2\pi\kappa z}e^{-\pi^2\Delta\kappa_\perp^2\rho^2/2(1\pm i\zeta)} \tag{5.23}$$

where $\zeta = \frac{\pi\Delta\kappa_\perp^2 z}{2\kappa}$. The amplitude point spread function thus corresponds to a Gaussian focus of waist $w_0 = \frac{\sqrt{2}}{\pi\Delta\kappa_\perp} = \frac{f}{\pi\kappa w_\xi}$, and Rayleigh length $z_R = \frac{2\kappa}{\pi\Delta\kappa_\perp^2} = \frac{f^2}{\pi\kappa w_\xi^2}$.

In both cases of a circular or Gaussian pupil, $H_\pm(\rho, z)$ undergoes an overall π phase shift as z traverses the focal plane (relative to the rapidly varying phase of $e^{\pm i2\pi\kappa z}$). This is again a manifestation of the Gouy phase shift which we have seen in Sections 2.4.1 and 5.1.1.

Useful Properties of 3D Amplitude Spread Functions

The following properties of $H_\pm(\vec{\rho}, z)$ are listed for reference:

- Normalization: $\int d^2\vec{\rho}\, H_\pm(\vec{\rho}, z) = e^{\pm i2\pi\kappa z}P(0)$.
- $H_+(\vec{\rho}, z) = H_-(\vec{\rho}, -z)$.
- If the pupil is real (i.e. $P(\vec{\xi}) = P^*(\vec{\xi})$), then $H_\pm(\vec{\rho}, z) = H_\pm^*(-\vec{\rho}, -z)$.
- If the pupil is symmetric (i.e. $P(\vec{\xi}) = P(-\vec{\xi})$), then $H_\pm(\vec{\rho}, z) = H_\pm(-\vec{\rho}, z)$.
- If the pupil is Hermitian (i.e. $P(\vec{\xi}) = P^*(-\vec{\xi})$), then $H_\pm(\vec{\rho}, z) = H_\pm^*(\vec{\rho}, -z)$.
- If the pupil is binary (i.e. $P(\vec{\xi}) = 0$ or 1) then:

$$\int d^2\vec{\rho}'\, H_\pm(\vec{\rho}_1 - \vec{\rho}', z_1)H_\pm(\vec{\rho}' - \vec{\rho}_2, z_2) = H_\pm(\vec{\rho}_1 - \vec{\rho}_2, z_1 + z_2) \tag{5.24}$$

$$\int d^2\vec{\rho}'\, H_\pm(\vec{\rho}' + \vec{\rho}_1, z_1)\, H_\pm^*(\vec{\rho}' + \vec{\rho}_2, z_2) = H_\pm(\vec{\rho}_1 - \vec{\rho}_2, z_1 - z_2) \tag{5.25}$$

$$\int d^2\vec{\rho}'\, H_\pm(\vec{\rho}' + \vec{\rho}_1, z_1)\, H_\mp(\vec{\rho}' + \vec{\rho}_2, z_2) = H_\pm(\vec{\rho}_1 - \vec{\rho}_2, z_1 - z_2) \tag{5.26}$$

$$\int d^2\vec{\rho}'\, H_\pm(\vec{\rho}' + \vec{\rho}_1, z_1)\, H_\mp^*(\vec{\rho}' + \vec{\rho}_2, z_2) = H_\pm(\vec{\rho}_1 - \vec{\rho}_2, z_1 + z_2) \tag{5.27}$$

5.2.3 3D Point Spread Function

Having derived the transfer functions for 3D imaging of monochromatic fields, we can now turn our attention to the case of incoherent fields. As discussed in Chapter 4, only the intensities in the object and image planes are of interest in this case, and accordingly we derive operational functions for 3D imaging of intensities. For ease of notation we consider only foward-propagating light.

From Eq. 4.70 we have

$$\text{PSF}(\vec{\rho}, z) = \frac{1}{\kappa^2} \left| H_+(\vec{\rho}, z) \right|^2 \tag{5.28}$$

which, from 5.15, leads to

$$\text{PSF}(\vec{\rho}, z) = \frac{1}{\kappa^2} \iint d^2\vec{\kappa}_\perp \, d^2\vec{\kappa}_\perp' \, \mathcal{H}_+(\vec{\kappa}_\perp; z) \mathcal{H}_+^*(\vec{\kappa}_\perp'; z) e^{i2\pi\vec{\rho} \cdot (\vec{\kappa}_\perp - \vec{\kappa}_\perp')} \tag{5.29}$$

Using the usual coordinate transformation defined in Eq. 4.13, we arrive at

$$\text{PSF}(\vec{\rho}, z) = \frac{1}{\kappa^2} \iint d^2\vec{\kappa}_{\perp c} \, d^2\vec{\kappa}_{\perp d} \, \mathcal{H}_+(\vec{\kappa}_{\perp c} + \tfrac{1}{2}\vec{\kappa}_{\perp d}; z) \mathcal{H}_+^*(\vec{\kappa}_{\perp c} - \tfrac{1}{2}\vec{\kappa}_{\perp d}; z) e^{i2\pi\vec{\rho} \cdot \vec{\kappa}_{\perp d}} \tag{5.30}$$

corresponding to a 3D PSF.

Incoherent intensity imaging in three dimensions is then described by

$$I_1(\vec{\rho}_1, z_1) = \int d^2\vec{\rho}_0 \, \text{PSF}(\vec{\rho}_1 - \vec{\rho}_0, \, z_1 - z_0) I_0(\vec{\rho}_0, z_0) \tag{5.31}$$

This equation is valid for unit magnification imaging. As before, the case of arbitrary magnification can be readily treated in the Fresnel approximation by making the replacement $I_1(\vec{\rho}_1, z_1) \to M^2 I_1(M\vec{\rho}_1, M^2 z_1)$ on the left-hand side. We note that Eq. 5.31 contains no integral over z_0, meaning that $I_1(\vec{\rho}_1, z_1)$ is tacitly a function of z_0. The 3D imaging described here thus remains a disguised version of 2D imaging. Bona-fide 3D imaging will be treated in the next section. We emphasize again that we are considering here only forward-propagating light. In the event of backward-propagating light, we must make the replacement $\text{PSF}(\vec{\rho}, z) \to \text{PSF}(\vec{\rho}, -z)$ in Eq. 5.31.

Following the same procedure as above, we consider the examples of circular and Gaussian pupils:

Circular Pupil

From Eq. 4.78 for a circular pupil, and adopting the Fresnel approximation, we arrive at

$$\frac{\text{PSF}(\rho, 0)}{\text{PSF}(0, 0)} = \text{jinc}^2 \left(\pi \Delta\kappa_\perp \rho \right) \tag{5.32}$$

$$\frac{\text{PSF}(0, z)}{\text{PSF}(0, 0)} = \text{sinc}^2 \left(\frac{\pi \Delta\kappa_\perp^2 z}{8\kappa} \right) \tag{5.33}$$

where the sinc function is defined by $\text{sinc}(x) = \frac{\sin(x)}{x}$. Denoting the solid angle spanned by the circular pupil as $\Omega_0 = \frac{\pi \Delta \kappa_\perp^2}{4\kappa^2}$ (see Eq. 3.41), we can write $\text{PSF}(0,0) = \kappa^2 \Omega_0^2$. Again, Eq. 5.32 is called an Airy pattern.

Gaussian Pupil

From Eq. 5.22 for a Gaussian pupil, also adopting the Fresnel approximation, we arrive at

$$\frac{\text{PSF}(\rho,z)}{\text{PSF}(0,0)} = \frac{1}{(1+\zeta^2)} e^{-\pi^2 \Delta \kappa_\perp^2 \rho^2 /(1+\zeta^2)} = \frac{1}{(1+\zeta^2)} e^{-2\rho^2 / w_0^2 (1+\zeta^2)} \tag{5.34}$$

where the solid angle spanned by the Gaussian pupil is $\Omega_0 = \frac{\pi \Delta \kappa_\perp^2}{4\kappa^2}$ (same as for the circular pupil), and $\text{PSF}(0,0) = 4\kappa^2 \Omega_0^2 = 1/z_R^2$. The profile of this PSF is Gaussian in the transverse direction (ρ) and Lorentzian in the axial direction (z), and as such it is referred to as a Gaussian–Lorentzian PSF.

While a Gaussian aperture is convenient in that it provides an analytic expression for H and PSF in three dimensions, it should be used with care. Its convenience can be offset by difficulties associated with the fact that a Gaussian pupil does not satisfy $P(\xi) = |P(\xi)|^2$ (recall that the pupil function as defined in Chapter 3 corresponds to a field transmission and not an intensity transmission), and hence many of the properties of amplitude point spread functions listed in Sections 3.3.5 and 5.2.2 do not apply for Gaussian pupils.

Useful Properties of 3D Point Spread Functions

The following properties of $\text{PSF}(\vec{\rho},z)$ are listed for reference:

- $\text{PSF}(\vec{\rho},z)$ is real.
- Normalization: $\int d^2\vec{\rho}\,\text{PSF}(\vec{\rho},z) = \frac{1}{f_0^2} \int d^2\vec{\xi}\,\left|P(\vec{\xi})\right|^2 = \frac{1}{\kappa^2} \int d^2\vec{\kappa}_\perp \left|\mathcal{H}(\vec{\kappa}_\perp)\right|^2 = \Omega_0$.
- If the pupil is real (i.e. $P(\vec{\xi}) = P^*(\vec{\xi})$), then $\text{PSF}(\vec{\rho},z) = \text{PSF}(-\vec{\rho},-z)$.
- If the pupil is symmetric (i.e. $P(\vec{\xi}) = P(-\vec{\xi})$), then $\text{PSF}(\vec{\rho},z) = \text{PSF}(-\vec{\rho},z)$.
- If the pupil is Hermitian (i.e. $P(\vec{\xi}) = P^*(-\vec{\xi})$), then $\text{PSF}(\vec{\rho},z) = \text{PSF}(\vec{\rho},-z)$.

5.2.4 3D Optical Transfer Function

Having derived the 3D point spread function, we can finally derive a 3D optical transfer function from

$$\text{OTF}(\vec{\kappa}_\perp;z) = \int d^2\vec{\rho}\,\text{PSF}(\vec{\rho},z) e^{-i2\pi \vec{\rho} \cdot \vec{\kappa}_\perp} \tag{5.35}$$

which, from Eq. 5.30, leads to

$$\text{OTF}(\vec{\kappa}_\perp;z) = \frac{1}{\kappa^2} \int d^2\vec{\kappa}_{\perp c}\,\mathcal{H}_+\left(\vec{\kappa}_{\perp c} + \tfrac{1}{2}\vec{\kappa}_\perp;z\right) \mathcal{H}_+^*\left(\vec{\kappa}_{\perp c} - \tfrac{1}{2}\vec{\kappa}_\perp;z\right) \tag{5.36}$$

In other words, what was true for in-focus imaging (Eq. 4.71) has now been generalized to out-of-focus imaging: $\text{OTF}(\vec{\kappa}_\perp;z)$ is simply equal to the autocorrelation of $\mathcal{H}_+(\vec{\kappa}_\perp;z)$.

Making use of the Fresnel approximation, the above equation may be expressed equivalently as

$$\mathrm{OTF}(\vec{\kappa}_\perp; z) = \frac{1}{\kappa^2} \int \mathrm{d}^2\vec{\kappa}_{\perp c}\, \mathcal{H}\left(\vec{\kappa}_{\perp c} + \tfrac{1}{2}\vec{\kappa}_\perp\right) \mathcal{H}^*\left(\vec{\kappa}_{\perp c} - \tfrac{1}{2}\vec{\kappa}_\perp\right) e^{-i2\pi\frac{z}{\kappa}\vec{\kappa}_{\perp c}\cdot\vec{\kappa}_\perp} \qquad (5.37)$$

It is important to note here that when the spatial frequency $\kappa_\perp$ is equal to zero, $\mathrm{OTF}(0; z) = \int \mathrm{d}^2\vec{\rho}\,\mathrm{PSF}(\vec{\rho}, z) = \Omega_0$ is a constant, independent of the defocus z. That is, light in free space incurs no loss in power as it propagates from one axial plane to another. This fundamental conservation principle has considerable ramifications when considering the problem of out-of-focus background rejection, a problem we will be faced with in future chapters.

In the meantime, we again consider the examples of circular and Gaussian pupils:

Circular Pupil

We derived an analytic expression for $\mathrm{OTF}(\vec{\kappa}_\perp; 0)$ for an unobstructed circular pupil in Section 4.4.1. Though it is not possible to derive an associated analytic expression for $\mathrm{OTF}(\vec{\kappa}_\perp; z)$, an excellent approximation is afforded by what is known as the Stokseth approximation [14]:

$$\mathrm{OTF}(\vec{\kappa}_\perp; z) \approx \mathrm{OTF}(\vec{\kappa}_\perp; 0)\, \mathrm{jinc}\left[\pi z \Delta\kappa_\perp \frac{\kappa_\perp}{\kappa}\left(1 - \frac{\kappa_\perp}{\Delta\kappa_\perp}\right)\right] \qquad (5.38)$$

where $\mathrm{OTF}(\vec{\kappa}_\perp; 0)$ is given by Eq. 4.82.

As illustrated in Fig. 5.3, if $\kappa_\perp = 0$ then $\mathrm{OTF}(0; z) = \Omega_0$, independent of z. In contrast, if $\kappa_\perp \neq 0$ then $\mathrm{OTF}(\vec{\kappa}_\perp; z)$ decays to zero in an oscillatory manner, where the envelope of the decay scales as $|z|^{-3/2}$ for large $|z|$. This scaling law will be useful, for example, when analyzing the capacity of optical sectioning microscopes to reject out-of-focus background. In particular, we will return to this scaling law when discussing structured illumination microscopes (Chapter 15).

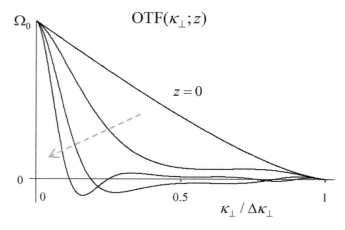

Figure 5.3. Stokseth approximations of a circular-pupil OTF for increasing defocus z.

Gaussian Pupil

In the case of a Gaussian pupil, Eqs. 5.34 and 5.35 lead to the simple relation

$$\text{OTF}(\vec{\kappa}_\perp; z) = \Omega_0 \exp\left[-\frac{\kappa_\perp^2}{\Delta\kappa_\perp^2}(1+\zeta^2)\right] \tag{5.39}$$

As before, if $\kappa_\perp = 0$ then $\text{OTF}(0; z) = \Omega_0$. If $\kappa_\perp \neq 0$ then $\text{OTF}(\vec{\kappa}_\perp; z)$ decays to zero, but this time exponentially and monotonically.

Useful Properties of 3D Optical Transfer Functions

The following properties of $\text{OTF}(\vec{\kappa}_\perp; z)$ are listed for reference:

- $\text{OTF}(\vec{\kappa}_\perp; z)$ is Hermitian (i.e. $\text{OTF}(\vec{\kappa}_\perp; z) = \text{OTF}^*(-\vec{\kappa}_\perp; z)$).
- Normalization: $\text{OTF}(0; z) = \frac{1}{f_0^2}\int d^2\vec{\xi}\,\left|P(\vec{\xi})\right|^2 = \frac{1}{\kappa^2}\int d^2\vec{\kappa}_\perp\,\left|\mathcal{H}(\vec{\kappa}_\perp)\right|^2 = \Omega_0$.
- If the pupil is real (i.e. $P(\vec{\xi}) = P^*(\vec{\xi})$), then $\text{OTF}(\vec{\kappa}_\perp; z) = \text{OTF}^*(\vec{\kappa}_\perp; -z)$.
- If the pupil is symmetric (i.e. $P(\vec{\xi}) = P(-\vec{\xi})$), then $\text{OTF}(\vec{\kappa}_\perp; z) = \text{OTF}(-\vec{\kappa}_\perp; z)$.
- If the pupil is Hermitian (i.e. $P(\vec{\xi}) = P^*(-\vec{\xi})$), then $\text{OTF}(\vec{\kappa}_\perp; z) = \text{OTF}^*(-\vec{\kappa}_\perp; -z)$.

5.3   3D IMAGING EXAMPLES

As emphasized throughout this chapter, the 3D imaging functions derived above are, in fact, generalized 2D imaging functions that characterize imaging from an object plane to an image plane. To apply these imaging functions to 3D volumes, certain conditions must be met. For example, let us consider the imaging of a thick object that is axially extended. By definition, most of this object will be out of focus. We therefore model this thick object as a stack of thin object planes of differential thicknesses dz_0. If the contributions of these thin planes to the final image intensity are independent, they can be evaluated separately and simply added to produce the final image intensity. However, two conditions must be met for such an independent (incoherent) summation of intensity contributions. The first is that the intensity generated at one plane does not affect the intensity generated at another. That is, the intensity at one plane must neither induce nor deplete the possibility of generation of intensity at another. The second condition is that the phases of the fields produced by each plane must be uncorrelated and rapidly varying, in which case the resulting intensities from each plane effectively add.

For the moment we will not go into details on how the above two conditions can be met in practice and simply assume them to be valid (more will be said about this in Chapters 9 and 10). In particular, we assume that the object consists of a distribution of discrete sources whose phases are indeed uncorrelated and rapidly varying in time. The local intensity generated inside the object can therefore be written as

$$I_0(\vec{\rho}_0, z_0) = dz_0\, I_{0z}(\vec{\rho}_0, z_0) \tag{5.40}$$

where $I_{0z}(\vec{\rho}_0, z_0)$ refers to an intensity per unit depth. Strictly speaking, the usage of the term intensity here is incorrect, since $I_0(\vec{\rho}_0, z_0)$ corresponds to a power per unit area *emitted* from a surface and as such should more properly be referred to as an emittance. Nevertheless, in the same way that we have already yielded to the use of the term intensity in place of irradiance (power per unit area *incident* on a surface), we will sidestep the issue of distinguishing intensity from emittance.

Because the final image intensity $I_1(\vec{\rho}_1, z_1)$ consists of an incoherent summation of object intensity contributions from different depths z_0, Eq. 5.31 generalizes to

$$I_1(\vec{\rho}_1, z_1) = \iint d^2\vec{\rho}_0 \, dz_0 \, \mathrm{PSF}(\vec{\rho}_1 - \vec{\rho}_0, z_1 - z_0) I_{0z}(\vec{\rho}_0, z_0) \tag{5.41}$$

or, equivalently, in frequency space,

$$\mathcal{I}_1(\vec{\kappa}_\perp; z_1) = \int dz_0 \, \mathrm{OTF}(\vec{\kappa}_\perp; z_1 - z_0) \mathcal{I}_{0z}(\vec{\kappa}_\perp; z_0) \tag{5.42}$$

These equations will play a central role throughout the remainder of this book. Specifically, let us consider the case where the sample consists of discrete sources of concentration $C(\vec{\rho}_0, z_0)$ (units: 1/volume), each emitting an elemental power ϕ_0 (units: power). The intensity generated inside the sample per unit depth is then given by

$$I_{0z}(\vec{\rho}_0, z_0) = \phi_0 C(\vec{\rho}_0, z_0) \tag{5.43}$$

which will serve as the basis for the examples below. For brevity, we consider the usual case where the image plane is at its nominal focus position. That is, we consider $z_1 = 0$, and for ease of notation we recast $I_1(\vec{\rho}_1, 0) \rightarrow I_1(\vec{\rho}_1)$.

Single Point Source

To begin, let us consider a single point source located at $(0, z_\sigma)$. The concentration associated with this source is

$$C(\vec{\rho}_0, z_0) = \delta^3(\vec{\rho}_0, z_0 - z_\sigma) \tag{5.44}$$

and from Eq. 5.41 the image intensity becomes

$$I_1(\vec{\rho}_1) = \phi_0 \iint d^2\vec{\rho}_0 \, dz_0 \, \mathrm{PSF}(\vec{\rho}_1 - \vec{\rho}_0, -z_0)\delta^3(\vec{\rho}_0, z_0 - z_\sigma) = \phi_0 \mathrm{PSF}(\vec{\rho}_1, -z_\sigma) \tag{5.45}$$

Here $I_1(\vec{\rho}_1)$ manifestly depends not only on $\vec{\rho}_1$ but also implicitly on the axial position z_σ of the point source.

It is instructive to evaluate the total power incident on the image plane. This is given by

$$\Phi = \int d^2\vec{\rho}_1 \, I_1(\vec{\rho}_1) = \phi_0 \Omega_0 \tag{5.46}$$

As expected from energy conservation, this total power is independent of z_σ. We conclude that, in addition to its other various interpretations (see Eq. 4.84), Ω_0 also corresponds to the collection efficiency associated with the transmission of power from object to image. More will be said about this collection efficiency in Chapter 6.

Planar Distribution of Sources

We consider now a distribution of point sources distributed in a thin plane such that

$$C(\vec{\rho}_0, z_0) = C_\rho(\vec{\rho}_0)\delta(z_0 - z_\rho) \tag{5.47}$$

where $C_\rho(\vec{\rho}_0)$ is the surface concentration of point sources. From Eq. 5.41 we obtain

$$I_1(\vec{\rho}_1) = \phi_0 \int d^2\vec{\rho}_0 \, \mathrm{PSF}(\vec{\rho}_1 - \vec{\rho}_0, -z_\rho)C_\rho(\vec{\rho}_0) \tag{5.48}$$

or, equivalently,

$$\mathcal{I}_1(\vec{\kappa}_\perp) = \phi_0 \mathrm{OTF}(\vec{\kappa}_\perp; -z_\rho)\mathcal{C}_\rho(\vec{\kappa}_\perp) \tag{5.49}$$

where $\mathcal{C}_\rho(\vec{\kappa}_\perp)$ denotes the Fourier transform of $C_\rho(\vec{\rho}_0)$.

In the very specific case where $C_\rho(\vec{\rho}_0)$ is a perfectly uniform distribution (i.e. $C_\rho(\vec{\rho}_0) = C_\rho$ or $\mathcal{C}_\rho(\vec{\kappa}_\perp) = C_\rho\delta^2(\vec{\kappa}_\perp)$) we obtain the result that $I_1(\vec{\rho}_1)$ and $\mathcal{I}_1(\vec{\kappa}_\perp)$ are now independent of z_ρ. That is, we conclude that it is impossible to infer the axial position of a uniform self-luminous plane based on information from the image intensity alone. Certainly a different conclusion was arrived at when we considered a point source, since, in contrast with Eq. 5.48, Eq. 5.45 was found to be dependent on z_σ. The reasons for this fundamental difference will be discussed in the following section.

Volume Distribution of Sources

Finally, an arbitrary volume distribution of sources leads to

$$I_1(\vec{\rho}_1) = \phi_0 \iint d^2\vec{\rho}_0 \, dz_0 \, \mathrm{PSF}(\vec{\rho}_1 - \vec{\rho}_0, -z_0)C(\vec{\rho}_0, z_0) \tag{5.50}$$

or, equivalently,

$$\mathcal{I}_1(\vec{\kappa}_\perp) = \phi_0 \int dz_0 \, \mathrm{OTF}(\vec{\kappa}_\perp; -z_0)\mathcal{C}(\vec{\kappa}_\perp; z_0) \tag{5.51}$$

These formulas are very useful in imaging applications and we will return to them repeatedly, particularly when dealing with fluorescence (see Chapter 13). At the risk of repetition, it is emphasized again that these formulas are valid only when the phases of the sources are completely uncorrelated, that is, when the sources are incoherent. More will be said about this notion of incoherence in Chapter 7.

5.4 FREQUENCY SUPPORT

To this point, we have considered only mixed representations of radiant fields which were derived from 2D Fourier transforms. In particular, we started from Eq. 5.2 and progressed from there to derive mixed representations of $\mathcal{H}$, H, PSF, and OTF. We can follow a similar procedure to derive fully 3D spatial-frequency representations. In this case, instead of starting with a mixed-representation of the defocus propagator $\mathcal{D}_{\pm}(\vec{\kappa}_{\perp}; z)$, we start with $\mathcal{D}_{\pm}(\vec{\kappa})$, which we have derived already in Chapter 2 and is given by (Eqs. 2.29 and 2.53)

$$\mathcal{D}_{\pm}(\vec{\kappa}) = \begin{cases} \delta\left(\kappa_z \mp \sqrt{\kappa^2 - \kappa_{\perp}^2}\right) & \text{(Rayleigh–Sommerfeld)} \\ \delta\left(\kappa_z \mp \kappa \pm \frac{\kappa_{\perp}^2}{2\kappa}\right) & \text{(Fresnel)} \end{cases} \tag{5.52}$$

The same arguments we followed at the beginning of Section 5.2 then lead to

$$\mathcal{H}_{\pm}\left(\vec{\kappa}\right) = \mathcal{H}\left(\vec{\kappa}_{\perp}\right) \mathcal{D}_{\pm}(\vec{\kappa}) \tag{5.53}$$

where, again, $\mathcal{H}\left(\vec{\kappa}_{\perp}\right)$ is the in-focus 2D coherent transfer function we are familiar with from Chapter 3. Equation 5.53 may be compared with Eq. 5.11. We note that the units of $\mathcal{H}_{\pm}\left(\vec{\kappa}\right)$ and $\mathcal{H}\left(\vec{\kappa}_{\perp}\right)$ differ. The former has unit of length while the latter is unitless.

It is worth spending a moment to consider the implications of Eq. 5.53. To begin, we observe that the Rayleigh-Sommerfeld approximation for $\mathcal{D}_{\pm}(\vec{\kappa})$ constrains $\vec{\kappa}$ to lie on a spherical shell of radius κ (the wavenumber). This shell is commonly referred to as the Ewald sphere. In contrast, the Fresnel approximation constrains $\vec{\kappa}$ to lie on a parabolic shell. In fact, it is deviations of this parabolic shell from the Ewald sphere that are at the origin of deviations of the Fresnel approximation from the more exact Rayleigh-Sommerfeld approximation. The effect of the prefactor $\mathcal{H}(\vec{\kappa}_{\perp})$ is to clip the magnitude of $\kappa_{\perp}$, thus constraining $\mathcal{D}_{\pm}(\vec{\kappa})$ to reside on spherical or parabolic caps, as illustrated in Fig. 5.4, and thus mitigating deviations of the Fresnel from the Rayleigh-Sommerfeld approximations. These caps constitute what is called the frequency support of $\mathcal{H}_{\pm}\left(\vec{\kappa}\right)$. In other words, only those spatial frequencies encompassed, or supported, by $\mathcal{H}_{\pm}\left(\vec{\kappa}\right)$ are allowed to be transferred by an imaging system from an object field to an image field.

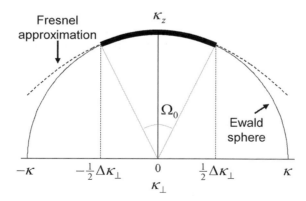

Figure 5.4. Frequency support of $\mathcal{H}$ defined by Ewald sphere or Fresnel approximation.

As before, $H_\pm(\vec{r})$ is equal to the inverse Fourier transform of $\mathcal{H}_\pm(\vec{\kappa})$, but now over three dimensions. That is,

$$H_\pm(\vec{r}) = \int d^3\vec{\kappa}\, e^{i2\pi\vec{\kappa}\cdot\vec{r}} \mathcal{H}_\pm(\vec{\kappa}) \tag{5.54}$$

leading to the same result as Eq. 5.15 (or Eq. 5.16 in the case of cylindrical symmetry).

Similarly, we can derive PSF and OTF from H_+ and $\mathcal{H}_+$; however, this time it is easier to derive $\mathrm{OTF}(\vec{\kappa})$ directly from $\mathcal{H}_+(\vec{\kappa})$ rather than by way of $\mathrm{PSF}(\vec{r})$. Indeed, $\mathrm{OTF}(\vec{\kappa})$ is equal to the normalized autocorrelation of $\mathcal{H}_+(\vec{\kappa})$, this time in 3D. That is,

$$\mathrm{OTF}(\vec{\kappa}) = \frac{1}{\kappa^2} \int d^3\vec{\kappa}_c\, \mathcal{H}_+\left(\vec{\kappa}_c + \tfrac{1}{2}\vec{\kappa}\right) \mathcal{H}_+^*\left(\vec{\kappa}_c - \tfrac{1}{2}\vec{\kappa}\right) \tag{5.55}$$

We can make headway if we adopt the Fresnel approximation, whereupon we find

$$\int d\kappa_{zc}\, \mathcal{D}_+\left(\vec{\kappa}_c + \tfrac{1}{2}\vec{\kappa}\right) \mathcal{D}_+^*\left(\vec{\kappa}_c - \tfrac{1}{2}\vec{\kappa}\right) = \delta\left(\kappa_z + \frac{1}{\kappa}\vec{\kappa}_{\perp c}\cdot\vec{\kappa}_\perp\right) \tag{5.56}$$

meaning Eq. 5.55 can be rewritten as

$$\mathrm{OTF}(\vec{\kappa}) = \frac{1}{\kappa^2} \int d^2\vec{\kappa}_{\perp c}\, \mathcal{H}\left(\vec{\kappa}_{\perp c} + \tfrac{1}{2}\vec{\kappa}_\perp\right) \mathcal{H}^*\left(\vec{\kappa}_{\perp c} - \tfrac{1}{2}\vec{\kappa}_\perp\right) \delta\left(\kappa_z + \frac{1}{\kappa}\vec{\kappa}_{\perp c}\cdot\vec{\kappa}_\perp\right) \tag{5.57}$$

where the $\mathcal{H}$'s inside the integral are now two-dimensional rather than three-dimensional.

The delta-function here defines the region of support for κ_z. However, in contrast to the delta function in Eq. 5.53, this region of support is no longer as strictly defined since κ_z is allowed to span a range of values for different values of $\vec{\kappa}_\perp$. To demonstrate this, we return to the example of a circular pupil where the $\mathcal{H}$ is defined by Eq. 4.78. The dot-product $\vec{\kappa}_{\perp c}\cdot\vec{\kappa}_\perp$ is then found to be bounded by

$$\left|\vec{\kappa}_{\perp c}\cdot\vec{\kappa}_\perp\right| \le \tfrac{1}{2}\kappa_\perp\left(\Delta\kappa_\perp - \kappa_\perp\right) \tag{5.58}$$

In turn, the delta-function in Eq. 5.57 imposes an upper bound on $|\kappa_z|$ given by

$$|\kappa_z| \le \frac{\kappa_\perp}{2\kappa}\left(\Delta\kappa_\perp - \kappa_\perp\right) \tag{5.59}$$

As we will see, this range of allowed values of $|\kappa_z|$ supported by $\mathrm{OTF}(\vec{\kappa})$ will have important consequences on axial resolution.

Remarkably, for the case of a circular pupil, Eq. 5.57 can be simplified into a compact analytical expression. The derivation of this expression is rather involved, and the reader is referred to the seminal paper by Frieden for details [6]. Only the final result is presented here:

$$\mathrm{OTF}(\vec{\kappa}) = \frac{\Delta\kappa_\perp}{\kappa\kappa_\perp}\sqrt{1 - \left(\frac{2\kappa\,|\kappa_z|}{\kappa_\perp\Delta\kappa_\perp} + \frac{\kappa_\perp}{\Delta\kappa_\perp}\right)^2} \tag{5.60}$$

This expression is valid for transverse spatial frequencies bounded by $\kappa_\perp \le \Delta\kappa_\perp$ and for axial spatial frequencies bounded by Eq. 5.59; otherwise it is zero. A plot of $\mathrm{OTF}(\vec{\kappa})$ is shown in Fig. 5.5. Two features are of note. First, $\mathrm{OTF}(\vec{\kappa})$ manifestly diverges when $\vec{\kappa} \to 0$. However, this divergence disappears when $\mathrm{OTF}(\vec{\kappa})$ is integrated along κ_z. For example, it

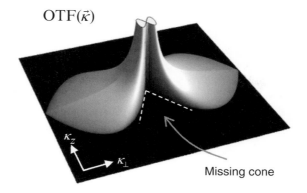

Figure 5.5. 3D OTF for unobstructed circular pupil (note divergence near $\vec{\kappa} \approx 0$ is clipped).

may be verified that $\int d\kappa_z \mathrm{OTF}\left(\vec{\kappa}\right)$ reduces to the in-focus 2D OTF given by Eq. 4.82, which is equal to Ω_0 when $\vec{\kappa}_\perp \to 0$.

A second feature of note is the pinching off of the axial frequency support as $\kappa_\perp \to 0$, leading to what is colloqially referred to as a "missing cone" [16]. We will examine this missing cone in detail in the next section. In the meantime, we complete our discussion by noting that the associated 3D PSF can be obtained from $\mathrm{OTF}\left(\vec{\kappa}\right)$ by a 3D Fourier transform. That is,

$$\mathrm{PSF}\left(\vec{r}\right) = \int \mathrm{d}^3\vec{\kappa}\, e^{i2\pi\vec{\kappa}\cdot\vec{r}}\,\mathrm{OTF}(\vec{\kappa}) \tag{5.61}$$

A summary of these results is presented in Fig. 5.6. This figure may be directly compared with Fig. 4.6.

5.4.1 Optical Sectioning

The missing cone in the $\mathrm{OTF}(\vec{\kappa})$ frequency support leads to one of the most fundamental problems in 3D imaging. Specifically, the frequency support of an imaging device governs the capacity of the device to perform optical sectioning. While the term optical sectioning is often used ambiguously in the literature, it is defined here as the capacity of an imaging device to provide axial resolution for *all* transverse spatial frequencies within the imaging bandwidth. Thus, for a microscope to perform full 3D imaging, it must, by definition, provide optical sectioning. However, as we will see below, restrictions to the frequency support of the microscope can severely undermine this capacity.

More specifically, let us return to the results of Section 5.3, which highlighted the difficulty in inferring the axial position of a uniform self-luminous plane. Recalling that the role of the OTF is to provide a link between the object and image intensity spectra

$$\mathcal{I}_1(\vec{\kappa}) = \mathrm{OTF}(\vec{\kappa})\mathcal{I}_0(\vec{\kappa}) \tag{5.62}$$

the underlying reason for this difficulty is now apparent. The intensity spectrum of a uniform plane is peaked at $\vec{\kappa}_\perp = 0$ and zero elsewhere. But, because of the missing cone, the frequency

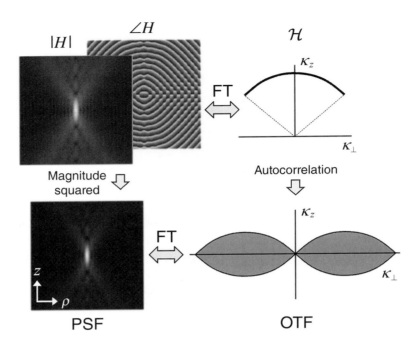

Figure 5.6. Relationships between 3D imaging functions.

support provided by $\text{OTF}(\vec{\kappa}_\perp = 0, \kappa_z)$ is in turn peaked at $\kappa_z = 0$ and zero for $\kappa_z \neq 0$, implying that regardless of whether $\mathcal{I}_0(\vec{\kappa})$ contains axial structure, the OTF causes $\mathcal{I}_0(\vec{\kappa})$ to *appear* to contain no axial structure. That is, a uniform plane cannot, upon imaging, be distinguished from a uniform featureless volume. Said differently, a uniform plane cannot, upon imaging, be axially resolved.

A somewhat more substantial argument comes from better defining what we mean by axial resolution. Let us return to a mixed representation of the image intensity spectrum given by

$$\mathcal{I}_1(\vec{\kappa}_\perp; z) = \int d^2\vec{\kappa}_\perp \, I_1(\vec{\rho}, z) e^{-i2\pi\vec{\rho}\cdot\vec{\kappa}_\perp} = \int d\kappa_z \, \mathcal{I}_1(\vec{\kappa}) e^{i2\pi\kappa_z z} \qquad (5.63)$$

From Eq. 5.62 this may be rewritten in the form

$$\mathcal{I}_1(\vec{\kappa}_\perp; z) = \int d\kappa_z \, \text{OTF}(\vec{\kappa}) \mathcal{I}_0(\vec{\kappa}) e^{i2\pi\kappa_z z} \qquad (5.64)$$

For the image intensity at a particular spatial frequency $\vec{\kappa}_\perp$ to be axially resolved, it is clear that $\mathcal{I}_1(\vec{\kappa}_\perp; z)$ must be localizable in z. However, from the Fourier transform uncertainty relation (see [4]) this means that $\mathcal{I}_1(\vec{\kappa})$ must exhibit an extended range of κ_z values. The broader this range, the more tightly $\mathcal{I}_1(\vec{\kappa}_\perp; z)$ can be localized. In Figs. 5.5 and 5.6, this range of κ_z values is constrained by the vertical span of $\text{OTF}(\vec{\kappa})$ along κ_z. Manifestly, the axial resolution capacity is null at $\kappa_\perp = 0$, meaning that we have violated the conditions for optical sectioning as defined above. Moreover, the axial resolution capacity is highly dependent on the transverse spatial frequency when $\kappa_\perp \neq 0$, meaning there is crosstalk between transverse and axial spatial frequencies. The crucial point, nevertheless, is that while axial resolution cannot occur for $\kappa_\perp = 0$, it is more or less restored for $\kappa_\perp > 0$ (up to the bandwidth limit $\Delta\kappa_\perp$). For example, it is maximum when $\kappa_\perp = \frac{1}{2}\Delta\kappa_\perp$. We saw a demonstration of this when imaging a single point source. Because $\mathcal{I}_0(\vec{\kappa})$ for a point source exhibits an extended range of transverse frequencies (as opposed to being peaked only at $\vec{\kappa}_\perp = 0$), full use can be made of the restored capacity for axial resolution in the range $0 < \kappa_\perp < \Delta\kappa_\perp$, ultimately enabling the possibility of depth discrimination for a point source.

Clearly, for an imaging device to be ideal it should exhibit an axial resolution capacity that is independent of transverse frequency. As is evident from Fig. 5.5, this is not the case here in part because of the missing cone, which precludes the possibility of optical sectioning, but also in part because of the limited spatial bandwidth permitted by the finite detection aperture, which forces the axial resolution capacity to eventually taper to zero. While a finite bandwidth is unavoidable in any physical imaging device, the difficulties arising from the missing cone in $\mathrm{OTF}(\vec{\kappa})$ are not. As we will see in future chapters, a variety of microscope techniques have been developed that exhibit no missing cone. These microscopes confer axial resolution for all $\kappa_\perp$ values within the imaging bandwidth (e.g. even when the sample is planar and uniform) and as such are called optically sectioning microscopes.

5.4.2 Extended Depth of Field

When imaging a volumetric sample with a standard microscope, in general there is only one in-focus plane within the volume. The rest of the planes are out of focus, producing blurred images that superpose onto the in-focus image and undermine its contrast. A variant of standard imaging is called extended-depth-of-field (EDOF) imaging. In this variant, the in-focus plane is axially swept through the sample (or, alternatively, the sample is swept through the in-focus plane). The effect of such a sweep is to produce an extended PSF, or EPSF, given by

$$\mathrm{EPSF}\left(\vec{\rho}\right) = \frac{1}{L}\int_{-L/2}^{L/2} dz \, \mathrm{PSF}\left(\vec{\rho}, z\right) \tag{5.65}$$

where L is the range of the focal sweep. In other words, the resultant image is a cumulative average of a continuous series of images of varying focal depth, all projected onto a single image plane. Whereas $\mathrm{PSF}(\vec{\rho}, z)$ is a function of three spatial dimensions, $\mathrm{EPSF}(\vec{\rho})$ is thus a function only of two. When L is large, for example when L is larger than the sample itself, then the associated extended OTF, or EOTF, can be approximated by its projection along z, obtaining

$$\mathrm{EOTF}\left(\vec{\kappa}_\perp\right) \approx \frac{1}{L}\mathrm{OTF}\left(\vec{\kappa}_\perp, \kappa_z = 0\right) \tag{5.66}$$

The apparent divergence at $\kappa_\perp = 0$ that comes from this approximation can be readily resolved by invoking energy conservation, which imposes the constraint $\mathrm{EOTF}(\vec{\kappa}_\perp \to 0) = \Omega_0$.

An example of $\mathrm{EOTF}(\vec{\kappa}_\perp)$ for a circular pupil is shown in Fig. 5.7, and reveals an interesting property of EDOF images, namely that all features in the images appear both in focus (the

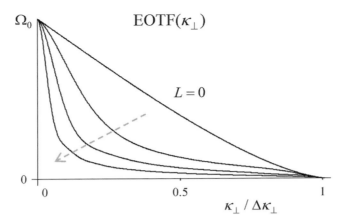

Figure 5.7. Circular-pupil EOTF for increasing depths of field L.

broad pedestal that extends out to the full bandwidth $\Delta\kappa_\perp$) and out of focus (the narrow peak about $\kappa_\perp = 0$) at the same time. Another interesting property of $\text{EOTF}(\vec{\kappa}_\perp)$ is that it exhibits no zero crossings within its bandwidth. This differs, for example, from standard defocused OTFs (e.g. see Fig. 5.3), which generally exhibit multiple zero crossings. Such an absence of zero crossings is particularly beneficial when compensating for out-of-focus blur by using a technique of deconvolution. We will not discuss deconvolution in this book, and the reader is referred instead to textbooks on the subject, such as [2].

5.4.3 Radiant Mutual Intensity

To conclude this chapter on 3D imaging, we briefly consider the full spatial-frequency representation of the radiant mutual intensity. Most of this discussion is drawn from the work by Streibl [15].

We consider here only forward-propagating light. From the definition of the radiant mutual intensity given by Eq. 4.11, the full spatial-frequency representation of the radiant mutual intensity is found to be

$$\mathcal{J}(\vec{\kappa}, \vec{\kappa}') = \mathcal{J}(\vec{\kappa}_\perp, \vec{\kappa}_\perp')\mathcal{D}_+\left(\vec{\kappa}\right)\mathcal{D}_+\left(\vec{\kappa}'\right) \tag{5.67}$$

where $\mathcal{D}_+\left(\vec{\kappa}\right)$ is either the forward-propagating Rayleigh–Sommerfeld or Fresnel free-space propagator, as defined by Eq. 5.52.

Applying our usual coordinate transformation (Eq. 4.13) specifically to the Fresnel approximation, we arrive at the full spatial-frequency representation of the intensity spectrum defined by Eq. 4.23, obtaining

$$\mathcal{I}(\vec{\kappa}_d) = \int d^2\vec{\kappa}_{\perp c}\, \mathcal{J}(\vec{\kappa}_{\perp c}, \vec{\kappa}_{\perp d})\, \delta\left(\kappa_{zd} + \frac{1}{\kappa}\vec{\kappa}_{\perp c} \cdot \vec{\kappa}_{\perp d}\right) \tag{5.68}$$

As before, the region of support for $|\kappa_{zd}|$ is bounded by $|\kappa_{zd}| \leq \frac{\kappa_{\perp d}}{2\kappa}\left(\Delta\kappa_\perp - \kappa_{\perp d}\right)$.

Finally, we can retrieve the mixed representation of the intensity spectrum from its full spatial-frequency representation using the definition

$$\mathcal{I}(\vec{\kappa}_\perp; z) = \int d\kappa_z \, \mathcal{I}(\vec{\kappa}) e^{i2\pi\kappa_z z} \tag{5.69}$$

From Eq. 5.68, we find then

$$\mathcal{I}(\vec{\kappa}_{\perp d}; z) = \int d^2\vec{\kappa}_{\perp c} \, \mathcal{J}(\vec{\kappa}_{\perp c}, \vec{\kappa}_{\perp d}) e^{-i2\pi\frac{z}{\kappa}\vec{\kappa}_{\perp c}\cdot\vec{\kappa}_{\perp d}} \tag{5.70}$$

This important result defines the axial propagation of the intensity spectrum based solely on a knowledge of the radiant mutual intensity at a given reference plane. Alternatively, this result could have been derived from the principle of conservation of radiant intensity given by 4.28, along with its relation to the intensity spectrum given by 4.23. In effect, Eq. 5.70 is a version of Eq. 4.29 in the frequency domain.

5.5 PROBLEMS

Problem 5.1
(a) Derive Eq. 5.24.
(b) What is the implication of the above relation? In particular, what does it say about the imaging properties of two identical, unit-magnification, binary-aperture imaging systems arranged in series?

Problem 5.2
Consider a unit-magnification 4f imaging system (all lenses of focal length f) with a square aperture defined by

$$P(\xi_x, \xi_y) = \begin{cases} 1 & |\xi_x| < a \text{ and } |\xi_y| < a \\ 0 & \text{elsewhere} \end{cases}$$

Based on the Fresnel approximation, derive analytically:

(a) $\mathcal{H}_+(\kappa_x, 0; 0)$ and $\mathcal{H}_+(0, 0; z)$
(b) $H_+(x, 0, 0)$ and $H_+(0, 0, z)$
(c) $\text{PSF}(x, 0, 0)$ and $\text{PSF}(0, 0, z)$
(d) $\text{OTF}(\kappa_x, 0; 0)$ and $\text{OTF}(0, 0; z)$

Note, it will be convenient to define a spatial bandwidth $\Delta\kappa_\perp = 2\kappa\frac{a}{f}$.

Note also, $\mathcal{H}_+$ and OTF are in mixed representations. You will run into special functions such as sinc(...) and erf(...). As such, this problem is best solved with the aid of integral tables or symbolic computing software such as *Mathematica*. Be careful with units and prefactors. For example, make sure the limits $x \to 0$ and $z \to 0$ converge to the same values!

Problem 5.3

Consider a unit-magnification 4f imaging system (of spatial frequency bandwidth $\Delta\kappa_\perp$) with a circular aperture. A planar object at a defocus position z_s emits a periodic, incoherent intensity distribution (per unit depth) given by

$$I_{0z}(x_0, y_0, z_0) = I_0 \left[1 + \cos\left(2\pi q_x x_0\right)\right] \delta(z_0 - z_s)$$

where I_0 is a constant.

(a) Based on Eq. 5.42, derive an expression for the imaged intensity distribution. This expression should look like

$$I_1(x_1, y_1) \propto \left[1 + M(q_x, z_s) \cos\left(2\pi q_x x_1\right)\right]$$

In other words, the imaged intensity is also periodic, but with a modulation contrast given by $M(q_x, z_s)$. What is $M(q_x, z_s)$?

(b) In the specific case where $q_x = \frac{1}{2}\Delta\kappa_\perp$, what is the modulation contrast when the object is in focus? At what defocus value does the modulation contrast fade to zero (express your result in terms of λ, n and NA)? What happens to the modulation contrast just beyond this defocus? (Hint: use the Stokseth approximation.)

Problem 5.4

(a) Verify that the second moment of an arbitrary function $F\left(\vec{\rho}\right)$ is given by

$$\int d^2\vec{\rho}\, \rho^2 F\left(\vec{\rho}\right) = -\frac{1}{4\pi^2}\nabla^2 \mathcal{F}\left(0\right)$$

where $\mathcal{F}\left(\vec{\kappa}_\perp\right)$ is the 2D Fourier transform of $F\left(\vec{\rho}\right)$.

(b) For a cylindrically symmetric imaging system whose OTF has a cusp at the origin, what does the above result tell you about the dependence of PSF on $\left|\vec{\rho}\right|$ for large $\left|\vec{\rho}\right|$? Verify this dependence for the case of an unobstructed circular pupil.

Problem 5.5

(a) Derive Eq. 5.8 from 5.7.

(b) Use this to re-express the amplitude point spread function $H_+\left(0, z\right)$ for a Gaussian pupil (Eq. 5.23) in a similar form as Eq. 5.21 for a circular pupil. For equal $\Delta\kappa_\perp$, which has the more rapidly varying carrier frequency in the vicinity of $z = 0$?

References

[1] Barrett, H. H. and Myers, K. J. *Foundations of Image Science*, Wiley-Interscience (2004).
[2] Bertero, M. and Boccacci, P. *Introduction to Inverse Problems in Imaging*, IOP Publishing (1998).
[3] Born, M. and Wolf, E. *Principles of Optics*, 6th edn, Cambridge University Press (1980).
[4] Bracewell, R. N. *The Fourier Transform and its Applications*, McGraw-Hill (1965).

[5] Gu, M. *Advanced Optical Imaging Theory*, Springer-Verlag (1999).

[6] Frieden, B. R. "Optical transfer of the three-dimensional object," *J. Opt. Soc. Am.* 57, 56–66 (1967).

[7] Gibson, S. F. and Lanni, F. "Diffraction by a circular aperture as a model for three-dimensional optical microscopy," *J. Opt. Soc. Am. A*, 6 1357–1367 (1989).

[8] Hopkins, H. H. "The frequency response of a defocused optical system," *Proc. R. Soc. London Ser. A* 231, 91–103 (1955).

[9] Kogelnik, H. and Li, T. "Laser beams and resonators," *Proc. IEEE* 54, 1312–1329 (1966).

[10] Mandel, L. and Wolf, E. *Optical Coherence and Quantum Optics*, Cambridge University Press (1995).

[11] McCutchen, C. W. "Generalized aperture and the three-dimensional diffraction image," *J. Opt. Soc. Am.* 54, 240–244 (1964).

[12] Saleh, B. E. A. and Teich, M. C. *Fundamentals of Photonics*, 2nd edn, Wiley Interscience (2007).

[13] Sheppard, C. J. R. and Mao, X. Q. "Three-dimensional imaging in a microscope," *J. Opt. Soc. Am. A* 6, 1260–1269 (1989).

[14] Stokseth, P. A. "Properties of a defocused optical system," *J. Opt. Soc. Am.* 59, 1314–1321 (1969).

[15] Streibl, N. "Fundamental restrictions for 3-D light distributions," *Optik* 66, 341–354 (1984).

[16] Streibl, N. "Depth transfer by an imaging system," *Opt. Acta* 31, 1233–1241 (1984).

6 | Radiometry

The term radiometry refers to the science of measurement of light. As such, radiometry plays an obligatory role in optical microscopy. Because light is detected as an intensity, an understanding of the relation between field propagation and intensity detection is essential. This relation was considered in Chapter 4 in the context of mutual intensity, which is a statistical property of a light beam defined for time averages longer than the coherence time (a quantity that will be defined in Chapter 7). The mutual intensity was used, in turn, to introduce the concept of spatial coherence, which we will re-examine here in the broader context of radiometry.

Efforts to establish an exact relation between radiometry and coherence have been long-standing, and were largely initiated by the seminal work of Walther [14, 15] and Wolf [16]. To date, most of these efforts have been limited to a scalar description of light, and to beams that are quasi-homogeneous. The reader is referred to Mandel and Wolf [7] for a comprehensive discussion of this link between radiometry and coherence, as well as [4, 6, 17], along with various articles such as [2, 5, 18]. Other discussions on radiometry and radiometric terms can be found in [1, 8, 9, 11]. Our aim in this chapter is to provide a cursory overview of some key results, specifically as they pertain to optical microscopy.

6.1 ENERGY FLUX DENSITY

The main premise in radiometry is that radiative energy flows in a manner that obeys certain conservation principles. We will derive such principles here. We begin with the general expression for mutual intensity given by Eq. 4.2. With our usual coordinate transformation, this is given by

$$J\left(\vec{r}_c, \vec{r}_d\right) = \left\langle E\left(\vec{r}_c + \tfrac{1}{2}\vec{r}_d\right) E^*\left(\vec{r}_c - \tfrac{1}{2}\vec{r}_d\right)\right\rangle \tag{6.1}$$

A particularly useful quantity when considering the flow of radiative energy is the derivative of this mutual intensity with respect to $\vec{r}_d$, namely

$$\vec{\nabla}_{r_d} J\left(\vec{r}_c, \vec{r}_d\right)\Big|_{r_d=0} = i2\pi\kappa\vec{F}\left(\vec{r}_c\right) \tag{6.2}$$

where

$$\vec{F}\left(\vec{r}\right) = \frac{i}{4\pi\kappa} \left\langle E\left(\vec{r}\right) \vec{\nabla} E^*\left(\vec{r}\right) - E^*\left(\vec{r}\right) \vec{\nabla} E\left(\vec{r}\right) \right\rangle \tag{6.3}$$

The vector $\vec{F}\left(r_c\right)$ is called the energy flux density, or flux density for short. As suggested by its name, this quantity provides a measure of the magnitude and direction of radiative energy flow, and will be key to our understanding of radiometry. We note that if the field $E\left(\vec{r}_c\right)$ satisfies the homogeneous Helmholtz equation (Eq. 2.11 with $S = 0$), then

$$\vec{\nabla} \cdot \vec{F}\left(\vec{r}\right) = 0 \tag{6.4}$$

meaning that radiative energy is conserved in homogeneous space.

6.1.1 Transport of Intensity

While Eq. 6.4 affords some degree of intuition, a much more informative relation comes from the separation of the gradient operator $\vec{\nabla}$ into its transverse and axial components. In particular, the transverse flux density is given by

$$\vec{F}_\perp\left(\vec{r}\right) = \frac{i}{4\pi\kappa} \left\langle E\left(\vec{r}\right) \vec{\nabla}_\perp E^*\left(\vec{r}\right) - E^*\left(\vec{r}\right) \vec{\nabla}_\perp E\left(\vec{r}\right) \right\rangle \tag{6.5}$$

leading to the transverse form of Eq. 6.2, given by

$$\vec{\nabla}_{\rho_d} J\left(\vec{\rho}_c, \vec{\rho}_d\right)\Big|_{\rho_d=0} = i2\pi\kappa \vec{F}_\perp\left(\vec{\rho}_c\right) \tag{6.6}$$

When the field $E\left(\vec{r}\right)$ is directional, that is when it is naturally written as $E\left(\vec{r}\right) \rightarrow E\left(\vec{r}\right) e^{i2\pi\kappa z}$ where the envelope field $E\left(\vec{r}\right)$ slowly varies as it propagates along z, the Helmholtz equation reduces to its well-known paraxial form given by

$$\nabla_\perp^2 E\left(\vec{r}\right) + i4\pi\kappa \frac{\partial}{\partial z} E\left(\vec{r}\right) = 0 \tag{6.7}$$

From Eqs. 6.5 and 6.7, we arrive at

$$\frac{\partial}{\partial z} I\left(\vec{r}\right) = -\vec{\nabla}_\perp \cdot \vec{F}_\perp\left(\vec{r}\right) \tag{6.8}$$

This is the so-called transport of intensity equation (TIE), which has found multiple applications in both optical and X-ray imaging (cf. [10]). The interpretation of this equation is clear. While Eq. 6.4 ensures that energy is neither created nor destroyed, Eq. 6.8 provides a more detailed description of where this energy goes. In particular, energy from neighboring points in the transverse plane can flow into or out of a beam as it propagates along the z axis. It should be noted that the transport of intensity equation was originally derived for coherent fields only [13]. As we see here, it also applies to partially coherent fields, provided they are paraxial.

6.2 RADIANCE

In addition to flux density, another important quantity in radiometry is called radiance (sometimes called brightness or specific intensity), defined by

$$\mathcal{L}\left(\vec{r}_c; \vec{\kappa}\right) = \kappa\,\kappa_z \int d^2\vec{\rho}_d\, J\left(\vec{r}_c, \vec{r}_d\right) e^{-i2\pi\vec{\kappa}_\perp \cdot \vec{\rho}_d} \tag{6.9}$$

Note that the integration here is only two-dimensional. Implicitly, we have $\kappa_z = \pm\sqrt{\kappa^2 - \kappa_\perp^2}$, depending on the propagation direction.

This is a fundamental quantity in radiometry, which defines not only where optical energy is concentrated but also where it is going. Indeed, from radiance, both the intensity and radiant intensity can be derived:

$$I\left(\vec{r}\right) = \int d^2\Omega_\kappa\, \mathcal{L}\left(\vec{r}; \vec{\kappa}\right) \tag{6.10}$$

$$\mathcal{R}\left(\vec{\kappa}_\perp\right) = \frac{\kappa_z}{\kappa} \int d^2\vec{\rho}\, \mathcal{L}\left(\vec{r}; \vec{\kappa}\right) \tag{6.11}$$

The first of these corresponds to an integration over solid angle (bearing in mind that $d^2\Omega_\kappa = \frac{d^2\vec{\kappa}_\perp}{\kappa|\kappa_z|}$ – see Eq. 3.39); the second corresponds to an integration over cross-sectional area. We can compare Eq. 6.11 with Eq. 4.21 and observe that the paraxial prefactor κ^2 has now been replaced with the exact prefactor κ_z^2.

Radiance is often used to characterize sources. As an example, let us consider an incoherent source located at plane 0 and characterized by a mutual intensity given by Eq. 4.43. From Eq. 6.9 we find that $\mathcal{L}_0\left(\vec{r}_c; \vec{\kappa}\right)$ varies as $\frac{\kappa_z}{\kappa}$ and hence $\mathcal{R}_0\left(\vec{\kappa}_\perp\right)$ varies as $\frac{\kappa_z^2}{\kappa^2}$. Accordingly, the far-field intensity from an incoherent source varies as

$$I_\infty(\vec{r}) = \frac{1}{r^2}\mathcal{R}_0\left(\frac{\kappa}{r}\vec{\rho}\right) = \frac{\cos^2\theta}{r^2}\mathcal{R}_0 \qquad \text{(incoherent)} \tag{6.12}$$

where $\mathcal{R}_0$ is the on-axis radiant intensity and θ is the off-axis tilt angle ($\cos\theta = \frac{z}{r}$).

Alternatively, for many sources the radiance appears to be isotropic instead, meaning it is independent of $\vec{\kappa}$. Such sources are called Lambertian. The far-field intensity from a Lambertian source thus varies as

$$I_\infty(\vec{r}) = \frac{\cos\theta}{r^2}\mathcal{R}_0 \qquad \text{(Lambertian)} \tag{6.13}$$

We arrive here at the surprising conclusion that a Lambertian source produces a far-field radiation pattern that has a *broader* angular spread than an incoherent source, despite the fact that an incoherent source has the narrowest spatial coherence possible. The answer to this seeming contradiction is that the coherence function of a Lambertian source produces non-local field correlations that precisely cause the field to be highly divergent.

Radiance is also used to characterize beams. For example, the radiances of plane-wave and Gaussian beams are given by

$$\mathcal{L}\left(\vec{r};\vec{\kappa}\right) = \kappa\,\kappa_z I_0\,\delta^2\left(\vec{\kappa}_\perp\right) \qquad \text{(plane wave)} \qquad (6.14)$$

$$\mathcal{L}\left(\vec{\rho}, z = 0;\vec{\kappa}\right) = 2\pi w_0^2\kappa\,\kappa_z I_0 e^{-2\left(\rho^2/w_0^2 + \pi^2 w_0^2 \kappa_\perp^2\right)} \qquad \text{(Gaussian)} \qquad (6.15)$$

In the second case, the radiance is defined at the focal plane of a Gaussian beam of waist w_0 (see Section 5.1.1). A generalization of this radiance to other planes is somewhat more complicated, but becomes trivial if we invoke the paraxial approximation $|\kappa_z| \rightarrow \kappa$. In this case, with our usual coordinate transformation, we find that the mutual and radiant mutual intensities (at plane z_c) are related to radiance by

$$J\left(\vec{\rho}_c, \vec{\rho}_d\right) = \frac{1}{\kappa^2}\int d^2\vec{\kappa}_\perp\,\mathcal{L}\left(\vec{r}_c;\vec{\kappa}\right)e^{i2\pi\vec{\kappa}_\perp\cdot\vec{\rho}_d} \qquad (6.16)$$

$$\mathcal{J}(\vec{\kappa}_\perp, \vec{\kappa}_{\perp d}) = \frac{1}{\kappa^2}\int d^2\vec{\rho}_c\,\mathcal{L}\left(\vec{r}_c;\vec{\kappa}\right)e^{-i2\pi\vec{\kappa}_{\perp d}\cdot\vec{\rho}_c} \qquad (6.17)$$

From Eq. 4.28, we then arrive at the remarkably simple free-space propagation law for radiance, given by

$$\mathcal{L}\left(\vec{\rho}, z;\vec{\kappa}\right) = \mathcal{L}\left(\vec{\rho} - \frac{z}{\kappa}\vec{\kappa}_\perp, 0;\vec{\kappa}\right) \qquad (6.18)$$

Manifestly, this law is in accord with our intuitive picture of light rays discussed in Chapter 1 (specifically Section 1.2 and Eq. 1.15), where propagation through free space involves the coupling of position and direction by way of a directional tilt angle $\sin\theta = \kappa_\perp/\kappa$. We will revisit this concept of radiance in much greater detail in Chapter 20, when we consider light propagation in scattering media.

6.2.1 Relation Between Flux Density and Radiance

What is still missing from our discussion is the link between flux density and radiance. While this link can be readily established if we invoke the paraxial approximation, it becomes more difficult when this approximation is relaxed, and the reader is referred to more authoritative treatises for detailed accounts and caveats (e.g. [7]). For brevity, we cite only some key results, which we will return to in future chapters. Specifically, the fundamental link between flux density and radiance is given by

$$\vec{F}\left(\vec{r}\right) = \int d^2\Omega_\kappa\,\mathcal{L}\left(\vec{r};\vec{\kappa}\right)\hat{\kappa} \qquad (6.19)$$

A consequence of this link is that the radiance-averaged wavevector direction, in turn, is given by (making use of Eq. 6.10)

$$\vec{\Theta}\left(\vec{r}\right) = \frac{\int d^2\Omega_\kappa\,\mathcal{L}\left(\vec{r};\vec{\kappa}\right)\hat{\kappa}}{\int d^2\Omega_\kappa\,\mathcal{L}\left(\vec{r};\vec{\kappa}\right)} = \frac{\vec{F}\left(\vec{r}\right)}{I\left(\vec{r}\right)} \qquad (6.20)$$

Another important result that may be cited, which follows from Eq. 6.19, is

$$\vec{F}_\infty\left(\vec{r}\right) = \frac{1}{r^2}\mathcal{R}_0\left(\frac{\kappa}{r}\vec{\rho}\right)\hat{r} = I_\infty\left(\vec{r}\right)\hat{r} \tag{6.21}$$

indicating that the far-field flux density points in the direction $\hat{r}$ with magnitude given by the far-field intensity.

These results apply to the full 3D flux density. Equivalent results for the transverse flux density $\vec{F}_\perp\left(\vec{r}\right)$ are obtained when $\hat{\kappa}$ and $\hat{r}$ are replaced by $\hat{\kappa}_\perp = \frac{\vec{\kappa}_\perp}{\kappa}$ and $\hat{\rho} = \frac{\vec{\rho}}{r}$ respectively (note that the vectors $\hat{\kappa}_\perp$ and $\hat{\rho}$ here do not have unit magnitude). From the transverse form of Eq. 6.20, the transport of intensity equation can then be recast into its more familiar form

$$\frac{\partial}{\partial z}I\left(\vec{r}\right) = -\vec{\nabla}_\perp \cdot \left(I\left(\vec{r}\right)\vec{\Theta}_\perp\left(\vec{r}\right)\right) \tag{6.22}$$

where $\vec{\Theta}_\perp\left(\vec{r}\right)$ corresponds to the local average of the wavevector off-axis tilt angle, assumed small (i.e. paraxial).

6.3 ÉTENDUE

In addition to flux density and radiance, another useful concept for characterizing energy flow is called étendue, a term borrowed from the French, meaning "extent" or "span." This concept seems to have fallen by the wayside in modern textbooks despite the fact that it is quite helpful in providing an intuitive notion of coherence propagation within optical systems. In its rare occurrences, the term étendue has been variously used to describe light beams or optical systems. To avoid ambiguity here, we apply it specifically to describe light beams. Another term, throughput, will be reserved for optical systems. For simplicity, we restrict ourselves to the paraxial approximation and assume, henceforth, that $\cos\theta \approx 1$.

The concept of étendue applies to beams that are quasi-homogeneous, as defined in Section 4.2.1. That is, the mutual intensity function, which we consider here in a 2D plane, must be separable into independent intensity and coherence functions such that

$$J(\vec{\rho}_c, \vec{\rho}_d) = I(\vec{\rho}_c)\mu(\vec{\rho}_d) \tag{6.23}$$

To each of these functions, we can associate a characteristic area. For example, the total beam area can be defined by

$$A = \frac{\left|\int \mathrm{d}^2\vec{\rho}_c\, I(\vec{\rho}_c)\right|^2}{\int \mathrm{d}^2\vec{\rho}_c\, I(\vec{\rho}_c)^2} = \frac{\Phi^2}{\int \mathrm{d}^2\vec{\rho}_c\, I(\vec{\rho}_c)^2} \tag{6.24}$$

where Φ is the total beam power, defined by Eq. 4.24.

The coherence area is conventionally defined by

$$A_\mu = \int \mathrm{d}^2\vec{\rho}_d\, \left|\mu(\vec{\rho}_d)\right|^2 \tag{6.25}$$

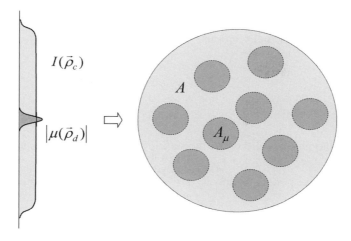

Figure 6.1. Schematic of a quasi-homogeneous beam.

The number of coherence areas comprised within the total beam area is then

$$N_\mu = \frac{A}{A_\mu} \qquad (6.26)$$

We refer to this as the number of spatial "modes" or "degrees of freedom" encompassed by the beam. If $N_\mu = 1$ the beam is said to be coherent; if $N_\mu > 1$ the beam is said to be partially coherent. The larger the number of modes N_μ, the less coherent the beam. A partially coherent beam of area A can therefore be thought of as a mosaic of coherent beamlets, each of area A_μ, as illustrated in Fig. 6.1. Conventionally, a beam is said to be fully incoherent when $N_\mu \gg 1$, but also when $A_\mu \approx \kappa^{-2}$.

By definition, the étendue of a beam is directly proportional to the number of modes and is given by

$$G = \kappa^{-2} N_\mu = \kappa^{-2} \frac{A}{A_\mu} \qquad (6.27)$$

(The variable G is generally used to denote étendue, and should not be confused with a Green function.)

As we will see, if the beam incurs no power loss as it propagates through an optical system, then its étendue remains conserved.

6.3.1   Conservation of Étendue

In Section 4.2 we considered the propagation of quasi-homogeneous fields between two planes in free space, and we established a scaled Fourier transform relationship between $I_1(\vec{\rho}_{1c})$ and $\mu_0(\vec{\rho}_{0d})$ (Eq. 4.38) and also between $I_0(\vec{\rho}_{0c})$ and $\mu_1(\vec{\rho}_{1d})$ (Eq. 4.39). These functions are thus scaled Fourier conjugates of one another. By applying Parseval's theorem (Eq. A.9) and making use of the definitions of intensity and coherence areas given above (Eqs. 6.24 and 6.25), we arrive at the fundamental relations

$$A_1 = \frac{z^2}{\kappa^2 A_{\mu 0}} \tag{6.28}$$

$$A_{\mu 1} = \frac{z^2}{\kappa^2 A_0} \tag{6.29}$$

These relations are valid under the same conditions as those required to derive Eqs. 4.38 and 4.39. Namely, we must have $\left(\vec{\rho}_{0d} \cdot \vec{\rho}_{0c}\right)_{max} \ll \lambda z$ and $\left(\vec{\rho}_{1d} \cdot \vec{\rho}_{1c}\right)_{max} \ll \lambda z$, which roughly translate to $A_{\mu 0} A_0 \ll \lambda^2 z^2$ and $A_{\mu 1} A_1 \ll \lambda^2 z^2$. As will be noted below, these conditions are quite restrictive.

Several conclusions can be drawn from Eqs. 6.28 and 6.29. First, both A_1 and $A_{\mu 1}$ expand with free space propagation in proportion to $\left(\frac{z}{\kappa}\right)^2$. This was already known from Section 4.2.2 (see Fig. 4.3). Second, a rearrangement of the above relations leads to

$$\frac{A_0}{A_{\mu 0}} = \frac{A_1}{A_{\mu 1}} \tag{6.30}$$

Based on the definition provided by Eq. 6.27, we conclude that the étendue G is conserved upon propagation through free space.

A closer inspection of Eqs. 6.28 and 6.29 also reveals that if we define the solid angles

$$\Omega_0 = \frac{A_1}{z^2} \tag{6.31}$$

$$\Omega_1 = \frac{A_0}{z^2} \tag{6.32}$$

then this law of conservation of étendue can be recast in the form

$$G = A_0 \Omega_0 = A_1 \Omega_1 \tag{6.33}$$

Equation 6.33 provides an intuitive interpretation of beam étendue as the product of the beam area and beam divergence, as illustrated in Fig. 6.2.

From Eqs. 6.33 and 6.27 we further obtain the relation

$$A_{\mu 0} \Omega_0 = A_{\mu 1} \Omega_1 = \kappa^{-2} \tag{6.34}$$

which holds generally, independently of the value of N_μ, and hence independently of the state of coherence of a beam. In words, Eq. 6.34 states that the far-field angular divergence of a partially coherent beam is governed by the width of the beam's local coherence function (which we knew already), but also that this global angular divergence is the same as that of a purely coherent beam of area A_μ. Examples of this result will be presented below.

Finally, there is a link between beam power and étendue, which is mediated by the source radiance $\mathcal{L}$. From Eq. 6.10, we have

$$\Phi = \mathcal{L} G = \mathcal{L} A \Omega \tag{6.35}$$

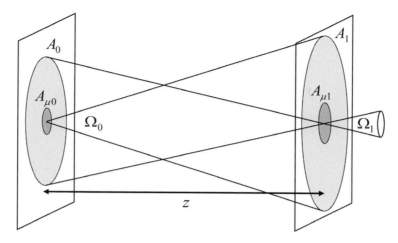

Figure 6.2. Free-space propagation relation between intensity and coherence functions.

where $\mathcal{L}$ is a characteristic average source radiance. We note that $\mathcal{L}$ here is an intrinsic property of the source independent of the source geometry (size, shape, etc.). The beam power in Eq. 6.35 has thus been separated into a term independent of beam geometry ($\mathcal{L}$) and a term dependent only on beam geometry (G).

6.3.2 Multimode Beam Propagation

The astute reader will remark that there is a conflict between the results derived above and the conditions assumed to derive them. On the one hand, we derived Eq. 6.34, but on the other hand we assumed $A_{\mu 0}\Omega_1 \ll \lambda^2$ and $A_{\mu 1}\Omega_0 \ll \lambda^2$ (and hence $A_{\mu 0}\Omega_0 A_{\mu 1}\Omega_1 \ll \kappa^{-4}$). Though this conflict will be largely resolved in the following section, it is clear we must be careful in drawing too many quantitative conclusions from our arguments. Nevertheless, despite its flaws, the concept of étendue remains a useful tool for providing a first-pass intuitive understanding into problems related to coherence propagation, particularly when considering propagation between conjugate planes, or between conjugate and Fourier planes, as will be demonstrated below.

Single-Mode Beam

As a first example, let us consider a perfectly coherent beam at plane 0. By definition, a coherent beam contains a single mode, meaning $N_\mu = 1$ and hence $A_{\mu 0} = A_0$. Technically, it would be difficult for such a single-mode beam to satisfy the condition of quasi-homogeneity defined in Section 4.2.1; however, as it turns out, a particular class of beam does satisfy this condition, namely a Gaussian field. This is given by $E_0(\vec{\rho}_0) = E_0 e^{-\rho_0^2/w_0^2}$, leading to a mutual intensity that can be written in a separated form as

$$J_0(\vec{\rho}_{0c}, \vec{\rho}_{0d}) = I_0(\rho_{0c})\mu_0(\rho_{0d}) = \left(I_0 e^{-2\rho_{0c}^2/w_0^2}\right)\left(e^{-\rho_{0d}^2/2w_0^2}\right) \tag{6.36}$$

thereby satisfying the condition of quasi-homogeneity. It can be verified from Eqs. 6.24, 6.25, and 6.26, that N_μ is indeed equal to 1.

The étendue of a Gaussian beam, from Eq. 6.27, is then simply

$$G = \kappa^{-2} \tag{6.37}$$

and, upon beam propagation, we have

$$A_1 = \frac{z^2}{\kappa^2 A_0} \tag{6.38}$$

$$A_{\mu 1} = \frac{z^2}{\kappa^2 A_0} \tag{6.39}$$

in agreement with our results obtained in Section 5.1.1. That is, the beam and coherence areas expand identically, and the beam remains coherent. The larger the initial beam area, the smaller the solid angle of the expansion. This is the reason laser beams expand so little upon propagation. For example, a 1 µm wavelength laser beam of waist $w_0 = 1$ mm expands with a solid angle of full width $\Delta\theta = 2/\pi w_0 \kappa \simeq 6 \times 10^{-4}$ radians.

A single-mode Gaussian beam is, in fact, a special case of a broader class of beams called Gaussian Schell-model fields, which are commonly used to describe partially coherent quasi-homogeneous beams. The reader is referred to [3, 12] for more details on this class of fields.

Multimode Beam

We turn now to the example of a quasi-homogeneous partially coherent beam, Schell-model or otherwise, which, by definition, contains multiple modes. The étendue of such a beam is given by

$$G = \kappa^{-2} N_\mu \tag{6.40}$$

and upon beam propagation we predict

$$A_1 = \frac{z^2}{\kappa^2 A_0} N_\mu \tag{6.41}$$

$$A_{\mu 1} = \frac{z^2}{\kappa^2 A_0} \tag{6.42}$$

The beam and coherence areas now do not expand identically (see Fig. 6.3). The beam area expands in a solid angle $\Omega_0 = 1/\kappa^2 A_{\mu 0}$, whereas the coherence area appears to expand with an effective solid angle N_μ times smaller. As was emphasized in Section 4.2.2 and is emphasized again here, the notion of an expanding coherence area is misleading because it suggests there is a one-to-one correspondence between a coherence area in plane 0 and a coherence area in plane 1. In reality, this is not the case. For a large enough propagation distance z, any given coherence area in plane 1 arises from the superposition of fields from *all* the coherence areas

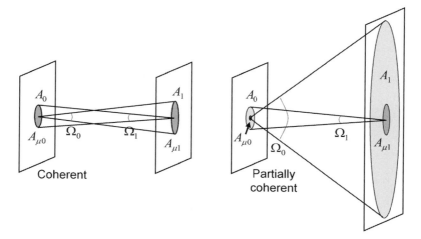

Figure 6.3. Free-space propagation of coherent versus partially coherent beam.

in plane 0. This is, indeed, the prescription for what we mean by a "large enough" propagation distance, as required for the derivations of Eqs. 6.28 and 6.29.

6.4 THROUGHPUT

As defined, the term étendue refers to a property of an optical beam. We turn our attention now to the related parameter of throughput, which refers to a property of an optical system. For clarification, an optical system is defined to have an input and an output, and in general consists of a collection of lenses and/or apertures. To define the throughput of an arbitrary optical system, let us imagine sending a fictitious beam of infinite area and maximum divergence angle through the system (i.e. an infinite-area incoherent beam). The input étendue of such a fictitious beam is, of course, infinite. However, when the beam is sent through the optical system, its output étendue becomes finite. The system throughput is thus defined to be equal to the resulting output étendue.

To better understand this concept of throughput, let us consider a simple optical system consisting of two apertures of areas A_0 and A_1 separated by a distance z, as illustrated in Fig. 6.4. By definition, the output étendue of a beam emerging from the system is given by the product of the beam area times the beam divergence solid angle. However, simple geometrical arguments tell us that, in the case of an incoherent input beam of infinite area, the output beam area must be equal to A_1 and the output beam divergence solid angle must be equal to $\Omega_1 = A_0/z^2$. By definition, then, the throughput of the two-aperture system is given by

$$G_T = \frac{A_0 A_1}{z^2} \qquad (6.43)$$

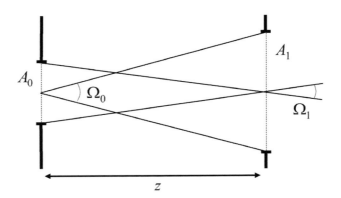

Figure 6.4. Geometry for throughput calculation of two apertures.

When rewritten in terms of the solid angles spanned by each aperture (as seen by the opposite aperture), this throughput becomes

$$G_T = A_0\Omega_0 = A_1\Omega_1 \tag{6.44}$$

where now $\Omega_0 = A_1/z^2$.

Equation 6.44 defining the system throughput bears a striking resemblance to Eq. 6.33 defining beam étendue. The difference between the two equations is that the throughput G_T depends only on system parameters whereas the étendue G depends only on beam parameters.

Two comments should be made here. First, as defined above, the system throughput is the same regardless of the orientation of the optical system. This is true in general. Second, our definition of system throughput is based on the use of a large, incoherent input beam to derive a resultant output étendue. We recall from the previous section that the concept of étendue is only well defined when certain conditions are met, namely $A_{\mu 0}A_0 \ll \lambda^2 z^2$ and $A_{\mu 1}A_1 \ll \lambda^2 z^2$. While these conditions posed a problem in the case of free space propagation, they can be readily met when A_0 and A_1 are constrained by apertures.

As an important second example, we consider an optical system consisting of a lens in a 2f configuration with an input and output aperture, as shown in Fig. 6.5. Following the same prescription for calculating the output étendue resulting from a large, incoherent input étendue, we find that the system throughput is given by

$$G_T = \frac{A_0 A_1}{f^2} \tag{6.45}$$

where A_0 and A_1 are the areas of the input and output apertures, and f is the lens focal length. The concept of throughput is well defined for this optical system since, as we saw in Section 4.3, the conditions of quasi-homogeneity which posed a problem in defining étendue for free space propagation are now relaxed. We further note that Eq. 6.45 is based on the presumption that the lens area is larger than both A_0 and A_1. If this is not the case, or if either of the apertures is absent, then the lens area itself plays the role of an aperture.

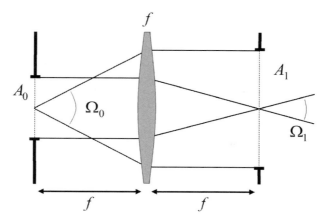

Figure 6.5. Geometry for throughput calculation of a 2f configuration with apertures.

In Section 6.3.1, it was concluded that étendue is conserved when a beam propagates through free space. Manifestly, this is no longer true when a beam propagates through an optical system, as demonstrated by the examples above where infinite input étendues are transformed into finite output étendues. But what happens when the input étendue is itself finite? A very useful principle governing the transmission of étendue through an optical system is given by

$$G_{\text{out}} \leq [G_T, G_{\text{in}}]_{\text{min}} \qquad (6.46)$$

which states that the output étendue G_{out} of a beam can be no greater than either the beam input étendue G_{in} or the system throughput G_T. The equality sign in Eq. 6.46 is a best-case scenario, which can be attained when the input beam is optimally coupled through the system. A conclusion that can be drawn from Eq. 6.46 is that if a beam is transmitted through a series of optical sub-systems, then the total throughput of the composite system is limited by the *smallest* throughput of the sub-systems comprising it. It suffices, therefore, to identify the throughput bottleneck in a composite system to define its overall throughput.

6.4.1 Apparent Source Coherence

Another conclusion can be drawn from Eq. 6.46. In particular, Eq. 6.46 suggests that by restricting throughput, we can reduce the étendue of a beam to smaller and smaller values. Recalling from Section 6.3 that a field becomes more coherent as its étendue becomes smaller, we arrive at the general conclusion that a partially coherent beam can be made more coherent upon propagation through an optical system. This general conclusion leads to the important concept of an *apparent* coherence.

Returning, for example, to the simple system illustrated in Fig. 6.4 comprising two apertures, let us again consider an input beam that is incoherent. By definition, the coherence area of an incoherent beam is $\approx \kappa^{-2}$. However, when the beam is sent through the optical system, for all intents and purposes the apparent input coherence area of the beam is no longer $\approx \kappa^{-2}$ but rather $A_{\mu 0} = \kappa^{-2}/\Omega_0$ (see Eq. 6.34) where Ω_0 is now restricted by the throughput of the optical system. From the definition of Ω_0 given by Eq. 6.44 we obtain

$$A_{\mu 0} = \kappa^{-2}\frac{z^2}{A_1} \tag{6.47}$$

That is, the apparent coherence of the input beam is no longer prescribed by the intrinsic properties of the input beam but rather by the optical system itself. Equation 6.47 is again valid only for $z^2 \gg A_1$, meaning that the apparent coherence area $A_{\mu 0}$ must, in general, be larger than κ^{-2}. However, it is clear that $A_{\mu 0}$ cannot increase without bound. In particular, $A_{\mu 0}$ is bounded by the area of the input aperture A_0 itself. In the event that z^2/A_1 is so large that this bound is attained, we arrive at $A_0/A_{\mu 0} = A_1/A_{\mu 1} = 1$, meaning that the beam has been rendered completely coherent by the optical system, both apparently at the system input and actually at the system output. Thus, as a general rule, a partially coherent input beam can be made more coherent upon propagation through an optical system, whereas a purely coherent input beam remains purely coherent.

It should be noted that we have been implicitly using this concept of apparent coherence from the start when describing incoherent beams. Indeed, the coherence area $\approx \kappa^{-2}$ that we have systematically ascribed to incoherent beams is in fact an apparent coherence area that arises from our prescription that the beam be radiative. The optical system in this case can be regarded as simple free space.

6.4.2 Transmission of Power

To demonstrate the practical utility of beam étendue and throughput, we apply these to estimate the maximum power that can be delivered through an arbitrary optical system.

To begin, the available power provided by a luminous source, from Eq. 6.35, can be written as

$$\Phi_{\mathcal{L}} = \mathcal{L}G_{\mathcal{L}} \tag{6.48}$$

where $\mathcal{L}$ and $G_{\mathcal{L}}$ are respectively the source radiance and étendue. From Eq. 6.46 the transmitted power through an optical system of throughput G_T is then limited by

$$\Phi \leq \mathcal{L}[G_T, G_{\mathcal{L}}]_{\min} \tag{6.49}$$

The net reduction in power is

$$\frac{\Phi}{\Phi_{\mathcal{L}}} \leq \frac{[G_T, G_{\mathcal{L}}]_{\min}}{G_{\mathcal{L}}} \tag{6.50}$$

which is independent of source radiance. In conclusion, to attain maximum power delivery we must match the system throughput to the source étendue. Some examples are presented below.

Power Cost to Render a Beam Coherent

The intuitive picture presented in Fig. 6.1 suggests that if we propagate a partially coherent quasi-homogeneous beam through a small aperture, or pinhole, of area $A_p \leq A_\mu$, then the number of modes transmitted through the pinhole reduces to one. The transmitted beam thus becomes purely coherent. We can verify this by using a throughput analysis where the optical system in question comprises an entrance aperture, defined by the area $A_{\mathcal{L}}$ of the beam at its source, and an exit aperture, defined by the area of the pinhole A_p, separated by a propagation distance z. From Eq. 6.43 we find that the pinhole area should be no larger than $z^2/(\kappa^2 A_{\mathcal{L}})$ to render the beam completely coherent, which, from Eq. 6.42, corresponds to the coherence area of the beam at the pinhole plane, as expected.

We emphasize that a partially coherent source can be made coherent only at the expense of power – an expense which depends on the actual étendue $G_{\mathcal{L}}$ of the source. For example, if the source is incoherent, we can adopt the estimate $G_{\mathcal{L}} \approx A_{\mathcal{L}}$ (from Eq. 6.27). The maximum power ratio transmitted through the pinhole from Eq. 6.50 is then $\approx A_p/z^2$, which can represent quite a significant reduction in power.

Power Delivery into an Optical Fiber

As a second example, we can estimate the maximum power delivered through an optical fiber. If the fiber has a core area A_f and an acceptance solid angle Ω_f, then its throughput is $G_T = A_f \Omega_f$.

For a fiber to be single-mode, this throughput must be no larger than κ^{-2}. Hence, for a single-mode fiber we must have $A_f \Omega_f \approx \lambda^2$, whereas for a multimode fiber $A_f \Omega_f > \lambda^2$. The number of modes that can be accommodated by a fiber, single-mode or multimode, is then roughly

$$N_\mu \approx \frac{A_f \Omega_f}{\lambda^2} \tag{6.51}$$

If an input beam has power Φ_{in} and étendue G_{in}, the output power Φ_{out} that can be delivered through the fiber becomes limited by

$$\Phi_{out} \leq \Phi_{in} \frac{[A_f \Omega_f, G_{in}]_{min}}{G_{in}} \tag{6.52}$$

The beam étendue and fiber throughput must therefore be matched to maximize power delivery. This means that the beam should be focused to an area no larger than the fiber core area A_f and to a focus angle no larger than the fiber acceptance angle Ω_f. These conclusions are intuitively obvious.

6.5 MICROSCOPE LAYOUT

Finally, we apply a throughput analysis to an optical microscope. In most cases an optical microscope consists of an illumination system and a detection system. We consider these separately. However, before doing this, we first introduce the general concepts of field and aperture stops.

6.5.1 Field and Aperture Stops

Any optical imaging system has an input and an output. Based on our definition of throughput as the resultant output étendue when the input étendue is infinite (i.e. the input beam is incoherent and of infinite area), we can formulate a simple recipe for estimating the throughput of an imaging system using ray tracing techniques. The recipe is as follows:

1. Determine Ω_{out} at the output plane of the system by back-projecting a diverging angular cone and identify the pupil in the system that most limits the angular spread of this cone. This pupil is called the aperture stop.
2. From the center of this aperture stop plane, project a new diverging angular cone forward and backward within the system to identify a new pupil in the system that most limits the angular spread of this cone. This second pupil is called the field stop.
3. Project this new cone to the system output plane and determine its area A_{out} at this plane.
4. The throughput of the system is then estimated to be $A_{out}\Omega_{out}$.

By definition, the field and aperture stops define A_{out} and Ω_{out} respectively, and in many cases are adjustable, allowing direct control of the system throughput. As examples of applications of the above recipe, we return to the optical systems illustrated in Figs. 6.4 and 6.5. From steps 1 and 2 we readily conclude that A_0 represents the aperture stop and A_1 represents the field stop in both Figs. 6.4 and 6.5 (provided the lens is larger than both apertures in Fig. 6.5), and indeed recover the values of throughput given by Eqs. 6.43 and 6.45.

As a point of clarification, it should be noted that what we call an optical system is allowed to extend into free space beyond any physical apertures in the system itself. As an example, let us consider the 4f imaging system illustrated in Fig. 3.4, and define the system input and output to be the object and image planes respectively, neither of which contain physical apertures. Step 1 from our recipe tells us that the pupil $P(\vec{\xi})$ is the aperture stop, and step 2 tells us that the field stop is located at one of the two lenses, whichever limits A_{out} most. Accordingly, the throughput of the 4f system is

$$G_T = \frac{A_\xi A_f}{f^2} \tag{6.53}$$

where f and A_f denote respectively the focal length and area of the field-stop lens, and A_ξ denotes the area of the aperture stop. This 4f system will be re-examined below in the context of standard microscope illumination and detection configurations.

6.5.2 Illumination Configurations

Let us examine the illumination and detection components of a standard microscope separately. We begin with the illumination component, and assume the illumination source is a lamp that produces a beam étendue larger than the illumination system throughput. Two types of illumination are considered:

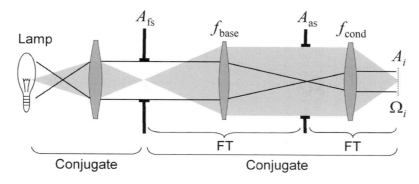

Figure 6.6. Critical illumination.

Critical Illumination

The first type of illumination is called critical illumination, where the lamp is located in a plane conjugate to the sample plane. That is, the source is imaged onto the sample by one or a set of lenses. An example of critical illumination is illustrated in Fig. 6.6, where, by convention, the lens closest to the source is called the collector, and the lens closest to the sample is called the condenser (f_{cond}). Between these is an intermediate lens which, for lack of a standardized term, we call a base lens (f_{base}). In most cases, the beam area at the sample plane is defined by either a field stop (A_{fs}) or by the base lens itself. We assume the former. Moreover, we assume that the field stop and sample planes are arranged in a 4f configuration. That is, the aperture stop (A_{as}) is located at a Fourier plane of both the condenser lens (usually referred to as the back aperture of the condenser lens) and the base lens.

By inspection of Fig. 6.6, a critical illumination system can be separated into a series of three sub-systems: an imaging system from the source to the field stop, a 2f system from the field stop to the aperture stop, and finally another 2f system from the aperture stop to the sample. The throughput bottleneck comes from the middle sub-system, meaning that, from Eq. 6.45, the throughput of the composite illumination system is given by

$$G_T = \frac{A_{as}A_{fs}}{f_{base}^2} \qquad (6.54)$$

Because the field stop is imaged onto the sample plane, the area of illumination at the sample plane is given by

$$A_i = A_{fs}\left(\frac{f_{cond}}{f_{base}}\right)^2 \qquad (6.55)$$

This is called the field of illumination, with its diameter (FOI) obeying the relation $A_i = \frac{\pi}{4}\mathrm{FOI}^2$.

In turn, the local convergence angle of the illumination beam at the sample plane is given by

$$\Omega_i = \frac{A_{as}}{f_{cond}^2} \qquad (6.56)$$

which can be re-expressed as $\Omega_i = \frac{\pi}{n^2}NA_i^2$ (see Eq. 4.84), where NA_i is the condenser numerical aperture and n the index of refraction at the sample plane.

As expected, we confirm $G_T = A_i\Omega_i$. Moreover, from Eq. 6.34, we conclude that the coherence area at the sample plane is

$$A_{\mu i} = \frac{f_{cond}^2}{\kappa^2 A_{as}} \tag{6.57}$$

which, from Section 4.3, we recognize to be the diffraction-limited resolution of a lens of aperture A_{as} and focal length f_{cond}. We will return to this coherence area below.

Finally, to estimate the power delivered to the sample, we recall that the beam étendue from the lamp is larger than the illumination throughput, by assumption. From Eq. 6.49, we then obtain

$$\Phi_i \leq \mathcal{L}G_T = \mathcal{L}\frac{A_{as}A_{fs}}{f_{base}^2} \tag{6.58}$$

where $\mathcal{L}$ is the lamp radiance.

Köhler Illumination

The second type of illumination we consider is Köhler illumination (which we have seen already in Section 4.3). Here the illumination source is located in a Fourier plane relative to the sample plane. An example of Köhler illumination is illustrated in Fig. 6.7. By inspection, the throughput bottleneck remains manifestly the same as before. The throughput of a Köhler illumination system is therefore

$$G_T = \frac{A_{as}A_{fs}}{f_{base}^2} \tag{6.59}$$

meaning that, despite having changed our illumination geometry, all our results concerning the sample-plane illumination area, coherence area, convergence angle, and power remain the same as in the case of critical illumination.

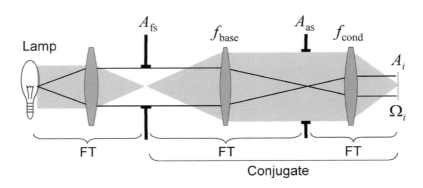

Figure 6.7. Köhler illumination.

We remark that our analysis of illumination throughput is somewhat idealized since we have tacitly assumed the lamp radiance $\mathcal{L}$ is uniform across the lamp area. If it is not uniform, as is usually the case in practice, then any variations in $\mathcal{L}$ become imaged onto the sample plane when critical illumination is applied. In contrast, these variations become homogenized (blurred) when Köhler illumination is applied. For this reason, while the two illuminations are formally equivalent, Köhler illumination is regarded as the preferred geometry, and utilized in all modern optical microscopes (provided these are not scanning microscopes – see Chapter 14).

Focal Spot Size and Coherence

A final comment should be made regarding illumination coherence. Though we assumed that the source was highly incoherent in our analyses of critical and Köhler illumination, the final illumination at the sample can, in fact, be highly coherent if the system throughput is small. For example, coherent illumination can be obtained at the sample by closing down either the field stop or the aperture stop. If the field stop is closed, the illumination is localized to a diffraction limited spot (we will return to this when examining confocal microscopy in Chapter 14). On the other hand, if the aperture stop is closed, the illumination becomes a plane wave (which is sometimes advantageous for phase contrast imaging, as will be seen in Chapter 10).

Alternatively, we could have started with a source that was coherent instead of incoherent. In this case, no matter what the throughput of the illumination system, the illumination étendue at the sample is given by $G_i = \kappa^{-2}$. Examples of coherent-source illumination include critical illumination from a point source or Köhler illumination from a plane-wave source. In both cases we find

$$A_i = A_{\mu i} = \frac{f_{\text{cond}}^2}{\kappa^2 A_{\text{as}}} = \frac{1}{\kappa^2 \Omega_i} \qquad \text{(coherent)} \qquad (6.60)$$

In contrast, if the illumination is partially coherent or incoherent we have instead

$$A_i = N_{\mu i} A_{\mu i} \qquad \text{(partially coherent)} \qquad (6.61)$$

As noted above, Eq. 6.60 corresponds to the diffraction-limited spot size associated with the condenser lens. Recalling from Eq. 4.84 its link with the condenser numerical aperture (NA_i), we can thus write

$$A_{\mu i} = \frac{\lambda^2}{\pi \text{NA}_i^2} \qquad (6.62)$$

which applies generally.

In summary, $A_{\mu i}$ is the coherence area at the sample plane and is independent of whether the illumination is coherent or incoherent, since it is independent of $N_{\mu i}$. On the other hand, from Eq. 6.61, the total beam area A_i at the sample plane manifestly depends on the number of illumination modes and hence on the state of illumination coherence. If there is only one mode, the illumination beam area is diffraction-limited and small; if there are many modes it is

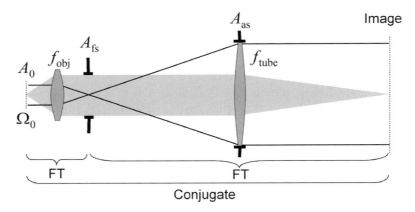

Figure 6.8. 4f detection configuration (often called infinity-corrected configuration).

larger. This is why it is possible to focus a coherent beam to a diffraction-limited spot but not an incoherent beam.

6.5.3 Detection Configuration

Almost all modern microscopes utilize a detection configuration based on a 4f geometry, as illustrated in Fig. 6.8. By convention, the lens closest to the sample is called the objective (f_{obj}) and the lens closest to the detector is called the tube lens (f_{tube}). The net magnification from the sample plane to the image plane is then

$$M = -\frac{f_{tube}}{f_{obj}} \tag{6.63}$$

Objectives that are designed for use in a 4f geometry are called "infinity-corrected." These usually comprise several lenses and can be quite thick. Objectives from the same manufacturer are usually designed so that the sum of twice their focal lengths and the spacing between their principal planes remains a fixed value, called the parfocal distance. A fixed parfocal distance allows one to change the microscope magnification by simply swapping different objectives without readjusting the position of the sample or the detector (at least in principle).

The detection throughput can be calculated two ways. In one method, the method we have used so far, the étendue at the image plane is calculated for a hypothetical infinite étendue at the sample plane. Alternatively, because a system throughput is symmetric, we can work backwards and calculate the étendue at the sample plane given a hypothetical infinite étendue at the image plane. In this second method, the detector at the image plane is effectively modeled as an incoherent source. We will use this second method to derive A_0 and Ω_0 at the sample plane, bearing in mind that these are not necessarily the same as the illumination A_i and Ω_i derived in the previous section.

In most microscopes, the aperture stop A_{as} is at the Fourier plane situated near, or even inside, the objective-lens housing. This Fourier plane is variously referred to as the objective back focal plane or back aperture ("back" referring to away from the sample). The field stop is generally defined by the physical size of the tube lens, though, in cases where the detector at the image plane has a small surface area, it could be defined by the detector size itself. Both cases lead to the same result: denoting as A_{fs} the tube lens or detector area, whichever appropriate, the throughput of the detection optics is found to be

$$G_T = \frac{A_{as}A_{fs}}{f_{tube}^2} \tag{6.64}$$

where

$$A_0 = A_{fs}\left(\frac{f_{obj}}{f_{tube}}\right)^2 = \frac{A_{fs}}{M^2} \tag{6.65}$$

and

$$\Omega_0 = \frac{A_{as}}{f_{obj}^2} \tag{6.66}$$

It is conventional to recast A_0 and Ω_0, respectively, in terms of the field-of-view (FOV) diameter and the objective numerical aperture (NA_0), such that

$$G_T = A_0\Omega_0 = \left(\frac{\pi}{4}FOV^2\right)\left(\frac{\pi}{n^2}NA_0^2\right) \tag{6.67}$$

where n is the index of refraction at the sample.

The *apparent* coherence area at the sample, from the detector point of view, is then

$$A_{\mu 0} = \frac{\lambda^2}{\pi NA_0^2} \tag{6.68}$$

corresponding to the diffraction-limited detection resolution. We remark that the illumination coherence area, defined by Eq. 6.62, and the detection coherence area defined above can be quite different. If $NA_i < NA_0$ the sample illumination appears coherent whereas if $NA_i \gg NA_0$ it appears incoherent (we will examine this in more detail in Section 10.1.2).

Practical Example

A detection throughput calculation is provided based on standard microscope parameters, as outlined below.

As far as detection is concerned, the most important parameters for a microscope chassis are the focal length and diameter of its tube lens. For many commercial microscopes, these are given by $f_{tube} = 180$ mm and $d_{tube} = 25$ mm.

The most important parameters for an objective lens are its focal length and numerical aperture. By convention, objectives are usually characterized by their magnification rather than their focal length. This magnification has meaning only if the tube lens focal length is specified in advance. For example, if an objective is specified by $NA_0 = 0.8$ and $|M| = 60\times$, then its focal length is $f_{obj} = f_{tube}/|M|$, or, in this particular case, 3 mm.

Assuming the tube lens plays the role of the field stop, the field-of-view diameter is given by $FOV = d_{tube}/|M| \approx 0.42$ mm, corresponding to the diameter of the sample area as seen by the detector. From Eq. 6.67, we obtain finally $G_T \approx 0.28$ mm², provided that the index of refraction at the sample is close to unity. For this calculation to be valid, the size of the detector must be at least as large as the tube lens, or, if not, some intermediate optics must be placed between the image plane and the detector plane to ensure that the detector size does not limit the FOV. If, in fact, the detector does limit the FOV, which is often the case in practice, the calculation must be modified accordingly.

6.6 PROBLEMS

Problem 6.1
(a) Derive Eq. 6.11 (i.e. $I_\infty\left(\vec{r}\right) = \left(\frac{1}{r}\right)^2 \mathcal{R}_0\left(\frac{\kappa}{r}\vec{\rho}\right)$) using the Fraunhofer approximation given by Eq. 2.62.

(b) Verify Eq. 6.16, using the paraxial approximation.

Problem 6.2
Assume that light emanating from an intensity distribution $I\left(\vec{\rho}_0, 0\right)$ obeys a paraxial angular distribution $\chi\left(\theta\right)$ everywhere (see figure). Based on purely geometrical arguments, one may write the convolutions

$$I\left(\vec{\rho}, z\right) = \frac{1}{z^2} \int d^2\vec{\rho}_0 \, \chi\left(\left|\vec{\rho} - \vec{\rho}_0\right|/z\right) I\left(\vec{\rho}_0, 0\right)$$

$$\vec{\Theta}_\perp\left(\vec{\rho}, z\right) = \frac{1}{z^3 I\left(\vec{\rho}, z\right)} \int d^2\vec{\rho}_0 \, \left(\vec{\rho} - \vec{\rho}_0\right) \chi\left(\left|\vec{\rho} - \vec{\rho}_0\right|/z\right) I\left(\vec{\rho}_0, 0\right)$$

Show that $I\left(\vec{\rho}, z\right)$ and $\vec{\Theta}_\perp\left(\vec{\rho}, z\right)$ constructed in this manner obey the transport of intensity equation (Eq. 6.22).

Hint: make use of $\vec{\nabla}|\vec{\rho}| = \vec{\rho}/|\vec{\rho}|$.

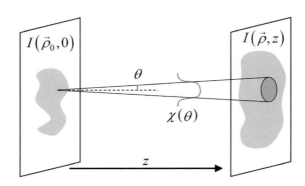

Problem 6.3

Consider the single-lens imaging system of arbitrary magnification M (see figure), which obeys the thin-lens formula. Assume the lens is large and $A_0 \approx A_1$.

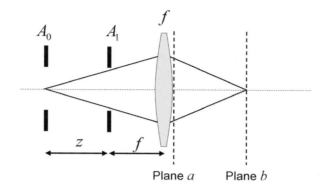

(a) Calculate the throughput of this system using the recipe outlined in Section 6.5.1, treating plane a as the output plane. Identify the aperture and field stops.

(b) Now do the same, but this time treating plane b as the output plane. Are the aperture and field stops the same?

Note: you should find that the throughput is independent of which plane a or b is treated as the output plane.

Problem 6.4

A lamp in a housing emits incoherent light through an aperture of area A_{lamp} (see figure). The emitted light power is Φ_{lamp}. This light illuminates an objective comprising a lens and an aperture at the back focal plane, both of area A_{obj} (assume $A_{obj} \lesssim A_{lamp}$). The lens has focal length f_{obj}. A variable distance z separates the lamp and the objective.

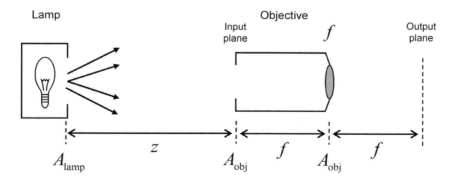

(a) In the case where the lamp touches the objective (i.e. $z = 0$), estimate the number of modes (coherence areas) that enter the objective at the input plane. What is maximum power of the beam at the output plane (i.e. the objective front focal plane)? What is the coherence area of the beam at the output plane? Estimate the beam spot size (total beam area) at the output plane.

(b) In the case where the lamp separated a large distance z from the objective, estimate the number of modes that enter the objective at the input plane. What is the maximum power of the beam at the output plane? What is the coherence area of the beam at the output plane? Estimate the beam spot size at the output plane.

(c) At what value of z does the beam at the output plane become a diffraction-limited spot (i.e. single mode)? At this value, what is the number of modes that enter the objective at the input plane?

Note: perform rough estimates only – that is, angular spreads of 2π steradians can be approximated as angular spreads of 1 steradian.

Problem 6.5

Consider a more general Gaussian–Schell beam whose mutual intensity is given by

$$J_0(\vec{\rho}_{0c}, \vec{\rho}_{0d}) = \left(I_0 e^{-2\rho_{0c}^2/w_c^2} \right) \left(e^{-\rho_{0d}^2/2w_d^2} \right)$$

(Note: this differs from the single-mode Gaussian beam described by Eq. 6.36 in that $w_c > w_d$.)

(a) Calculate the number of modes in this beam.

(b) Calculate the area and coherence area of this beam upon propagation a large distance z. Show explicitly that the number of modes is conserved.

(c) Consider using a lens of numerical aperture NA_i to focus this beam. If the beam just fills the lens (roughly speaking), estimate the size the the resultant focal spot.

(d) If instead the beam overfills the lens such that only 1% of the beam power is focused, estimate the size of the resultant focal spot.

References

[1] Boyd, R. W. *Radiometry and the Detection of Optical Radiation*, Wiley Series in Pure and Applied Optics, Wiley-Interscience (1983).

[2] Carter, W. H. and Wolf, E. "Coherence and radiometry with quasihomogeneous planar sources," *J. Opt. Soc. Am.* 67, 785–796 (1977).

[3] Foley, J. T. and Zubairy, M. S. "The directionality of Gaussian Schell-model beams," *Opt. Commun.* 26, 297–300 (1978).

[4] Friberg, A. T. and Thompson, B. J. Eds., *Selected Papers on Coherence and Radiometry*, SPIE Milestone Series, MS69, SPIE Press (1993).

[5] Friberg, A. T. "Propagation of a generalized radiance in paraxial optical systems," *Appl. Opt.* 30, 2443–2446 (1991).

[6] Goodman, J. W. *Statistical Optics*, 2nd edn, Wiley (2015).

[7] Mandel, L. and Wolf, E. *Optical Coherence and Quantum Optics*, Cambridge University Press (1995).

[8] McCluney, R. *Introduction to Radiometry and Photometry*, Artec House Publishers (1994).

[9] Miller, J. L. and Friedman, E. *Photonics Rules of Thumb: Optics, Electro-Optics, Fiber Optics and Lasers*, 2nd edn, McGraw-Hill (2003).

[10] Nugent, K. A. "The measurement of phase through the propagation of intensity: an introduction," *Contemp. Phys.* 52, 55–69 (2011).

[11] Smith, W. J. *Modern Optical Engineering*, 3rd edn, SPIE Press and McGraw-Hill (2000).

[12] Starikov, A. and Wolf, E. "Coherent-mode representation of Gaussian Schell-model sources and of their radiation fields," *J. Opt. Soc. Am.* 72, 923–928 (1982).

[13] Teague, M. R. "Deterministic phase retrieval: a Green's function solution," *J. Opt. Soc. Am.* 73, 1434–1441 (1983).

[14] Walther, A. "Radiometry and coherence," *J. Opt. Soc. Am.* 58, 1256–1259 (1968).

[15] Walther, A. "Radiometry and coherence," *J. Opt. Soc. Am.* 63, 1622–1623 (1973).

[16] Wolf, E. "Coherence and radiometry," *J. Opt. Soc. Am.* 68, 6–17 (1978).

[17] Wolf, E. *Introduction to the Theory of Coherence and Polarization of Light*, Cambridge University Press (2007).

[18] Yoshimori, K. and Itoh, K. "Interferometry and radiometry," *J. Opt. Soc. Am.* A 14, 3379–3387 (1997).

7 Intensity Fluctuations

So far we have largely sidestepped the issue of temporal variations in optical fields. In Chapter 2, it was noted that optical fields vary with frequencies that are quite high, typically on the order of petahertz (10^{15} Hz), which is much faster than can be directly observed by standard detectors. These harmonic variations were hidden in Chapter 3, which was focused on monochromatic fields, allowing us to consider only the complex field amplitudes. They were also neglected in Chapter 4, which invoked temporal averages $\langle \ldots \rangle$, allowing us to generalize our treatment to non-monochromatic fields. However, little discussion went into specifying what exactly was meant by a non-monochromatic field. An aim of this chapter will be to tie up loose ends and better define the concepts of temporal averaging and non-monochromaticity. As in Chapter 4, we will allow for the possibility of temporal field fluctuations that go beyond those of a harmonic variation. Borrowing from the parlance of radio communications, these additional temporal fluctuations can be thought of in the frequency domain as residing in sidebands about a primary harmonic frequency, or carrier frequency, thus imparting a spectral bandwidth to the field. The span of this bandwidth depends on the dynamics of the temporal fluctuations, and will systematically be assumed to be much smaller than the carrier frequency itself, meaning that the field, while non-monochromatic, will nevertheless be regarded as quasi-monochromatic. As we will see, a general conclusion that can be derived from this chapter is that temporal fluctuations can be treated in an analogous way as spatial fluctuations, and many of the principles we obtained previously in the space domain can be directly translated into the time domain.

Further reading on the vast subject of intensity fluctuations can be found in textbooks by Goodman [5], Mandel and Wolf [14], Loudon [13], and Ishimaru [9], along with a fascinating review by Harris [7]. Additionally, a wide selection of original papers can be found in [15].

7.1 TEMPORAL COHERENCE

In Chapter 4 the spatial fluctuations of a field were characterized by a mutual intensity (see Eq. 4.2). Such a characterization can be extended to the temporal domain by what is called a mutual coherence function,

$$\Gamma(\vec{r}, \vec{r}'; t, \tau) = \left\langle E(\vec{r}, t + \tau) E^*(\vec{r}', t) \right\rangle \tag{7.1}$$

where the time dependence of $E(\vec{r}, t)$ is now explicit. The brackets $\langle ... \rangle$ again denote a long-time average, specified as longer than the field coherence time, which will be defined below, though short relative to any coarse-grained temporal dynamics in the field that might be of interest.

To begin, let us investigate the temporal fluctuations at a given location in space. That is, we set $\vec{r} = \vec{r}'$ in Eq. 7.1 and momentarily drop the spatial coordinates from our notation. Moreover, we adopt a slight elaboration in language. Thus far, the local intensity of a light beam has been defined by Eq. 1.10, meaning that intensity is well defined only when considered over a long-time average. Since we will examine some finer details of intensity fluctuations throughout this chapter, we introduce what is called an instantaneous intensity, defined by

$$I(t) = E(t)E^*(t) \tag{7.2}$$

In terms of this new instantaneous intensity (without averaging), our previous definition of intensity (with averaging) becomes $\langle I(t) \rangle$.

The concept of an instantaneous intensity is, of course, idealized since any physical detector must possess a finite bandwidth and thus cannot perform an intensity measurement with infinitely fine time resolution. Nevertheless, just as the brackets $\langle ... \rangle$ refer to a time average longer than a beam coherence time, the term instantaneous will refer to an intensity measurement with temporal resolution much finer than the beam coherence time, regardless of whether such a measurement is practically feasible. The critical parameter here is the coherence time, whose definition we will finally establish.

7.1.1 Quasi-Stationary Beams

For many fields of interest, the statistics of the fluctuations do not significantly change over time, or if they change at all it is over time scales that are slow compared to the scale specified by the long-time average $\langle ... \rangle$. The mutual coherence of such fields is then roughly separable, and can be written as

$$\Gamma(\bar{t}, \tau) = \langle I(t) \rangle \, \gamma(\tau) \tag{7.3}$$

where $\bar{t}$ is slowly varying.

Such fields are called quasi-stationary and are the temporal analogs of quasi-homogeneous fields defined by Eq. 4.31. Just as $\mu(\vec{r}_d)$ was called a spatial coherence function, $\gamma(\tau)$ here is called a temporal coherence function and obeys the same normalization property

$$\gamma(0) = 1 \tag{7.4}$$

Moreover, just as the coherence area was defined by Eq. 6.25, the coherence time here is defined by

$$\tau_\gamma = \int d\tau \, |\gamma(\tau)|^2 \tag{7.5}$$

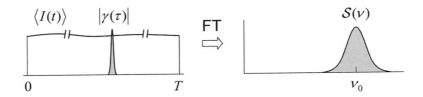

Figure 7.1. Quasi-stationary beam.

Continuing with this analogy between $\mu(\vec{r}_d)$ and $\gamma(\tau)$ (see Section 4.2.1), the spectral density of the field is defined by

$$S(\nu) = \int d\tau\, \gamma(\tau) e^{i2\pi\nu\tau} \tag{7.6}$$

which is normalized such that

$$\int d\nu\, S(\nu) = \gamma(0) = 1 \tag{7.7}$$

Parseval's theorem (Eq. A.9) then provides an alternative definition of coherence time given by

$$\tau_\gamma = \int d\nu\, |S(\nu)|^2 \tag{7.8}$$

from which the spectral bandwidth becomes

$$\Delta\nu = \frac{\left|\int d\nu\, S(\nu)\right|^2}{\int d\nu\, |S(\nu)|^2} = \frac{1}{\tau_\gamma} \tag{7.9}$$

An important remark must be made here. Throughout this book we have characterized light as a complex field $E(t)$. Moreover, we have imposed the constraint, albeit arbitrary, that $E(t)$ is analytic, meaning that $\mathcal{E}(\nu) = 0$ for $\nu < 0$ (see Section 1.1). As a consequence, $\Gamma(\bar{t}, \tau)$ too must be analytic, meaning that $S(\nu) = 0$ for $\nu < 0$, as illustrated in Fig 7.1. That is, even though the integrals in Eqs. 7.7 and 7.8 formally extend over all frequencies, it should be kept in mind that they are effectively one-sided.

The condition of analyticity of $E(t)$, and hence of $\Gamma(\bar{t}, \tau)$, is thoroughly discussed in [3] and [5].

7.2   COHERENCE FUNCTIONS

Thus far, we have treated free-space propagation mostly in the context of Huygens wavelets that arise from virtual, or secondary, sources. In this chapter, we turn our attention to the physical, or primary, sources themselves. These might be externally driven, as in the case of particles that scatter light, or self-luminous, as in the case of excited fluorescent molecules. Both cases will be treated in more detail in Chapters 9 and 13, respectively. For the moment the exact nature of these primary sources will not be specified, and they are referred to simply as radiators.

To illustrate the concept of temporal coherence, let us consider an arbitrary source consisting of an ensemble of radiators oscillating at a carrier frequency $\bar{\nu}$ and fluctuating temporally in both amplitude and phase. The field from a single elemental radiator is written as $E_r(t)e^{-i2\pi\bar{\nu}t+i\phi(t)}$. In turn, the total field from a source results from the superposition of fields generated by many such radiators. If the fluctuations of the radiator fields are temporally uncorrelated, the superposition is said to be incoherent.

7.2.1 Interrupted Phase Model

We begin with a standard model wherein each elemental radiator emits a field whose amplitude is relatively constant, but whose phase fluctuates randomly in time (see, for example, [13]). The statistics of these phase fluctuations depend on the type of radiator in question, and as a specific example, we consider phase fluctuations that arise from random phase interruptions. That is, we adopt a model where each elemental radiator phase $\phi(t)$ remains constant for a certain amount of time until it is interrupted, at which point it assumes another random value between 0 and 2π, and so forth (see Fig. 7.2). If the probability of occurrence of a phase interruption is constant per unit time, then the interruption statistics are Poissonian and the probability that no phase interruption occurs in a time interval τ is

$$p(\tau) = \frac{1}{\tau_\gamma}e^{-\tau/\tau_\gamma} \tag{7.10}$$

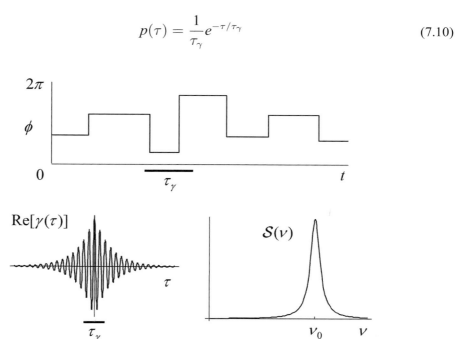

Figure 7.2. Interrupted phase model with Lorentzian spectral density.

where τ_γ is the average time interval between phase interruptions. The coherence function for such a phase-interrupted source is readily found to be

$$\gamma(\tau) = e^{-i2\pi\bar{\nu}\tau} e^{-|\tau|/\tau_\gamma} \tag{7.11}$$

which, from Eq. 7.6, leads to

$$S(\nu) = \frac{2\tau_\gamma}{1 + 4\pi^2\tau_\gamma^2(\nu - \bar{\nu})^2} \tag{7.12}$$

It can be verified that τ_γ introduced in Eq. 7.10 to characterize the phase interruption statistics indeed corresponds to the coherence time of the field, as defined by Eqs. 7.5 or 7.8.

We note from Eq. 7.12 that the frequency spectrum is centered about the carrier frequency $\bar{\nu}$ but has broad wings characterized by a Lorentzian profile. This simple model is often used to describe radiation from fluorescing atoms or molecules undergoing random collisions with their environment, in which case τ_γ is generally quite short. For example, for a gas of atoms in vacuum τ_γ is typically on the order of 10^{-11} s (i.e. $\Delta\lambda = \lambda_0^2\Delta\nu/c \approx 0.1$ nm); for molecules in solution it is even shorter, typically on the order of 10^{-13} s (i.e. $\Delta\lambda \approx 10$ nm).

To evaluate the intensity fluctuations arising from this interrupted phase model, we begin by considering a single radiator at a source plane and evaluate the resulting intensity at a distant observation plane, so distant that field curvature can be neglected. The field at the observation plane is then

$$E(t) = E_r e^{-i2\pi\bar{\nu}t + i\phi(t)} \tag{7.13}$$

where E_r is the field amplitude resulting from the single radiator, which, in the interrupted phase model, is taken to be constant. The instantaneous intensity at the observation plane is then

$$I(t) = |E_r|^2 = I_r \tag{7.14}$$

which is also manifestly constant and exhibits no fluctuations. This is an uninteresting example.

However, let us now consider the more complicated scenario wherein several elemental radiators reside at exactly the same location at the source plane (the concept of "same position" will be clarified in a moment).

The field at the observation plane then becomes the superposition

$$E(t) = E_r e^{-i2\pi\bar{\nu}t} \sum_n^{N_r} e^{i\phi_n(t)} \tag{7.15}$$

where the summation is over the number of radiators N_r. Accordingly, the intensity is given by

$$I(t) = I_r \left| \sum_n^{N_r} e^{i\phi_n(t)} \right|^2 = I_r \sum_{n,m}^{N_r} e^{i(\phi_n(t) - \phi_m(t))} = I_r \left(N_r + 2\mathrm{Re}\left[\sum_{n<m}^{N_r} e^{i(\phi_n(t) - \phi_m(t))} \right] \right) \tag{7.16}$$

The first term on the right-hand side is fixed in time and represents the sum of the intensities from the individual radiators. The second term arises from the interference between distinct radiators and fluctuates in time because of the relative fluctuations in the phases of these fields.

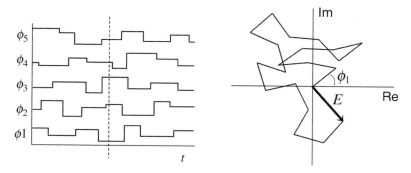

Figure 7.3. Interrupted phase model corresponds to an instantaneous random walk.

If the phases are uncorrelated in time, the second term vanishes on average. The qualifier "on average" must be emphasized here. That is, if we were to sample the instantaneous intensity at an arbitrary instant in time, this second term would not vanish in general, but instead would vary from measurement to measurement.

To evaluate the statistics of the instantaneous intensity fluctuations we use $\{\phi_n\}$ to denote a particular realization of the phases at an arbitrary time t. The instantaneous field, from Eq. 7.15, is then given by the summation

$$E = E_r e^{-i2\pi\bar{\nu}t} \sum_n^{N_r} e^{i\phi_n} \tag{7.17}$$

As is evident from Fig. 7.3, the summation can be thought of as a classical 2D random walk of fixed step size E_r. An abundance of literature is devoted to the subject of random walks, and the reader is referred to [5] and [18] for more details. Briefly, if N_r is presumed to be large and the instantaneous radiator phases are presumed to be uniformly distributed in the interval $[0, 2\pi]$, then the probability densities for $\text{Re}[E]$ and $\text{Im}[E]$ are found to obey independent Gaussian statistics with zero mean and equal variance. That is, the probability distribution governing the instantaneous complex field E is found to obey what are known as circular Gaussian statistics (see, for example, [5]). Accordingly, the probability density for the instantaneous intensity, given by $I = |E|^2$, converges to a negative exponential distribution

$$p_I(I) = \frac{1}{\langle I \rangle} \exp\left(-\frac{I}{\langle I \rangle}\right) \tag{7.18}$$

where the average value of I is given by $\langle I \rangle = N_r I_r$, in agreement with the first term in Eq. 7.16 (see Fig. 7.4). This distribution is manifestly spread over a wide range of intensity values, confirming that the second term in Eq. 7.16, while vanishing on average, does indeed cause significant fluctuations in the sampled realizations of I. To characterize the extent of these fluctuations, we can calculate their variance, obtaining

$$\sigma_I^2 = \langle I^2 \rangle - \langle I \rangle^2 = \int_0^\infty dI\, I^2 p_I(I) - \langle I \rangle^2 = \langle I \rangle^2 \tag{7.19}$$

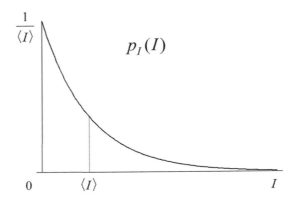

Figure 7.4. Negative-exponential probability density.

Notably, the standard deviation of these fluctuations, defined as the square root of the variance, is found to be as large as the average of I itself! Conventionally, the ratio of standard deviation to average is referred to as a contrast. Thus, from Eq. 7.19, the contrast of the instantaneous intensity fluctuations is given by

$$\frac{\sigma_I}{\langle I \rangle} = 1 \tag{7.20}$$

7.2.2 Gaussian Light

Though we have evaluated the statistics of the distribution of the instantaneous intensity $I(t)$, we have not yet considered how $I(t)$ actually varies in time. We turn our attention now to these temporal dynamics. Specifically, in the same way that temporal field fluctuations were characterized by

$$\Gamma(\bar{t}, \tau) = \langle E(t + \tau)E^*(t) \rangle \tag{7.21}$$

in Section 7.1, temporal intensity fluctuations are characterized here by

$$\Gamma_I(\bar{t}, \tau) = \langle I(t + \tau)I(t) \rangle = \langle E(t + \tau)E^*(t + \tau)E(t)E^*(t) \rangle \tag{7.22}$$

where, as always, $\langle ... \rangle$ signifies a time average much longer than the coherence time, and $\bar{t}$ makes allowances for the possibility of even slower variations in the mutual coherence.

A fundamental relation in statistical optics is called the Siegert relation, which states that for fields obeying quasi-stationary circular Gaussian statistics Eq. 7.22 reduces to

$$\Gamma_I(\tau) = \langle I \rangle^2 \left(1 + |\gamma(\tau)|^2 \right) \tag{7.23}$$

The Siegert relation is quite general and can be used to describe many types of light ranging from thermally driven incandescence, to fluorescence generated by atoms or molecules, to laser light scattering from macroscopic particles in Brownian motion, etc. Because the range of light types is so broad, it has garnered numerous designations, including Gaussian light [10], thermal light [5], chaotic light [13], etc. We will adopt the first of these designations to emphasize the reliance of the Siegert relation on circular Gaussian field statistics. Ultimately, the reason the Siegert relation applies so generally stems from what is known as the central limit theorem, which will be discussed in more detail below.

In the meantime, several conclusions can be drawn from Eq. 7.23. First, we can confirm that the variance of the instantaneous intensity fluctuations is indeed given by

$$\sigma_I^2 = \Gamma_I(0) - \langle I \rangle^2 = \langle I \rangle^2 \tag{7.24}$$

which we already knew from Eq. 7.20. However, the Siegert relation provides information not only about the variance of the instantaneous intensity fluctuations but also about their temporal dynamics. In particular, Eqs. 7.23 and 7.5 indicate that the fluctuations in $I(t)$ subsist for an average duration τ_γ. This is a remarkable result. In effect, the Siegert relation enables us to infer the *field* coherence time of the individual radiators based on a measurement of the *intensity* coherence time from a large number of these radiators. This conclusion presupposes that the radiator fields are temporally uncorrelated, as is the case in our simple random phase model. It also presupposes that they possess an identical frequency spectrum centered about a same carrier frequency. When these presuppositions are valid, the frequency spectrum is said to be homogeneously broadened.

In practice, however, it is often the case that the spectra of the individual radiators are in fact not centered on the same carrier frequencies. For example, radiators that undergo thermal motion emit fields with carrier frequencies that are Doppler shifted depending on their velocities. Assuming these velocities obey a Boltzmann distribution, the resulting frequency spectrum of the total field produced by a large number of radiators in thermal motion becomes Gaussian (as opposed to Lorentzian), and we have

$$\mathcal{S}(\nu) = \frac{1}{\sqrt{2\pi}\sigma_\nu} e^{-(\nu - \bar{\nu})^2/2\sigma_\nu^2} \tag{7.25}$$

where $\bar{\nu}$ is the average radiator carrier frequency and σ_ν defines the width of the spectrum. From Eq. 7.6, the coherence function associated with this spectrum is

$$\gamma(\tau) = e^{-i2\pi\bar{\nu}\tau} e^{-\pi\tau^2/2\tau_\gamma^2} \tag{7.26}$$

and the coherence time is

$$\tau_\gamma = \frac{1}{2\sqrt{\pi}\sigma_\nu} \tag{7.27}$$

Such a frequency spectrum is said to be inhomogeneously broadened. The Siegert relation still holds but $\gamma(\tau)$ now corresponds to a coherence function that characterizes the temporal dynamics not of the individual radiators but of the radiators as an ensemble. To recover the temporal dynamics of individual radiators from a knowledge of $\gamma(\tau)$ alone requires additional information about the actual distribution of carrier frequencies. In the specific

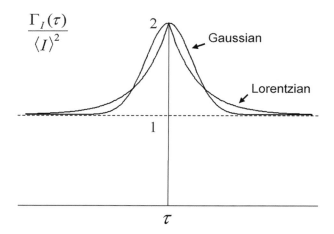

Figure 7.5. Intensity autocorrelation functions associated with Lorentzian and Gaussian spectral densities.

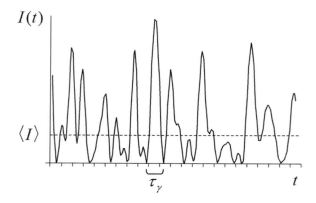

Figure 7.6. Instantaneous intensity associated with Gaussian spectral density.

example provided above involving thermal motion, this distribution is known to be the Boltzmann distribution; however, it could easily be otherwise depending on a variety of environmental factors or physical constraints on the radiators. The principle of using $\gamma(\tau)$ to recover the dynamics of individual radiators is the basis of an entire field of study called dynamic light scattering (DLS). For more information on this field the reader is referred to [2, 4, 16, 19, 21].

Figure 7.5 illustrates $\Gamma_I(\tau)$ for both Lorentzian and Gaussian frequency spectra. A representative trace of the intensity fluctuations for a Gaussian spectrum is shown in Fig. 7.6. As expected, the variations in this trace are quite large and subsist for a characteristic time τ_γ. Note that the most probable intensity value is zero. This is a direct consequence of the negative-exponential probability distribution (Eq. 7.18). A fluctuating intensity of the type depicted in Fig. 7.6 is sometimes referred to as temporal speckle (as opposed to spatial speckle, which we will examine below). Any source comprising a large number of radiators that emit light with temporally uncorrelated phases will produce such temporal speckle; however, in standard imaging applications involving self-luminous radiators (particularly fluorescent molecules) this temporal speckle is generally not observed, for reasons that will be explained in Section 7.3.

7.2.3 Köhler Illumination

Thus far, we have looked at the instantaneous intensity produced by a population of elemental radiators located at a same point in a source plane. Though this instantaneous intensity was found to be spatially uniform across the entire observation plane (assuming the observation plane was far enough that we could neglect field curvature), the actual value of the instantaneous intensity was found to vary significantly in time. We generalize our treatment now to the more realistic case where the source radiators are not confined to a single point but rather distributed over space. Moreover, instead of considering an observation plane far from the source plane, we turn to a more common geometry found in microscopy applications, namely the Köhler illumination configuration where the source and observation planes are separated by a lens in a 2f geometry. As we have already seen in Sections 3.2 and 4.3, this is essentially equivalent to considering planes that are far apart, but where the large separation distance z is replaced by the more tractable focal length f. Again, our discussion is based on the assumption that the fields produced by each radiator are uncorrelated in phase.

To begin, let us suppose that the physical size of each elemental radiator is much smaller than a wavelength and that the radiators are distributed over an area $A_0 \gg \kappa^{-2}$, corresponding to the total source area (see Fig. 7.7). The light produced at the source plane is then, by definition, incoherent. However, the resulting light at the observation (or Fourier) plane is, in general, partially coherent since the coherence area $A_{\mu 1}$ at this plane is larger than κ^{-2}. As we know

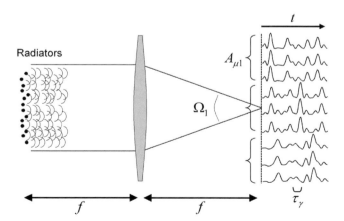

Figure 7.7. Field produced by many independent radiators in a 2f configuration.

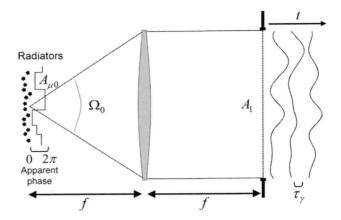

Figure 7.8. Field produced by many independent radiators in a 2f configuration.

from Sections 4.3 and 6.5.2, $A_{\mu 1}$ is defined by the solid angle Ω_1 incident on the observation and is roughly given by

$$A_{\mu 1} = \frac{f^2}{\kappa^2 A_0} \tag{7.28}$$

Moreover, from the reciprocity relations established in Section 4.3, if only a portion of the light is allowed to be recovered from the Fourier plane, say a portion that passes through an aperture of area A_1, then the *apparent* coherence area $A_{\mu 0}$ at the source plane is also larger than κ^{-2} (see Fig. 7.8). As a result, the phases of the radiator fields *appear* to be correlated over areas larger than the physical size of the radiators themselves, such that

$$A_{\mu 0} = \frac{f^2}{\kappa^2 A_1} \tag{7.29}$$

Again, it may seem surprising that many radiators whose phases are a priori uncorrelated can suddenly appear to be emitting light with a correlated phase; however, this is simply a consequence of the fact that the radiators, upon observation through the aperture A_1, cannot be resolved individually. The net phase they exhibit as a smallest resolvable group is that which results from the random-walk summation of each radiator phase encompassed in that group (i.e. in an area $A_{\mu 0}$). From Section 7.2.1, we found that if the coherence time of each individual, uncorrelated radiator phase is τ_γ, then so too is the coherence time of the net phase resulting from their random walk summation. As such, the radiators appear to be emitting with a same phase over an area $A_{\mu 0}$ and for a duration τ_γ.

Figure 7.8 suggests that the apparent source field can be described by what we refer to as a phase mosaic model. This model is the spatial analog of the interrupted phase model described in Section 7.2.1. That is, the instantaneous source amplitude is relatively uniform across the entire source area, but the instantaneous source phase is subdivided into a mosaic of correlation areas $A_{\mu 0}$. Within each of these correlation areas the phase is roughly uniform; however, across

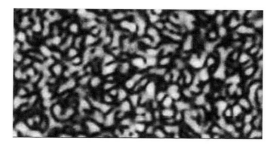

Figure 7.9. Speckle pattern.

different correlation areas the value of this phase is randomly distributed in the range $[0, 2\pi]$. If we assume that the values of the instantaneous phases in each correlation area are uncorrelated and fluctuate with a coherence time τ_γ, then Fig. 7.8 can be thought of as depicting a snapshot of the phase profile in time. This snapshot subsists for an average duration τ_γ, after which it becomes randomized again. The phase mosaic model presented here essentially reproduces the same results as the quasi-homogeneous beam model presented in Section 4.2.1, except that now it includes even instantaneous time scales, whereas in Section 4.2.1 a quasi-homogeneous beam was defined only over coarse time scales (i.e. averaged over times longer than τ_γ).

Based on a phase mosaic model for the source, we can now infer a resultant instantaneous intensity pattern at the observation plane. Since in a 2f configuration all the source correlation areas $A_{\mu 0}$ contribute with roughly equal weight to this plane, the instantaneous intensity at any arbitrary point in the observation plane can be calculated from a summation of $N_{\mu 0} = A_0/A_{\mu 0}$ fields, each with random phase, as was done in Eq. 7.17. Again, provided $N_{\mu 0}$ is large, we arrive at an instantaneous intensity that obeys the negative-exponential probability distribution given by Eq. 7.18, leading to the same conclusions regarding intensity variance and contrast as derived above. The difference with this present calculation is that we are now considering an instantaneous intensity that varies in space rather than in time. An example of an intensity pattern at the observation plane is depicted in Fig. 7.9. This is called a spatial speckle pattern, or speckle for short. The correlation area $A_{\mu 1}$ corresponds to the characteristic area of a speckle grain.

It should be emphasized that the speckle pattern illustrated in Fig. 7.9 corresponds to an instantaneous intensity resulting from a particular realization of a phase mosaic pattern at the source plane. This is not, in general, what would be observed with a standard detector at the observation plane since the lifetime of this pattern is only τ_γ, which can be quite a bit shorter than the detector response time (often by several orders of magnitude!). The effect of temporal filtering on intensity observation will be discussed below.

To tie our results with those from Chapter 4, we recall that upon time averaging over times longer than τ_γ, the phase mosaic model reduces to the quasi-homogeneous beam model in Section 4.2.1. That is, we can express the source mutual intensity as

$$J_0(\vec{\rho}_{0c}, \vec{\rho}_{0d}) = \langle I_0(\vec{\rho}_{0c}) \rangle \, \mu_0(\vec{\rho}_{0d}) \tag{7.30}$$

where the coherence function $\mu_0(\vec{\rho}_d)$ spans the correlation area $A_{0\mu}$ over which the phase front is spatially uniform but temporally fluctuating (time averaging has now been made explicit). Following the same formalism developed in Chapter 4, we can also derive a corresponding $\mu_1(\vec{r}_{1d})$ at the observation plane.

Moreover, just as we defined an intensity correlation function in time (Eq. 7.22), we can define an intensity correlation function in space, here generalized to 3D:

$$J_I(\vec{r}_c, \vec{r}_d) = \left\langle I(\vec{r}_c + \tfrac{1}{2}\vec{r}_d) I(\vec{r}_c - \tfrac{1}{2}\vec{r}_d) \right\rangle \tag{7.31}$$

Correspondingly, provided the instantaneous field throughout the observation plane obeys quasi-stationary circular Gaussian statistics, we arrive at a spatial analog of the Siegert relation (Eq. 7.23) given by

$$J_I(\vec{r}_d) = \langle I \rangle^2 \left(1 + \left|\mu(\vec{r}_d)\right|^2\right) \tag{7.32}$$

where the subscript 1 referring to the observation plane has been dropped. Note that $J_I(\vec{r}_d)$ here is an intensity correlation function and should not be confused with a mutual intensity, which is a field correlation function. Nevertheless, $J_I(\vec{r}_d)$ depends on the field correlations through $\left|\mu(\vec{r}_d)\right|^2$. In effect, the volume encompassed by $\left|\mu(\vec{r}_d)\right|^2$ is the characteristic intensity correlation volume of a speckle grain.

Spatiotemporal Coherence

From the preceding discussions, we conclude that correlations in light subsist both in time and space. Both of these are, of course, coupled by the wave function that governs the propagation of light (see Chapter 2). To gain an intuitive picture of this coupling, we can define a coherence span in time and space. Since the former, when multiplied by the speed of light, leads to an axial coherence length, this is equivalent to defining the axial and transverse dimensions of a 3D coherence volume [20].

In fact, we have already derived an expression for the 3D mutual coherence of light arising from an incoherent source distribution in a 2f configuration, which is given in Eq. 4.63. But this expression was derived under the assumption that the light was quasi-monochromatic, with a well-defined wavenumber κ. In this chapter, we specifically relax this assumption of quasi-monochromaticity and allow the light to feature a range of wavenumbers $\Delta\kappa$, centered about an average wavenumber $\bar{\kappa}$. The relationship between $\bar{\kappa}$ and the average frequency $\bar{\nu}$ is straightforward and given by Eq. 1.3, or, written differently, $\bar{\nu} = \upsilon_p \bar{\kappa}$, where $\upsilon_p = \frac{c}{n(\nu)}$ is called the phase velocity of light. As always, the medium through which the light is travelling is characterized by its index of refraction $n(\nu)$, where we have made allowances that the index of refraction might be frequency dependent. The relation between $\Delta\kappa$ and $\Delta\nu$ is somewhat more involved, and given by $\Delta\nu = \upsilon_g \Delta\kappa$, where υ_g is called the group velocity of light, which obeys

$$\frac{\upsilon_g}{\upsilon_p} = 1 - \bar{\nu} \left.\frac{dn}{d\nu}\right|_{\bar{\nu}} \tag{7.33}$$

In most cases, $\frac{dn}{d\nu}$ is a very small quantity that is positive when the light propagation through the medium obeys what is called "normal" dispersion. In other words, in most cases v_g is almost the same as v_p, though slightly smaller (see [8] for a more detailed discussion on the concept of group velocity).

With these definitions of $\bar{\kappa}$ and $\Delta\kappa$, we can gain a picture of the frequency support associated with Köhler illumination. For monochromatic light of wavenumber $\bar{\kappa}$, this can be determined by taking a 3D Fourier transform of $\mu(\vec{r}_d)$, which, from Eq. 4.63 and replacing κ by $\bar{\kappa}$, leads to

$$\hat{\mu}\left(\vec{\kappa}\right) = \frac{1}{\Phi_0} \int d^2\vec{\rho}_{0c}\, \delta\left(\bar{\kappa} - \kappa_z - \frac{\bar{\kappa}}{2f^2}\rho_{0c}^2\right) \delta^2\left(\vec{\kappa}_\perp + \frac{\bar{\kappa}}{f}\vec{\rho}_{0c}\right) I_0\left(\vec{\rho}_{0c}\right) \tag{7.34}$$

We observe here that $\vec{\kappa}$ is restricted to a parabolic cap defined by $\kappa_\perp = -\frac{\bar{\kappa}}{f}\rho_{0c}$ and $\kappa_z = \bar{\kappa}\left(1 - \frac{1}{2}\frac{\rho_{0c}^2}{f^2}\right)$. In fact, this parabolic cap is an approximation (Fresnel) to the more accurate surface defined by a spherical cap (Rayleigh–Sommerfeld), defined by $\kappa_\perp = -\bar{\kappa}\sin\theta_1$ and $\kappa_z = \bar{\kappa}\cos\theta_1 \approx \bar{\kappa}\left(1 - \frac{1}{2}\sin^2\theta_1\right)$, where θ_1 is the half-angle spanned by Ω_1 in Fig. 7.7. This spherical cap corresponds to the frequency support associated with monochromatic Köhler illumination of wavenumber $\bar{\kappa}$. But here we are considering non-monochromatic Köhler illumination comprising a range of wavenumbers $\Delta\kappa$. Making the key assumption that the frequency supports associated with each wavenumber κ are uncorrelated, the overall non-monochromatic frequency support consists simply of an incoherent superposition of the constituent monochromatic frequency supports, as illustrated in Fig. 7.10. In other words, the non-monochromatic frequency support comprises a full volume in κ-space, rather than simply a surface. In turn, this volume defines the spatiotemporal coherence of the light, where by spatial coherence we mean the coherence associated with the diversity of wavevector directions, and by temporal coherence we mean the coherence associated with the diversity in wavevector magnitudes. This distinction is made clear when we consider the overall spans of the non-monochromatic frequency support in the transverse and axial directions, given by

$$\Delta\kappa_\perp = 2\bar{\kappa}\sin\theta_1 + \Delta\kappa\sin\theta_1 \tag{7.35}$$

$$\Delta\kappa_z = 2\bar{\kappa}\sin^2\frac{\theta_1}{2} + \Delta\kappa\cos^2\frac{\theta_1}{2} \tag{7.36}$$

We observe that each of these spans consists of two parts, the first associated with spatial coherence and the second with temporal coherence. We can loosely define transverse and axial coherence lengths of the light to be given by the inverse of these spans, that is, $l_\perp = \frac{1}{\Delta\kappa_\perp}$ and $l_z = \frac{1}{\Delta\kappa_z}$ respectively. In particular, temporally coherent Köhler illumination ($\Delta\kappa = 0$) exhibits a finite coherence length (partial coherence) in both the transverse and axial directions. As expected, the transverse coherence length is the same we have seen before in Eqs. 3.31 and 4.84. When a spectral bandwidth is introduced to this Köhler illumination ($\Delta\kappa > 0$) these coherence lengths become smaller. For example, in the case of Köhler illumination with very

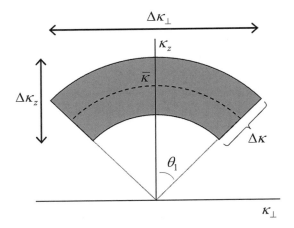

Figure 7.10. Frequency support for spatially incoherent monochromatic illumination.

low NA ($\theta_1 \approx 0$) the axial coherence length collapses from a very long length to one that scales inversely with $\Delta\kappa$. It should be noted that the frequency support depicted in Fig. 7.10 is that associated with $\mu(\vec{r}_d)$, which is a correlation function involving fields. The intensity correlation function obtained by the spatial Siegert relation (Eq. 7.32) is defined by $\left|\mu(\vec{r}_d)\right|^2$ and thus exhibits an even larger frequency support.

7.2.4 Central Limit Theorem

The Siegert relations expressed in Eqs. 7.23 and 7.32 are tremendously useful when considering randomly fluctuating fields, in part because of their broad ranges of validity. So far, we have concentrated on fields produced by radiating sources that randomly fluctuate in phase but not in amplitude. That is, in calculating the summation in Eq. 7.17, we took the amplitude of each radiator field to be fixed (i.e. E_r). As it turns out, this condition of fixed amplitude is more stringent than required and we can generalize our random phase model to also include statistically varying amplitudes, provided certain conditions are met. These conditions are presented below.

A well-known principle in statistics is the central limit theorem, which states that the fluctuations of many independent random variables, when added, exhibit Gaussian statistics (see, for example, [1] and [5]). A key feature of this theorem is that the statistics of the individual random variables need not themselves be Gaussian for the theorem to hold – they need only be quasi-stationary and independent. In fact, we tacitly made use of this theorem in the derivation of Eq. 7.18 from Eq. 7.17. Indeed, the projections of each radiator field in Eq. 7.17 along the real or imaginary axes are given by $E_r \cos\phi_n$ and $E_r \sin\phi_n$ respectively, with E_r taken to be fixed and ϕ_n to be uniformly random. Clearly these projections do not individually obey Gaussian statistics, and yet upon summation of a large number of these projections, the distributions of the summed projections $\mathrm{Re}[E]$ and $\mathrm{Im}[E]$ do converge to

Gaussian statistics, by virtue of the central limit theorem. Moreover, since these distributions are independent, of zero mean and of equal variance, the final distribution of the summed complex field E converges to circular Gaussian statistics, from which the Siegert relations (Eq. 7.23 and 7.32) were ultimately derived.

It is important to note, however, that a convergence to circular Gaussian statistics would have been attained even if E_r had not been fixed but instead had been allowed to vary independently from radiator to radiator. In other words, the Siegert relations derived for the interrupted phase model are also valid when this model is generalized to include varying amplitudes, regardless of the statistics of the amplitude variations, provided only that these are quasi-stationary and independent, and that the number of radiators is large. For this reason, the Siegert relations apply quite generally.

Nevertheless, to underline the importance of the conditions of quasi-stationarity and independence, let us consider a source for which the Siegert relations fail. In particular, let us consider a source consisting of a large number of radiators whose phases are random but whose amplitudes fluctuate in a perfectly correlated manner. That is, the total field produced by such radiators is given by

$$E(t) = E_r(t)e^{-i2\pi\bar{\nu}t}\sum_{n}^{N_r} e^{i\phi_n(t)} \tag{7.37}$$

where $E_r(t)$ is no longer constant but varies in time, randomly or otherwise. Thus, at any instant in time, the same $E_r(t)$ is associated with all radiators. Such a situation is encountered, for example, if the radiators are driven by an incident field that is itself time-varying. The resulting instantaneous intensity produced by the radiators is then

$$I(t) = I_r(t)\left|\sum_{n}^{N_r} e^{i\phi_n(t)}\right|^2 \tag{7.38}$$

which contains two time dependences, one arising from the incident field intensity ($I_r(t)$), and another arising from the summation of random radiator phases. We know how to calculate the statistics of the latter from our 2D random walk model. On the other hand, the statistics of the former depend on the incident field. For example, if $I_r(t)$ randomly fluctuates in time independently of the phase fluctuations, then Eq. 7.38 is a product of two statistically independent random variables. From Eq. B.8, we find then

$$\frac{\sigma_I}{\langle I\rangle} = \sqrt{1 + 2\frac{\sigma_{I_r}^2}{\langle I_r\rangle^2}} \tag{7.39}$$

which clearly does not satisfy the Siegert relation since $\frac{\sigma_I}{\langle I\rangle}$ is now greater than 1. In the still more specific case where $I_r(t)$ is itself governed by negative-exponential statistics (Eq. 7.18), the contrast of the instantaneous intensity fluctuations reduces to $\sqrt{3}$. Such a case is sometimes referred to as "speckled speckle," which is discussed in more detail in [6]. Various other cases for which the Siegert relations do not apply are discussed, for example, in [11] and [12].

7.3   FILTERED FLUCTUATIONS

The instantaneous intensity statistics derived above are idealized in the sense that they do not take into account the practical limitations of actual intensity measurements. In particular, physical detectors are neither infinitely fast nor infinitely small in size. Because detectors have finite electrical bandwidths and non-zero surface areas, intensity measurements are inevitably filtered both in time and space. The effects of such filtering can be negligible or severe depending on the scales of the intensity variations of interest. The purpose of this section is to derive modified Siegert relations that takes such filtering into account. In particular, we will examine the effects of filtering on the measured contrast of the intensity fluctuations.

7.3.1 Temporal Filtering

We begin by considering the effect of a finite detector bandwidth. To isolate this effect, we assume for the moment that the detector is infinitely small in area, meaning that it provides a measure of localized intensity fluctuations (as opposed to power fluctuations, which will be considered in the next section). A finite detector bandwidth leads to a temporal filtering of the instantaneous intensity $I(t)$ such that

$$I_T(t) = \int dt' \, R_T(t - t')I(t') \tag{7.40}$$

where $I_T(t)$ is the measured intensity and $R_T(t)$ is the detector response function. This function is assumed to be time invariant (stationary). Moreover, it is assumed to be real and normalized so that

$$\int dt \, R_T(t) = 1 \tag{7.41}$$

Finally, for the response function to be causal, it must obey the constraint $R_T(t < 0) = 0$.

Following our usual convention, the detector integration time associated with the response function is defined to be

$$T = \frac{\left| \int dt \, R_T(t) \right|^2}{\int dt \, R_T^2(t)} = \frac{1}{\int dt \, R_T^2(t)} \tag{7.42}$$

Having established the relationship between measured and instantaneous intensities (Eq. 7.40), we can now derive the relationship between measured and instantaneous intensity correlation functions, the former being denoted by

$$\bar{\Gamma}_I(\tau) = \langle I_T(t + \tau)I_T(t) \rangle \tag{7.43}$$

A straightforward calculation leads to

$$\bar{\Gamma}_I(\tau) = \int d\tau' \, Q_T(\tau - \tau')\Gamma_I(\tau') \tag{7.44}$$

where we have introduced the response autocorrelation function

$$Q_T(\tau) = \int dt'\, R_T(t' + \tau) R_T(t') \tag{7.45}$$

We note that $Q_T(\tau)$ is a symmetric function $(Q_T(\tau) = Q_T(-\tau))$, and obeys the normalization conditions

$$\int d\tau\, Q_T(\tau) = 1 \tag{7.46}$$

$$Q_T(0) = \frac{1}{T} \tag{7.47}$$

Equations 7.44 and 7.45 apply in general. However, we are interested here in their effect on the temporal Siegert relation. In other words, we specifically consider the case where $\Gamma_I(\tau)$ satisfies Eq. 7.23. We find then

$$\bar{\Gamma}_I(\tau) = \langle I_T \rangle^2 \left[1 + \int d\tau'\, Q_T(\tau - \tau')\, |\gamma(\tau')|^2 \right] \tag{7.48}$$

where we have made use of the relation $\langle I_T(t) \rangle = \langle I(t) \rangle$ that comes from Eq. 7.41.

Equation 7.48 represents the modification of the temporal Siegert relation that comes from performing actual measurements. We observe that the function $|\gamma(\tau)|^2$ is filtered here, not by the detector response function $R_T(t)$ but rather by the autocorrelation of this function $Q_T(\tau)$.

We can go a step further by deriving the contrast of the measured intensity fluctuations. We start by defining the variance

$$\sigma_{I_T}^2 = \langle I_T^2 \rangle - \langle I_T \rangle^2 = \langle I_T \rangle^2 \int d\tau'\, Q_T(\tau')\, |\gamma(\tau')|^2 \tag{7.49}$$

The functions $Q_T(\tau)$ and $|\gamma(\tau)|^2$ specify different time scales. The characteristic width of $Q_T(\tau)$ is the detector integration time T; the characteristic width of $|\gamma(\tau)|^2$ is the coherence time τ_γ (from Eq. 7.5). Defining N_γ to be the number of coherence times effectively "seen" by the detector response, such that

$$N_\gamma = \begin{cases} 1 & \text{if } \frac{T}{\tau_\gamma} \leq 1 \\ \frac{T}{\tau_\gamma} & \text{if } \frac{T}{\tau_\gamma} > 1 \end{cases} \tag{7.50}$$

the contrast of the measured fluctuations is finally given by

$$\frac{\sigma_{I_T}}{\langle I_T \rangle} \simeq \frac{1}{\sqrt{N_\gamma}} \tag{7.51}$$

We conclude that in the extreme case where the detector is indeed very fast and able to track intensity fluctuations instantaneously, then $N_\gamma = 1$ and we recover the contrast associated with temporal speckle given by Eq. 7.20. On the other hand, if the detector is slow, meaning that N_γ is large, then the measured intensity fluctuations become severely filtered compared to

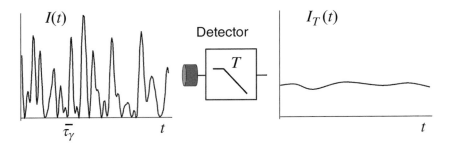

Figure 7.11. Instantaneous intensity and temporally filtered measured intensity.

the instantaneous intensity fluctuations incident on the detector. The latter case is illustrated in Fig. 7.11.

As a practical example, a detector bandwidth of the order 1 GHz is quite fast. This corresponds to a detector integration time $T \approx 10^{-9}$ s which, nevertheless, is several orders of magnitude slower than the typical coherence times associated with radiation from atoms or molecules (see Section 7.1). As such, the rapid intensity fluctuations intrinsic to atomic or molecular radiation are, in practice, so severely filtered as to be undetectable. In contrast, the coherence times associated with light scattering from macroscopic particles in Brownian motion are typically on the order of 10^{-2} to 10^{-4} s, which can easily be tracked by standard detectors.

For reference, a standard example of temporal filtering is with a boxcar gate defined by

$$R_T(t) = \begin{cases} 0 & t \le 0 \\ \frac{1}{T} & 0 < t \le T \\ 0 & t > T \end{cases} \tag{7.52}$$

leading to

$$Q_T(\tau) = \begin{cases} \frac{1}{T}\left(1 - \frac{|\tau|}{T}\right) & \text{for } |\tau| \le T \\ 0 & \text{otherwise} \end{cases} \tag{7.53}$$

7.3.2 Spatial Integration

Following the same procedure we used to take into account temporal filtering, we can now take into account spatial integration of the measured intensity caused by a non-zero detector size, noting that, in effect, such integration is equivalent to spatial filtering. Introducing the spatial window function of the detector area, again assumed to be real and normalized, the total power measured by the detector becomes

$$\Phi_A(t) = A \int d^2\vec{\rho}\, R_A(\vec{\rho}) I(\vec{\rho}, t) \tag{7.54}$$

where $I(\vec{\rho}, t)$ is the instantaneous local intensity at the detector plane, A is the detector area, and $R_A(\vec{\rho})$ is a (real) detector window function. In analogy with Eqs. 7.42 and 7.45, the detector area is defined to be

$$A = \frac{\left|\int d^2\vec{\rho}\, R_A(\vec{\rho})\right|^2}{\int d^2\vec{\rho}\, R_A^2(\vec{\rho})} = \frac{1}{\int d^2\vec{\rho}\, R_A^2(\vec{\rho})} \tag{7.55}$$

and the detector spatial autocorrelation function becomes

$$Q_A(\vec{\rho}_d) = \int d^2\vec{\rho}_c\, R_A(\vec{\rho}_c + \tfrac{1}{2}\vec{\rho}_d) R_A(\vec{\rho}_c - \tfrac{1}{2}\vec{\rho}_d) \tag{7.56}$$

which is symmetric ($Q_A(\vec{\rho}_d) = Q_A(-\vec{\rho}_d)$) and obeys the normalization conditions

$$\int d^2\vec{\rho}_d\, Q_A(\vec{\rho}_d) = 1 \tag{7.57}$$

$$Q_A(0) = \frac{1}{A} \tag{7.58}$$

In the specific case where the beam is quasi-homogeneous such that its local intensity across the detector area is approximately uniform when averaged over time (i.e. $\langle \Phi_A \rangle = A \langle I \rangle$), and where the spatial Siegert relation (Eq. 7.32) is presumed valid, we obtain finally

$$\bar{J}_{\Phi_A}(\vec{\rho}_d) = A^2 \langle I \rangle^2 \left[1 + \int d^2\vec{\rho}'\, Q_A(\vec{\rho}_d - \vec{\rho}') \left| \mu(\vec{\rho}') \right|^2 \right] \tag{7.59}$$

Note that the spatial variations are considered here only in two dimensions, in accord with the detector geometry. Equation 7.59 is the spatial equivalent of Eq. 7.48 where now the detector area A plays the role of T, and the coherence area A_μ plays the role of τ_γ. Correspondingly, we can define N_μ to be the number of coherence areas (or speckle-grain areas) "seen" by the detector area, such that

$$N_\mu = \begin{cases} 1 & \text{if } \frac{A}{A_\mu} \leq 1 \\ \frac{A}{A_\mu} & \text{if } \frac{A}{A_\mu} > 1 \end{cases} \tag{7.60}$$

leading to a contrast of the measured power fluctuations given by

$$\frac{\sigma_{\Phi_A}}{\langle \Phi_A \rangle} \simeq \frac{1}{\sqrt{N_\mu}} \tag{7.61}$$

We conclude from Eq. 7.61 that if the detector area is so small as to record only from a single beam coherence area, then the measured power fluctuations exhibit the same contrast as the local intensity fluctuations, namely those associated with temporal speckle. On the other hand, if the detector area is much larger than the beam coherence area, then the intensity fluctuations associated with each coherence area become filtered when they are detected as power fluctuations, even in the idealized case where the detector is instantaneously fast (see Fig. 7.12). This conclusion, however, rests on the assumption that the intensity fluctuations

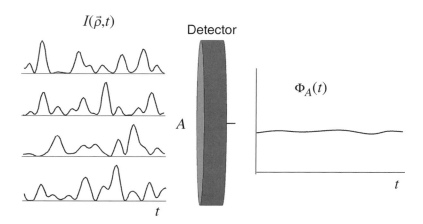

$I(\vec{\rho},t)$

Figure 7.12. Local instantaneous intensity and spatially integrated measured power.

associated with each coherence area are *uncorrelated*, as is required for the spatial Siegert relation to be valid. More will be said about the relation between coherence and correlation below.

Finally, if the measured power Φ_T is both temporally filtered and spatially integrated we have

$$\Phi_T(t) = A \iint d^2\vec{\rho}\, dt'\, R_T(t - t')R_A(\vec{\rho})I(\vec{\rho}, t') \tag{7.62}$$

leading to, ultimately,

$$\frac{\sigma_{\Phi_T}}{\langle \Phi_T \rangle} \simeq \frac{1}{\sqrt{N_\gamma N_\mu}} \tag{7.63}$$

which again rests on the assumption that the temporal and spatial fluctuations are uncorrelated.

7.4   COHERENCE EXAMPLES

Liberal use has been made throughout this chapter of the term coherence. To gain a more intuitive notion of what is meant by this term, let us consider two arbitrary points $\vec{r}$ and $\vec{r}'$ within a light beam. The fields at these points are said to be coherent if they are correlated with one another over the course of a measurement. That is, the quantity $\langle E(\vec{r}) E^*(\vec{r}') \rangle$ should be significantly different from zero, and the phrase "over the course of a measurement" is codified by the brackets $\langle ... \rangle$, which indicate a temporal averaging. In particular, coherence is ensured if the phase *difference* between the fields $E(\vec{r})$ and $E(\vec{r}')$ remains the same over the course of a measurement. This is made clear with examples.

As a first example, let us consider a strictly monochromatic beam, such as an idealized laser beam, and let us consider two points $\vec{\rho}$ and $\vec{\rho}'$ in a cross-sectional area of the beam. The phases at these points are governed by $e^{-i2\pi\bar{\nu}t}$, meaning that the phase difference between these two points is identically zero. These two points are coherent with one another. In fact, the coherence area of the beam spans the entire beam cross-section.

Now let us send this beam through a thin diffuser plate that is fixed, and consider the two points to be on the diffuser surface. The phases imparted on the two fields by the diffuser are given by $\varphi\left(\vec{\rho}\right)$ and $\varphi\left(\vec{\rho}'\right)$. The phase difference is thus $\varphi\left(\vec{\rho}\right) - \varphi\left(\vec{\rho}'\right)$, always. These two points remain coherent with one another, even though the phases themselves are unknown, and, since the same can be said for any two points in the beam, the coherence area again spans the entire beam cross-section.

Now let us laterally translate the diffuser plate such that the phases $\varphi\left(\vec{\rho}\right)$ and $\varphi\left(\vec{\rho}'\right)$ become periodically randomized. In this case, the phase difference $\varphi\left(\vec{\rho}\right) - \varphi\left(\vec{\rho}'\right)$ fluctuates randomly in time and the two points now become incoherent. However, one can imagine that if $\vec{\rho}$ and $\vec{\rho}'$ are in close proximity, closer than the characteristic grit size of the diffuser, then $\varphi\left(\vec{\rho}\right) \approx \varphi\left(\vec{\rho}'\right)$, and the fields regain coherence. In other words, the coherence area of the beam has collapsed now to roughly the diffuser grit area (actually, smaller than this if the grit pitch is large – see [6]). In effect, we have reproduced here the phase mosaic model described in Section 7.2.3. Strictly speaking, the beam emerging from the diffuser is no longer monochromatic since it exhibits local phase variations that vary in time. Instead, it is more correctly referred to as quasi-monochromatic (see [17] for a highly pedaogical demonstration). We will make ample use of such beams throughout this book.

Let us complicate matters still further by replacing the monochromatic laser with a distant non-monochromatic point source of the type described in Section 7.2.1. The light from such a source is no longer temporally coherent. Nevertheless, provided the source is far way, the light is spatially coherent because its phase front, even though it is rapidly and randomly fluctuating with time, remains flat. Upon transmission through the diffuser, the phase difference between the two points remains $\varphi\left(\vec{\rho}\right) - \varphi\left(\vec{\rho}'\right)$, independent of the fluctuating phase of the incident light. That is, if the diffuser is fixed, the beam emerging from the diffuser remains fully spatially coherent across the entire diffuser surface. On the other hand, if the diffuser is translating, the coherence area collapses again to a small size.

It should be clear from the above examples that a temporally coherent (e.g. monochromatic) beam can be spatially incoherent, while a temporally incoherent (e.g. non-monochromatic) beam can be spatially coherent. In general, beams exhibit partial coherence, both temporal and spatial. But it should also be clear that the coherence of light depends on how it is measured. We saw this in Sections 6.4.1 and 7.2.3, where apertures were found to play a critical role in defining an apparent coherence area. We see this here as well, where the measurement bandwidth affects the coherence area. If the measurement is fast, with an integration time shorter than τ_γ, then the phase fluctuations are effectively frozen in time and the coherence area appears large. On the other hand, if the measurement detector is slow, then areas that appear spatially coherent with a fast detector begin to appear spatially incoherent. One might say that coherence is in the eye of the beholder.

7.5   PROBLEMS

Problem 7.1

Non-monochromatic fields can be described by explicitly taking into account their time dependence. It can be shown that when the time dependence of a field is made explicit, the radiative Rayleigh–Sommerfeld diffraction integral (Eq. 2.45 and 2.47) can be re-written in the form

$$E(\vec{\rho}, z, t) = -i\bar{\kappa} \int d^2\vec{\rho}_0 \frac{\cos\theta}{r} E(\vec{\rho}_0, 0, t - r/c)$$

which is valid for narrowband fields whose wavenumber is centered around $\bar{\kappa}$ (assuming propagation in vacuum). This expression can be simplified using the Fresnel approximation (Section 2.4.1). Based on this expression, evaluate the intensity distribution $I(\vec{\rho}, z)$ a distance z from two pinholes irradiated by a beam $I_0(\vec{\rho}_0, 0)$ that is partially coherent both in space and time. In particular, assume that the irradiating beam is both quasi-homogeneous and quasi-stationary, with a separable mutual coherence function given by

$$\Gamma(\vec{\rho}, \vec{\rho}'; t, t + \tau) = \langle I_0 \rangle \, \mu(\rho_d)\gamma(\tau)$$

where $\rho_d = |\vec{\rho} - \vec{\rho}'|$, and $\mu(\rho_d)$ and $\gamma(\tau)$ are Gaussian. That is, we have

$$\mu(\rho_d) = e^{-\rho_d^2/2\rho_\mu^2}$$

$$\gamma(\tau) = e^{-i2\pi\bar{\nu}\tau}e^{-\pi\tau^2/2\tau_\gamma^2}$$

where $\bar{\nu} = \bar{\kappa}c$.

The pinholes are separated by a distance a along the x direction (see figure below).

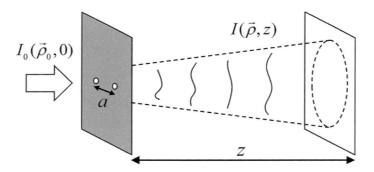

(a) Consider only the x direction and derive an expression for $I(x, z)$. Your expression should look something like

$$I(x, z) \propto \frac{1}{z^2} \langle I_0 \rangle \left[1 + M(x) \cos 2\pi x/p\right]$$

representing a fringe pattern of modulation $M(x)$ and period p.

(b) What is the maximum modulation strength $M(x)_{max}$? What happens to this strength as ρ_μ or τ_γ tends toward infinity? Does this strength depend on z?

(c) What is the period p of the fringes? Express your answer in terms of λ_0 and $\theta = \frac{a}{z}$, corresponding to the angle subtended by the pinholes.

(d) How far do the fringes extend in x? Specifically, at what value $x_{1/e}$ does the modulation strength decrease by a factor of $1/e$ relative to its maximum? Express your answer in terms of θ and the coherence length $l_\gamma = \tau_\gamma c$. Does $x_{1/e}$ depend on ρ_μ?

Problem 7.2

Consider a light beam that randomly switches between two states of intensities I_A and I_B. Let $P_A(t)$ and $P_B(t)$ be the probabilities that the beam is in states A and B respectively, such that

$$\frac{d}{dt}P_A(t) = -\lambda P_A(t) + \mu P_B(t)$$

where λ and μ are switching rate constants. Such a beam is called a random telegraph wave.

(a) Show that the mean intensity of the beam is given by

$$\langle I \rangle = \frac{\mu I_A + \lambda I_B}{\mu + \lambda}$$

(b) Show that the intensity variance of the beam is given by

$$\sigma_I^2 = \frac{\mu \lambda (I_A - I_B)^2}{(\mu + \lambda)^2}$$

Problem 7.3

A technique of laser speckle contrast analysis can be used to assess blood flow within tissue. In this technique, laser light is back-scattered from tissue, and a camera is used to record the resultant speckle pattern (assumed to obey circular Gaussian field statistics). Any motion in the tissue causes the speckle pattern to fluctuate in time. By measuring the contrast of these fluctuations as a function of the camera exposure time T one can deduce a temporal coherence time τ_γ. The local blood flow velocity can then be inferred from τ_γ, provided one is equipped with a theoretical model relating the two.

(a) The coherence function of light scattered from randomly flowing particles is often assumed to obey the statistics of a phase-interrupted source (see Eq. 7.11). Show that the expected contrast of the measured speckle fluctuations is given by

$$\frac{\sigma_{I_T}}{\langle I \rangle} = \sqrt{\frac{\tau_\gamma}{T}\left(1 - \frac{\tau_\gamma}{2T}\left(1 - e^{-2T/\tau_\gamma}\right)\right)}$$

(b) Verify that when $\tau_\gamma \ll T$ the contrast obeys the relation given by Eq. 7.51.

Problem 7.4

Consider the intensity distribution $I(\vec{\rho})$ at the image plane of a unit-magnification imaging system whose point spread function is written $\mathrm{PSF}(\vec{\rho})$. This intensity distribution is detected by a camera, which consists of a 2D array of detectors (pixels), each of area $A = p \times p$. As a result, $I(\vec{\rho})$ becomes integrated upon detection, and then sampled. The detected power, prior to sampling, can thus be written as

$$\Phi_A(\vec{\rho}) = A \int \mathrm{d}^2\vec{\rho}' \, R_A(\vec{\rho} - \vec{\rho}')I(\vec{\rho}')$$

(a) Provide expressions for $R_A(\vec{\rho})$ and its Fourier transform $\mathcal{R}_A(\vec{\kappa}_\perp)$.

(b) Let the intensity distribution at the object plane $I_0(\vec{\rho}_0)$ have the statistics of a "fully developed" speckle pattern. It can be shown (e.g. see Eq. 4.62) that the coherence function of a such a speckle pattern is given by

$$\left| \mu_0(\vec{\rho}_{0d}) \right|^2 = \frac{\mathrm{PSF}_s(\vec{\rho}_{0d})}{\mathrm{PSF}_s(0)}$$

where PSF_s is the point spread function associated with the speckle generation (not necessarily the same as PSF).

Express the spatial contrast of the imaged speckle pattern recorded by the camera in terms of $\mathcal{R}_A(\vec{\kappa}_\perp)$, $\mathrm{OTF}(\vec{\kappa}_\perp)$ and $\mathrm{OTF}_s(\vec{\kappa}_\perp)$.

(c) What happens to the above contrast as the size of the camera pixels becomes much larger than the spans of both $\mathrm{PSF}(\vec{\rho})$ and $\mathrm{PSF}_s(\vec{\rho})$?

Problem 7.5

Derive Eqs. 7.35 and 7.36 from Fig. 7.10.

References

[1] Barrett, H. H. and Myers, K. J. *Foundations of Image Science*, Wiley-Interscience (2004).

[2] Berne, B. J. and Pecora, R. *Dynamic Light Scattering: with Applications to Chemistry, Biology, and Physics*, Dover (2000).

[3] Born, M. and Wolf, E. *Principles of Optics*, 6th edn, Cambridge University Press (1980).

[4] Clark, N. A., Lunacek, J. H. and Benedek, G. B. "A study of Brownian motion using light scattering," *Am. J. Phys.* 38, 575–585 (1970).

[5] Goodman, J. W. *Statistical Optics*, 2nd edn, Wiley (2015).

[6] Goodman, J. W. *Speckle Phenomena in Optics: Theory and Applications*, Roberts and Company (2007).

[7] Harris, M. "Light-field fluctuations in space and time," *Contemp. Phys.* 36, 215–233 (1995).

[8] Hecht, E. *Optics*, 4th edn, Addison Wesley (2002).

[9] Ishimaru, A. *Wave Propagation and Scattering in Random Media*, IEEE and Oxford University Press (1997).

[10] Jakeman, E. and Pike, E. R. "The intensity-fluctuation distribution of Gaussian light," *J. Phys.* A 1, 128–138 (1968).

[11] Jakeman, E. "On the statistics of K-distributed noise," *J. Phys. A: Math. Gen.* 13, 31–48 (1980).

[12] Lemieux, P.-A. and Durian, D. J. "Investigating non-Gaussian scattering processes by using nth-order intensity correlation functions," *J. Opt. Soc. Am. A* 16, 1651–1664 (1999).

[13] Loudon, R. *The Quantum Theory of Light*, 3rd edn, Oxford University Press (2000).

[14] Mandel, L. and Wolf, E. *Optical Coherence and Quantum Optics*, Cambridge University Press (1995).

[15] Mandel, L. and Wolf, E. Eds., *Selected Papers on Coherence and Fluctuations of Light, Vols. 1 and 2*, Dover (1970).

[16] Maret, G. "Recent experiments on multiple scattering and localization of light," in *Les Houches Session LXI 1994, Physique Quantique Mésoscopique*, Eds. E. Akkermans, G. Montambaux, J.-L. Pichard, and J. Zinn-Justin, Elsevier Science (1995).

[17] Martienssen, W. and Spiller, E. "Coherence and fluctuations in light beams," *Am. J. Phys.* 32, 919–926 (1964).

[18] Montroll, E. W. and West, B. J. "On an enriched collection of stochastic processes," in *Fluctuation Phenomena*, Eds. E. W. Montroll and J. L. Lebowitz, North-Holland Publishing (1979).

[19] Pine, D. J., Weitz, D. A., Zhu, J. X. and Herbolzheimer, E. "Diffusing wave spectroscopy: dynamic light scattering in the multiple scattering limit," *J. Phys. France* 51, 2101–2127 (1990).

[20] Ryabukho, V. P., Lyakin, D. V., Grebenyuk, A. A. and Klykov, S. S. "Wiener-Khintchin theorem for spatial coherence of optical wave field," *J. Opt.* 15, 025405 (2013).

[21] Weitz, D. A. and Pine, D. J. "Diffusive wave spectroscopy," in *Dynamic Light Scattering: the Method and some Applications* Ed., W. Brown. Oxford University Press (1993).

8 | Detection Noise

Whenever performing a measurement, we must bear in mind the noise that is inevitably introduced by the measurement device. In Chapter 7 we looked at temporal and spatial filtering caused by the limited speed and non-zero size of a detector. While this filtering had the effect of quenching the intensity fluctuations recorded by the detector, it did not, in itself, introduce noise into the measurement. In this chapter, we specifically examine the issue of detection noise. In particular, we consider various detection noise sources and discuss their effects on the temporal variance of an intensity measurement. This will allow us to characterize the intensity measurement in terms of a signal-to-noise ratio (SNR), which, needless to say, is of fundamental importance in any experiment involving light detection.

As always, the variance of a fluctuating variable X is defined to be

$$\sigma_X^2 = \langle X^2 \rangle - \langle X \rangle^2 \tag{8.1}$$

From now on, however, the brackets $\langle ... \rangle$ will be used to signify a time average much longer than both the coherence time of the light (as before) and the integration time of the detector. Moreover, both the mean and variance of the fluctuating variable X will be assumed to be independent of time, or alternatively, if they do depend of time, to vary on a time scale that is much longer than the coherence and measurement times. This last condition is tantamount to our definition of quasi-stationarity adopted previously. Some rules governing the variances of quasi-stationary independently fluctuating variables when they are added or multiplied are listed in Appendix B.

Throughout this chapter, for reasons that will be made clear below, we will work with relative variances, defined by $\sigma_X^2 / \langle X \rangle^2$. These have the advantage of being unitless, which considerably facilitates the problem of unit conversion. For example, if X is expressed in a different unit as Y such that $Y = \xi X$, where ξ is a constant scaling factor, we have (obviously),

$$\frac{\sigma_Y^2}{\langle Y \rangle^2} = \frac{\sigma_X^2}{\langle X \rangle^2} \tag{8.2}$$

In other words, the relative variance is independent of units, or of multiplication by a constant scaling factor. We will make ample use of this property below.

Moreover, there is a link between relative variance and SNR, though this link depends on convention. If we are considering the SNR in terms of signal and noise *powers*, the power SNR is given by

$$\text{SNR} = \frac{\langle X \rangle^2}{\sigma_X^2} \tag{8.3}$$

corresponding to the inverse of the relative variance. Alternatively, the amplitude SNR is defined as the square root of the power SNR given above. That is, the amplitude SNR corresponds to an inverse contrast (see Section 7.2.1).

Different noise sources can influence the relative variance of a measured signal (and hence SNR). We consider only the most important of these.

8.1 SHOT NOISE

Up to now, light has been described as a superposition of propagating waves. This description is "classical" in the sense that light is treated as an entity whose attributes are independent of the physical device used to measure it. Upon an actual measurement of light, however, we must adopt a "quantum" description wherein the presence of light can only be recorded as discrete particle-like units, called photons. Thus we adopt the general semi-classical rule: light propagates as waves but is detected as particles. This rule, which can be attributed to Dirac [1], is almost always valid. For information on classes of light for which this semi-classical rule no longer applies, the reader is referred to [2, 4, 5].

In a standard optical detector, light power is converted to an electrical current. The instantaneous photoelectron conversion rate $\alpha(t)$ is proportional to the instantaneous incident power $\Phi(t)$ such that

$$\alpha(t) = \frac{\eta(\nu)}{h\nu} \Phi(t) \tag{8.4}$$

where $\eta(\nu)$ is the quantum efficiency of the detector that characterizes its spectral sensitivity, and h is the well-known Planck constant ($h = 6.63 \times 10^{-34}$ J s). As defined, $\alpha(t)$ is expressed in units of photoelectrons/s.

While the instantaneous conversion rate $\alpha(t)$ is a well defined function of $\Phi(t)$, the actual instances when the photoelectron conversions take place are not. These instances are randomly distributed in time, leading to an actual realization of the photoelectric conversions given by

$$k(t) = \sum_n \delta(t - t_n) \tag{8.5}$$

where t_n is the actual conversion instance of the nth photoelectron (see Fig 8.1).

As we saw in Section 7.3, the use of any physical detector leads to a filtering of the measured power, both temporal and spatial. That is, for a detector with response function $R_T(t)$ corresponding to an integration time T, the measured power is expected to be

$$\Phi_T(t) = \int dt' \, R_T(t - t') \Phi(t') \tag{8.6}$$

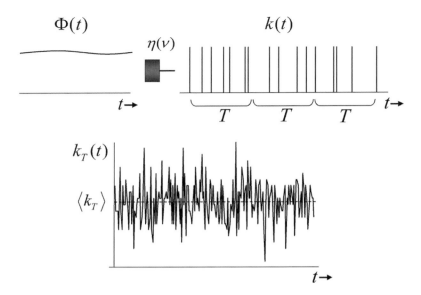

Figure 8.1. The number of photoelectrons per detector integration time T varies from measurement to measurement, leading to shot noise in the photocount K. Note: $\langle k_T \rangle = \langle \alpha_T \rangle = \langle \alpha \rangle$.

where, presumably, any spatial integration over the detector surface area is taken into account in $R_T(t)$ (see Eq. 7.62).

The same filtering applies to both the instantaneous photoelectron conversion rate and the actual photoelectron conversion signal, leading to, respectively,

$$\alpha_T(t) = \int dt' \, R_T(t - t') \alpha(t') \tag{8.7}$$

and

$$k_T(t) = \int dt' \, R_T(t - t') k(t') \tag{8.8}$$

Finally, we define K to be the measured number of photoelectron conversions that occur within an arbitrary detector integration time T. That is

$$K = \int_T dt \, k_T(t) \tag{8.9}$$

Let us begin by considering the case where the light power $\Phi(t)$ is perfectly stable in time, meaning that the photoelectron conversion rate $\alpha_T(t)$ is also stable. Despite the photoelectron conversion rate being a constant, the actual conversion events occur at random times and the value of K is thus found to fluctuate from measurement to measurement. These fluctuations are characterized by Poisson statistics. That is, the distribution of K for a given α_T (i.e. conditioned upon α_T) is given by

$$P_K(K \mid \bar{K}) = \frac{\bar{K}^K e^{-\bar{K}}}{K!} \tag{8.10}$$

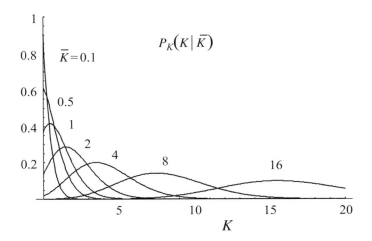

Figure 8.2. Poisson distribution for different values of $\bar{K}$. Note that K takes on only integer values.

where $\bar{K} = T\alpha_T$. Equation 8.10 is called a Poisson distribution, and is illustrated in Fig. 8.2. We observe from this distribution that even when Φ_T is perfectly stable in time, the measurement of K appears noisy (see Fig 8.1). This noise is called shot noise.

Equation 8.10 can be generalized to include the case where $\Phi_T(t)$ itself varies in time with fluctuation statistics that are quasi-stationary. Correspondingly, $\alpha_T(t)$ varies in time and the probability distribution for K becomes

$$P_K(K) = \int_0^\infty d\alpha_T \, P_K(K \mid \alpha_T) p_{\alpha_T}(\alpha_T) \tag{8.11}$$

where $p_{\alpha_T}(\alpha_T)$ is the probability density function for the occurrence of α_T. Note that P_K is a probability distribution whereas $p_{\alpha_T}(\alpha_T)$ is a probability density because K is a discrete variable whereas α_T is a continuous variable. Equation 8.11 is called a Poisson transform (or sometimes a Mandel transform, in deference to its originator [5]). It results in a mean and relative variance for K given by, respectively,

$$\langle K \rangle = \bar{K} = T\langle \alpha_T \rangle = T\langle \alpha \rangle \tag{8.12}$$

and

$$\frac{\sigma_K^2}{\langle K \rangle^2} = \frac{1}{\langle K \rangle} + \frac{\sigma_{\alpha_T}^2}{\langle \alpha_T \rangle^2} \tag{8.13}$$

Equation 8.12 indicates that the mean number of photoelectrons generated in an integration time T remains directly proportional to the average light power, as was the case when Φ_T was fixed in time.

Equation 8.13 is more interesting. The first term in Eq. 8.13 stems directly from the quantum nature of the detection process, that is, from the conversion of the continuous variable Φ_T to the discrete variable K. This extra quantum noise term is what we called shot noise above.

The second term in Eq. 8.13 stems from the usual classical fluctuations in the (filtered) light power, which we studied in Chapter 7. Indeed, Eq. 8.13 can be rewritten directly in terms of this filtered light power, obtaining

$$\frac{\sigma_K^2}{\langle K \rangle^2} = \frac{1}{\langle K \rangle} + \frac{\sigma_{\Phi_T}^2}{\langle \Phi_T \rangle^2} \tag{8.14}$$

where, again, $\Phi_T(t)$ is given by Eq. 8.6.

These classical fluctuations are sometimes called relative-intensity-noise (RIN) when they constitute unwanted intensity variations, such as in laser beams that are intended to be stable. On the other hand, these classical fluctuations may represent signal, and consitute precisely the intensity measurements of interest!

8.1.1 Power Spectral Density

In many cases, it is more convenient to work with a Fourier description of light detection. To this end, we define the power spectral density of the photoelectron conversion function to be

$$S_k(\nu) = \frac{1}{T_\infty} \left| \hat{k}(\nu) \right|^2 \tag{8.15}$$

where

$$\hat{k}(\nu) = \int_{-T_\infty/2}^{T_\infty/2} dt \, k(t) e^{i2\pi\nu t} \tag{8.16}$$

and T_∞ is a time taken to be much longer than any temporal scale of interest. This is the same time used here to specify the temporal averages denoted by $\langle ... \rangle$.

Detailed discussions of $S_k(\nu)$ can be found in textbooks such as [3] and [5]. In particular, a derivation of $S_k(\nu)$ starting from Eq. 8.5 is found to lead to the important relation

$$S_k(\nu) = \langle \alpha \rangle + S_\alpha(\nu) \tag{8.17}$$

where $S_\alpha(\nu)$ is the power spectral density of the photoelectron rate function $\alpha(t)$. That is,

$$S_\alpha(\nu) = \frac{1}{T_\infty} |\hat{\alpha}(\nu)|^2 \tag{8.18}$$

where

$$\hat{\alpha}(\nu) = \int_{-T_\infty/2}^{T_\infty/2} dt \, \alpha(t) e^{i2\pi\nu t} \tag{8.19}$$

As expected, we observe from Eq. 8.17 that $S_k(\nu)$ contains both a quantum shot noise contribution (first term) and a classical fluctuation contribution (second term).

Once again, the use of a physical detector leads to an inevitable filtering of the photoelectron conversion function, which we described previously in Eq. 8.8 by a convolution operation. The corresponding power spectral density of the actually measured photoelectron conversion signal is then obtained from a straightforward application of the convolution theorem (Eq. A.6), leading to

$$S_{k_T}(\nu) = |\mathcal{R}_T(\nu)|^2 \, S_k(\nu) \tag{8.20}$$

where $\mathcal{R}_T(\nu)$ is the Fourier transform of the detector response function

$$\mathcal{R}_T(\nu) = \int_{-T_\infty/2}^{T_\infty/2} dt \, R_T(t) e^{i2\pi\nu t} \tag{8.21}$$

As a side comment, it should be noted that $|\mathcal{R}_T(\nu)|^2$ provides a direct measure of the detector bandwidth B. In particular, the relation between B and the detector integration time T is set by the Nyquist criterion to be $B = \frac{1}{2T}$ (recall that, as opposed to spatial bandwidths, temporal bandwidths are defined here as one-sided, hence the factor of $1/2$ – for a didactic explanation the Nyquist criterion, see [6]). From Eq. 7.47 and Parseval's theorem (Eq. A.9), then

$$\int d\nu \, |\mathcal{R}_T(\nu)|^2 = \frac{1}{T} = 2B \tag{8.22}$$

We will return to this definition of detection bandwidth on several occasions.

Continuing our discussion, Eq. 8.20 can be expanded in a more explicit form

$$S_{k_T}(\nu) = |\mathcal{R}_T(\nu)|^2 \, \langle\alpha\rangle + S_{\alpha_T}(\nu) \tag{8.23}$$

where the power spectral density of the filtered photoelectron rate function is given by

$$S_{\alpha_T}(\nu) = |\mathcal{R}_T(\nu)|^2 \, S_\alpha(\nu) \tag{8.24}$$

Equation 8.23 is the main result of this section. A schematic of this result is presented in Fig. 8.3, which depicts the distribution of spectral power at various frequencies. In particular, we note that shot noise produces uniform spectral power at all frequencies. That is, the larger the bandwidth of the detector (i.e. the larger the width of $|\mathcal{R}_T(\nu)|^2$, or equivalently the smaller

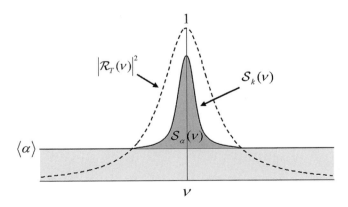

Figure 8.3. Power spectral density $S_k(\nu)$ before filtering. Shaded areas depict shot noise and classical fluctuation contributions. Note: $S_k(\nu)$ has units s^{-1} whereas the filtering function $|\mathcal{R}_T(\nu)|^2$ is unitless.

the detector integration time T), the proportionally larger the contribution of shot noise to the total detected power. To maximize SNR, it is clear that the bandwidth of the detector should be adjusted so as to minimize this shot-noise contribution while not undermining the spectral power in the classical signal. That is, the bandwidth of the detector should be adjusted to the width of $\mathcal{S}_\alpha(\nu)$, and no larger.

We note that a link between Eq. 8.23 and Eq. 8.14 is recovered when we integrate both sides of Eq. 8.23 over all frequencies. Again making ample use of Parseval's theorem, we obtain

$$\left\langle k_T^2(t) \right\rangle = \frac{1}{T}\langle \alpha \rangle + \left\langle \alpha_T^2(t) \right\rangle \qquad (8.25)$$

Noting further that $\langle k_T \rangle = \langle \alpha_T \rangle = \langle \alpha \rangle$, we recast Eq. 8.25 in the form

$$\sigma_{k_T}^2 = \frac{1}{T}\langle \alpha \rangle + \sigma_{\alpha_T}^2 \qquad (8.26)$$

which, upon normalization, leads finally to Eq. 8.13, or equivalently to Eq. 8.14.

8.1.2 Gaussian Light Example

The question of whether shot noise is dominant over classical fluctuations is generally of critical importance in any optical measurement. An answer to this question depends on the relative weights of the two terms in Eq. 8.14. For example, if the light wave is stable in time, as in the case of a laser beam, then $\sigma_{\Phi_T}^2 \to 0$ and shot noise is dominant by default. On the other hand, if the light wave fluctuates, one must look more closely at the different time scales that come into play.

In particular, let us again examine the specific case of the Gaussian light model presented in Section 7.2.2. Moreover, let us assume that the detector area is smaller than the beam coherence area, meaning that the detector provides a measure of local intensity fluctuations. If the detector integration time T is less than the coherence time τ_γ then from Eqs. 8.12, 8.14, and 7.20 we have

$$\frac{\sigma_K^2}{\langle K \rangle^2} \to \frac{1}{\langle \alpha \rangle T} + 1 \qquad (T \lesssim \tau_\gamma) \qquad (8.27)$$

In this case, shot noise becomes dominant when $\langle \alpha \rangle T < 1$, corresponding to an average of less than one photoelectron conversion per integration time. For light powers this feeble, or detector integration times this short, the photoelectron conversion statistics take on Poissonian statistics independently of the classical fluctuation statistics (see Fig. 8.4).

On the other extreme, if the integration time is much greater than the coherence time, then from Eq. 7.51 we have

$$\frac{\sigma_K^2}{\langle K \rangle^2} \to \frac{1}{\langle \alpha \rangle T} + \frac{\tau_\gamma}{T} \qquad (T \gg \tau_\gamma) \qquad (8.28)$$

and in this case shot noise becomes dominant when $\langle \alpha \rangle \tau_\gamma < 1$, corresponding to an average of less than one photoelectron conversion per coherence time (as opposed to per

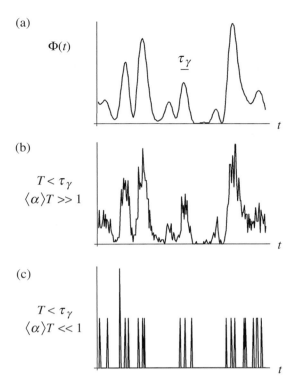

(a)

$\Phi(t)$

τ_γ

t

(b)

$T < \tau_\gamma$
$\langle \alpha \rangle T \gg 1$

t

(c)

$T < \tau_\gamma$
$\langle \alpha \rangle T \ll 1$

t

Figure 8.4. Instantaneous power (a) and detected power (b, c).

integration time above). The parameter $\langle \alpha \rangle \tau_\gamma$ is called the photon degeneracy factor. When this photon degeneracy factor is much smaller than 1, the photoelectron conversion statistics again take on Poissonian statistics independently of the classical fluctuation statistics. For more detailed information on shot noise versus classical fluctuation statistics, the reader is referred to [3].

Finally, if the detector area is larger than the coherence area, then the classical fluctuation terms in Eqs. 8.27 and 8.28 are subject to further filtering (see Eq. 7.63) and must be modified accordingly.

8.1.3 Unit Conversion

As noted above, an advantage of characterizing noise by its relative variance is that this facilitates the problem of unit conversion. For example, let us imagine that the signal X provided by the detector is in some arbitrary unit, such as amperes, volts, etc., and that the link between X and the number of detected photoelectrons K is defined by a generalized conversion gain G_X. That is, we write

$$X = G_X K \qquad (8.29)$$

Because K is unitless, the unit of G_X must be the same as that of X. In fact, G_X can be interpreted as the minimum signal arising from a single photoelectron, or, alternatively, as the "quantum" associated with the unit in question. In this manner, the relative variance of shot noise is given by, from Eq. 8.2,

$$\frac{\sigma_X^2}{\langle X \rangle^2} = \frac{1}{\langle K \rangle} = \frac{G_X}{\langle X \rangle} \qquad \text{(shot noise)} \qquad (8.30)$$

The "quantum" G_X associated with charge is the electron charge q in coulombs ($q = 1.60 \times 10^{-19}$ C); with current it is $\frac{q}{T}$ in amperes; with voltage it is $\frac{q}{T}R$ in volts, where R is the transimpedance linking current to voltage; and so forth. For example, if the signal produced by a detector is in the form of a current I_α, then shot noise is manifested as a current noise, and from Eq. 8.30 we obtain

$$\frac{\sigma_{I_\alpha}^2}{\langle I_\alpha \rangle^2} = \frac{q/T}{\langle I_\alpha \rangle} \qquad (8.31)$$

Making use of Eq. 8.22, this expression can be recast in the perhaps more familiar form

$$\sigma_{I_\alpha} = \sqrt{2qB \langle I_\alpha \rangle} \qquad \text{(shot noise)} \qquad (8.32)$$

where B is the detector bandwidth. This is the standard deviation of the current fluctuations arising from shot noise alone.

Alternatively, the parameter G_X may not be readily available, and instead the detector may be characterized by a generalized sensitivity S_X, such that

$$X = S_X \Phi \qquad (8.33)$$

where Φ is the incident light power. That is, S_X is expressed in units of X per Watt (W), where, again, X can be in units of coulombs, amperes, volts, etc. In this case, G_X is derived from S_X using the relation

$$G_X = 2 \frac{h\nu}{\eta(\nu)} BS_X \qquad (8.34)$$

where we have made use of Eq. 8.4. We thus obtain

$$\sigma_X = \sqrt{\frac{2h\nu}{\eta(\nu)} BS_X \langle X \rangle} \qquad \text{(shot noise)} \qquad (8.35)$$

This is the standard deviation of shot-noise-induced fluctuations expressed in any unit.

8.2 OTHER NOISE SOURCES

Shot noise is unavoidable when making intensity measurements (almost always). However, shot noise is not the only noise source. In many instances other types of noise can dominate shot noise and play an even more significant role in undermining SNR. These other noise sources are more technical in the sense that they can be controlled to some degree, usually by lowering the temperature of the detector or by better designing its internal electronics. We examine some of these noise sources below.

8.2.1 Dark-Current Noise

Because detectors operate at non-zero temperatures, the thermal energy at the detector active area produces a background current I_d. This current subsists even when there is no light incident on the detector, and for this reason is called a dark current. For semiconductor-based detector active areas, dark current is thermionic in origin and roughly scales as

$$I_d \sim e^{-E_g/2k_BT} \tag{8.36}$$

where E_g is the detector-cathode bandgap energy, k_B is Boltzmann's constant ($k_B = 1.38 \times 10^{-23}$ J K^{-1}), and T here is temperature in kelvin. For example, as a rule of thumb for a silicon-based detector, $\langle I_d \rangle$ decreases by a factor of about 2 for every 7–8 K drop in temperature, meaning that dark current can be quite substantially reduced by cooling the detector active area even a few tens of degrees.

It might appear that such cooling should not be necessary because the dark current can easily be measured and hence corrected for (i.e. by simple subtraction). While this is true for the *average* dark current, it is not true for the noise associated with the dark current. Inasmuch as dark-current noise obeys the same statistics as photocurrent noise, the two are indistinguishable. The total average current generated at the detector is thus

$$I_q = I_\alpha + I_d \tag{8.37}$$

and the noises associated with these currents are independent and additive, leading to $\sigma_{I_q}^2 = \sigma_{I_\alpha}^2 + \sigma_{I_d}^2$ from Eq. B.5, obtaining finally

$$\frac{\sigma_{I_q}^2}{\langle I_q \rangle^2} = \frac{2qB}{\langle I_q \rangle} \tag{8.38}$$

8.2.2 Avalanche Noise

Let us imagine now that an amplification gain M is directly applied to the detector current. This is typically performed through a current avalanche mechanism, examples of which are discussed below. For the moment, we assume this gain is noiseless, such that the amplified current is

$$I_m = MI_q \tag{8.39}$$

From Eq. 8.30 we can derive the resulting shot noise associated with this amplified current, noting that the "quantum" associated with the current measurement is also amplified by M. We obtain then

$$\frac{\sigma_{I_m}^2}{\langle I_m \rangle^2} = \frac{\sigma_{I_q}^2}{\langle I_q \rangle^2} \tag{8.40}$$

arriving at the unsurprising result that noiseless current amplification does not affect overall SNR.

The situation is different, however, if the avalanche amplification process is itself noisy, in which case it introduces what is known as excess noise to the measurement, and Eq. 8.40 is no longer valid. This is conventionally characterized in terms of an excess noise factor, defined by

$$F^2 = \frac{\sigma_{I_m}^2}{\langle M \rangle^2 \sigma_{I_q}^2} \tag{8.41}$$

where $\langle M \rangle$ is the average gain. When $F = 1$, the amplification is noiseless.

The detailed mechanisms underlying excess noise tend to be fairly involved, and depend to a large degree on the specific detector in question. An example of a detector that performs avalanche amplification of photocurrent is the photomultiplier tube (PMT) whose gain is typically $M \approx 10^6$–10^7 with an excess noise factor $F \approx 2$. Another example is the avalanche photodiode (APD) whose gain is typically $M \approx 10$–1000 with an excess noise factor $F \approx M^{0.15}$. More information on the excess noise associated with various avalanche-based detectors can be found in [8, 9, 10].

8.2.3 Johnson Noise

In cases where current is converted to voltage, the detection electronics must include an overall transimpedance R. However, this transimpedance inevitably dissipates heat, and hence, from the fluctuation-dissipation theorem, introduces an additive current noise

$$\sigma_{I_j}^2 = \frac{4k_B T B}{R} \tag{8.42}$$

where T here is temperature and B is the detector bandwidth. This noise is called Johnson noise.

We can directly compare Johnson noise with shot noise (Eq. 8.38) to determine the conditions where one dominates the other. Allowing for the possibility of a noiseless current amplification, the average measured voltage becomes

$$\langle V \rangle = \langle I_m \rangle R = M \langle I_q \rangle R \tag{8.43}$$

Thus, from Eqs. 8.38, 8.40, and 8.42 we conclude that shot noise dominates Johnson noise when $\langle V \rangle > 2k_B T / Mq$. When no current amplification is applied ($M = 1$), this threshold voltage is equal to ≈ 52 mV at room temperature. Remarkably, this threshold voltage is independent of the detection bandwidth.

In principle, shot noise can be rendered dominant over Johnson noise simply by using a large enough M or R; however, in practice this may not be so easy. For example, restrictions occur when M introduces unacceptable excess noise or when R is so large that stray capacitances in the detector electronics lead to measurement response times that are unacceptably slow.

For more information on Johnson noise and its relation to the fluctuation-dissipation theorem, the reader is referred to [7].

8.2.4 Noise Equivalent Power

To maximize SNR, it is always better for shot noise to be dominant over all other detector-related noises. When this is the case, the detection is said to be shot-noise limited. Unfortunately, this is often not the case. A common metric used to assess detector noise is called a noise equivalent power, or NEP. This defines the effective light power that is required to obtain a SNR equal to one. NEP is generally expressed in units of $W/\sqrt{Hz}$, and is a catch-all that embodies any additive noise sources associated with the detection process. To understand how NEP works, let us first derive the effective NEP associated with shot noise alone. For example, let us consider a detector that converts an optical power Φ to a signal X of arbitrary unit, with a sensitivity S_X defined according to Eq. 8.33. Typically this sensitivity is in units of A/W or V/W, and is a manufacturer specification of the detector.

From Eq. 8.2 the apparent power fluctuations resulting from shot noise are characterized by

$$\frac{\sigma_\Phi^2}{\langle \Phi \rangle^2} = \frac{\sigma_X^2}{\langle X \rangle^2} \tag{8.44}$$

which, from Eq. 8.35, lead to

$$\frac{\sigma_\Phi}{\sqrt{B}} = \sqrt{\frac{2h\nu}{\eta S_X} \langle X \rangle} \tag{8.45}$$

This is an effective shot-noise NEP associated with shot noise alone. If it is smaller than the specified detector NEP, the signal $\langle X \rangle$ is too weak and detector noise is dominant over shot noise. On the other hand, if the light power (and hence $\langle X \rangle$) is increased to the point that this shot-noise NEP becomes larger than the specified detector NEP, then the detection becomes shot-noise limited.

8.2.5 Example: Split Detector

As an example of how to manipulate noises when calculating the SNR of a measurement, let us consider a simple technique to estimate the position of a beam using a split detector. Here, a light beam straddles two independent semi-detectors that are juxtaposed, as shown in Fig. 8.5. For the sake of argument, let as assume these semi-detectors report voltage signals.

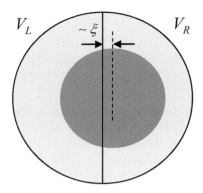

Figure 8.5. Beam incident on split detector.

The voltage imbalance produced by these semi-detectors can thus provide an estimate of the horizontal beam position, which we call here ξ, defined as

$$\xi = \frac{V_R - V_L}{V_T} \tag{8.46}$$

where $V_{R,L}$ are the right/left voltages respectively, and $V_T = V_R + V_L$ is the total voltage produced by the split detector. Note that ξ here is unitless, and the conversion of ξ to a true position requires an additional scaling factor on the order of half the beam width. Moreover, this scaling factor depends on ξ itself, to take into account the beam profile, etc. In general, ξ is a good (i.e. linear) surrogate for beam position when its value is much smaller than the beam width. But we overlook such details here. The question we are interested is, what is the SNR associated with the measurement of ξ?

To begin, let us ascribe noises to the measurements of V_R and V_L. That is, we write $V_R = \langle V_R \rangle + \delta V_R$ and $V_L = \langle V_L \rangle + \delta V_L$, where the noises, by definition, average to zero (i.e. $\langle \delta V_{R,L} \rangle = 0$). Similarly we can separate ξ into signal and noise components by writing $\xi = \langle \xi \rangle + \delta \xi$. Henceforth, we assume that the voltage noises are small relative to the voltage averages, meaning we need only keep terms to first order in these noises. We thus readily establish the correspondences

$$\langle \xi \rangle = \frac{\langle V_R \rangle - \langle V_L \rangle}{\langle V_T \rangle} \tag{8.47}$$

$$\delta \xi = \frac{1 - \langle \xi \rangle}{\langle V_T \rangle} \delta V_R - \frac{1 + \langle \xi \rangle}{\langle V_T \rangle} \delta V_L \tag{8.48}$$

To calculate the SNR associated with the measurement of ξ, we must calculate the variance of $\delta \xi$. Assuming the voltage noises $\delta V_{R,L}$ are uncorrelated, we obtain

$$\sigma_\xi^2 = \left(\frac{1 - \langle \xi \rangle}{\langle V_T \rangle} \right)^2 \sigma_R^2 + \left(\frac{1 + \langle \xi \rangle}{\langle V_T \rangle} \right)^2 \sigma_L^2 \tag{8.49}$$

Assuming further that the voltage noises are comprised of shot noise and additive detector electronic noise, we have

$$\sigma_{R,L}^2 = G_V \langle V_{R,L} \rangle + \sigma_D^2 \tag{8.50}$$

where we have made use of Eq. 8.30 and the variance of the detector noise σ_D^2 is taken to be the same for both semi-detectors.

Finally, inserting Eq. 8.50 into Eq. 8.49, we arrive at

$$\frac{\langle \xi \rangle^2}{\sigma_\xi^2} = \frac{\langle V_T \rangle}{G_V} \left[\frac{\langle \xi \rangle^2}{1 - \langle \xi \rangle^2 + \frac{2\sigma_D^2}{G_V \langle V_T \rangle} \left(1 + \langle \xi \rangle^2 \right)} \right] \tag{8.51}$$

This is the power SNR associated with the measurement of ξ. If we are interested in the amplitude SNR, we must take the square root of this. In any case, two conclusions can be drawn. First, an increase in $\langle V_T \rangle$, which corresponds to an increase in the total optical power incident on the split detector, leads to an increase in SNR, as expected. Second, the SNR is not a constant, but depends on the value of $\langle \xi \rangle$ itself. In fact, when the detector noise is small, the SNR appears to increase violently when $\langle \xi \rangle$ approaches its limit of ± 1. But we must remember that we have derived the SNR associated with ξ as defined by Eq. 8.46. This is not, strictly speaking, the horizontal beam position, and only becomes linearly related to the beam position for small values of ξ.

Equation 8.51 can be recast in an alternative form if we make the association $\sigma_D^2 = S_V^2 \text{NEP}^2 B$, which leads to

$$\frac{2\sigma_D^2}{G_V} = \frac{\eta S_V}{h\nu} \text{NEP}^2 \tag{8.52}$$

in the denominator. The parameters S_V and NEP tend to be more convenient since they are usually readily available.

It should be mentioned that this same procedure for determining a beam position along a horizontal axis can, of course, be extended to the vertical axis as well, using a detector that is split into four quadrants.

8.3 CAMERAS

We turn now to another type of detector that is routinely used in microscopy applications: a camera. Basically, a camera is a pixellated array of detectors that reports the numbers of photoelectrons accumulated in each pixel during an exposure time. Two main classes of cameras are currently available, based on charge-coupled device (CCD) or complementary metal-oxide-semiconductor (CMOS) technologies. The main difference between these is how the photoelectrons are amplified and converted to signal. In a CCD camera, the photoelectrons are shuttled from one pixel to the next before they are amplified by a common amplifier; in a

CMOS camera the amplifications are performed directly at every pixel. In either case, the pixel signals are digitized, such that

$$N_n = GK_n \qquad (8.53)$$

where K_n is the number of accumulated photoelectrons in pixel n, G is the camera readout gain, and N_n is the resulting readout number reported by the camera. Both N_n and K_n cannot be arbitrarily large. N_n is limited by the digitizer range; K_n is limited by what is known as the pixel well capacity. In most cases, the readout gain G is set to approximately the ratio of these limits. Neglecting readout noise for now, Eq. 8.53 can be regarded as a simple re-scaling of the number of photoelectrons electrons per pixel, from which we conclude (omitting the index n for ease of notation)

$$\frac{\sigma_N^2}{\langle N \rangle^2} = \frac{\sigma_K^2}{\langle K \rangle^2} \qquad (8.54)$$

In the event the optical power incident on the particular pixel of interest is fluctuating in time, Eq. 8.54 becomes

$$\frac{\sigma_N^2}{\langle N \rangle^2} = \frac{G}{\langle N \rangle} + \frac{\sigma_{\Phi_T}^2}{\langle \Phi_T \rangle^2} \qquad (8.55)$$

where we have made use of Eqs. 8.14 and 8.30, and the subscript T denotes the camera exposure (integration) time.

Again, the first term on the right-hand side of Eq. 8.55 represents quantum shot noise inherent in the light detection process, while the second term represents classical fluctuations in the measured light power.

8.3.1 Readout Noise

Equation 8.55 neglected the possibility of other detector noise contributions in addition to shot noise. Such additional noise contributions can be particularly problematic when the number of accumulated photoelectrons is small, which occurs when the light level is low or when the exposure time is short. In the former case, dark-current noise is usually the dominant additional noise, and in Section 8.2.1 we discussed how cooling the camera active area can mitigate this problem. In this section, we concentrate on the latter case where the exposure time is short, corresponding to fast imaging. The dominant additional noise in this latter case is usually the noise associated with the photoelectron digitization process itself and is called readout noise.

The introduction of readout noise into the photoelectron digitization process is schematically depicted in Fig. 8.6. Because readout noise is independent and additive, we have from Eqs. B.5 and 8.54

$$\frac{\sigma_N^2}{\langle N \rangle^2} = \frac{1}{\langle K \rangle} + \frac{\sigma_r^2}{\langle K \rangle^2} = \frac{G}{\langle N \rangle} + \frac{G^2 \sigma_r^2}{\langle N \rangle^2} \qquad (8.56)$$

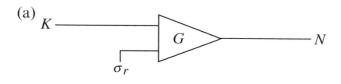

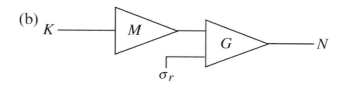

Figure 8.6. Schematic of readout in camera without (a) and with (b) electron preamplification. Bottom panel corresponds to an ICCD or EM-CCD.

where, to simplify our discussion, we have neglected the possibility of classical intensity fluctuations. The readout noise is usually expressed as an equivalent standard deviation in photoelectron number, denoted by the unitless quantity σ_r. Whether or not readout noise dominates shot noise depends on the relative weights of each in Eq. 8.56. In particular, if the mean number of photoelectrons accumulated per exposure time satisfies the condition $\langle K \rangle > \sigma_r^2$, then shot noise is greater than readout noise. For scientific grade CCD cameras whose readout frequencies are slow (1 MHz or slower), typical readout noises tend to be small enough that $\sigma_r < 1$, meaning that shot noise is dominant for light levels as low as one detected photoelectron per exposure time. On the other hand, for fast CCD cameras (readout frequencies in the tens of MHz), then readout noise is invariably exacerbated, typically to the point where $\sigma_r \gg 1$. As a result, readout noise usually dominates shot noise when performing fast imaging with CCD cameras. Recently, scientific grade CMOS (or sCMOS) cameras have become available that circumvent this problem and provide both high-speed readout frequencies and near shot-noise-limited readout noise with $\sigma_r \approx 1$. These cameras are fast supplanting CCD cameras and becoming the industry standard.

8.3.2 Photoelectron Preamplification

But all is not lost for CCD cameras. From the above discussion, one might conclude that a CCD camera cannot attain shot-noise-limited detection when performing fast imaging at low light levels. This is not the case. A standard strategy to sidestep the problem of readout noise is to preamplify the number of photoelectrons *prior to* their digitization (this is essentially the same strategy as prescribed to sidestep the problem of Johnson noise in Section 8.2.3). A variety of strategies are available for such preamplication. One strategy involves the introduction of a microchannel plate (MCP) in front of the camera active area, providing an avalanche mechanism of preamplification similar to that of a PMT. CCD cameras based on this strategy are called intensified CCD (or ICCD) cameras. We will revisit this type of camera in Section 13.3.

Another more recent strategy is based on on-chip avalanche preamplification. CCD cameras based on this second strategy are called electron-multiplication CCD (or EM-CCD) cameras. Schematics of both cases are depicted in Fig. 8.6b.

The basic idea of photoelectron preamplification is to amplify the detected photoelectrons to the point where their associated shot noise becomes dominant over readout noise. Taking into account both photoelectron preamplification and readout gain, we write

$$\langle N \rangle = MG \langle K \rangle \tag{8.57}$$

where M denotes the photoelectron preamplification gain (generally called "multiplication" gain in this context), leading to

$$\frac{\sigma_N^2}{\langle N \rangle^2} = \frac{1}{\langle K \rangle} + \frac{\sigma_r^2}{M^2 \langle K \rangle^2} \tag{8.58}$$

The benefit of photoelectron preamplification is apparent. As the multiplication gain M is increased, the relative contribution of readout noise to the overall relative variance becomes smaller and eventually drops below that of shot noise, even for small values of $\langle K \rangle$.

In the event that the multiplication gain M introduces its own avalanche noise characterized by an excess noise factor F, then the first term on the right-hand side of Eq. 8.58 becomes multiplied by F^2. Recasting Eq. 8.58 in terms of $\langle N \rangle$ rather than $\langle K \rangle$, we obtain finally

$$\frac{\sigma_N^2}{\langle N \rangle^2} = \frac{F^2 MG}{\langle N \rangle} + \frac{G^2 \sigma_r^2}{\langle N \rangle^2} \tag{8.59}$$

Manifestly, a multiplication gain is only beneficial if the excess noise introduced by the multiplication gain is small. For ICCDs, typically $M \approx 10^3$–10^5 and $F \approx 1.6$–2. For EM-CCDs, typically $M \approx 10^2$–10^4 and $F \approx 1.3$. Both technologies present advantages and disadvantages. An advantage of an ICCD is that an MCP permits very fast gain gating, on the order of nanoseconds, while disadvantages are that it is noisier, exhibits poor quantum efficiency ($\eta \approx 20\%$ or less), and does not provide high pixel densities. In contrast, advantages of EM-CCD are that it can be quieter (with proper temperature stabilization!), and operates with the higher quantum efficiency and pixel density of the CCD itself, since no MCP is required.

8.4   ELECTRONIC GAIN VERSUS OPTICAL GAIN

In closing this chapter, an important distinction must be made between electronic gain and optical gain. As we have seen above, a noiseless electronic gain does not affect the relative noise variance associated with the light detection process. In particular, it does not affect the relative shot-noise variance. By definition, such electronic gain is applied post detection.

In contrast, when a gain is applied directly to the light power prior to detection, the results become quite different. For example, if the light power is (somehow) increased by a factor M, then from Eq. 8.4 the photoelectron conversion rate increases by the same factor M. From Eq. 8.30, this leads to a net decrease in the relative shot-noise variance by a factor M, no matter what the unit of measurement. Hence, while an electronic gain does not increase SNR, an optical gain certainly does, provided the SNR is shot-noise limited.

8.5 PROBLEMS

Problem 8.1

For photoelectron arrival times governed by Poissonian statistics, it can be shown that the wait time τ between successive photoelectrons is governed by the probability distribution (see Eq. 7.10):

$$p(\tau) = \frac{1}{\bar{\tau}} e^{-\tau/\bar{\tau}}$$

where $\bar{\tau}$ is the average wait time. It can also be shown that, starting at an *arbitrary* time, the wait time for the next photoelectron is given by the same probability distribution with the same $\bar{\tau}$. The same is true for the "wait time" (going backward in time) for the previous photoelectron. But from these last two statements, it appears that the average wait time between successive photoelectrons should be $2\bar{\tau}$ and not $\bar{\tau}$. How can one reconcile all these statements? This is a classic problem in probability theory. (Hint: the arbitrary start time is more likely to fall within photoelectron intervals that are large.)

Problem 8.2

(a) Show that if the instantaneous power Φ of a light beam obeys a negative-exponential probability density, then, upon detection, the number of photoelectron conversions per detector integration time T obeys a probability distribution given by

$$P_K(K) = \frac{1}{1 + \langle K \rangle} \left(\frac{\langle K \rangle}{1 + \langle K \rangle} \right)^K$$

where $\langle K \rangle = \frac{\eta}{h\nu} \langle \Phi \rangle T$.

This is called a Bose–Einstein probability distribution (in probability theory it is called a geometric distribution).

(b) Based on the above result, verify that the variance in the detected number of photoelectron conversions is

$$\sigma_K^2 = \langle K \rangle + \langle K \rangle^2$$

Note: for part (b), you will find the following identity to be useful:

$$\sum_{k=0}^{\infty} k^n \gamma^k = \begin{cases} \frac{1}{1-\gamma} & (n=0) \\ \frac{\gamma^n (n+1)!}{(1-\gamma)^{n+1}} \sum_{m=0}^{n-1} \sum_{j=0}^{m+1} (-1)^j \frac{(m-j+1)^n}{j!(n-j+1)!} \gamma^{-m} & (n \geq 1) \end{cases}$$

Problem 8.3
Derive Eq. 8.26 from Eq. 8.23.

Problem 8.4
Consider a detector voltage measured through an impedance $R = 10^5 \ \Omega$ (this is a typical value). Assume that the detector is at room temperature, but that dark current is negligible. The charge of a single electron is 1.6×10^{-19} C.

(a) Say a single photoelectron is generated at the detector cathode (i.e. input). What is the minimum detector bandwidth B required for the measurement of this photoelectron to be shot-noise limited?

(b) The bandwidth derived above is found to be unrealistic. In fact, the detector bandwidth is known to be 10 MHz (also a typical value). What is the minimum current preamplification M required for the measurement of the single photoelectron to be shot-noise limited? (assuming this preamplification to be noiseless).

Problem 8.5
Consider a camera with a 12-bit dynamic range and a pixel well capacity of 10,000e$^-$. Assume that the camera gain G is properly set to accommodate these ranges. The camera amplifier produces a readout noise of 10e$^-$ (i.e. $\sigma_r = 10$; note that the readout noise is in units of *number* of electrons as opposed to electron charge). Assume the illumination light is stable (i.e. exhibits no classical fluctuations). Dark noise and Johnson noise are negligible.

(a) What is the minimum average readout value $\langle N \rangle$ for the measured signal to be shot-noise limited?

(b) This is not good enough. Let us say we want to measure a signal as low as $\langle N \rangle = 1$. To do this, we will incorporate an electron multiplication stage in our camera. What electron multiplication gain M is required to guarantee that the measurement will be shot-noise limited even at this low signal? (Consider the electron multiplication stage to be noiseless.)

References

[1] Dirac, P. A. M. *The Principles of Quantum Mechanics*, 4th edn, Clarendon Press (1982).
[2] Glauber, R. J. *Quantum Theory of Optical Coherence: Selected Papers and Lectures*, Wiley-VCH (2007).
[3] Goodman, J. W. *Statistical Optics*, 2nd edn, Wiley (2015).
[4] Loudon, R. *The Quantum Theory of Light*, 3rd edn, Oxford University Press (2000).

[5] Mandel, L. and Wolf, E. *Optical Coherence and Quantum Optics*, Cambridge University Press (1995).

[6] Press, W. H., Teukolsky, S. A., Vetterling, W. T. and Flannery, B. P. *Numerical Recipes: the Art of Scientific Computing*, 3rd edn, Cambridge University Press (2007).

[7] Reif, F. *Fundamentals of Statistical and Thermal Physics*, McGraw-Hill (1965).

[8] Saleh, B. E. A. and Teich, M. C. *Fundamentals of Photonics*, 2nd edn, Wiley Interscience (2007).

[9] Sze, S. M. and Ng, K. K. *The Physics of Semiconductor Devices*, 3rd edn, Wiley-Interscience (2006).

[10] Yariv, A. *Quantum Electronics*, 3rd edn, Wiley (1989).

9 Absorption and Scattering

In Chapters 1 through 8, we developed a formalism for the treatment of light propagation through lenses and apertures, as well as the detection of light. This formalism provides the basis for the description of optical microscopes; however, one crucial component has been largely missing: the sample. The remainder of this book is intended to complete our formalism by specifically examining different types of samples and different varieties of optical microscopes designed to image these samples. While the range of microscopes we will consider is certainly not exhaustive, it will be representative of the state of the art, and hopefully will provide the reader with some foundations upon which to build.

Ultimately, the goal of any optical microscope is to produce a spatial map of a sample based on the interaction of the sample with light. We very briefly considered the possibility of imaging self-luminous samples in Section 5.3, which we will return to in future chapters. Here we consider samples that are not self-luminous. Optical fields traversing such samples incur modifications in their amplitude and phase, which, in turn, cause light absorption and scattering. While these phenomena can be treated within the same formalism, as we will see below they do not yield the same optical contrasts.

The first part of this chapter will be limited to a 2D formalism, appropriate for treating beam propagation that is more or less directional. The second part will extend this to a more general 3D formalism that is applicable to non-paraxial propagation, as well as to vectorial fields. In general, our goal will be to determine the distal far field that results from proximal interactions between an incident field and a sample. Throughout this chapter, we consider only monochromatic fields.

The subject of absorption and scattering is quite vast and only an overview is presented here. For more comprehensive treatments, the reader is referred to several textbooks dedicated to the subject, including [2, 12, 14], along with more general references [3, 7, 11].

9.1 2D FORMALISM

To begin, let us consider a planar sample located at $z_0 = 0$. Let us further assume the sample thickness δz is so small that a field traversing it does not have a chance to be displaced in space. We model such a sample by a transmission function $t(\vec{\rho}_0)$ such that

$$E_0(\vec{\rho}_0) = t(\vec{\rho}_0)E_i(\vec{\rho}_0) \tag{9.1}$$

where $E_i\left(\vec{\rho}_0\right)$ is the field incident on the sample, and $E_0(\vec{\rho}_0)$ is the field directly emanating from the sample at the same transverse location $\vec{\rho}_0$.

If the sample consists only of this 2D plane, then the remainder of the problem is straightforward since the field then continues to propagate unimpeded through free space. That is, the far field $E(\vec{\rho}, z)$ at a large distance z is given by

$$E(\vec{\rho}, z) = \int d^2\vec{\rho}_0 \, D_+(\vec{\rho} - \vec{\rho}_0, z) t(\vec{\rho}_0) E_i\left(\vec{\rho}_0\right) \tag{9.2}$$

where $D_+(\vec{\rho})$ is the forward-directed free-space propagator associated with the Rayleigh–Sommerfeld or Fresnel approximations, whichever is appropriate (see Chapter 2).

The question remains how to model the transmission function $t(\vec{\rho}_0)$ that characterizes the sample. A standard model treats the 2D sample as a phase screen that simply imparts a phase to the transmitted field. That is, we write

$$t(\vec{\rho}_0) = e^{i2\pi\kappa\delta n\left(\vec{\rho}_0\right)\delta z} \tag{9.3}$$

where $\delta n\left(\vec{\rho}_0\right) = \frac{1}{n}\left(n\left(\vec{\rho}_0\right) - n\right)$ characterizes the variations of the sample index of refraction $n\left(\vec{\rho}_0\right)$ relative to the surrounding homogenous index of refraction n. The term phase screen is a bit of a misnomer here since we will make allowances for the possibility that $\delta n\left(\vec{\rho}_0\right)$ is complex. Strictly speaking, only the real part of $\delta n\left(\vec{\rho}_0\right)$ induces a phase shift, while the imaginary part induces instead an attenuation by causing the magnitude of $t(\vec{\rho}_0)$ to become smaller than one. Nevertheless, we will continue to use the term phase screen, by convention.

9.1.1 Thin Sample Approximation

The key advantage to treating our sample as thin is that δz is small, meaning that, without significant error, we can expand the transmission function to first order and write

$$t(\vec{\rho}_0) = 1 + i2\pi\kappa\delta z\delta n\left(\vec{\rho}_0\right) \tag{9.4}$$

The far field becomes, from Eq. 9.2,

$$E(\vec{\rho}, z) = E_i(\vec{\rho}, z) + E_s(\vec{\rho}, z) \tag{9.5}$$

where

$$E_i(\vec{\rho}, z) = \int d^2\vec{\rho}_0 \, D_+(\vec{\rho} - \vec{\rho}_0, z) E_i\left(\vec{\rho}_0\right) \tag{9.6}$$

$$E_s(\vec{\rho}, z) = i2\pi\kappa\delta z \int d^2\vec{\rho}_0 \, D_+(\vec{\rho} - \vec{\rho}_0, z) \delta n\left(\vec{\rho}_0\right) E_i\left(\vec{\rho}_0\right) \tag{9.7}$$

The first term (Eq. 9.6) represents the far field that would arise if the sample were absent. This comes from the unimpeded free-space propagation of the incident field (sometimes called the ballistic field). The second term (Eq. 9.7) represents the additional field component that results from the sample presence, and arises from scattering. We note the prefactor i associated with $\delta n\left(\vec{\rho}_0\right)$, indicating that our phase screen model inherently treats the sample as a primary source despite the fact that the free-space propagator D_+ is normally associated with secondary sources (this point will be made clearer in our comparison with 3D models).

The above equations can be expressed equivalently in the transverse spatial frequency domain:

$$\mathcal{E}(\vec{\kappa}_\perp; z) = \mathcal{E}_i(\vec{\kappa}_\perp; z) + \mathcal{E}_s(\vec{\kappa}_\perp; z) \tag{9.8}$$

where

$$\mathcal{E}_i(\vec{\kappa}_\perp; z) = \mathcal{D}_+(\vec{\kappa}_\perp; z)\mathcal{E}_i(\vec{\kappa}_\perp) \tag{9.9}$$

$$\mathcal{E}_s(\vec{\kappa}_\perp; z) = i2\pi\kappa\delta z \mathcal{D}_+(\vec{\kappa}_\perp; z) \int d^2\vec{\kappa}'_\perp \, \delta\hat{n}(\vec{\kappa}_\perp - \vec{\kappa}'_\perp)\mathcal{E}_i(\vec{\kappa}'_\perp) \tag{9.10}$$

and $\delta\hat{n}(\vec{\kappa}_\perp)$ is the 2D Fourier transform of $\delta n(\vec{\rho})$. Depending on the approximation, $\mathcal{D}_+(\vec{\kappa}_\perp; z)$ is given by Eq. 2.28 or Eq. 2.54.

9.1.2 Plane-Wave Illumination

Let us consider the simplest case of on-axis plane-wave illumination, meaning that $E_i(\vec{\rho}_0)$ becomes a constant E_i, and $\mathcal{E}_i(\vec{\kappa}_\perp) = E_i\delta^2(\vec{\kappa}_\perp)$. The ballistic and scattered radiant fields then reduce to

$$\mathcal{E}_i(\vec{\kappa}_\perp; z) = E_i e^{i2\pi\kappa z}\delta^2(\vec{\kappa}_\perp) \tag{9.11}$$

$$\mathcal{E}_s(\vec{\kappa}_\perp; z) = i2\pi\kappa\delta z E_i \mathcal{D}_+(\vec{\kappa}_\perp; z)\delta\hat{n}(\vec{\kappa}_\perp) \tag{9.12}$$

Several conclusions can be drawn from this.

Scattering Angle Versus Object Size

For one, Eq. 9.12 provides a clear picture of the range of wavevector angles that propagate to the far field. This range is defined by either $\mathcal{D}_+(\vec{\kappa}_\perp; z)$ or $\delta\hat{n}(\vec{\kappa}_\perp)$, whichever limits the span of $|\vec{\kappa}_\perp|$ the most. In the case of Rayleigh–Sommerfeld propagation, $\mathcal{D}_+(\vec{\kappa}_\perp; z)$ imposes the limit $|\vec{\kappa}_\perp| \leq \kappa$, and thus permits a wide range of wavevector angles spanning the full 2π steradians in the forward direction. In general, it is not $\mathcal{D}_+(\vec{\kappa}_\perp; z)$ that limits the range of wavevector angles, but rather $\delta\hat{n}(\vec{\kappa}_\perp)$. As an example, let us consider a disk-shaped object of radius a, defined by

$$\delta n(\rho_0) = \begin{cases} \delta n & \rho_0 \leq a \\ 0 & \rho_0 > a \end{cases} \tag{9.13}$$

$$\delta\hat{n}(\kappa_\perp) = \pi a^2 \delta n \, \mathrm{jinc}\,(2\pi\kappa_\perp a) \tag{9.14}$$

A plot of the jinc function is shown in Appendix A, illustrating that the product $\kappa_\perp a$ is confined to roughly the value of one. In other words, there is a reciprocal relationship between

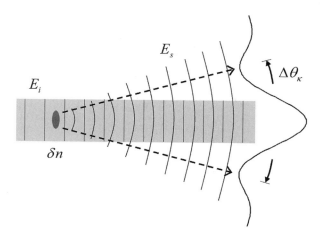

Figure 9.1. Scattering geometry.

the size of the object and the angular range of the scattered wavevectors, leading to the general rule of thumb for the span of wavevector tilt angles given by

$$\Delta\theta_\kappa \approx \frac{\lambda}{a} \tag{9.15}$$

as we have previously seen when dealing with propagation from a small aperture (see Section 2.4.2)

The above relation, illustrated in Fig. 9.1, is valid regardless of whether δn is real or imaginary, provided only that $a \gtrsim \lambda$.

Scattering From a Grating

Let us consider now a grating, defined by a cosinusoidally varying index of refraction. That is, we write $\delta n \left(\vec{\rho}_0 \right) = \delta n \cos \left(2\pi \vec{q}_\perp \cdot \vec{\rho}_0 + \phi \right)$ where $\vec{q}_\perp$ defines the grating spatial frequency and direction, and we make an allowance for an arbitrary phase ϕ. In the case where δn is purely real, the object is called a phase grating, since it induces pure phase shifts at the sample plane. In the case where δn is purely imaginary, the object is called an absorption grating, since it causes a net loss in radiative energy.

In both cases, we can write

$$\delta\hat{n}(\vec{\kappa}_\perp) = \frac{\delta n}{2} \left[e^{i\phi}\delta^2(\vec{\kappa}_\perp - \vec{q}_\perp) + e^{-i\phi}\delta^2(\vec{\kappa}_\perp + \vec{q}_\perp) \right] \tag{9.16}$$

The scattered field from the grating defined is thus found to be confined to two off-axis wavevectors propagating at off-axis angle θ_κ defined by $\sin\theta_\kappa = \pm\frac{q_\perp}{\kappa}$. We note that a lateral translation of the grating caused by a change in phase ϕ does not alter the direction of these wavevectors, but only alters their relative phases, as illustrated in Fig. 9.2. An intuitive explanation for why a grating produces two off-axis rays that are symmetrically distributed comes from the interpretation sample structure as comprised of spatial momentum components.

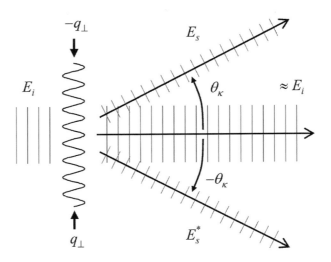

Figure 9.2. Scattering from a phase grating.

The stationary sinusoidal profile of the grating can be thought of as the summation of two counterpropagating transverse momentum components $\pm \vec{q}_{\perp}$, in accord with Eq. 9.16. These transverse momentum components effectively impart two transverse momenta to the incident field, causing parts of it to deflect off-axis.

9.1.3 Beam Propagation Method

So far we have considered only infinitesimally thin 2D samples. The question remains how to generalize our 2D formalism so that it is applicable to volumetric samples. A powerful technique to do this is called the beam propagation method (BPM), originally developed for the modeling of propagation in waveguides. The basic idea of this method is to consider the sample as a succession of phase screens, as illustrated in Fig. 9.3. Each phase screen is defined by a transmission function given by Eq. 9.3, but instead of propagating the field from the phase screen to a distant far zone, as was done above, we propagate it to the next phase screen in the sequence. The separation between screens is δz, where δz is now treated as a small distance, but not infinitesimally small. Formally, the beam propagation method amounts to iteratively calculating the field incident on each phase screen by applying the operations

$$E\left(\vec{\rho}_0, z_0 + \delta z\right) = D_+\left(\vec{\rho}_0, \delta z\right) * \left(t\left(\vec{\rho}_0, z_0\right) \times E\left(\vec{\rho}_0, z_0\right)\right) \qquad (9.17)$$

where $t\left(\vec{\rho}_0, z_0\right)$ denotes the phase screen located at axial position z_0. That is, the method involves alternating the applications of multiplication and convolution operators. The iteration procedure is initiated at a position $z_0 = z_{in}$ in front of the sample, where $E\left(\vec{\rho}_0, z_{in}\right)$ is presumably known and given by $E_i\left(\vec{\rho}_0\right)$, and then proceeds step by step through the sample until the end of the sample has been attained at $z_0 = z_{out}$. The resulting field at $E\left(\vec{\rho}_0, z_{out}\right)$ is

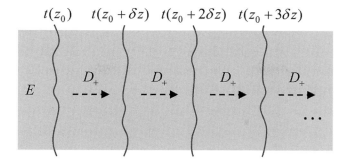

Figure 9.3. Beam propagation method.

then equivalent to the sample field and may be propagated through free space to a far zone if desired.

Since δz remains a small quantity, the expansion given by Eq. 9.4 remains valid, and the beam propagation method may be recast as an iterative two-step procedure

$$E_i\left(\vec{\rho}_0, z_0\right) \leftarrow \left(1 + i2\pi\kappa\delta z\delta n\left(\vec{\rho}_0, z_0\right)\right) E_i\left(\vec{\rho}_0, z_0\right) \tag{9.18}$$

$$E_i\left(\vec{\rho}_0, z_0 + \delta z\right) = \int \mathrm{d}^2\vec{\rho}_0' D_+\left(\vec{\rho}_0 - \vec{\rho}_0', \delta z\right) E_i\left(\vec{\rho}_0', z_0\right) \tag{9.19}$$

making explicit the interpretation that the incident field at each phase screen is progressively updated based on the interactions with previous phase screens.

The question remains what value to choose for δz. If the details of the sample are known, then δz can be arbitrarily small. On the other hand, if the sample can be characterized only by mesoscopic statistical properties, the assignment of δz must be made carefully. In particular, it should be no larger than what is called the scattering mean free path. This will be discussed in detail in Chapter 20.

9.2   3D FORMALISM

In the previous section, a 3D sample was modeled as a series of 2D phase screens, defined by Eq. 9.3. The physical origin of this model, however, was sidestepped. We turn now to a brief derivation of how this model naturally arises from the Helmholtz equation when we consider *inhomogeneities* in the sample medium. This derivation is based on a rigorous formalism for scattering or absorption, in which sample inhomogeneities are treated as primary radiating sources that are driven by an incident field $E_i(\vec{\rho}_0)$. The phases of the these primary sources is thus defined by the phase of the incident field, meaning that the scattered fields generated by the primary sources are coherent both with one another and with the incident field. As we will see, the coherent superposition of such scattered fields can lead to highly structured radiation patterns. We begin by developing a simple scalar description of these fields, which we later generalize to vector fields. While a scalar description is usually sufficient for most imaging applications, a vector description is required when field polarization plays an important role, as it does, for example, in multiharmonic or pump-probe microscopies (see Chapters 17 and 18).

As we saw in Chapter 2, the propagation of a monochromatic field through a medium is governed by the Helmholtz equation

$$\left(\nabla^2 + 4\pi^2\kappa^2\right) E(\vec{r}) = 0 \tag{9.20}$$

where, as a reminder, $\kappa = n\frac{\nu}{c}$ and an overall time dependence $e^{i2\pi\nu t}$ is understood. As in our 2D formalism, we consider the index of refraction $n\left(\vec{r}\right)$ of the sample to be spatially varying about the homogenous value n of its surrounding medium. When $n\left(\vec{r}\right)$ is inserted into the Helmholtz equation, and terms are rearranged, we obtain

$$\left(\nabla^2 + 4\pi^2\kappa^2\right) E(\vec{r}) = -4\pi^2\kappa^2\delta\varepsilon\left(\vec{r}\right) E(\vec{r}) \tag{9.21}$$

where $\delta\varepsilon\left(\vec{r}\right) = \frac{1}{n^2}\left(n^2\left(\vec{r}\right) - n^2\right)$. We note that $\delta\varepsilon\left(\vec{r}\right)$, which we will use to characterize the sample scattering strength, is closely related to the sample dielectric constant defined by $\epsilon\left(\vec{r}\right) = n^2\left(\vec{r}\right)$. The quantity $4\pi^2\kappa^2\delta\varepsilon\left(\vec{r}\right)$ is sometimes called a scattering potential. We also observe that in the usual case where $n\left(\vec{r}\right) \approx n$, we have $\delta\varepsilon\left(\vec{r}\right) \approx 2\delta n\left(\vec{r}\right)$.

From Chapter 2, the right-hand side of Eq. 9.21 is recognized as corresponding to an effective source term in the Helmholtz equation (see Eq. 2.11). This source term is manifestly governed by the total field $E(\vec{r})$, as is made explicit with the association

$$S(\vec{r}) = -4\pi^2\kappa^2\delta\varepsilon\left(\vec{r}\right) E(\vec{r}) \tag{9.22}$$

Moreover, the total field $E(\vec{r})$ is itself a superposition of the incident field $E_i(\vec{r})$, corresponding to the field in the absence of any induced sources, and an induced scattered field $E_s(\vec{r})$. That is, we have

$$E(\vec{r}) = E_i(\vec{r}) + E_s(\vec{r}) \tag{9.23}$$

(the same as Eq. 9.5).

By inserting Eqs. 9.22 and 9.23 into Eq. 2.11 and noting that $E_i(\vec{r})$ must obey Eq. 9.20, we conclude

$$\left(\nabla^2 + 4\pi^2\kappa^2\right) E_s(\vec{r}) = -4\pi^2\kappa^2\delta\varepsilon\left(\vec{r}\right) E(\vec{r}) \tag{9.24}$$

Our goal here is to solve for $E_s(\vec{r})$. But we are already equipped with the Green function for the Helmholtz equation, and since we are only considering forward-directed light, we can write

$$E_s(\vec{r}) = -4\pi^2\kappa^2 \int d^3\vec{r}_0\, G_+(\vec{r} - \vec{r}_0)\delta\varepsilon\left(\vec{r}_0\right) E(\vec{r}_0) \tag{9.25}$$

or, more succinctly,

$$E_s(\vec{r}) = \int d^3\vec{r}_0\, G_+(\vec{r} - \vec{r}_0)S(\vec{r}_0) \tag{9.26}$$

So far, we have made no approximations, and Eq. 9.25 is thus exact. However, it is also somewhat problematic in that $E(\vec{r}_0)$ in the integrand represents the total field which itself contains $E_s(\vec{r})$. In other words, Eq. 9.25 provides only a formal solution to $E_s(\vec{r})$ which must be solved self-consistently.

9.2.1 Born Approximation

While an exact self-consistent solution of Eq. 9.25 is generally intractable, a considerable simplification is obtained when $\delta\varepsilon(\vec{r}_0)$ is assumed to be small and hence the scattering is assumed to be weak, in which case the approximation $E(\vec{r}_0) \approx E_i(\vec{r}_0)$ becomes valid. This is called the first Born approximation, or Born approximation for short (or, in the present context, sometimes the Rayleigh–Debye or Rayleigh–Gans approximation). Equation 9.25 then reduces to

$$E_s(\vec{r}) = -4\pi^2\kappa^2 \int d^3\vec{r}_0\, G_+(\vec{r} - \vec{r}_0)\delta\varepsilon(\vec{r}_0)\, E_i(\vec{r}_0) \qquad (9.27)$$

The correspondence between Eqs. 9.27 and 9.7 is immediate, particularly bearing in mind that $\delta\varepsilon(\vec{r}) \approx 2\delta n(\vec{r})$, and, in the case where the paraxial approximation is valid, $G_+(\vec{r}) = -\frac{i}{4\pi\kappa}D_+(\vec{r})$ (see Eq. 2.35 with $\cos\theta \to 1$). The main difference between the two is that Eq. 9.27 is fully three-dimensional, and the sample $\delta\varepsilon(\vec{r}_0)$ is treated fully volumetrically rather than as a succession of phase screens. In other words, the total field $E(\vec{r})$ can be calculated in a single calculation (through Eqs. 9.23 and 9.27) rather than requiring an iterative application of many calculations, as in the beam propagation method. This is not to say, however, that the Born approximation is superior to the beam propagation method. A critical requirement of the Born approximation is that $\delta\varepsilon(\vec{r}_0)$ be small enough that $E(\vec{r}_0) \approx E_i(\vec{r}_0)$ meaning that the incident field is only weakly perturbed by the sample. In contrast, the requirement for the beam propagation method is only that δz be small. By progressively updating the incident field in the beam propagation method, the incident field can end up becoming quite strongly perturbed. Said differently, the Born approximation only allows the sample to interact with the unperturbed incident field rather than the total field, meaning it only allows the possibility of single scattering. It does not allow the possibility of multiple scattering, such as, for example, when the scattered field from one location becomes scattered again at another location. In contrast, the beam propagation method inherently incorporates the possibility of multiple scattering. It can readily be shown that in the limit where the sample $\delta n(\vec{r}_0)$ is so weak that multiple scattering can be neglected, and the fields are mostly directional (i.e. paraxial), then the beam propagation method yields identical results as the Born approximation.

Fourier Diffraction Theorem

Despite the restriction of the Born approximation to problems involving single scattering only, it remains one of the mainstays of imaging theory. A classic application is in the reconstruction of a 3D sample distribution based on planar measurements of the scattered far field obtained with plane-wave illumination. This application was first described by Wolf [15]. For this, let us rewrite Eq. 9.27 in terms of radiant fields, leading to

$$\mathcal{E}_s(\vec{\kappa}) = -4\pi^2\kappa^2\mathcal{G}_+(\vec{\kappa})\int d^3\vec{\kappa}'\,\delta\hat{\varepsilon}(\vec{\kappa}-\vec{\kappa}')\mathcal{E}_i(\vec{\kappa}') \tag{9.28}$$

where $\delta\hat{\varepsilon}(\vec{\kappa})$ is the 3D Fourier transform of $\delta\varepsilon(\vec{r})$.

In three dimensions, an incident plane wave in the direction $\vec{\kappa}_i$ is given by $E_i(\vec{r}_0) = E_i e^{i2\pi\vec{\kappa}_i\cdot\vec{r}_0}$, or, equivalently, $\mathcal{E}_i(\vec{\kappa}) = E_i\delta^3(\vec{\kappa}-\vec{\kappa}_i)$. For plane-wave illumination, we have then

$$\mathcal{E}_s(\vec{\kappa}) = -4\pi^2\kappa^2 E_i\mathcal{G}_+(\vec{\kappa})\delta\hat{\varepsilon}(\vec{\kappa}-\vec{\kappa}_i) \tag{9.29}$$

Our goal is to derive an estimate of $\delta\hat{\varepsilon}(\vec{\kappa})$ based on a measurement of the scattered field $\mathcal{E}_s(\vec{\kappa}_\perp;z)$ in a cross-sectional plane located at z. By taking the Fourier transform of both sides of Eq. 9.29 with respect to κ_z, and making use of Eq. 2.21, we obtain

$$\mathcal{E}_s(\vec{\kappa}_\perp;z) = i\pi\kappa^2 E_i\frac{e^{i2\pi\sqrt{\kappa^2-\kappa_\perp^2}\,z}}{\sqrt{\kappa^2-\kappa_\perp^2}}\delta\hat{\varepsilon}\left(\vec{\kappa}_\perp-\vec{\kappa}_{\perp i},\sqrt{\kappa^2-\kappa_\perp^2}-\kappa_{zi}\right) \tag{9.30}$$

or, rewritten with $\kappa_z = \sqrt{\kappa^2-\kappa_\perp^2}$,

$$\delta\hat{\varepsilon}(\vec{\kappa}-\vec{\kappa}_i) = \frac{\kappa_z}{i\pi\kappa^2 E_i}e^{-i2\pi\kappa_z z}\mathcal{E}_s(\vec{\kappa}_\perp;z) \tag{9.31}$$

This is the well-known Fourier diffraction theorem. Its interpretation is illustrated in Fig. 9.4 for the case of on-axis illumination. Though $\mathcal{E}_s(\vec{\kappa}_\perp;z)$ is measured in a plane, the information it provides about $\delta\hat{\varepsilon}(\vec{\kappa})$ is confined to a frequency support in the shape of a thin spherical cap. For example, if $\delta\varepsilon(\vec{r})$ is point-like, the axial location of this point can be determined with a resolution given by the inverse height of the cap. If instead $\delta\varepsilon(\vec{r})$ is planar, for example if it is a uniform plane, the axial resolution is given by the inverse thickness of the cap – that is, a uniform planar sample cannot be axially resolved at all.

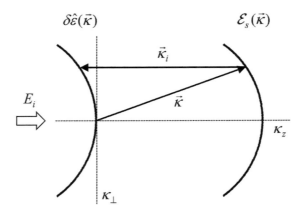

Figure 9.4. Frequency support of reconstructed $\delta\hat{\varepsilon}(\vec{\kappa})$.

Fraunhofer Approximation

When the sample is localized to a small region and the observation point is distant, the Fraunhofer approximation can be applied in addition to the Born approximation. We have already seen this approximation in Section 2.4.2. Making use of Eq. 2.61, the forward-propagating Green function becomes (for $z > z_0$)

$$G_+(\vec{r} - \vec{r}_0) = -\frac{e^{i2\pi\kappa r}}{4\pi r} e^{-i2\pi\frac{\kappa}{r}\vec{r}\cdot\vec{r}_0} \qquad \text{(Fraunhofer)} \qquad (9.32)$$

and the scattered field reduces to

$$E_s(\vec{r}) = \pi\kappa^2 \frac{e^{i2\pi\kappa r}}{r} \int d^3\vec{r}_0\, e^{-i2\pi\frac{\kappa}{r}\vec{r}\cdot\vec{r}_0} \delta\varepsilon(\vec{r}_0) E_i(\vec{r}_0) \qquad (9.33)$$

This relation will be useful in later chapters, for example when we consider multiharmonic microscopies (Chapter 17). In the particular case of illumination with a plane wave of wavevector $\vec{\kappa}_i$, we have

$$E_s(\vec{r}) = \pi\kappa^2 E_i \frac{e^{i2\pi\kappa r}}{r} \delta\hat{\varepsilon}(\kappa\hat{r} - \vec{\kappa}_i) \qquad (9.34)$$

where $\hat{r} = \frac{\vec{r}}{r}$. In other words, the scattered far field is given by an outgoing spherical wave whose amplitude along any direction $\hat{r}$ is directly determined by the Fourier transform of the sample at the difference between outgoing and incoming wavevectors. Upon inversion, we write

$$\delta\hat{\varepsilon}(\kappa\hat{r} - \vec{\kappa}_i) = \frac{r}{\pi\kappa^2 E_i} e^{-i2\pi\kappa r} E_s(\vec{r}) \qquad (9.35)$$

indicating that when the Fraunhofer approximation is valid, $\delta\hat{\varepsilon}(\vec{\kappa})$ can be reconstructed directly from $E_s(\vec{r})$ rather than from $\mathcal{E}_s(\vec{\kappa}_\perp; z)$ as it was above (Eq. 9.31).

9.2.2 Rytov Approximation

In the Born approximation, scattering from the sample produces an additive perturbation to the incident field. There exists another approximation, called the Rytov approximation, where scattering produces instead a phase perturbation to the incident field. That is, the total field is given by

$$E(\vec{r}) = e^{i\Psi(\vec{r})} E_i(\vec{r}) \qquad (9.36)$$

A derivation of $\Psi(\vec{r})$ would take us too far afield, and the reader is referred instead to the original reference (e.g. [13]). Suffice it to say that in the Rytov approximation we arrive at

$$\Psi(\vec{r}) = i\frac{4\pi^2\kappa^2}{E_i(\vec{r})} \int d^3\vec{r}_0\, G_+(\vec{r} - \vec{r}_0) \delta\varepsilon(\vec{r}_0) E_i(\vec{r}_0) \qquad (9.37)$$

We note that if $\Psi(\vec{r})$ is small, then the Rytov approximation is equivalent to the Born approximation. However, the advantage of the Rytov approximation is that $\Psi(\vec{r})$ need not be small. What is required is only that the spatial derivatives of $\Psi(\vec{r})$ be small. While the exact regimes of validity of the Born versus Rytov approximations are difficult to pin down, a rough rule of thumb is that the Born approximation works well when the scatterers are sparsely distributed and weak, or small in size, even if they exhibit sharp edges. The Rytov approximation works better with larger, slowly varying scatterers that exhibit soft edges. In general, the Rytov approximation is useful in problems where the total field remains largely directional. In this regard, it resembles the beam propagation method and features the same advantage of making allowances for multiple scattering.

9.3   CROSS-SECTIONS

As described above, a light beam interacting with a sample can be partially absorbed or scattered. In either case, power is extracted from the beam. If the sample is a discrete particle, this extracted power Φ_e is characterized in terms of the particle extinction cross-section σ_e, such that

$$\Phi_e = \sigma_e I_i \tag{9.38}$$

where I_i is the incident beam intensity, assumed to be uniform across the particle. The extinction cross-section corresponds then to the effective area of the particle as seen by the light beam (σ_e has units of area). Because the extraction of power can result from both absorption and scattering, σ_e in general contains both absorption and scattering components. Our goal in this section is to derive these components. The procedure we follow is mostly described in [4].

Let us consider the scenario depicted in Fig. 9.5. A particle is surrounded by a volume V bounded by a surface S. A field $E_i\left(\vec{r}\right)$ originating from outside the volume is incident on the particle. According to Eq. 9.23, this incident field propagates unimpeded through the volume *and* also produces a scattered field given by Eq. 9.25. One may wonder how it is possible to maintain energy conservation in such a scenario since it seems, at first glance, that the energy carried by the incident beam remains unchanged leaving no energy possible for the scattered

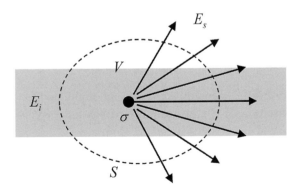

Figure 9.5. Scattering from particle.

field. However, this first glance neglects the interference between incident and scattered fields. It is precisely this interference that leads to energy conservation, as we will see below.

We begin, as usual, with the Helmholtz equation, which we used to derive Eq. 9.25. In particular, we begin with Eq. 9.21. When we multiply both sides of this equation by $E^*(\vec{r})$, and multiply both sides of the complex conjugate of this equation by $E(\vec{r})$, and subtract the two, we obtain

$$E^*(\vec{r})\nabla^2 E(\vec{r}) - E(\vec{r})\nabla^2 E^*(\vec{r}) = -i8\pi^2\kappa^2 \text{Im}\left[\delta\varepsilon(\vec{r}) E^*(\vec{r})E(\vec{r})\right] \qquad (9.39)$$

If we now integrate both sides over the volume V, and apply Green's theorem (see Appendix B.1) to the left side, we obtain

$$\int_S d^2\vec{r} \left(E^*(\vec{r})\vec{\nabla}E(\vec{r}) - E(\vec{r})\vec{\nabla}E^*(\vec{r})\right) \cdot \hat{s} = -i8\pi^2\kappa^2 \int_V d^3\vec{r}\,\text{Im}\left[\delta\varepsilon(\vec{r}) E^*(\vec{r})E(\vec{r})\right]$$
$$(9.40)$$

where the left side has now become a surface integral and $\hat{s}$ is the unit vector pointing normally outward from the surface. But we have seen the integrand on the left before (see Section 6.1). Applying a time-average to both sides, we obtain then

$$\int_S d^2\vec{r}\,\vec{F}(\vec{r}) \cdot \hat{s} = -2\pi\kappa \int_V d^3\vec{r}\,\text{Im}\left[\delta\varepsilon(\vec{r}) \langle E^*(\vec{r})E(\vec{r})\rangle\right] \qquad (9.41)$$

where $\vec{F}(\vec{r})$ is the energy flux defined by Eq. 6.3. If the integrals in Eq. 9.41 vanish, the net energy flux entering or exiting the volume V is zero, and energy is conserved. On the other hand, if the integrals do not vanish, it is an indication that energy must be created or destroyed within the volume. Since, in our case, we only allow the possibility of the latter via absorption, we conclude that the net absorbed power within the volume is given by

$$\Phi_a = 2\pi\kappa \int_V d^3\vec{r}\,\text{Im}\left[\delta\varepsilon(\vec{r})\right] I(\vec{r}) \qquad (9.42)$$

leading to the expected result that absorption occurs only if $\delta\varepsilon(\vec{r})$ contains an imaginary component.

A similar procedure can be followed to derive the net scattered power that exits the volume V. This time, we start with Eq. 9.24 and multiply by $E_s^*(\vec{r})$, and so forth, to obtain

$$\Phi_s = \int_S d^2\vec{r}\,\vec{F}_s(\vec{r}) \cdot \hat{s} = -2\pi\kappa \int_V d^3\vec{r}\,\text{Im}\left[\delta\varepsilon(\vec{r}) \langle E_s^*(\vec{r})E(\vec{r})\rangle\right] \qquad (9.43)$$

Finally, the sum of the absorbed and scattered powers must correspond to the depleted power from the incident field, called the extinction power Φ_e. From Eq. 9.23, we thus arrive at

$$\Phi_e = \Phi_a + \Phi_s = 2\pi\kappa \int_V d^3\vec{r}\,\text{Im}\left[\delta\varepsilon(\vec{r}) \langle E_i^*(\vec{r})E(\vec{r})\rangle\right] \qquad (9.44)$$

We observe that even if there is no absorption, the extinction power is given by

$$\Phi_e = 2\pi\kappa \int_V d^3\vec{r}\,\text{Im}\left[\delta\varepsilon(\vec{r}) \langle E_i^*(\vec{r})E_s(\vec{r})\rangle\right] \qquad \text{(no absorption)} \qquad (9.45)$$

It is clear here that the interference between incident and scattered fields carries power, resolving the apparent conundrum with energy conservation we had above.

9.3.1 Scattering Amplitude

Let us consider a small, localized particle, such that we can readily apply the Fraunhofer approximation, and make use of Eq. 2.61. The total field, upon illumination of the particle by an incident plane wave wavevector $\vec{\kappa}_i$, is then written as

$$E(\vec{r}) = E_i(\vec{r}) + \frac{e^{i2\pi\kappa r}}{r} f(\hat{\kappa}, \hat{\kappa}_i) \tag{9.46}$$

where $f(\hat{\kappa}, \hat{\kappa}_i)$ is called the scattering amplitude, defined by

$$f(\hat{\kappa}, \hat{\kappa}_i) = \pi\kappa^2 \int d^3\vec{r}_0 \, e^{-i2\pi\vec{\kappa}\cdot\vec{r}_0} \delta\varepsilon\left(\vec{r}_0\right) E(\vec{r}_0) \tag{9.47}$$

and $\hat{\kappa}_i$ and $\hat{\kappa}$ are input and output direction vectors associated with the incident and scattered fields respectively ($E(\vec{r}_0)$ is implicitly a function of $\vec{\kappa}_i$ and we have made the far-field association $\hat{\kappa} = \hat{r}$). Note that we have not relied on the Born approximation, as we did to arrive at Eq. 9.33. In other words, $E(\vec{r}_0)$ inside the integrand of Eq. 9.47 is the total field and not just the incident field.

From our connection between intensity and energy flux in the far field (Eq. 6.21) we readily find

$$|f(\hat{\kappa}, \hat{\kappa}_i)|^2 = r^2 I_s(\vec{r}) = r^2 \left| \vec{F}_s(\vec{r}) \right| \qquad \text{(for } r \to \infty) \tag{9.48}$$

In other words, $|f(\hat{\kappa}, \hat{\kappa}_i)|^2$ can be thought of as deflecting incident power from the direction $\hat{\kappa}_i$ into the direction $\hat{\kappa} = \hat{r}$. We may use this interpretation to define the differential scattering cross-section associated with the particle, given by

$$\frac{d\sigma_s}{d\Omega} = \frac{|f(\hat{\kappa}, \hat{\kappa}_i)|^2}{I_i} \tag{9.49}$$

where, in the far field, we have $\Omega_\kappa = \Omega$ (synonymous with $\hat{\kappa} = \hat{r}$).

The total scattering cross-section of the particle is then

$$\sigma_s = \frac{1}{I_i} \int d^2\Omega \, |f(\hat{\kappa}, \hat{\kappa}_i)|^2 \tag{9.50}$$

and the total scattered power is expressed in terms of this scattering cross-section by

$$\Phi_s = \sigma_s I_i \tag{9.51}$$

9.3.2 Optical Theorem

We now derive the extinction cross-section of the particle. This can be obtained from Eq. 9.44 by replacing $E_i^*(\vec{r})$ with an incident (complex-conjugated) plane wave $\sqrt{I_i}e^{-i2\pi\vec{\kappa}_i\cdot\vec{r}_0}$, leading to

$$\Phi_e = 2\pi\kappa\sqrt{I_i}\int d^3\vec{r}_0\,\mathrm{Im}\left[\delta\varepsilon(\vec{r}_0)\,e^{-i2\pi\vec{\kappa}_i\cdot\vec{r}_0}E(\vec{r}_0)\right] \tag{9.52}$$

A comparison of Eqs. 9.52 and 9.47 leads immediately to

$$\Phi_e = \sigma_e I_i = \frac{2\sqrt{I_i}}{\kappa}\mathrm{Im}\left[f(\hat{\kappa}_i,\hat{\kappa}_i)\right] \tag{9.53}$$

from which we determine that

$$\sigma_e = \frac{2}{\kappa\sqrt{I_i}}\mathrm{Im}\left[f(\hat{\kappa}_i,\hat{\kappa}_i)\right] \tag{9.54}$$

This remarkable equation is called the optical theorem. It states that the extinction cross-section is defined by the imaginary part of the scattering amplitude in the *same direction* as the incident wave.

Finally, for completeness we write

$$\Phi_a = \sigma_a I_i \tag{9.55}$$

which defines the absorption cross-section, and we have $\sigma_e = \sigma_a + \sigma_s$.

9.4 VECTOR DIPOLE SCATTERING

While in principle Eq. 9.25 need not be restricted to near-axis (paraxial) scattering, its applicability to larger scattering angles becomes problematic because it does not properly take into account the true vectorial nature of light. When the scattering angles are large (i.e. the scattering centers are small), a full vectorial treatment of light fields is indispensable. The same is true if the scatterers are themselves tensorial in nature, producing polarization directions that may differ from that of the driving incident field $\vec{E}_i$. Because we rarely stray from the paraxial approximation throughout this book, only a cursory discussion of vector scattering is presented here and the reader is referred is to textbooks such as [2, 3, 7, 12] for much more complete accounts. In particular, we will limit ourselves to conditions where the scattering is observed in the far field as defined in Chapter 2, and is primarily electric dipole in nature. When these conditions are met, the scattered radiation generated by primary sources $\vec{S}(\vec{r}_0)$, now taken to be vectorial, can be expressed by a simple modification of Eq. 9.26 given by

$$\vec{E}_s(\vec{r}) = \int d^3\vec{r}_0\, G_+(\vec{r}-\vec{r}_0)\vec{S}^{(t)}(\vec{r}_0) \tag{9.56}$$

where we have defined $\vec{S}^{(t)}(\vec{r}_0)$ to be the component of $\vec{S}(\vec{r}_0)$ transverse to the far-field propagation direction $\hat{r}$. That is

$$\vec{S}^{(t)}(\vec{r}_0) = \vec{S}(\vec{r}_0) - \hat{r}\left(\hat{r} \cdot \vec{S}(\vec{r}_0)\right) = -\hat{r} \times \left(\hat{r} \times \vec{S}(\vec{r}_0)\right) \tag{9.57}$$

As a result, the field $\vec{E}_s(\vec{r})$ is also transverse to the propagation direction $\hat{r}$. This condition of being transverse to the propagation direction is one of the defining characteristics of far-field radiation and suggests that a coordinate system aligned along this propagation direction would considerably simplify our representation of $\vec{E}_s(\vec{r})$. To this end, we adopt the spherical coordinate system defined by

$$\begin{pmatrix} \hat{r} \\ \hat{\theta} \\ \hat{\varphi} \end{pmatrix} = \begin{pmatrix} \sin\theta\cos\varphi & \sin\theta\sin\varphi & \cos\theta \\ \cos\theta\cos\varphi & \cos\theta\sin\varphi & -\sin\theta \\ -\sin\varphi & \cos\varphi & 0 \end{pmatrix} \begin{pmatrix} \hat{x} \\ \hat{y} \\ \hat{z} \end{pmatrix} \tag{9.58}$$

where the carets denote unit vectors, θ is the tilt of the propagation direction $\hat{r}$ from the optical axis $\hat{z}$, and φ is the tilt of the plane comprising vectors $\hat{r}$ and $\hat{z}$ from the $\hat{x}$ axis (see Fig. 9.6). This plane is called the scattering plane; polarization directions parallel and perpendicular to this scattering plane are conventionally referred to as (p) and (s) polarization directions respectively. In summary, then, we have

$$\begin{aligned} \hat{r} &= \quad \text{propagation direction} \\ \hat{\theta} &= \quad \text{p-polarization direction} \\ \hat{\varphi} &= \quad \text{s-polarization direction} \end{aligned} \tag{9.59}$$

where we note that the coordinate transformation matrix in Eq. 9.58 is orthogonal, meaning that the corresponding inverse transformation is simply given by the transpose of this matrix [1].

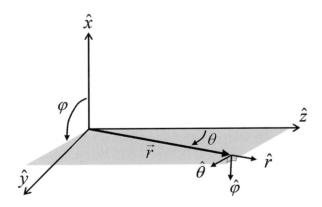

Figure 9.6. Scattering plane and associated coordinate system.

From Eq. 9.57 we obtain the relations

$$\hat{r} \cdot \vec{S}^{(t)}(\vec{r}_0) = 0$$
$$\hat{\theta} \cdot \vec{S}^{(t)}(\vec{r}_0) = \hat{\theta} \cdot \vec{S}(\vec{r}_0) = S^{(p)}(\vec{r}_0) \qquad (9.60)$$
$$\hat{\varphi} \cdot \vec{S}^{(t)}(\vec{r}_0) = \hat{\varphi} \cdot \vec{S}(\vec{r}_0) = S^{(s)}(\vec{r}_0)$$

leading to

$$\hat{r} \cdot \vec{E}_s(\vec{r}) = 0$$
$$\hat{\theta} \cdot \vec{E}_s(\vec{r}) = E_s^{(p)}(\vec{r}) \qquad (9.61)$$
$$\hat{\varphi} \cdot \vec{E}_s(\vec{r}) = E_s^{(s)}(\vec{r})$$

As anticipated, the three-dimensional vector $\vec{E}_s(\vec{r})$ in $\hat{x}$-$\hat{y}$-$\hat{z}$ space is now expressed more succinctly in two dimensions, namely in $\hat{\theta}$–$\hat{\varphi}$ space. We thus adopt the shorthand notation

$$\begin{pmatrix} S^{(p)}(\vec{r}_0) \\ S^{(s)}(\vec{r}_0) \end{pmatrix} = \begin{pmatrix} \hat{\theta} \\ \hat{\varphi} \end{pmatrix} \cdot \vec{S}^{(t)}(\vec{r}_0) = \begin{pmatrix} \hat{\theta} \\ \hat{\varphi} \end{pmatrix} \cdot \vec{S}(\vec{r}_0) \qquad (9.62)$$

$$\begin{pmatrix} E_s^{(p)}(\vec{r}) \\ E_s^{(s)}(\vec{r}) \end{pmatrix} = \begin{pmatrix} \hat{\theta} \\ \hat{\varphi} \end{pmatrix} \cdot \vec{E}_s(\vec{r}) \qquad (9.63)$$

Because the directions $\hat{\theta}$ and $\hat{\varphi}$ are orthogonal, the scattered intensity at the observation point $\vec{r}$ then reduces to

$$I_s(\vec{r}) = \left| E_s^{(p)}(\vec{r}) \right|^2 + \left| E_s^{(s)}(\vec{r}) \right|^2 \qquad (9.64)$$

Equations 9.56, 9.63, and 9.64 provide the basis for our treatment of radiation from a localized distribution of sources. When these sources are induced by an incident driving field, as we are assuming here, then this radiation can be interpreted as scattering. It should be emphasized again that the relation between $\vec{S}(\vec{r}_0)$ and $\vec{E}(\vec{r}_0)$ must be well defined for our interpretation to be accurate. In most cases this relation is not well defined and the Born approximation must be invoked.

In the event the Fraunhofer approximation is applicable, the scattered field $\vec{E}_s(\vec{r})$ becomes

$$\vec{E}_s(\vec{r}) = \frac{1}{4\pi r} e^{i2\pi\kappa r} \int d^3\vec{r}_0\, e^{-i2\pi\kappa\hat{r}\cdot\vec{r}_0} \vec{S}^{(t)}(\vec{r}_0) \qquad (9.65)$$

or, in the coordinate system of the scattering plane,

$$E_s^{(p,s)}(\vec{r}) = \frac{1}{4\pi r} e^{i2\pi\kappa r} \int d^3\vec{r}_0\, e^{-i2\pi\kappa\hat{r}\cdot\vec{r}_0} S^{(p,s)}(\vec{r}_0) \qquad (9.66)$$

We will find these results to be useful in later chapters. For reference, from Eq. 9.58 the dot product $\hat{r} \cdot \vec{r}_0$ can be recast as

$$\hat{r} \cdot \vec{r}_0 = r_0 \left(\cos\theta \cos\theta_0 + \sin\theta \sin\theta_0 \cos(\varphi - \varphi_0) \right) \qquad (9.67)$$

Radiation From a Single Dipole

As an example, let us consider an induced point source directed along the $\hat{x}_0$ axis. That is, let us consider $\vec{S}(\vec{r}_0) = V_s s_x \delta^3(\vec{r}_0)\hat{x}_0$, where V_s is the infinitesimally small source volume. Because our source is discrete here, s_x may be interpreted as a simple dipole moment, units notwithstanding. Equations 9.66, 9.62, 9.58, and 9.64 then lead to

$$\begin{pmatrix} E_s^{(p)}(\vec{r}) \\ E_s^{(s)}(\vec{r}) \end{pmatrix} = \begin{pmatrix} \frac{V_s}{4\pi r} e^{i2\pi\kappa r} s_x \cos\theta \cos\varphi \\ -\frac{V_s}{4\pi r} e^{i2\pi\kappa r} s_x \sin\varphi \end{pmatrix} \tag{9.68}$$

and

$$I_s(\vec{r}) = \left(\frac{V_s}{4\pi r}\right)^2 |s_x|^2 \left(1 - \sin^2\theta \cos^2\varphi\right) \tag{9.69}$$

Note that $\left(1 - \sin^2\theta \cos^2\varphi\right)$ can be expressed as $\sin^2\psi$, where ψ is the tilt angle between $\hat{r}$ and $\hat{x}$. The radiation distribution produced by a single dipole thus exhibits a well-known donut pattern that is symmetric about the $\hat{x}$ axis, as illustrated in Fig. 9.7. Radiation along the $\hat{x}$ axis is impossible because, as defined here, $\vec{S}(\vec{r}_0)$ possesses no component that is transverse to this propagation direction.

The total power radiated by a single dipole is calculated by integrating Eq. 9.69 over a sphere of large radius. This total power is found to be

$$\Phi_s = \frac{1}{6\pi} V_s^2 |s_x|^2 \tag{9.70}$$

As before, a link can be made between a point source s_x and a point inhomogeneity given by $\delta\varepsilon(\vec{r}_0) = V_s \delta\varepsilon \, \delta^3(\vec{r}_0)$. Referring to the association provided by Eq. 9.22, we arrive at

$$\Phi_s = \frac{8\pi^3}{3} V_s^2 \kappa^4 \delta\varepsilon^2 \, I_i(\vec{r}_0) = \sigma_s I_i \tag{9.71}$$

commonly referred to as the Clausius–Mossotti relation when V_s is a sphere.

We observe that Φ_s scales here as κ^4, which is a characteristic of dipole radiation from a source that is much smaller than the optical wavelength. When such a point-like source is treated as a scattering center, it produces what is known as Rayleigh scattering.

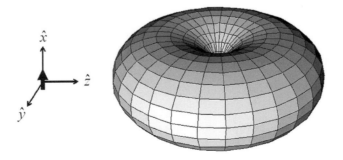

Figure 9.7. Radiation pattern from a single dipole.

Radiation From Two Dipoles

While the example of radiation from a single induced point source is instructive, it does not underline the importance of phase when calculating radiation patterns produced by distributed sources. In particular, induced sources can be distributed over volumes that are large relative to the optical wavelength. Inasmuch as the phase of a driving field $\vec{E}_i(\vec{r}_0)$ is spatially dependent, then so too must be the phase of the induced $\vec{S}(\vec{r}_0)$, meaning that different volume elements of $\vec{S}(\vec{r}_0)$, in general, radiate with different phases. Not only can $\vec{S}(\vec{r}_0)$ exhibit a spatially dependent phase, but to properly evaluate $\vec{E}_s(\vec{r})$ we must also take into account the additional phase shifts imparted by $G_+(\vec{r} - \vec{r}_0)$, which in turn depend on propagation direction and polarization. These latter phase shifts by themselves can cause a dramatic structuring of the radiation pattern. As a simple example, let us extend our discussion of radiation from a single point dipole to a pair of point dipoles. In particular, let us consider two point dipoles separated by a distance Δ_x along the $\hat{x}$ axis. If these dipoles are driven by an $\hat{x}$-polarized plane wave propagating in the $\hat{z}$ direction, then both dipoles are driven with identical amplitudes and phases, and we write

$$\vec{S}(\vec{r}_0) = V_s s_x \left[\delta(x_0 - \tfrac{1}{2}\Delta_x) + \delta(x_0 + \tfrac{1}{2}\Delta_x) \right] \delta(y_0)\delta(z_0)\hat{x}_0 \tag{9.72}$$

where V_s is again the infinitesimal volume associated with each dipole.

For simplicity, let us assume that the observation distance r is much greater than Δ_x, allowing us to continue using the Fraunhofer approximation. From Eqs. 9.66, 9.62, and 9.64 we find

$$I_s^{(2)}(\vec{r}) = 2I_s^{(1)}(\vec{r}) \left[1 + \cos\left(\xi_x \sin\theta \cos\varphi \right) \right] \tag{9.73}$$

where $\xi_x = 2\pi\kappa\Delta_x$ and $I_s^{(1)}(\vec{r})$ refers to the radiated intensity pattern from a single dipole (Eq. 9.69) located at the origin. Figure 9.8 provides an illustration of this radiation pattern for increasing Δ_x. When $\Delta_x = 0$, the pattern reduces to that of a single dipole, as expected. However, as Δ_x increases the radiation pattern becomes increasingly structured, exhibiting multiple lobes in all directions (with the exception of the $\hat{x}$ direction).

The total power radiated by the two dipoles can be evaluated by integrating Eq. 9.73 over all directions, obtaining

$$\Phi_s^{(2)} = 2\Phi_s^{(1)} \left[1 - 3 \left(\frac{\cos\xi_x}{\xi_x^2} - \frac{\sin\xi_x}{\xi_x^3} \right) \right] \tag{9.74}$$

where $\Phi_s^{(1)}$ refers to the total power radiated by a single isolated dipole (i.e. Eq. 9.71).

Similar results are obtained for dipoles separated along the $\hat{y}$ axis, where now $\xi_y = 2\pi\kappa\Delta_y$ and

$$I_s^{(2)}(\vec{r}) = 2I_s^{(1)}(\vec{r}) \left[1 + \cos\left(\xi_y \sin\theta \sin\varphi \right) \right] \tag{9.75}$$

leading to

$$\Phi_s^{(2)} = 2\Phi_s^{(1)} \left[1 + \frac{3}{2} \left(\frac{\sin\xi_y}{\xi_y} + \frac{\cos\xi_y}{\xi_y^2} - \frac{\sin\xi_y}{\xi_y^3} \right) \right] \tag{9.76}$$

Plots of $\Phi_s^{(2)}$ in both cases are illustrated in Fig. 9.9. The plots are remarkable in that they reveal that the total power radiated by two dipoles is not constant, but depends critically on their

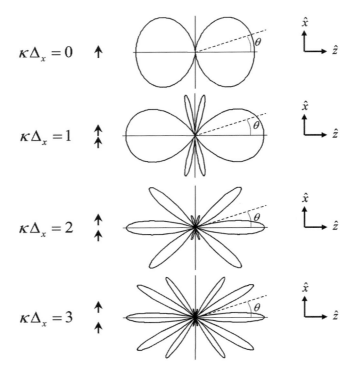

Figure 9.8. Radiation pattern from a dipole pair, for different dipole separations.

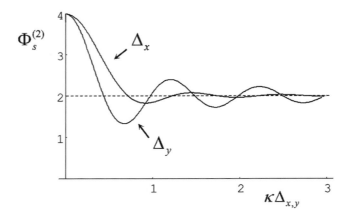

Figure 9.9. Total power radiated by a dipole pair with fixed dipole moments, as a function of dipole separation.

location relative to one another other, despite the fact that the dipole amplitudes and phases are held fixed! In particular, when the dipoles are more closely spaced than an optical wavelength, they are found to radiate four times the power of a single dipole, rather than simply twice the power. This result might have been anticipated given that two overlapping dipoles correspond effectively to a single dipole of source strength $2s_x$, which, from Eq. 9.70, leads to a factor of four increase in power. Nevertheless, one may wonder where this increase in power comes from. Ultimately, of course, this power must come from the incident driving field E_i. That is, the work extracted from this driving field in order to maintain fixed source strengths s_x at each dipole must depend on the distance between these dipoles. Borrowing terminology from antenna theory, the dipoles, even though they have been taken here to be non-interacting, in fact do interact and exhibit what is known as a mutual impedance [8]. This interaction, it turns out, is partially mediated by the near-field produced by the dipoles, something we have neglected altogether. Nevertheless, our omission of the near field remains quite legitimate here because we are considering only radiation in the far field. Remarkably, the effects of near-field interactions, whose calculation by direct means using a full Green function formalism would have been quite tedious, can be inferred indirectly from a simple far-field calculation. The intimate relation between near-field and far-field is well known, and is discussed in considerable depth elsewhere (see, for example, [12]). Suffice it to say here that when the two dipoles in our example become separated by a distance beyond the reach of their respective near-fields, then the power they radiate essentially reduces to that of two independent dipoles. As observed in Fig. 9.9, this return to independence of the dipoles is particularly rapid when the dipoles are separated along the $\hat{x}$ direction since, along this direction, far-field coupling is prohibited and only near-field coupling is possible.

It should be noted that similar enhancements in radiated power occur when a single dipole is placed in close proximity to a dielectric interface. A very clear discussion of this is presented by Lukosz [9, 10], making use of a full Green function formalism that includes near-field light. Other references on the subject include [5] and [6].

As a closing remark, the reader is reminded that the radiated powers calculated above are those of two dipoles whose source strengths s_x (interpreted as dipole moments) are held fixed. These powers do not necessarily correspond to the radiated powers scattered from two dipoles driven by a fixed incident field E_i, since the respective source strengths, in reality, are driven by the total field E and not by the incident field E_i. Since E does indeed include both near and far-field scattering contributions, the source strengths s_x must be calculated self-consistently. Nevertheless, the fixed source strength (as opposed to fixed incident field) calculations presented above are instructive in that they highlight the dramatic influence of phase both on the radiation patterns produced by distributed sources, and also on the corresponding radiation powers.

9.5 PROBLEMS

Problem 9.1

Equation 9.34 indicates that, in the Born approximation, when a plane wave is transmitted through a sample, the scattered far-field is related to the Fourier transform of the sample distribution $\delta\varepsilon\left(\vec{r}_0\right)$. This is true under conditions of the Fraunhofer approximation, but it is not true under conditions of the Fresnel approximation.

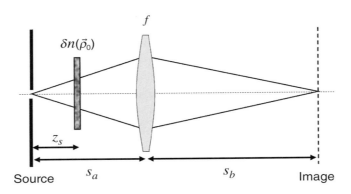

Show that by instead using point-source illumination and a single lens, as shown in the figure, then the Fourier transform relationship can be restored under the Fresnel approximation. For simplicity, consider only a thin sample, and use a 2D formalism, where the sample is characterized by index-of-refraction variations $\delta n\left(\vec{\rho}_0\right)$. Specifically, derive the resulting field at the image plane and show that it is directly proportional to the Fourier transform of $\delta n\left(\vec{\rho}_0\right)$ when the sample is located a distance $z_s = s_a - f$ from the source. Assume that s_a and s_b obey the thin-lens formula.

Note: There are several ways to solve this problem. Use the fact that a forward projection of the field from the sample plane to the image plane is equivalent to a backward projection of this field to the source plane (without the sample), followed by a forward projection to the image plane. This last projection, in the case of perfect imaging, is given by Eq. 3.17.

Problem 9.2

The beam propagation method (BPM) involves first evaluating the transmitted field though a phase screen and then propagating this transmitted field to the next phase screen.

(a) Show that the transmitted mutual intensity through a phase screen can be written as

$$J_{z_0}\left(\vec{\rho}_c, \vec{\rho}_d\right) \leftarrow K_{\delta z}\left(\vec{\rho}_d\right) J_{z_0}\left(\vec{\rho}_c, \vec{\rho}_d\right)$$

where the index of refraction variations in the phase screens are assumed statistically homogeneous. Derive an expression for $K_{\delta z}(\vec{\rho}_d)$? Hint: use results from Appendix B.5.

(b) Use Eqs. 6.16 and 6.18 to show that the radiance from one phase screen to the next transforms as

$$\mathcal{L}_{z_0+\delta z}\left(\vec{\rho}_c; \vec{\kappa}_\perp\right) = \int d^2\vec{\kappa}'_\perp \mathcal{K}_{\delta z}\left(\vec{\kappa}'_\perp\right) \mathcal{L}_{z_0}\left(\vec{\rho}_c - \frac{\delta z}{\kappa}\vec{\kappa}_\perp; \vec{\kappa}_\perp - \vec{\kappa}'_\perp\right)$$

where $\mathcal{K}_{\delta z}(\vec{\kappa}_\perp)$ is the Fourier transform of $K_{\delta z}(\vec{\rho}_d)$

Problem 9.3

Consider illuminating a sample with a plane wave directed along $\hat{z}$, and recording the resultant scattered far field in the transmission direction within a cone of half-angle θ_{max}. Based on the information provided by this scattered far field, how well can a sample $\delta\varepsilon$ be axially resolved if the sample is:

(a) a thin, uniform plane?
(b) a point?

Use the Fourier diffraction theorem and in particular Fig. 9.4 to provide estimates as a function of $\theta_{\max}$.

Problem 9.4

A wave traveling through a slowly spatially varying index of refraction $n\left(\vec{r}\right)$ can be written as

$$E\left(\vec{r}\right) = A\left(\vec{r}\right) e^{i2\pi W(\vec{r})/\lambda}$$

where λ is the free-space wavelength. This expression is similar to the Rytov approximation except that $A\left(\vec{r}\right)$ does not represent the incident field, but rather represents a slowly varying amplitude (real). Surfaces of constant $W\left(\vec{r}\right)$ are called wavefronts of the field.

Show that when the above expression is inserted into the Helmholtz equation (Eq. 9.20), and in the geometrical-optics limit where λ becomes vanishingly small, one arrives at the so-called Eikonal equation:

$$\left|\vec{\nabla} W\left(\vec{r}\right)\right|^2 = n^2(\vec{r})$$

This equation serves to define $\vec{\nabla} W\left(\vec{r}\right)$, which can be interpreted as a light ray direction in geometrical optics.

Problem 9.5

Derive Eqs. 9.73 and 9.75.

References

[1] Arfken, G. B. and Weber, H. J. *Mathematical Methods for Physicists*, 4th edn, Academic Press (1995).
[2] Bohren, C. F. and Huffman, D. R. *Absorption and Scattering of Light by Small Particles*, Wiley-Interscience (1998).
[3] Born, M. and Wolf, E. *Principles of Optics*, 6th edn, Cambridge University Press (1980).
[4] Carney, P. S., Schotland, J. C. and Wolf, E. "Generalized optical theorem for reflection, transmission, and extinction of power for scalar fields," *Phys. Rev. E* 70, 036611 (2004).
[5] Chance, R. R., Prock, A. and Silbey, R. "Molecular fluorescence and energy transfer near interfaces," *Adv. Chem. Phys.*, 2184 (1978).
[6] Drexhage, K. H. "Interaction of light with monomolecular dye layers," in *Progress in Optics Vol. XII*, North-Holland (1974).
[7] Jackson, J. D. *Classical Electrodynamics*, 3rd edn, Wiley (1998).
[8] Kraus, J. D. and Marhefka, R. J. *Antennas*, 3rd edn, McGraw-Hill Education Singapore (2007).

[9] Lukosz, W. and Kunz, R. E. "Light emission by magnetic and electric dipoles close to a plane interface. I. Total radiated power," *J. Opt. Soc. Am.* 67, 1607–1615 (1977).

[10] Lukosz, W. and Kunz, R. E. "Light emission by magnetic and electric dipoles close to a plane dielectric interface. II. Radiation patterns of perpendicular oriented dipoles," *J. Opt. Soc. Am.* 67, 1615–1619 (1977).

[11] Marion, J. B. and Heald, M. A. *Classical Electromagnetic Radiation*, 2nd edn, Academic Press (1980).

[12] Nieto-Vesperinas, M. *Scattering and Diffraction in Physical Optics*, 2nd edn, World Scientific (2006).

[13] Rytov, S. M., Kravtsov, Y. A. and Tatarski, V. I. *Principles of Statistical Radiophysics 4: Wave Propagation Through Random Media*, Springer (2011)

[14] van de Hulst, H. C. *Light Scattering by Small Particles*, Dover (1981).

[15] Wolf, E. "Three-dimensional structure determination of semi-transparent objects from holographic data," *Opt. Commun.* 1, 153–156 (1969).

10   Widefield Microscopy

We have finally established the toolbox required to analyze how a microscope forms an image of a sample. In this chapter, we will limit ourselves to microscopes based on camera detection. That is, we consider the capture of the detected field over a wide area, namely a camera sensor. Such microscopes are called widefield microscopes. Two major problems arise when performing widefield microscopy. The first is that the camera sensor is generally planar in geometry, meaning that even though we may be looking at a three-dimensional sample, the captured information is intrinsically two-dimensional. The second major problem is that cameras are sensitive to intensity, not field, which significantly exacerbates the problem of sample reconstruction. To gain an understanding of the process of image formation, we begin by considering thin 2D samples, as we did in Section 9.1. Toward the end of the chapter we will proceed to more general 3D imaging, since, ultimately, most samples are volumetric.

Widefield microscopes can be globally divided into two categories, those based on a transmission geometry, where the illumination source and detection camera are on opposite sides of the sample, and those based on a reflection geometry, where they are on the same side. Both geometries reveal fundamentally different sample structures, as we will see below.

The theory of widefield microscopy is well established, and can be found in a variety of classic textbooks, some of the more recent being Goodman [6], though, as always, Born and Wolf [4] remains a standard. More specialized books examine 3D imaging in particular, such as [7], or phase-contrast imaging, such as [14]. As for the original papers, there are many, too many to cite here, though some indispensable papers cannot be overlooked, such as Frieden [5] and Streibl [18], as well as the body of work by Sheppard (e.g. [15, 16]).

10.1   TRANSMISSION MICROSCOPY

We first examine microscopes based on a transmission geometry, where incident light from a source transilluminates a sample before being collected by imaging optics and ultimately detected by a camera. Though our camera detects only intensities, we nonetheless consider the propagation of fields through the microscope – intensities will only be calculated at the very end, at the moment they are detected. This procedure has the advantage of being applicable to both monochromatic and non-monochromatic fields (albeit still narrowband). In the latter case, we consider non-monochromatic fields as a superposition of independent monochromatic fields, and simply sum their final intensities upon detection. In other words, without loss of generality, we confine our analysis here to monochromatic or quasi-monochromatic fields.

10.1.1 Thin Samples

To analyze images obtained with a thin sample, we proceed in much the same way as in Section 9.1, except that instead of propagating the sample field through free space to the far field, we propagate it through the microscope optics to a camera. For simplicity, let us assume that the sample is located at the in-focus, or focal, plane of our microscope (we will generalize to out-of-focus imaging later). If we treat the sample as a thin phase screen, then from Eqs. 9.1 and 9.4 we have

$$E_0\left(\vec{\rho}_0\right) = E_i\left(\vec{\rho}_0\right) + E_{0s}\left(\vec{\rho}_0\right) \tag{10.1}$$

where $E_i\left(\vec{\rho}_0\right)$ is the field incident on the sample, assumed known, $E_{0s}\left(\vec{\rho}_0\right)$ is the additive perturbation caused by the sample, and $E_0\left(\vec{\rho}_0\right)$ is the resultant field transmitted through the sample, all located at the focal plane. In terms of radiant fields, we start with

$$\mathcal{E}_0\left(\vec{\kappa}_\perp\right) = \mathcal{E}_i\left(\vec{\kappa}_\perp\right) + \mathcal{E}_{0s}\left(\vec{\kappa}_\perp\right) \tag{10.2}$$

Our general strategy will be to first propagate this sample radiant field through the microscope optics. Since the propagation takes the radiant field from the focal to image plane, it is defined by the in-focus coherent transfer function $\mathcal{H}\left(\vec{\kappa}_\perp\right)$, given by Eq. 3.27, obtaining

$$\mathcal{E}\left(\vec{\kappa}_\perp\right) = \mathcal{H}\left(\vec{\kappa}_\perp\right)\mathcal{E}_0\left(\vec{\kappa}_\perp\right) \tag{10.3}$$

Next, we calculate the radiant mutual intensity at the camera, given by (Eq. 4.11)

$$\mathcal{J}\left(\vec{\kappa}_\perp, \vec{\kappa}_\perp'\right) = \left\langle \mathcal{E}\left(\vec{\kappa}_\perp\right)\mathcal{E}^*\left(\vec{\kappa}_\perp'\right)\right\rangle \tag{10.4}$$

Finally, following our usual coordinate transformation, the intensity spectrum at the camera is obtained from (Eq. 4.23)

$$\mathcal{I}\left(\vec{\kappa}_{\perp d}\right) = \int \mathrm{d}^2\vec{\kappa}_{\perp c}\,\mathcal{J}\left(\vec{\kappa}_{\perp c}, \vec{\kappa}_{\perp d}\right) \tag{10.5}$$

This is our final answer, since the intensity distribution at the camera (i.e. the image) can be directly obtained from the intensity spectrum by an inverse Fourier transform. Having established this general strategy, we now consider its details.

10.1.2 Brightfield Microscopy

The most common transmission microscope delivers incident light to the sample by way of Köhler illumination (as illustrated in Fig. 6.7) and performs imaging with an objective and tube lens in a 4f configuration (as illustrated in Fig. 6.8). Such a microscope is called a brightfield microscope since, even in the absence of a sample, light is incident on the camera. In other words, the signal arising from the sample appears on a bright background.

An application of the strategy outlined above to brightfield microscopy requires first calculating $\mathcal{E}_{0s}\left(\vec{\kappa}_\perp\right)$, the scattered radiant field at the focal plane. Using our phase screen model (Eqs. 9.1 and 9.4), and applying a Fourier transform, we arrive at

$$\mathcal{E}_{0s}(\vec{\kappa}_\perp) = i\pi\kappa\delta z \int d^2\vec{\kappa}_\perp'' \, \delta\hat{\varepsilon}\left(\vec{\kappa}_\perp - \vec{\kappa}_\perp''\right) \mathcal{E}_i(\vec{\kappa}_\perp'') \tag{10.6}$$

where we have modeled the sample in terms of $\delta\hat{\varepsilon}\left(\vec{\kappa}_\perp\right)$ rather than $\delta\hat{n}\left(\vec{\kappa}_\perp\right)$ to ultimately make our transition to volumetric imaging a bit more streamlined (recall $\delta\hat{\varepsilon}\left(\vec{\kappa}_\perp\right) \approx 2\delta\hat{n}\left(\vec{\kappa}_\perp\right)$). The reader may wonder here why we have chosen to work with radiant fields rather than fields, since Eq. 10.6 appears unnecessarily complicated. This will become clear below.

Following our strategy outlined above, the radiant mutual intensity at the camera $\mathcal{J}\left(\vec{\kappa}_\perp, \vec{\kappa}_\perp'\right)$ can be separated into four terms

$$\mathcal{J}\left(\vec{\kappa}_\perp, \vec{\kappa}_\perp'\right) = \mathcal{H}\left(\vec{\kappa}_\perp\right)\mathcal{H}^*\left(\vec{\kappa}_\perp'\right)\left(T_1 + T_2 + T_3 + T_4\right) \tag{10.7}$$

The first term $T_1 = \left\langle \mathcal{E}_i\left(\vec{\kappa}_\perp\right)\mathcal{E}_i^*(\vec{\kappa}_\perp')\right\rangle$ is the background radiant mutual intensity obtained in the absence of a sample. The second and third terms, $T_2 = \left\langle \mathcal{E}_{0s}\left(\vec{\kappa}_\perp\right)\mathcal{E}_i^*(\vec{\kappa}_\perp')\right\rangle$ and $T_3 = \left\langle \mathcal{E}_i\left(\vec{\kappa}_\perp\right)\mathcal{E}_{0s}^*(\vec{\kappa}_\perp')\right\rangle$, are the main signals arising from the sample. The fourth term $T_4 = \left\langle \mathcal{E}_{0s}\left(\vec{\kappa}_\perp\right)\mathcal{E}_{0s}^*(\vec{\kappa}_\perp')\right\rangle$ we will neglect for now since it is presumed to be much smaller than the rest, by virtue of our thin-sample approximation. As it happens, all the terms can be recast as a function of $\mathcal{J}_i\left(\vec{\kappa}_\perp, \vec{\kappa}_\perp'\right)$, the radiant mutual intensity incident on the sample at the focal plane. Because $\mathcal{J}_i\left(\vec{\kappa}_\perp, \vec{\kappa}_\perp'\right)$ is presumed known, this will serve as the crux of our calculation. In particular, for Köhler illumination, we have from Eq. 4.58

$$\mathcal{J}_i\left(\vec{\kappa}_\perp, \vec{\kappa}_\perp'\right) = \kappa^{-2}I_\xi\left|\mathcal{H}_i\left(-\vec{\kappa}_\perp\right)\right|^2 \delta^2\left(\vec{\kappa}_\perp - \vec{\kappa}_\perp'\right) \tag{10.8}$$

where I_ξ is the homogeneous illumination intensity incident on the pupil of the Köhler illumination optics and, for ease of notation, we have rewritten the pupil function in terms of an effective illumination coherent transfer function $\mathcal{H}_i$ (see Eq. 3.27).

We can now evaluate the terms in Eq. 10.7. The first term is already calculated, since it is given by $\mathcal{J}_i\left(\vec{\kappa}_\perp, \vec{\kappa}_\perp'\right)$ itself. The second term is more interesting, leading to, from Eq. 10.6

$$\left\langle \mathcal{E}_{0s}\left(\vec{\kappa}_\perp\right)\mathcal{E}_i^*(\vec{\kappa}_\perp')\right\rangle = i\pi\kappa\delta z \int d^2\vec{\kappa}_\perp'' \, \delta\hat{\varepsilon}\left(\vec{\kappa}_\perp - \vec{\kappa}_\perp''\right)\left\langle \mathcal{E}_i(\vec{\kappa}_\perp'')\mathcal{E}_i^*(\vec{\kappa}_\perp')\right\rangle \tag{10.9}$$

which, making use of Eq. 10.8, becomes

$$\left\langle \mathcal{E}_{0s}\left(\vec{\kappa}_\perp\right)\mathcal{E}_i^*(\vec{\kappa}_\perp')\right\rangle = i\pi\frac{\delta z}{\kappa}I_\xi\delta\hat{\varepsilon}\left(\vec{\kappa}_\perp - \vec{\kappa}_\perp'\right)\left|\mathcal{H}_i\left(-\vec{\kappa}_\perp'\right)\right|^2 \tag{10.10}$$

Similarly, the third term becomes

$$\left\langle \mathcal{E}_i\left(\vec{\kappa}_\perp\right)\mathcal{E}_{0s}^*(\vec{\kappa}_\perp')\right\rangle = -i\pi\frac{\delta z}{\kappa}I_\xi\delta\hat{\varepsilon}^*\left(\vec{\kappa}_\perp' - \vec{\kappa}_\perp\right)\left|\mathcal{H}_i\left(-\vec{\kappa}_\perp\right)\right|^2 \tag{10.11}$$

Putting these terms together, and effecting our coordinate transformation (Eq. 4.13), we arrive at

$$\mathcal{J}\left(\vec{\kappa}_{\perp c}, \vec{\kappa}_{\perp d}\right) = I_{\xi}\left[\kappa^{-2}\mathcal{H}_{\cap}\left(\vec{\kappa}_{\perp c}, 0\right)\delta^2\left(\vec{\kappa}_{\perp d}\right) + i\pi\frac{\delta z}{\kappa}\delta\hat{\varepsilon}\left(\vec{\kappa}_{\perp d}\right)\mathcal{H}_{\cap}^*\left(\vec{\kappa}_{\perp c}, -\vec{\kappa}_{\perp d}\right)\right.$$
$$\left.-i\pi\frac{\delta z}{\kappa}\delta\hat{\varepsilon}^*\left(-\vec{\kappa}_{\perp d}\right)\mathcal{H}_{\cap}\left(\vec{\kappa}_{\perp c}, \vec{\kappa}_{\perp d}\right)\right]$$

(10.12)

where we have introduced an effective transmission coherent transfer function, defined by

$$\mathcal{H}_{\cap}\left(\vec{\kappa}_{\perp c}, \vec{\kappa}_{\perp d}\right) = \mathcal{H}\left(\vec{\kappa}_{\perp c} + \tfrac{1}{2}\vec{\kappa}_{\perp d}\right)\mathcal{H}^*\left(\vec{\kappa}_{\perp c} - \tfrac{1}{2}\vec{\kappa}_{\perp d}\right)\left|\mathcal{H}_i\left(-\vec{\kappa}_{\perp c} - \tfrac{1}{2}\vec{\kappa}_{\perp d}\right)\right|^2 \quad (10.13)$$

As a reminder, $\mathcal{H}_i$ and $\mathcal{H}$ are the coherent transfer functions associated with the microscope illumination and detection optics respectively. The transmission coherent transfer function represents the overlap of these functions when displaced relative one another by $\pm\tfrac{1}{2}\vec{\kappa}_{\perp d}$. An example with centered circular pupils is illustrated in Fig. 10.1.

According to Eq. 10.5, to calculate the intensity spectrum we must integrate Eq. 10.12 over all $\vec{\kappa}_{\perp c}$. We finally arrive at

$$\mathcal{I}\left(\vec{\kappa}_{\perp d}\right) = I_b\delta^2\left(\vec{\kappa}_{\perp d}\right) + \mathcal{I}_s\left(\vec{\kappa}_{\perp d}\right) \quad (10.14)$$

where I_b is the featureless intensity background that is incident on the camera even when no sample is present (often referred to as ballistic light), and $\mathcal{I}_s\left(\vec{\kappa}_{\perp d}\right)$ is the additive intensity spectrum resulting from the sample, given by

$$\mathcal{I}_s\left(\vec{\kappa}_{\perp d}\right) = i\pi\frac{\delta z}{\kappa}I_{\xi}\int d^2\vec{\kappa}_{\perp c}\left[\delta\hat{\varepsilon}\left(\vec{\kappa}_{\perp d}\right)\mathcal{H}_{\cap}^*\left(\vec{\kappa}_{\perp c}, -\vec{\kappa}_{\perp d}\right) - \delta\hat{\varepsilon}^*\left(-\vec{\kappa}_{\perp d}\right)\mathcal{H}_{\cap}\left(\vec{\kappa}_{\perp c}, \vec{\kappa}_{\perp d}\right)\right]$$

(10.15)

Equations 10.14 and 10.15 are the main results of our analysis of transmission microscopy with thin samples.

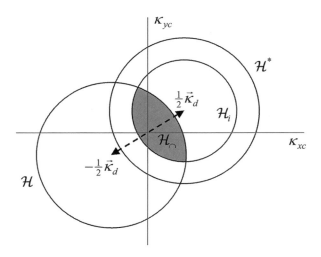

Figure 10.1. $\mathcal{H}_{\cap}\left(\vec{\kappa}_{\perp c}, \vec{\kappa}_{\perp d}\right)$ for centered circular pupils.

We can push our analysis a little further by considering different kinds of samples. For example, if the sample is a pure phase object, then $\delta\varepsilon\left(\vec{\rho}_0\right)$ is a real function and $\delta\hat{\varepsilon}\left(\vec{\kappa}_\perp\right)$ is Hermitian (i.e. $\delta\hat{\varepsilon}^*\left(\vec{\kappa}_\perp\right) = \delta\hat{\varepsilon}\left(-\vec{\kappa}_\perp\right)$). We find in this case

$$\mathcal{I}_s\left(\vec{\kappa}_{\perp d}\right) = 2\pi\kappa\delta z I_\xi \mathcal{H}_\phi\left(\vec{\kappa}_{\perp d}\right)\delta\hat{\varepsilon}\left(\vec{\kappa}_{\perp d}\right) \qquad \text{(phase object)} \qquad (10.16)$$

where

$$\mathcal{H}_\phi\left(\vec{\kappa}_{\perp d}\right) = \frac{1}{i2\kappa^2}\int d^2\vec{\kappa}_{\perp c}\left[\mathcal{H}_\cap\left(\vec{\kappa}_{\perp c},\vec{\kappa}_{\perp d}\right) - \mathcal{H}_\cap^*\left(\vec{\kappa}_{\perp c},-\vec{\kappa}_{\perp d}\right)\right] \qquad (10.17)$$

is the so-called phase transfer function.

On the other hand, if the sample is a pure absorption object, then $\delta\varepsilon\left(\vec{\rho}_0\right)$ is imaginary and $\delta\hat{\varepsilon}\left(\vec{\kappa}_\perp\right)$ is anti-Hermitian (i.e. $\delta\hat{\varepsilon}^*\left(\vec{\kappa}_\perp\right) = -\delta\hat{\varepsilon}\left(-\vec{\kappa}_\perp\right)$), in which case

$$\mathcal{I}_s\left(\vec{\kappa}_{\perp d}\right) = i2\pi\kappa\delta z I_\xi \mathcal{H}_A\left(\vec{\kappa}_{\perp d}\right)\delta\hat{\varepsilon}\left(\vec{\kappa}_{\perp d}\right) \qquad \text{(absorption object)} \qquad (10.18)$$

where

$$\mathcal{H}_A\left(\vec{\kappa}_{\perp d}\right) = \frac{1}{2\kappa^2}\int d^2\vec{\kappa}_{\perp c}\left[\mathcal{H}_\cap\left(\vec{\kappa}_{\perp c},\vec{\kappa}_{\perp d}\right) + \mathcal{H}_\cap^*\left(\vec{\kappa}_{\perp c},-\vec{\kappa}_{\perp d}\right)\right] \qquad (10.19)$$

is the so-called absorption transfer function. With these definitions, both $\mathcal{H}_\phi\left(\vec{\kappa}_\perp\right)$ and $\mathcal{H}_A\left(\vec{\kappa}_\perp\right)$ are Hermitian, meaning that their inverse Fourier transforms $H_\phi\left(\vec{\rho}\right)$ and $H_A\left(\vec{\rho}\right)$ are real.

An important conclusion can be drawn from Eqs. 10.17 and 10.19: an in-focus brightfield microscope can reveal the zero-frequency component ($\vec{\kappa}_{\perp d} = 0$) of an absorption object but not of a phase object. In other words, in-focus brightfield microscopy is insensitive to average phase.

Coherence Parameter

A useful parameter to characterize the effective transmission coherent transfer function introduced in Eq. 10.12 is called the coherence parameter, defined by

$$\varsigma = \frac{\Delta\kappa_{\perp i}}{\Delta\kappa_\perp} \qquad (10.20)$$

where $\Delta\kappa_{\perp i}$ and $\Delta\kappa_\perp$ are the spatial frequency bandwidths of $\mathcal{H}_i$ and $\mathcal{H}$ respectively. That is, from Eq. 3.35, ς corresponds to the ratio of illumination and detection numerical apertures. To gain a better appreciation of the significance of this parameter, let us consider imaging an absorption object in the simplified case where both illumination and detection pupils are centered on axis, and where the illumination pupil is binary (i.e. $\left|\mathcal{H}_i\left(\vec{\kappa}_\perp\right)\right|^2 = \mathcal{H}_i\left(\vec{\kappa}_\perp\right)$ is equal to 0 or 1).

When $\varsigma \to 0$, the bandwidth of $\mathcal{H}_i\left(\vec{\kappa}_\perp\right)$ becomes vanishingly narrow. This is called a coherent illumination configuration, corresponding to illumination with an on-axis plane wave, and leads to

$$\mathcal{H}_A\left(\vec{\kappa}_{\perp d}\right) \to \tfrac{1}{2}\left[\mathcal{H}\left(0\right)\mathcal{H}^*\left(-\vec{\kappa}_{\perp d}\right) + \mathcal{H}^*\left(0\right)\mathcal{H}\left(\vec{\kappa}_{\perp d}\right)\right] \qquad \text{(coherent)} \qquad (10.21)$$

which, for a real and symmetric detection pupil, simplifies to $\mathcal{H}_A\left(\vec{\kappa}_{\perp d}\right) \to \mathcal{H}\left(0\right)\mathcal{H}\left(\vec{\kappa}_{\perp d}\right)$.

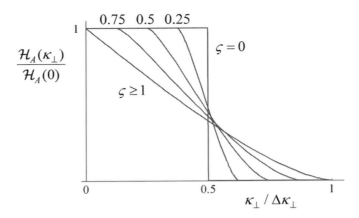

Figure 10.2. $\mathcal{H}_A\left(\vec{\kappa}_{\perp d}\right)$ for circular pupil, as a function of coherence parameter ς.

When $\varsigma \geq 1$, the illumination pupil plays no role in defining $\mathcal{H}_\cap\left(\vec{\kappa}_{\perp c}, \vec{\kappa}_{\perp d}\right)$. This is called an incoherent illumination configuration, and leads instead to

$$\mathcal{H}_A\left(\vec{\kappa}_{\perp d}\right) \rightarrow \mathrm{OTF}\left(\vec{\kappa}_{\perp d}\right) \qquad \text{(incoherent)} \qquad (10.22)$$

where we have made use of Eq. 4.71. Plots of $\mathcal{H}_A\left(\vec{\kappa}_{\perp d}\right)$ for circular pupils and for various values of ς are illustrated in Fig. 10.2. It is apparent that as ς increases from zero, $\mathcal{H}_A\left(\vec{\kappa}_{\perp d}\right)$ makes a smooth transition from a coherent transfer function to an optical transfer function (see Fig. 4.6), thus leading to an increase in its frequency support by a factor of two.

10.1.3 Phase Imaging

In many cases, for example in biological microscopy, samples are not very absorptive and instead impart only phase shifts to the transmitted incident light. Such samples, befittingly, are called phase objects. The problem of phase imaging is one of the most well known in microscopy and has been longstanding for decades. Many solutions are available for phase imaging, only a few of which will be described here. The solutions depend, to a large extent, on the degree of coherence of the illumination, as parametrized by ς, and we will consider scenarios where the illumination is coherent, partially coherent, and completely incoherent.

For ease of notation, we introduce the sample phase function $\phi\left(\vec{\rho}_0\right) = \pi\kappa\delta z\delta\varepsilon\left(\vec{\rho}_0\right)$, enabling us to recast the phase screen model of our 2D sample (Eq. 9.3) into the simpler form

$$t\left(\vec{\rho}_0\right) = \left|t\left(\vec{\rho}_0\right)\right|e^{i\phi\left(\vec{\rho}_0\right)} \qquad (10.23)$$

To be clear here, the term phase denoted by $\phi\left(\vec{\rho}_0\right)$, refers to the phase of the sample itself and not of the light. Indeed, as noted above, though we are confining our analysis to quasi-monochromatic fields, our analysis is generalizable to non-monochromatic fields whose phases, in general, are ill defined.

Coherent Illumination

The simplest scenario for phase imaging involves coherent illumination with an on-axis plane wave. This occurs, for example, if the diameter of a binary illumination pupil is reduced to zero (while nevertheless retaining sufficient power to allow imaging). In this case, from Eq. 10.17, we have

$$\mathcal{H}_\phi\left(\vec{\kappa}_{\perp d}\right) = \tfrac{1}{i2}\left[\mathcal{H}\left(0\right)\mathcal{H}^*\left(-\vec{\kappa}_{\perp d}\right) - \mathcal{H}^*\left(0\right)\mathcal{H}\left(\vec{\kappa}_{\perp d}\right)\right] \qquad (10.24)$$

And herein lies a problem: if the detection pupil is both real and symmetric, as is the case with standard brightfield microscopy, then $\mathcal{H}_\phi\left(\vec{\kappa}_{\perp d}\right)$ vanishes identically and phase imaging is impossible. Clearly, in-focus coherent phase imaging with on-axis illumination requires the detection pupil to be either not completely real or not completely symmetric (or both).

Zernike Phase Contrast

The first of these solutions, namely where the detection pupil is symmetric but not completely real, is known as Zernike phase contrast, in deference to its inventor [20]. In this solution, almost the entire pupil is unobstructed except for the very center, wherein a tiny optical element is inserted that induces a $\frac{\pi}{2}$ phase shift. As a result, $\mathcal{H}\left(\vec{\kappa}_{\perp d} \neq 0\right)$ is real while $\mathcal{H}\left(\vec{\kappa}_{\perp d} = 0\right)$ is imaginary, and, from Eq. 10.24, $\mathcal{H}_\phi\left(\vec{\kappa}_{\perp d}\right) \approx i\,\mathrm{Im}\left[\mathcal{H}\left(0\right)\right]\mathcal{H}\left(\vec{\kappa}_{\perp d}\right)$. An illustration of a Zernike phase contrast microscope is shown in Fig. 10.3, which provides an intuitive interpretation of the technique. Indeed, from Eq. 10.7, we observe that phase information is carried from the sample to the camera by the terms depending exclusively on the interference between incident and scattered fields (terms T_2 and T_3, neglecting term T_4). The unscattered incident field propagates through the center of the detection pupil, while the scattered field propagates mostly away from the center. When the incident and scattered fields are $\frac{\pi}{2}$ out of phase, as in the case of phase imaging, then their interference can produce no intensity variations at the camera. On the other hand, by introducing an additional $\frac{\pi}{2}$ phase shift in the center of the pupil, the incident and scattered light fields are brought back into phase, leading to detectable intensity variations. It should be noted that Fig. 10.3 represents an overly simplistic and impractical version of Zernike phase contrast, since a vanishingly small illumination pupil

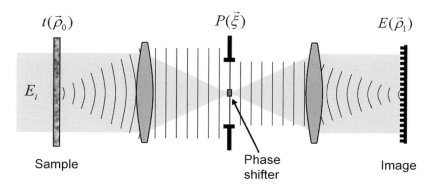

Figure 10.3. Zernike phase contrast with on-axis phase mask.

is not very light efficient. In practice, Zernike phase contrast is usually obtained instead with an annular illumination pupil for improved light efficiency, combined with an annular phase shifter in the detection pupil. The basic principle, however, remains the same.

Schlieren Microscopy

The second of the solutions suggested above for coherent in-focus phase imaging involves breaking the symmetry of the detection aperture. A simple way to achieve this is to use a knife edge to block half of an otherwise centered, open detection pupil (leaving the center just barely unblocked, so that $\mathcal{H}(0)$ is not zero). The detection pupil in this case can be decomposed into even and odd components, denoted by $\mathcal{H}_{\text{even}}$ and $\mathcal{H}_{\text{odd}}$ respectively, such that $\mathcal{H}\left(\vec{\kappa}_{\perp d}\right) = \mathcal{H}_{\text{even}}\left(\vec{\kappa}_{\perp d}\right) + \mathcal{H}_{\text{odd}}\left(\vec{\kappa}_{\perp d}\right)$. From Eq. 10.24, this leads to $\mathcal{H}_\phi\left(\vec{\kappa}_{\perp d}\right) \approx i\mathcal{H}(0)\mathcal{H}_{\text{odd}}\left(\vec{\kappa}_{\perp d}\right)$. In other words, the odd component of the pupil carries phase information while the even component does not. This knife-edge technique is called Schlieren microscopy, and is illustrated in Fig. 10.4. It should be noted that here too the figure is overly simplistic, for the same reason that a closed illumination pupil is not very light efficient. In practice, illumination in Schlieren microscopy is usually performed through a slit pupil, where the slit is oriented along the same direction as the knife edge.

A feature of Schlieren microscopy is that because $\mathcal{H}_{\text{odd}}\left(\vec{\kappa}_{\perp d}\right)$ is odd, so too must be $\mathcal{H}_\phi\left(\vec{\kappa}_{\perp d}\right)$. As a result, Schlieren microscopy reveals not the phase object itself but rather the *gradient* of the phase object along the direction perpendicular to the knife edge. This becomes apparent if we take the inverse Fourier transform of Eq. 10.16 to recover the light intensity at the camera, obtaining

$$I_s\left(\vec{\rho}_1\right) = 2I_\xi \int \mathrm{d}^2\vec{\rho}_0 H_\phi\left(\vec{\rho}_1 - \vec{\rho}_0\right) \phi\left(\vec{\rho}_0\right) \tag{10.25}$$

From the properties of Fourier transforms (see Appendix A), $H_\phi\left(\vec{\rho}\right)$ must also be odd along the direction perpendicular to the knife edge. A schematic illustration of $H_{\text{even}}\left(\vec{\rho}\right)$ and $H_{\text{odd}}\left(\vec{\rho}\right)$ is shown in Fig. 10.4. If the imaging is reasonably well resolved, we can assume $H_\phi\left(\vec{\rho}\right)$ is locally peaked about $\vec{\rho} = 0$. Moreover, if we assume $\phi\left(\vec{\rho}_0\right)$ is slowly varying on the length scale of the width of this peak, we can make the variable change $\vec{\rho}' = \vec{\rho}_1 - \vec{\rho}_0$ in Eq. 10.25, and perform a Taylor expansion of $\phi\left(\vec{\rho}_0\right)$ about $\vec{\rho}_1$. The zeroth order of this expansion vanishes owing to the odd symmetry of $H_\phi\left(\vec{\rho}\right)$, and so, to first order we have

$$I_s\left(\vec{\rho}_1\right) \approx 2I_\xi \int \mathrm{d}^2\vec{\rho}' H_\phi\left(\vec{\rho}'\right) \left[\vec{\rho}' \cdot \vec{\nabla}_\perp \phi\left(\vec{\rho}'\right)\Big|_{\vec{\rho}_1}\right] \tag{10.26}$$

bringing us to the conclusion that $I_s\left(\vec{\rho}_1\right)$ is dependent on the gradient of $\phi\left(\vec{\rho}_0\right)$.

An interesting variation of Schlieren microscopy comes with the use of a phase knife-edge rather than an amplitude knife-edge. This too leads to phase-gradient contrast by way of a Hilbert transform [10]. A knife-edge in phase can even be applied in a radially symmetric manner, leading to what is called spiral-phase contrast [3, 9].

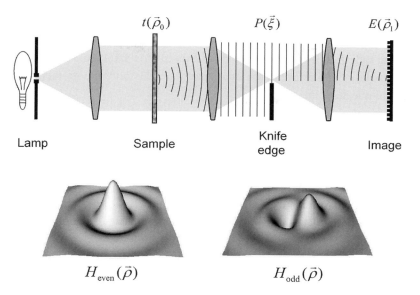

Figure 10.4. Configuration for Schlieren microscopy, and even and odd compoents of H.

Partially Coherent Illumination

Phase imaging is not restricted to coherent illumination ($\varsigma = 0$), and indeed some of the most effective techniques for phase imaging are based on non-coherent illumination ($\varsigma > 0$). But phase imaging is not always straightforward. To appreciate some of the difficulties involved, let us consider again the usual case of unobstructed (i.e. binary) pupils both in the illumination and detection optics. From Eqs. 10.17 and 10.13, we have then

$$\mathcal{H}_\phi\left(\vec{\kappa}_{\perp d}\right) = \frac{1}{i2\kappa^2} \int d^2\vec{\kappa}_{\perp c}\, \mathcal{H}\left(\vec{\kappa}_{\perp c} + \tfrac{1}{2}\vec{\kappa}_{\perp d}\right) \mathcal{H}\left(\vec{\kappa}_{\perp c} - \tfrac{1}{2}\vec{\kappa}_{\perp d}\right)$$
$$\times \left[\mathcal{H}_i\left(-\vec{\kappa}_{\perp c} - \tfrac{1}{2}\vec{\kappa}_{\perp d}\right) - \mathcal{H}_i\left(-\vec{\kappa}_{\perp c} + \tfrac{1}{2}\vec{\kappa}_{\perp d}\right)\right] \quad (10.27)$$

Let us further consider the case, also usual, where both $\mathcal{H}_i\left(\vec{\kappa}_\perp\right)$ and $\mathcal{H}\left(\vec{\kappa}_\perp\right)$ are symmetric. An examination of Eq. 10.27 reveals that, by symmetry, the integral must vanish, and hence $\mathcal{H}_\phi\left(\vec{\kappa}_{\perp d}\right) = 0$. In other words, in-focus brightfield imaging with symmetric, unobstructed pupils cannot reveal phase objects, independent of the coherence parameter ς.

Oblique-Field Microscopy

But Eq. 10.27 is not restricted to symmetric pupils. In the previous section we considered the effect of imparting an asymmetry to the detection pupil by introducing a knife-edge. A similar asymmetry can be imparted simply by offsetting either pupil. If the detection pupil is offset, the light from the sample is forced to be detected in an oblique manner. Similarly, if the illumination pupil is offset, it is the illumination that is oblique. Either case enables the possibility of phase imaging. This can be verified by adding a displacement vector to $\vec{\kappa}_{\perp c}$ in either $\mathcal{H}$ or $\mathcal{H}_i$ (recalling that a shift in position in the pupil plane corresponds to a shift in angle in the sample plane), and noting that the integral in Eq. 10.27 no longer vanishes. Such phase imaging based on oblique detection or illumination falls under the general rubric of oblique-field imaging. Just as Schlieren microscopy reveals only the gradient of phase objects, so too does oblique-field imaging.

A particularly intuitive interpretation of oblique-field imaging comes from considering the effect of the sample on the illumination flux density. To calculate the flux density into and out of a sample, we begin by evaluating the transmission of the incident mutual intensity through the sample. This is given by

$$J_0\left(\vec{\rho}_{0c}, \vec{\rho}_{0d}\right) = T\left(\vec{\rho}_{0c}, \vec{\rho}_{0d}\right) J_i\left(\vec{\rho}_{0c}, \vec{\rho}_{0d}\right) \tag{10.28}$$

where $T\left(\vec{\rho}_{0c}, \vec{\rho}_{0d}\right) = t\left(\vec{\rho}_{0c} + \frac{1}{2}\vec{\rho}_{0d}\right) t^*\left(\vec{\rho}_{0c} - \frac{1}{2}\vec{\rho}_{0d}\right)$ denotes the sample mutual transmittance. Assuming the illumination is only partially coherent, such that $J_i\left(\vec{\rho}_{0c}, \vec{\rho}_{0d}\right)$ is narrowly peaked about $\vec{\rho}_{0d} = 0$, we can confine our analysis to small values of $\vec{\rho}_{0d}$ and adopt the first-order approximation

$$T\left(\vec{\rho}_{0c}, \vec{\rho}_{0d}\right) = \left|t\left(\vec{\rho}_{0c}\right)\right|^2 \left[1 + i\vec{\rho}_{0d} \cdot \vec{\nabla}\phi\left(\vec{\rho}_{0c}\right)\right] \tag{10.29}$$

Combining Eqs. 10.28 and 10.29, and making use of the definition of transverse flux density provided by Eq. 6.6, we arrive at

$$\vec{F}_{0\perp}\left(\vec{\rho}_{0c}\right) = \left|t\left(\rho_{0c}\right)\right|^2 \left[\vec{F}_{i\perp}\left(\vec{\rho}_{0c}\right) + \frac{1}{2\pi\kappa}I_i\left(\vec{\rho}_{0c}\right)\vec{\nabla}\phi\left(\vec{\rho}_{0c}\right)\right] \tag{10.30}$$

In other words, when the flux density is transmitted through the sample it incurs not only an attenuation in regions where $\left|t\left(\rho_{0c}\right)\right|^2$ is less than 1, but also a transverse deflection where the sample phase exhibits a gradient. This occurrence of local deflections of flux density provides a simple mechanisms for phase imaging with oblique fields, as illustrated in Fig. 10.5. Indeed, as is the principle of all phase imaging mechanisms, the basic idea is to convert phase variations in the sample into intensity variations at the camera. In the case of oblique field microscopy, the flux density is clipped by the detection pupil, either because the detection pupil is offset or because the illumination is oblique. Local phase gradients in the sample cause this clipping to increase or decrease depending on the sign and strength of the phase gradients, thus leading to measurable intensity variations at the imaging camera. Note that the angular distribution of light incident upon any sample location in Fig. 10.5 is defined by the illumination numerical aperture.

Examples of oblique field microscopy were proposed early on involving offset pupils [2], or partially attenuating pupils with specialized optics, as in the case of Hoffman contrast [8]. More recent examples of microscopes utilizing oblique illumination can be found in [11], or oblique detection in [13], both of which provide quantitative phase contrast. These can be combined in various ways, yielding images such as shown in Fig. 10.6.

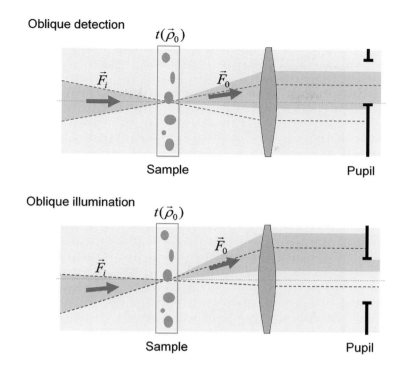

Figure 10.5. Configurations for oblique field microscopy. A component of light traversing a particular location in the sample is shown in darker gray. In both cases, local phase gradients in the sample lead to local intensity changes in the image.

Incoherent Illumination

So far, we have examined phase imaging strategies that require coherent or partially coherent incident illumination. We turn now to a remarkable strategy based on completely incoherent illumination. We recall that the coherence area of spatially incoherent illumination is defined by a 2D delta function (see Section 4.2.3), meaning that the phase of the illumination is uncorrelated and random at every point. The possibility of measuring a deterministic sample phase using incident light that is completely indeterministic seems, at first glance, quite hopeless. Nevertheless, such a possibility exists and was invented decades ago by Nomarski [12]. It is called differential interference contrast, or DIC.

Shear Interferometry

Before considering the details of DIC, let us examine the more general concept of shear interferometry, on which DIC is based. We start with a standard brightfield microscope. Let us imagine, as a thought experiment, that two identical copies of an incident field $E_i\left(\vec{\rho}_0\right)$ are separated from one another by a small transverse offset $\Delta\vec{\rho}_d$ before being transmitted through a sample. Once transmitted, they are then combined into a single field with their offset annulled.

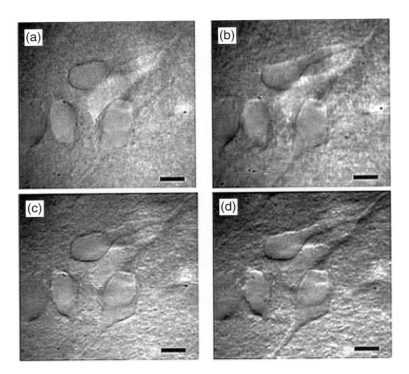

Figure 10.6. Images of neurons in a brain slice acquired with standard imaging (a), a partial beam block in the illumination (b) or detection (c) aperture, and partial beam blocks in both apertures (d). Scale bar = 10 μm. Adapted from [19].

That is, the radiant field after the sample is written as the superposition of two components

$$\mathcal{E}_0(\vec{\kappa}_\perp)^\pm = \mathcal{E}_i(\vec{\kappa}_\perp) + \mathcal{E}_{0s}(\vec{\kappa}_\perp)^\pm \tag{10.31}$$

where the superscripts $+$ and $-$ refer to the different paths through the sample, and, from Eq. 10.6 and the Fourier shift theorem (Eq. A.3), we have

$$\mathcal{E}_{0s}(\vec{\kappa}_\perp)^\pm = ie^{\pm i\pi\vec{\kappa}_\perp \cdot \Delta\vec{\rho}_d} \int d^2\vec{\kappa}_\perp'' \, \hat{\phi}\left(\vec{\kappa}_\perp - \vec{\kappa}_\perp''\right) e^{\mp i\pi\vec{\kappa}_\perp'' \cdot \Delta\vec{\rho}_d} \mathcal{E}_i(\vec{\kappa}_\perp'') \tag{10.32}$$

where $\hat{\phi}\left(\vec{\kappa}_\perp\right)$ is the 2D Fourier transform of $\phi\left(\vec{\rho}_0\right)$. The interpretation of the above equation is clear. The rightmost exponential offsets the incoming incident field by the vectors $\pm\frac{1}{2}\Delta\vec{\rho}_d$, while the leftmost exponential offsets the outgoing fields by the opposite vectors $\mp\frac{1}{2}\Delta\vec{\rho}_d$, thus annulling the offsets.

Continuing with our thought experiment, let us further imagine that these radiant field components incur additional phase shifts $\pm\psi$ upon propagation through the microscope to the camera. That is, the two radiant field components incident on the camera are given by

$$\mathcal{E}(\vec{\kappa}_\perp)^\pm = \mathcal{H}\left(\vec{\kappa}_\perp\right) e^{\pm i\psi} \left[\mathcal{E}_i(\vec{\kappa}_\perp) + \mathcal{E}_{0s}(\vec{\kappa}_\perp)^\pm\right] \tag{10.33}$$

and the total radiant field is thus $\mathcal{E}(\vec{\kappa}_\perp) = \mathcal{E}(\vec{\kappa}_\perp)^+ + \mathcal{E}(\vec{\kappa}_\perp)^-$.

We can proceed now in the same manner as above. The total radiant mutual intensity at the camera can be separated into four terms

$$\mathcal{J}\left(\vec{\kappa}_\perp, \vec{\kappa}_\perp'\right) = \mathcal{J}\left(\vec{\kappa}_\perp, \vec{\kappa}_\perp'\right)^{++} + \mathcal{J}\left(\vec{\kappa}_\perp, \vec{\kappa}_\perp'\right)^{--} + \mathcal{J}\left(\vec{\kappa}_\perp, \vec{\kappa}_\perp'\right)^{+-} + \mathcal{J}\left(\vec{\kappa}_\perp, \vec{\kappa}_\perp'\right)^{-+}$$

(10.34)

The first two of these correspond to the interferences of each radiant field component with itself. The second two correspond to cross interferences between different components. Following our usual coordinate transformation, each of these terms produces an intensity spectrum at the camera, according to Eq. 10.5.

At this point, we can make assumptions about our illumination. Specifically, let us consider the case of fully incoherent Köhler illumination, and write $\mathcal{J}_i\left(\vec{\kappa}_{\perp c}, \vec{\kappa}_{\perp d}\right) = \kappa^{-2} I_\xi \delta^2\left(\vec{\kappa}_{\perp d}\right)$ (see Eq. 4.58). A tedious but straightforward calculation then leads to

$$\mathcal{I}\left(\vec{\kappa}_{\perp d}\right)^{++} = I_\xi \mathrm{OTF}\left(\vec{\kappa}_{\perp d}\right)\left[\delta^2\left(\vec{\kappa}_{\perp d}\right) + i\hat{\phi}\left(\vec{\kappa}_{\perp d}\right) e^{i\pi \vec{\kappa}_{\perp d}\cdot\Delta\vec{\rho}_d} - i\hat{\phi}^*\left(-\vec{\kappa}_{\perp d}\right) e^{i\pi \vec{\kappa}_{\perp d}\cdot\Delta\vec{\rho}_d}\right]$$

(10.35)

$$\mathcal{I}\left(\vec{\kappa}_{\perp d}\right)^{--} = I_\xi \mathrm{OTF}\left(\vec{\kappa}_{\perp d}\right)\left[\delta^2\left(\vec{\kappa}_{\perp d}\right) + i\hat{\phi}\left(\vec{\kappa}_{\perp d}\right) e^{-i\pi \vec{\kappa}_{\perp d}\cdot\Delta\vec{\rho}_d} - i\hat{\phi}^*\left(-\vec{\kappa}_{\perp d}\right) e^{-i\pi \vec{\kappa}_{\perp d}\cdot\Delta\vec{\rho}_d}\right]$$

(10.36)

$$\mathcal{I}\left(\vec{\kappa}_{\perp d}\right)^{+-} = I_\xi \mathrm{OTF}\left(\vec{\kappa}_{\perp d}\right) e^{i2\psi}\left[\delta^2\left(\vec{\kappa}_{\perp d}\right) + i\hat{\phi}\left(\vec{\kappa}_{\perp d}\right) e^{i\pi \vec{\kappa}_{\perp d}\cdot\Delta\vec{\rho}_d} - i\hat{\phi}^*\left(-\vec{\kappa}_{\perp d}\right) e^{-i\pi \vec{\kappa}_{\perp d}\cdot\Delta\vec{\rho}_d}\right]$$

(10.37)

$$\mathcal{I}\left(\vec{\kappa}_{\perp d}\right)^{-+} = I_\xi \mathrm{OTF}\left(\vec{\kappa}_{\perp d}\right) e^{-i2\psi}\left[\delta^2\left(\vec{\kappa}_{\perp d}\right) + i\hat{\phi}\left(\vec{\kappa}_{\perp d}\right) e^{-i\pi \vec{\kappa}_{\perp d}\cdot\Delta\vec{\rho}_d} - i\hat{\phi}^*\left(-\vec{\kappa}_{\perp d}\right) e^{i\pi \vec{\kappa}_{\perp d}\cdot\Delta\vec{\rho}_d}\right]$$

(10.38)

We note that $\mathcal{I}\left(\vec{\kappa}_{\perp d}\right)^{++}$ and $\mathcal{I}\left(\vec{\kappa}_{\perp d}\right)^{--}$ are individually Hermitian while $\mathcal{I}\left(\vec{\kappa}_{\perp d}\right)^{+-} + \mathcal{I}\left(\vec{\kappa}_{\perp d}\right)^{-+}$ is only Hermitian in combination. The total intensity spectrum is given by the sum of Eqs. 10.35 to 10.38. Yet another round of calculations brings us to

$$\mathcal{I}\left(\vec{\kappa}_{\perp d}\right) = 4 I_\xi \mathrm{OTF}\left(\vec{\kappa}_{\perp d}\right)\left[\cos^2\psi\, \delta^2\left(\vec{\kappa}_{\perp d}\right) + i\cos^2\psi\left(\hat{\phi}\left(\vec{\kappa}_{\perp d}\right) - \hat{\phi}^*\left(-\vec{\kappa}_{\perp d}\right)\right)\right.$$
$$\left. + i\pi \sin\psi\cos\psi\left(\Delta\vec{\rho}_d\cdot\vec{\kappa}_{\perp d}\right)\left(\hat{\phi}\left(\vec{\kappa}_{\perp d}\right) + \hat{\phi}^*\left(-\vec{\kappa}_{\perp d}\right)\right)\right]$$

(10.39)

where we have made the assumption that $\Delta\rho_d$ is small, allowing us to expand the exponentials to first order.

This result is more general than we need since it makes allowances for the possibility that $\phi\left(\vec{\rho}_0\right)$ is complex. In the case we are considering a pure phase sample, then $\phi\left(\vec{\rho}_0\right)$ is real (meaning $\hat{\phi}\left(\vec{\kappa}_{\perp d}\right)$ is Hermitian) and we arrive at finally

$$\mathcal{I}\left(\vec{\kappa}_{\perp d}\right) = 4 I_\xi \mathrm{OTF}\left(\vec{\kappa}_{\perp d}\right)\left[\cos^2\psi\, \delta^2\left(\vec{\kappa}_{\perp d}\right) + i2\pi \sin\psi\cos\psi\left(\Delta\vec{\rho}_d\cdot\vec{\kappa}_{\perp d}\right)\hat{\phi}\left(\vec{\kappa}_{\perp d}\right)\right]$$

(10.40)

According to this result, our thought experiment has led us to a mechanism for measuring sample phase based on the use of completely incoherent Köhler illumination. But, as we have seen before, it is not the sample phase that is measured here, but rather the gradient of the sample phase along the direction of $\Delta\vec{\rho}_d$. This is apparent upon taking the Fourier transform of Eq. 10.40, leading to

$$I\left(\vec{\rho}_1\right) = I_b\left[\Omega_0\cos^2\psi + \sin\psi\cos\psi\int d^2\rho_0\,\mathrm{PSF}\left(\vec{\rho}_1-\vec{\rho}_0\right)\Delta\vec{\rho}_d\cdot\vec{\nabla}\phi\left(\vec{\rho}_0\right)\right] \quad (10.41)$$

where $I_b = 4I_\xi$ (see Eq. 4.84) and $I_b\Omega_0\cos^2\psi$ is the homogeneous background illumination intensity incident on the camera without the presence of the sample.

Differential Interference Contrast

The reader may wonder how it is possible to put our thought experiment into practice. Again, the basic principle of shear interferometry is to split an incident field into identical copies, shear these copies relative to one another before they interact with the sample and recombine them after they have interacted with the sample. This can be accomplished by the technique of Nomarski DIC [1, 12], shown in Fig. 10.7.

As it happens, DIC is based on something we have not yet considered in this chapter: polarization. The reader is referred to Appendix C for a brief summary of the Jones matrix formalism to treat light polarization. Thus, as we have already seen in Section 9.4, a field in a transverse plane can be projected into horizontal (x) and vertical (y) polarization components, written in column matrix form as

$$\vec{E}(\vec{\rho}) = \begin{pmatrix} E^{(p)}\left(\vec{\rho}\right) \\ E^{(s)}\left(\vec{\rho}\right) \end{pmatrix} \quad (10.42)$$

where the superscripts (p) and (s) label these projections. In its basic configuration, a DIC microscope makes use of two polarizers and two Nomarski prisms placed in the illumination and detection pupils (Fig. 10.7). The effect of polarizers on the different polarization components of a light beam is described in Appendix C. For our purposes, the effect of the Nomarski prisms will be simply to impart a small angular separation between the horizontal and vertical polarization components.

In brief, from left to right, the first polarizer, oriented at $+45°$, creates two identical (albeit cross polarized) copies of the field $E_\xi\left(\vec{\rho}_\xi\right)$ incident on the illumination pupil. This illumination

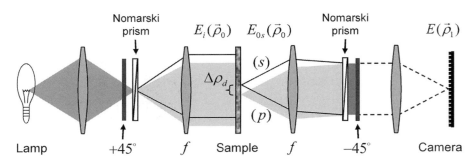

Figure 10.7. DIC microscope configuration.

pupil is presumed to be infinitely large, and so it is not shown in the figure. The first Normarski prism then imparts an angular separation to these beams. Because this angular separation occurs at the pupil plane, it results in a spatial separation at the focal plane, corresponding in our case to the shear vector $\Delta\vec{\rho}_d$ described above. The second Nomarski prism, identical to the first, then recombines the sheared fields after they have traversed the sample. Finally, bearing in mind that orthogonal polarization components do not interfere, a final polarizer typically oriented at $-45°$ is required to project these components onto a common polarization axis so that they can interfere, which they do at the camera.

In more detail, if the field just in front of the first polarizer is given by

$$\vec{E}_\xi(\vec{\xi}) = \begin{pmatrix} E_\xi^{(p)}(\vec{\xi}) \\ E_\xi^{(s)}(\vec{\xi}) \end{pmatrix} \tag{10.43}$$

then, from Eq. C.8, the field just after the polarizer but before the first Nomarski prism is given by

$$\vec{E}_\xi'(\vec{\xi}) = \frac{1}{2}\begin{pmatrix} 1 & 1 \\ 1 & 1 \end{pmatrix}\begin{pmatrix} E_\xi^{(p)}(\vec{\xi}) \\ E_\xi^{(s)}(\vec{\xi}) \end{pmatrix} = \frac{1}{2}\begin{pmatrix} E_\xi^{(p)}(\vec{\xi}) + E_\xi^{(s)}(\vec{\xi}) \\ E_\xi^{(p)}(\vec{\xi}) + E_\xi^{(s)}(\vec{\xi}) \end{pmatrix} \tag{10.44}$$

And herein lies the crux of DIC. Even though the field polarization components in Eq. 10.43 are separately fully incoherent and completely independent of one another, after they pass through the polarizer they become rigorously identical.

To establish the connection between this section and the last, we note that the first Nomarski prism and the focal plane are arranged in a 2f lens configuration (see Section 4.3). We can thus make the association, neglecting an irrelevant phase factor,

$$\mathcal{E}_i\left(\vec{\kappa}_\perp\right) = \frac{1}{2}\left[E_\xi^{(p)}\left(-\frac{f}{\kappa}\vec{\kappa}_\perp\right) + E_\xi^{(s)}\left(-\frac{f}{\kappa}\vec{\kappa}_\perp\right)\right] \tag{10.45}$$

The only remaining item that is missing from our connection is the additional phase shifts $\pm\psi$ imparted on the fields. In practice, this is usually accomplished by a lateral translation of the second Nomarski prism. The relative phase difference 2ψ is called a bias phase. From Eq. 10.41, this is observed to play a critical role in tuning the signal-to-background ratio of a DIC microscope. If $\psi = \frac{\pi}{2}$, then both the background and phase-gradient signal vanish. In this case, a more involved calculation taken to higher order in signal strength reveals that DIC microscopy becomes sensitive to the second-derivative, or Laplacian, of the sample phase. If instead $\psi = 0$, then the background becomes relatively quite large, whereas the signal remains vanishingly second-order in strength. In practice, the bias is usually chosen such that $\psi \approx \frac{\pi}{4}$, which maximizes the product $\sin\psi\cos\psi$. The intensity at the camera then varies linearly with phase-gradient strength, increasing or decreasing relative to the background intensity depending on the sign of the phase-gradient. As with Schlieren or oblique-field microscopy, such a linear dependence on phase gradients leads to shadowing effects that confer an apparent 3D relief structure to the sample (see Fig. 10.8).

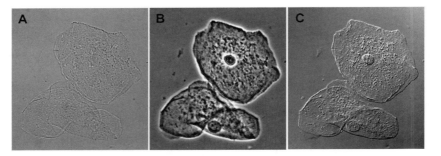

Figure 10.8. Images of cheek cells acquired with (A) standard widefield illumination, (B) Zernike phase contrast microscopy with annular illumination, and (C) Nomarski differential interference microscopy.

It should be emphasized again that the illumination in DIC microscopy is completely incoherent. No pinhole, annulus, or slit is required in the illumination pupil plane, and indeed no pupil is required at all, making the technique highly light efficient and capable of working with an ordinary low-power lamp. Moreover, the technique requires only a small modification to standard brightfield microscopy, namely the addition of two polarizers and two Nomarski prisms. As such, DIC microscopy remains one of the most popular phase imaging techniques available. A comparison of Zernike phase contrast and DIC images is shown in Fig. 10.8.

10.1.4 Volume Samples

We have confined our analysis of transmission microscopy to thin samples. While such samples can be readily found in practice, a more general treatment of transmission microscopy should allow for the possibility of thick samples. The main strategy we have used thus far is summarized in four basic equations, namely Eqs. 10.2–10.5. This strategy has the particularity of being based on radiant fields instead of fields, with the motivation that this simplifies calculations significantly. For example, the expression for $\mathcal{H}_\cap \left(\vec{\kappa}_{\perp c}, \vec{\kappa}_{\perp d} \right)$ given by Eq. 10.13, which involves a simple product of pupil functions, would be much more complicated had we worked directly with fields.

Yet another benefit that comes from working with radiant fields is in the generalization of our strategy to volume samples. As it happens, only the first two of the four basic equations in our strategy need be slightly modified. In particular, for volume samples, we must add a third dimension to our calculations. We continue to base our analysis on radiant fields, but now explicitly include z_0 in our representation of the radiant fields about the focal plane (defined by $z_0 = 0$). Equations 10.2 and 10.3 become

$$\mathcal{E}_0 \left(\vec{\kappa}_\perp; 0 \right) = \mathcal{E}_i \left(\vec{\kappa}_\perp; 0 \right) + \mathcal{E}_{0s} \left(\vec{\kappa}_\perp; 0 \right) \tag{10.46}$$

$$\mathcal{E} \left(\vec{\kappa}_\perp \right) = \mathcal{H} \left(\vec{\kappa}_\perp \right) \mathcal{E}_0 \left(\vec{\kappa}_\perp; 0 \right) \tag{10.47}$$

where $\mathcal{H} \left(\vec{\kappa}_\perp \right)$ continues to be the same in-focus coherent transfer function we used for thin samples.

The key difference in extending to the third dimension resides is how we calculate $\mathcal{E}_{0s}\left(\vec{\kappa}_\perp; 0\right)$ in Eq. 10.46. Since the sample is now volumetric, we must add a third dimension to its model, and we write $\delta\hat{\varepsilon}\left(\vec{\kappa}_\perp; z_0\right)$. As always, the major simplification comes from the use of the Born approximation, which neglects multiple scattering. From Eq. 9.28, we arrive at

$$\mathcal{E}_{0s}(\vec{\kappa}_\perp; 0) = -4\pi^2\kappa^2 \iint \mathrm{d}^2\vec{\kappa}_\perp'' \, \mathrm{d}z_0 \, \mathcal{G}_+(\vec{\kappa}_\perp; -z_0)\delta\hat{\varepsilon}\left(\vec{\kappa}_\perp - \vec{\kappa}_\perp''; z_0\right) \mathcal{D}_+(\vec{\kappa}_\perp''; z_0)\mathcal{E}_i(\vec{\kappa}_\perp''; 0)$$

$$(10.48)$$

This equation warrants a closer examination. Its interpretation becomes clear when read from right to left. The in-focus incident radiant field $\mathcal{E}_i(\vec{\kappa}_\perp''; 0)$ is propagated as if the sample were absent to an out-of-focus plane by $\mathcal{D}_+(\vec{\kappa}_\perp''; z_0)$, where it interacts with the sample $\delta\hat{\varepsilon}\left(\vec{\kappa}_\perp; z_0\right)$ at this plane. The resulting field from this interaction is then propagated back to focal plane by $\mathcal{G}_+(\vec{\kappa}_\perp; -z_0)$, again as if the sample were absent. Note that because we are using $\mathcal{G}_+$ rather than $\mathcal{G}$, the sample can be distributed upstream or downstream from the focal plane, provided we are considering only forward-directed wavevectors (see discussion in Section 2.3.2). We recall that, by definition, $\mathcal{E}_i(\vec{\kappa}_\perp''; 0)$ is the incident in-focus radiant field in the absence of the sample. Moreover, the propagations to and from the out-of-focus plane can be considered to be free-space propagations because we have neglected multiple scattering, meaning that each out-of-focus sample plane is treated independently.

A simplification is afforded if we make use of the paraxial form of $\mathcal{G}_+$ (Eq. 2.35 with $\cos\theta \to 1$), such that Eq. 10.48 becomes

$$\mathcal{E}_{0s}(\vec{\kappa}_\perp; 0) = i\pi\kappa \iint \mathrm{d}^2\vec{\kappa}_\perp'' \, \mathrm{d}z_0 \, \mathcal{D}_+(\vec{\kappa}_\perp; -z_0)\delta\hat{\varepsilon}\left(\vec{\kappa}_\perp - \vec{\kappa}_\perp''; z_0\right) \mathcal{D}_+(\vec{\kappa}_\perp''; z_0)\mathcal{E}_i(\vec{\kappa}_\perp''; 0)$$

$$(10.49)$$

Equation 10.49 represents a 3D generalization of Eq. 10.6, and as expected, in the case of a thin sample where $\delta\hat{\varepsilon}\left(\vec{\kappa}_\perp; z_0\right) \to \delta z\, \delta\hat{\varepsilon}\left(\vec{\kappa}_\perp\right)\delta(z_0)$, the two become equivalent. The remainder of our strategy continues as before, arriving finally at the calculation of the intensity spectrum $\mathcal{I}\left(\vec{\kappa}_\perp\right)$ at the camera.

Phase Imaging Revisited

The addition of a third dimension to our analysis has many ramifications. As an example, let us revisit the possibility of phase imaging. In particular, it was asserted above that in-focus brightfield imaging with symmetric, unobstructed pupils cannot reveal the phase of a thin sample, regardless of the coherence parameter ς. This turns out to be no longer true when the sample is displaced out of focus, and indeed, since the invention of transmission microscopy, it has been well known that transparent objects become more visible when they are imaged out of focus. We will not delve deeply into this phenomenon, but rather illustrate it with the example of point object.

In particular, let us derive the image produced by a 3D point object located at an axial position z_σ. That is, we write $\delta\varepsilon\left(\vec{r}_0\right) = \delta\varepsilon_\sigma\delta^2\left(\vec{\rho}_0\right)\delta\left(z_0 - z_\sigma\right)$, or equivalently $\delta\hat{\varepsilon}\left(\vec{\kappa}_\perp; z_0\right) = \delta\varepsilon_\sigma\delta\left(z_0 - z_\sigma\right)$. Equation 10.49 then simplifies in the Fresnel approximation to

$$\mathcal{E}_{0s}(\vec{\kappa}_\perp; 0) = i\pi\kappa\delta\varepsilon_\sigma \int \mathrm{d}^2\vec{\kappa}_\perp'' \, e^{i\pi\frac{z_\sigma}{\kappa}\left(\kappa_\perp^2 - \kappa_\perp''^2\right)}\mathcal{E}_i(\vec{\kappa}_\perp''; 0) \qquad (10.50)$$

Following the procedure outlined in Section 10.1.2, we find that Eq. 10.15 becomes modified to

$$\mathcal{I}_s\left(\vec{\kappa}_{\perp d}\right) = i\pi\kappa^{-1}I_\varepsilon \int d^2\vec{\kappa}_{\perp c}\, e^{i2\pi\frac{z_\sigma}{\kappa}\vec{\kappa}_{\perp c}\cdot\vec{\kappa}_{\perp d}}\left[\delta\varepsilon_\sigma\mathcal{H}_\cap^*\left(\vec{\kappa}_{\perp c}, -\vec{\kappa}_{\perp d}\right) - \delta\varepsilon_\sigma^*\mathcal{H}_\cap\left(\vec{\kappa}_{\perp c},\vec{\kappa}_{\perp d}\right)\right]$$

(10.51)

where $\mathcal{H}_\cap\left(\vec{\kappa}_{\perp c},\vec{\kappa}_{\perp d}\right)$ is the effective transmission coherent transfer function defined by Eq. 10.13. The addition of an exponential in the integrand has significant consequences.

Let us simplify the problem further my considering only unobstructed (binary) and symmetric pupils. We can start by examining the case of incoherent illumination where $\varsigma \geq 1$. We have then

$$\mathcal{H}_\cap\left(\vec{\kappa}_{\perp c},\vec{\kappa}_{\perp d}\right) = \mathcal{H}_\cap^*\left(\vec{\kappa}_{\perp c}, -\vec{\kappa}_{\perp d}\right) = \mathcal{H}\left(\vec{\kappa}_{\perp c} + \tfrac{1}{2}\vec{\kappa}_{\perp d}\right)\mathcal{H}\left(\vec{\kappa}_{\perp c} - \tfrac{1}{2}\vec{\kappa}_{\perp d}\right)$$

(10.52)

Inserting this into Eq. 10.51 and making use of Eq. 5.37 we obtain

$$\mathcal{I}_s\left(\vec{\kappa}_{\perp d}\right) = -2\pi\kappa I_\varepsilon \text{OTF}\left(\kappa_{\perp d}; -z_\sigma\right)\text{Im}\left[\delta\varepsilon_\sigma\right]$$

(10.53)

arriving at the conclusion that we still cannot image the phase of our point object.

On the other hand, the situation changes in the case of partially coherent illumination ($\varsigma < 1$). Let us look, for example, at the extreme case of completely coherent illumination where the illumination pupil is essentially a 2D delta function, such that

$$\mathcal{H}_\cap\left(\vec{\kappa}_{\perp c},\vec{\kappa}_{\perp d}\right) \rightarrow \kappa^2\mathcal{H}\left(\vec{\kappa}_{\perp d}\right)\delta^2\left(\vec{\kappa}_{\perp c} + \tfrac{1}{2}\vec{\kappa}_{\perp d}\right)$$

(10.54)

which leads to

$$\mathcal{I}_s\left(\vec{\kappa}_{\perp d}\right) = -2\pi\kappa I_\varepsilon\mathcal{H}\left(\vec{\kappa}_{\perp d}\right)\left[\text{Re}\left[\delta\varepsilon_\sigma\right]\sin\left(\pi\frac{z_\sigma}{\kappa}\kappa_{\perp d}^2\right) + \text{Im}\left[\delta\varepsilon_\sigma\right]\cos\left(\pi\frac{z_\sigma}{\kappa}\kappa_{\perp d}^2\right)\right]$$

(10.55)

We thus conclude that the phase of a point object can now be imaged, but only when the object is out of focus ($z_\sigma \neq 0$).

Frequency Support

Throughout this chapter, we have treated transmission microscopy as a so-called forward problem. That is, we have derived the fields (and hence the intensity) arriving at the camera as a function of the sample being imaged. In practice, however, it is rarely the fields we are interested in. Instead, it is the sample itself we are interested in. The so-called inverse problem of reconstructing the sample based on the measurements of fields is a complicated one, exacerbated by the fact that cameras do not measure fields but rather measure intensities. We encountered such an inverse problem in our derivation of the Fourier diffraction theorem (Section 9.2.1). According to this theorem, the sample information that can be reconstructed from a scattered field measurement at a distant plane is confined to a frequency support in the shape of a thin spherical cap. This is scant information indeed. However, this theorem was derived for the specific case of plane wave illumination. In transmission microscopy, the illumination is generally not a plane wave (unless $\varsigma = 0$), but rather it is Köhler illumination, which can be thought of as an incoherent superposition of multiple uncorrelated plane waves propagating in different directions. As such, the sample information that can be recovered from transmission microscopy is expected to be more intricate. Though we will not engage in 3D sample reconstruction here, we can obtain an intuitive picture of the information that can be contained in this reconstruction by deriving its 3D frequency support. This is best achieved by considering the Born approximation using fully 3D wavevectors, in which case Eqs. 10.2 and 10.3 (or 10.46 and 10.47) become

$$\mathcal{E}_0\left(\vec{\kappa}\right) = \mathcal{E}_i\left(\vec{\kappa}\right) - 4\pi^2\kappa^2\mathcal{G}_+\left(\vec{\kappa}\right)\int d^3\vec{\kappa}_i\,\delta\hat{\varepsilon}\left(\vec{\kappa}-\vec{\kappa}_i\right)\mathcal{E}_i\left(\vec{\kappa}_i\right) \tag{10.56}$$

$$\mathcal{E}_0\left(\vec{\kappa}_\perp;0\right) = \int d\kappa_z\,\mathcal{E}_0\left(\vec{\kappa}\right) \tag{10.57}$$

$$\mathcal{E}\left(\vec{\kappa}_\perp\right) = \mathcal{H}\left(\vec{\kappa}_\perp\right)\mathcal{E}_0\left(\vec{\kappa}_\perp;0\right) \tag{10.58}$$

where Eq. 10.56 is the radiant field in the sample region, Eq. 10.57 projects this radiant field onto the focal plane, and, as before, Eq. 10.58 transfers the radiant field from the focal plane to the camera.

From Eq. 10.56 it is clear that the information that can recovered about the sample $\delta\hat{\varepsilon}$ depends on the range of wavevectors $\vec{\kappa}_i$ and the range of wavevectors $\vec{\kappa}$, or, more precisely, the range of the *difference* between these two wavevectors $\vec{\kappa} - \vec{\kappa}_i$ that is allowed through the microscope system. The range, or frequency support, associated with the input wavevectors $\vec{\kappa}_i$ is defined by both the span of the illumination pupil, characterized by $\mathcal{H}_i\left(\vec{\kappa}_\perp\right)$ (see Eq. 3.27), and the fact that these wavevectors must radiatively forward propagate, characterized by $\mathcal{D}_+\left(\vec{\kappa}\right)$ (Eq. 2.29). Similarly, the frequency support associated with the output wavevectors $\vec{\kappa}$ is characterized by $\mathcal{H}\left(\vec{\kappa}_\perp\right)$ and $\mathcal{G}_+\left(\vec{\kappa}\right)$ (Eq. 2.21). Both frequency supports correspond to thin spherical caps defined by $\delta\left(\kappa_z - \sqrt{\kappa^2 - \kappa_\perp^2}\right)$, with spans of $\vec{\kappa}_\perp$ bounded by their associated pupils. In general, transmission microscopes are operated with coherence factors ς in the range 0 to 1 (coherent to incoherent), meaning that the illumination and detection pupils are generally different in size. Once the two spherical caps defined by $\mathcal{H}_i\left(\vec{\kappa}_\perp\right)\mathcal{D}_+\left(\vec{\kappa}\right)$ and $\mathcal{H}\left(\vec{\kappa}_\perp\right)\mathcal{D}_+\left(\vec{\kappa}\right)$ are known, a calculation of the frequency support associated with the difference wavevector $\vec{\kappa} - \vec{\kappa}_i$ becomes straightforward. An illustration of this is shown in Fig. 10.9.

Several conclusions can be drawn from this illustration. First, the frequency support associated with the sample bears resemblance to the frequency support associated with a microscope OTF (Fig. 5.6), though the two are associated with physically different quantities. Whereas the two fields contributing to an OTF arise from the same detection pupil, they arise from generally different illumination and detection pupils in transmission microscopy. For example, the smaller the illumination pupil relative to the detection pupil, the thinner the frequency support becomes along the axial direction and the more it curves downward, approaching Fig. 5.6 in the limit $\varsigma \rightarrow 0$. The axial width of the frequency support is inversely related to axial resolution associated with the sample, indicating that, as with the

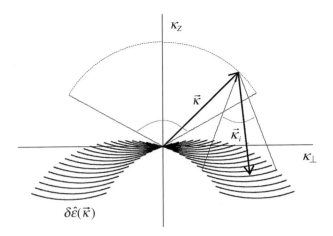

Figure 10.9. Frequency support of $\delta\hat{\varepsilon}(\vec{\kappa})$ for a widefield transmission configuration.

OTF, a laterally uniform sample cannot be axially resolved because of a missing cone. Similarly, the transverse width of the frequency support is inversely related to the transverse resolution associated with the sample. This transverse width is given by $\Delta\kappa_{\perp} + \Delta\kappa_{\perp i}$, corresponding to the sum of the spatial frequency bandwidths associated with the illumination and detection pupils. The transverse resolution of a transmission microscope is thus commonly defined to be

$$\delta\rho = \frac{\lambda}{\mathrm{NA} + \mathrm{NA}_i} \tag{10.59}$$

Finally, it should be noted that we used Eq. 10.56 to derive the frequency support for sample reconstruction. This equation involves fields and not intensities, and it may be unclear how it relates to measurements obtained by a camera, which measures only intensities. The key point is that, as we have seen above, the intensity component bearing sample information in a transmission microscope comes precisely from the interference between the scattered field (sample dependent) the incident field (sample independent). Thus, the measure of intensity in a transmission microscope does indeed provide an indirect measure of the field arising from the sample, but unfortunately this measure is incomplete. More precisely, referring to Eq. 10.7, the frequency support illustrated in Fig. 10.9 corresponds to that obtained from the second term T_2 alone. In standard transmission microscopy, this term is inevitably accompanied by the third term T_3, which is the complex conjugate of T_2, and thus leads to a complex-conjugated version of the frequency support shown in Fig. 10.9 (i.e. reflected about axial and lateral axes). Said differently, the frequency support provided by a standard transmission microscope comes not from the term T_2, but rather from only the real part of this term, leading to a loss of information compared to what would be obtained from a fully complex measure of T_2. For now, we will simply accept this loss as inevitable. In Chapters 11 and 12 we will discuss strategies to recover this loss.

10.2 REFLECTION MICROSCOPY

So far, we have considered widefield microscopes where the illumination source is on one side of the sample, and the camera on the other. Often such microscopes are difficult to put into practice because of constraints due to the sample. For example, the sample may be too thick, or does not provide easy access to more than one side. In such cases, a reflection microscope, where the illumination source and camera are on the same side of the sample, may be the only option.

The sample information obtained with a reflection microscope is fundamentally different from that obtained with a transmission microscope. The main reason for this is that the change of direction between the incident and scattered wavevectors is inherently much more severe. Another reason is that reflection microscopes are often darkfield in nature (more on this below), leading to a nonlinear dependence of the detected intensity on the sample structure. These points will only be touched upon here, and made clearer in Chapter 11.

10.2.1 Brightfield Microscopy

We first address a semantic difficulty that often arises when considering reflection microscopy regarding how one defines the sample itself. For example, let us consider a sample consisting of a single mirror-like surface of reflectance $r\left(\vec{\rho}_0\right)$ located at the microscope focus. In a similar manner as we used for a transmitting sample (Eq. 10.23), we can recast this reflectance as

$$r\left(\vec{\rho}_0\right) = \left|r\left(\vec{\rho}_0\right)\right| e^{i\phi\left(\vec{\rho}_0\right)} \tag{10.60}$$

where $\phi\left(\vec{\rho}_0\right)$ is the local phase shift caused by the sample. This phase shift could be a property of the sample itself, or it could result from a small height variations in the sample (small enough that the sample may still be regarded as in focus).

Based on Eq. 10.60, one could argue that if the sample is absent, then $r\left(\vec{\rho}_0\right) \to 0$ and no intensity can be measured at the camera at all. Alternatively, one could argue that what is interesting is not the sample reflectance per se, but rather the local variations of this reflectance about an average, in which case $\left|r\left(\vec{\rho}_0\right)\right| \to \bar{r}_0$ can represent this average (assumed, without loss of generality, to induce zero phase shift), and the sample of interest can be considered to be $\phi\left(\vec{\rho}_0\right)$, with the possibility that $\phi\left(\vec{\rho}_0\right)$ might be complex if the reflectance is accompanied by absorption. This latter interpretation is similar to the interpretation we adopted above for transmission microscopy. Indeed we find in the weak sample-phase limit, that

$$\mathcal{E}_0\left(\vec{\kappa}_\perp\right) = \bar{r}_0 \mathcal{E}_i\left(\vec{\kappa}_\perp\right) + \mathcal{E}_{0s}\left(\vec{\kappa}_\perp\right) \tag{10.61}$$

$$\mathcal{E}_{0s}\left(\vec{\kappa}_\perp\right) = i\bar{r}_0 \int d^2\vec{\kappa}_\perp'' \, \hat{\phi}\left(\vec{\kappa}_\perp - \vec{\kappa}_\perp''\right) \mathcal{E}_i\left(\vec{\kappa}_\perp''\right) \tag{10.62}$$

which, compared with Eqs. 10.2 and 10.6, is essentially identical. In other words, the strategy summarized in Eqs. 10.2 to 10.5 is equally applicable in reflection microscopy, provided the sample is in focus.

When the sample is displaced out of focus, however, the situation is no longer the same. In this case Eq. 10.62 becomes, in the paraxial approximation,

$$\mathcal{E}_{0s}(\vec{\kappa}_\perp;0) = i\bar{r}_0 \int d^2\vec{\kappa}_\perp'' \, \mathcal{D}_-(\vec{\kappa}_\perp;-z_0)\hat{\phi}\left(\vec{\kappa}_\perp - \vec{\kappa}_\perp''\right) \mathcal{D}_+(\vec{\kappa}_\perp'';z_0)\mathcal{E}_i(\vec{\kappa}_\perp'';0) \quad (10.63)$$

When comparing this with Eq. 10.49, some differences become apparent. For one, we do not integrate in z_0. That is, though we have added a third dimension, we are still considering the displacement of a single surface only. Indeed, we cannot think of a volumetric reflecting sample in the same context of brightfield microscopy, since the incident field cannot be reflected in an unaffected manner from multiple depths at once.

But there is another key difference. One of the free-space transfer functions has been replaced by $\mathcal{D}_-$. Though seemingly innocuous, this modification has significant ramifications regarding image formation. These ramifications will be examined in detail when we consider alternative microscopy techniques, such as interference (Chapter 11) and confocal (Chapter 14) microscopies.

10.2.2 Darkfield Microscopy

An altogether different class of widefield microscope than we have considered so far is the so-called darkfield microscope, which is commonly configured in a reflection configuration. By definition, when the sample is removed in a darkfield microscope, no background light is incident on the camera and only darkness is recorded. The image produced by a darkfield microscope thus originates exclusively from the scattered light from the sample. This scattered light is typically in the backward (i.e. reflection) direction, though as we will see below, other directions are possible.

To analyze the process of imaging with a darkfield microscope, we can return to our general strategy summarized by Eqs. 10.2 to 10.5, though this time with the omission of $\mathcal{E}_i\left(\vec{\kappa}_\perp\right)$ in Eq. 10.2. While at first glance this omission might appear to be a simplification, it actually leads to a complication. In particular, if we carry out our derivation of the radiant mutual intensity at the camera, we find that all the terms in Eq. 10.7 vanish, except one, the last term T_4. This was precisely the term we previously neglected because it was only second order in sample strength. In other words, a darkfield microscope image is no longer linearly dependent on sample strength, as was a brightfield microscope image, and we cannot, for example, derive a corresponding frequency support as we did in Fig. 10.9.

This problem becomes apparent when we calculate the intensity recorded by the camera. To obtain a tractable solution, we consider a thin sample only (of thickness δz), though located at an arbitrary axial position z_ρ. We also work with fields rather than radiant fields, largely for the sake of variety.

If we replace $t\left(\vec{\rho}_0\right)$ with $r\left(\vec{\rho}_0\right) = 2\pi\kappa\delta n \, \delta z$ and treat our sample in a similar manner as in Section 9.1, the field at the camera becomes

$$E\left(\vec{\rho}_1\right) = \int d^2\vec{\rho}_0 \, H_-\left(\vec{\rho}_1 - \vec{\rho}_0, -z_\rho\right) r\left(\vec{\rho}_0\right) E_i\left(\vec{\rho}_0, z_\rho\right) \quad (10.64)$$

where $E_i\left(\vec{\rho}_0, z_\rho\right)$ is the forward-propagating incident field at plane z_p, and $H_-\left(\vec{\rho}, z\right)$ is the backward-propagating 3D amplitude point spread function (see Section 5.2.2). The intensity at the camera is, accordingly

$$I\left(\vec{\rho}_1\right) = \iint d^2\vec{\rho}_0\, d^2\vec{\rho}_0'\, H_-\left(\vec{\rho}_1 - \vec{\rho}_0, -z_\rho\right) H_-^*\left(\vec{\rho}_1 - \vec{\rho}_0', -z_\rho\right) r\left(\vec{\rho}_0\right) r^*\left(\vec{\rho}_0'\right) J_{i,z_\rho}(\vec{\rho}_0, \vec{\rho}_0')$$

$$(10.65)$$

where $J_{i,z_\rho}(\vec{\rho}_0, \vec{\rho}_0')$ is the incident mutual intensity at plane z_ρ, which can be derived from the in-focus mutual intensity $J_i(\vec{\rho}_0, \vec{\rho}_0')$ using Eq. 4.25.

As is clear from Eq. 10.65, the image produced by the camera depends nonlinearly on the sample reflectance, and in general we can make little headway beyond this equation. Even in the case of perfectly coherent illumination with a plane wave, such that $J_{i,z_\rho}(\vec{\rho}_0, \vec{\rho}_0') = J_i(\vec{\rho}_0, \vec{\rho}_0') = I_i$, Eq. 10.65 remains at a sticking point, particularly when the imaging is out of focus, in which case $H_-\left(\vec{\rho}, z\right)$ becomes broadened and $I\left(\vec{\rho}_1\right)$ depends on points in the sample $r\left(\vec{\rho}_0\right)$ and $r\left(\vec{\rho}_0'\right)$ that can be far apart.

There is a scenario, however, in which simplicity is restored. This is the scenario of fully incoherent illumination. From our argument in Section 4.3, the mutual intensity in this case remains incoherent throughout a volume, obeying the same statistics regardless of z_ρ (for z_ρ smaller than the beam size). We thus have $J_{i,z_\rho}(\vec{\rho}_0, \vec{\rho}_0') = J_i(\vec{\rho}_0, \vec{\rho}_0') = \kappa^{-2} I_i \delta^2(\vec{\rho}_0 - \vec{\rho}_0')$, and Eq. 10.65 readily simplifies to

$$I\left(\vec{\rho}_1\right) = I_i \int d^2\vec{\rho}_0\, \text{PSF}\left(\vec{\rho}_1 - \vec{\rho}_0, z_\rho\right) \left|r\left(\vec{\rho}_0\right)\right|^2 \qquad (10.66)$$

where we have made use of Eq. 5.30 and accounted for the fact that the PSF here is backward rather than forward propagating (hence the change of sign in z_ρ). In other words, we have recovered a linear dependence of the image on the sample, but the dependence is on $\left|r\left(\vec{\rho}_0\right)\right|^2$ and not $r\left(\vec{\rho}_0\right)$. We have lost all information about the sample phase.

Darkfield Microscopy Configurations

Because of the inherent difficulties associated with sample reconstruction, darkfield microscopy is rarely used to perform quantitative measurements. Instead, it is used more to obtain qualitative information. For example, it is particularly effective at highlighting sharp or punctate features in a sample, such as edges or defects. Because it depends on sample strength only to second order, the signals it provides are generally weak. On the other hand, whatever signals are detected benefit from a very high signal to background ratio, owing to the absence of background.

Most often, darkfield microscopy is used to examine surfaces, such as semiconductor wafers, or thin samples such as a blood smears. In such cases, to avoid the direct backreflection of the incident light into the microscope, the illumination is often delivered obliquely onto the sample using a ring illuminator. The light scattered from the sample is then detected within the dark region of the illumination cone (see Fig. 10.10).

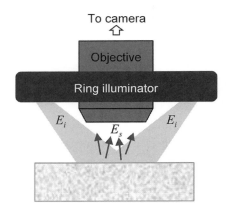

Figure 10.10. Darkfield microscopy configuration.

Alternatively, the illumination can be so oblique as to be side-on. For example, a cylindrical lens can be used to weakly focus an illumination beam along one axis but not the other thus creating a "sheet" of light, which can be delivered orthogonal to the microscope axis. This configuration has the advantage that it does not illuminate the sample above or below the focal plane, meaning that it too effectively probes only a thin sample region even if the sample is volumetric. Such a side-on illumination configuration is over a century old, when it was modestly referred to as ultramicroscopy [17]. A more modern appellation is light-sheet microscopy. We will revisit this configuration in the context of fluorescence imaging in Chapter 14.

10.3 PROBLEMS

Problem 10.1

Consider a thin sample that induces both phase shifts $\varphi(\vec{\rho}_0)$ and attenuation $\alpha(\vec{\rho}_0)$. The local sample transmittance can be written as $t(\vec{\rho}_0) = e^{i\phi(\vec{\rho}_0)}$, where $\phi(\vec{\rho}_0) = \varphi(\vec{\rho}_0) + i\alpha(\vec{\rho}_0)$ is a generalized complex phase function ($\varphi(\vec{\rho}_0)$ and $\alpha(\vec{\rho}_0)$ are real). Show that this complex phase function can be effectively imaged with a modified Zernike phase microscope.

Specifically, consider a Zernike phase contrast microscope whose pupil function can be controlled so that

$$P(\xi) = \begin{cases} e^{i\psi} & \xi \leq \varepsilon \\ 1 & \varepsilon < \xi \leq a \\ 0 & \xi > a \end{cases}$$

where ψ is an adjustable phase shift that is user-defined (assume $\varepsilon \ll a$).

The sample is illuminated with an on-axis plane wave of amplitude E_i. The resultant intensity recorded at the image plane, for a given ψ, is written as $I^{(\psi)}(\vec{\rho})$.

(a) Show that by acquiring a sequence of four images with $\psi = \left\{0, \frac{\pi}{2}, \pi, \frac{3\pi}{2}\right\}$, and by processing these four images using the algorithm

$$\widetilde{I}(\vec{\rho}) = \frac{1}{4}\left[\left(I^{(0)}(\vec{\rho}) - I^{(\pi)}(\vec{\rho})\right) + i\left(I^{(\pi/2)}(\vec{\rho}) - I^{(3\pi/2)}(\vec{\rho})\right)\right]$$

we obtain

$$\widetilde{I}(\vec{\rho}) = iI_i \int d^2\vec{\rho}_0\, H(\vec{\rho} - \vec{\rho}_0)\phi(\vec{\rho}_0)$$

where $I_i = |E_i|^2$.

That is, the constructed complex "intensity" $\widetilde{I}(\vec{\rho})$ is effectively an image of the complex phase function of the sample, from which we can infer both $\varphi(\vec{\rho}_0)$ and $\alpha(\vec{\rho}_0)$. The imaging response function is given by the microscope amplitude point spread function. Use the Born approximation in real space and assume unit magnification.

(b) Derive a similar algorithm that achieves the same result but with a sequence of only three images.

Problem 10.2

Consider a modified Schlieren microscope where the knife edge, instead of blocking light, produces a π phase shift. Compare this modified Schlieren microscope with the standard Schlieren microscope described in Section 10.1.3 (all other imaging conditions being equal).

(a) Which microscope is more sensitive to samples that are purely phase shifting? (Assume weak phase shifts and perform calculations in real space.)

(b) Which microscope is more sensitive to samples that are purely absorbing? (Assume weak attenuation and perform calculations in real space.)

Problem 10.3

(a) Derive Eq. 10.30 from Eqs. 10.28 and 10.29.

(b) Rewrite Eq. 10.30 in terms of local tilt angles $\Theta_i\left(\vec{\rho}_{0c}\right)$ and $\Theta_0\left(\vec{\rho}_{0c}\right)$ going into and out of the sample (see Eq. 6.20 for definition of tilt angles). Express $\vec{\nabla}\phi\left(\vec{\rho}_{0c}\right)$ in terms of $\Delta\Theta\left(\vec{\rho}_{0c}\right) = \Theta_0\left(\vec{\rho}_{0c}\right) - \Theta_i\left(\vec{\rho}_{0c}\right)$.

Problem 10.4

Write Eq. 10.32 in terms of fields rather than radiant fields, and use this to derive Eq. 10.41 more directly (at least, for thin samples that are in focus). Assume, as in the text, that $\phi\left(\vec{\rho}_0\right)$ is both real and small.

Problem 10.5

In DIC microscopy, a bias is used to adjust the relative phase between the cross-polarized fields. Such a bias can be obtained by introducing a quarter wave plate (QWP) between the Nomarski prism and the polarizer in the DIC detection optics. When the fast axis of the QWP is set to $45°$ from vertical (or horizontal), then the bias phase $\Delta\theta$ can be adjusted by rotating the polarizer angle ϕ. The Jones matrix for a QWP whose fast axis is aligned in the vertical direction is given by

$$\mathbf{M}_{\mathrm{QWP}}^{(0°)} = e^{i\pi/4} \begin{pmatrix} 1 & 0 \\ 0 & -i \end{pmatrix}$$

Show that the relation between $\Delta\theta$ and ϕ is given by $\Delta\theta = 2\phi + \frac{\pi}{2}$.

References

[1] Allen, R. D., David, G. B. and Nomarski, G. "The Zeiss-Nomarski differential interference equipment for transmitted-light microscopy," *Z. Wiss. Mikrosk.* 69, 193–221 (1969).

[2] Axelrod, D. "Zero-cost modification of bright field microscopes for imaging phase gradient on cells: Schlieren optics," *Cell Biophys.* 3, 167–173 (1981).

[3] Bernet, S., Jesacher, A., Fürhapter, S., Maurer, C. and Ritsch-Marte, M. "Quantitative imaging of complex samples by spiral phase contrast microscopy," *Opt. Express* 14, 3792–3805 (2006).

[4] Born, M. and Wolf, E. *Principles of Optics*, Cambridge University Press (1999).

[5] Frieden, B. R. "Optical transfer of the three-dimensional object," *J. Opt. Soc. Am.* 57, 56–66 (1967).

[6] Goodman, J. W. *Introduction to Fourier Optics*, 4th edn, W. H. Freeman (2017).

[7] Gu, M. *Advanced Optical Imaging Theory*, Springer (1999).

[8] Hoffman, R. and Gross, L. "Modulation contrast microscopy," *Appl. Opt.* 14, 1169–1176 (1975).

[9] Khonina, S. N., Kotlyar, V. V., Shinkaryev, M. V., Soifer, V. A. and Uspleniev, G. V. "The phase rotor filter," *J. Mod. Opt.* 39, 1147–1154 (1992).

[10] Lowenthal, S. and Belvaux, Y. "Observation of phase objects by optically processed Hilbert transform," *Appl. Phys. Lett.* 11, 49–51 (1967).

[11] Mehta, S. B. and Sheppard, C. J. R. "Quantitative phase-gradient imaging at high resolution with asymmetric illumination-based differential phase contrast," *Opt. Lett.* 34, 1924–1926 (2009).

[12] Nomarski, G. "Microinterféromètre différentiel à ondes polarisées [in French]," *J. Phys. Radium* 16, S9 (1955).

[13] Parthasarathy, A. B., Chu, K. K., Ford, T. N. and Mertz, J. "Quantitative phase imaging using a partitioned detection aperture," *Opt. Lett.* 37, 4062–4064 (2012).

[14] Popescu, G. *Quantitative Phase Imaging of Cells and Tissues*, McGraw-Hill (2011).

[15] Sheppard, C. J. R. and Mao, X. Q. "Three-dimensional imaging in a microscope," *J. Opt. Soc. Am. A* 6, 1260–1269 (1989).

[16] Sheppard, C. J. R. "The optics of microscopy," *J. Opt. A: Pure Appl. Opt.* 9, S1–S6 (2007).

[17] Siedentopf, H. and Zsigmondy, R. "Über sichtbarmachung und grössenbestimmung ultramikroskopischer teilchen, mit besonder anwendung auf goldrubingläsern [in German]," *Ann. Phys.* 10, 1–39 (1903).

[18] Streibl, N. "Three-dimensional imaging by a microscope," *J. Opt. Soc. Am. A* 2, 121–127 (1985).

[19] Yi, R., Chu, K. K. and Mertz, J. "Graded-field microscopy with white light," *Opt. Express* 14, 5191–5200 (2006).

[20] Zernike, F. "Das Phasenkontrastverfahren bei der mikroskopischen Beobachtung [in German]," *Z. Tech. Phys.* 16, 454 (1935).

11 | Interference Microscopy

In the previous chapter, we examined various techniques to image a sample with a camera. The general principle consisted of illuminating the sample with a field, coherent or incoherent, and then monitoring the effect of the sample on this field. A difficulty came from the fact that a camera cannot measure fields, which are complex-valued, directly. Instead, it can only measure intensities, which are real-valued. This was particularly problematic when performing phase imaging, and many of the techniques we examined required mechanisms to convert phase variations caused by the sample into intensity variations detectable by the camera. Some of these, such as Zernike phase contrast and Nomarski differential interference contrast, have gained widespread popularity in biological applications involving tissue slice or cell culture imaging. However, their uses remain limited because they provide only qualitative depictions of the phase variations from which it is difficult to extract quantitative measurements.

We turn now to microscopy techniques specifically designed to produce a quantitative mapping of the complex sample field, while continuing to be based on cameras. These techniques also rely on a mechanism to convert phase variations into intensity variations, but in a well-defined manner. The basic idea is to measure the complex variations in the sample field relative to a stable reference field. The reference field is supplied in the form of a external reference beam (sometimes called a local oscillator) and the relative phase difference between the two fields is revealed by interfering these on the camera. As in the previous chapter, the phase of the sample field is inferred from intensity variations at the camera; however, now because the reference field is externally defined and controllable, the phase of the sample field, along with its amplitude, can be determined quantitatively.

The technique of reconstructing a complex sample field from its 2D interference map with a reference field is generally referred to as interference imaging. A classic example of this is holography, which was invented by Gabor in 1948 [6]. At the time, the 2D interference maps, or holograms, were stored on photographic film, and the complex field reconstruction was performed optically. A requirement of holograms is that they exhibit high spatial resolution and for many years photographic film was the only recording medium that provided sufficient resolution. However, technological advances both in camera resolution and computer speed have enabled the possibility of digital holography, where a hologram is recorded directly onto a camera chip and reconstruction is performed numerically rather than optically, as originally suggested by Goodman [7], Lohmann [17], and Yaroslavski [13]. The reader is referred to textbooks [8, 11] for a full account of holography and its various applications. This chapter will be concerned with digital (as opposed to optical) holography and will describe various holographic microscopy configurations and reconstruction techniques, many of which are reviewed in [20]. We will also consider more general interference microscopy configurations involving low-coherence reference fields, with emphasis on the application of interference microscopy to quantitative phase imaging. Throughout this chapter, we will consider only monochromatic, or near-monochromatic, illumination. For a much more general treatment of interference microscopes that extends to broadband and vectorial fields, the reader is referred, for example, to [22].

11.1 PRINCIPLE

A basic layout of an interference microscope is shown in Fig. 11.1. A complex field produced by a sample is denoted by $E_{0s}(\vec{\rho}_0)$, and propagates a distance z to a camera. The sample field at the camera plane is then given by

$$E_s(\vec{\rho}) = \int d^2\vec{\rho}_0 \, D_+(\vec{\rho} - \vec{\rho}_0, z) E_{0s}(\vec{\rho}_0) \tag{11.1}$$

where $D_+(\vec{\rho} - \vec{\rho}_0, z)$ is the forward-propagating free-space propagator, Rayleigh–Sommerfeld or Fresnel (see Chapter 2).

The goal of interference microscopy is to reconstruct the (complex) sample based on images recorded at the camera plane. Because a camera records only intensity, it loses all phase information unless somehow this phase information is encoded into intensity. Such encoding is performed in interference microscopy by directing an additional reference field onto the camera, denoted by $E_r(\vec{\rho})$, which interferes with $E_s(\vec{\rho})$ and leads to phase-dependent intensity variations. A crucial requirement for the sample and reference fields to interfere is that they be mutually coherent, and, as noted above, we have assumed that $E_s(\vec{\rho})$ and $E_r(\vec{\rho})$ are monochromatic or near-monochromatic. The total field incident on the camera is

$$E(\vec{\rho}) = E_r(\vec{\rho}) + E_s(\vec{\rho}) \tag{11.2}$$

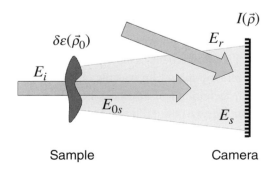

Figure 11.1. Layout of an interference microscope.

and the recorded intensity $I(\vec{\rho}) = \langle |E(\vec{\rho})|^2 \rangle$ is then

$$I(\vec{\rho}) = I_r(\vec{\rho}) + I_s(\vec{\rho}) + \langle E_s(\vec{\rho})E_r^*(\vec{\rho}) \rangle + \langle E_r(\vec{\rho})E_s^*(\vec{\rho}) \rangle \qquad (11.3)$$

The first and second terms are, respectively, the intensities arising from the reference and sample fields alone. The third and fourth terms arise from the interference between these two fields and are complex conjugates of one another. Taken together the last two terms are called an interferogram, and their sum is real. And herein lies the problem. Because the interferogram is real, we cannot in a simple manner recover the complex field $E_s(\vec{\rho})$, which is of interest to us since it contains the information about the sample.

The underlying principle behind interference microscopy is that it provides a mechanism to isolate the components within the interferogram and to measure them separately. In other words, the cross-term $\langle E_s(\vec{\rho})E_r^*(\vec{\rho}) \rangle$ can be measured separately from $\langle E_r(\vec{\rho})E_s^*(\vec{\rho}) \rangle$. We will not consider the details of how this mechanism works until later in this chapter. For the moment, we simply assume this mechanism exists, and that it allows us to measure one of the cross-terms, say the third term in Eq. 11.3, which we denote explicitly as

$$\widetilde{I}_{sr}(\vec{\rho}) = \langle E_s(\vec{\rho})E_r^*(\vec{\rho}) \rangle \qquad (11.4)$$

We emphasize that $\widetilde{I}_{sr}(\vec{\rho})$ is not a true intensity because it is complex (hence the tilde). Once $\widetilde{I}_{sr}(\vec{\rho})$ is measured, and assuming the reference field is known, then the complex sample field $E_s(\vec{\rho})$ can be readily determined.

As in the previous chapter, we will find that it is simpler to work with radiant fields rather than fields, and to derive Eq. 11.4 by way of a radiant mutual intensity. We thus replace Eq. 11.3 with

$$\mathcal{J}(\vec{\kappa}_\perp, \vec{\kappa}_\perp') = \langle \mathcal{E}_r(\vec{\kappa}_\perp)\mathcal{E}_r^*(\vec{\kappa}_\perp') \rangle + \langle \mathcal{E}_s(\vec{\kappa}_\perp)\mathcal{E}_s^*(\vec{\kappa}_\perp') \rangle + \langle \mathcal{E}_s(\vec{\kappa}_\perp)\mathcal{E}_r^*(\vec{\kappa}_\perp') \rangle + \langle \mathcal{E}_r(\vec{\kappa}_\perp)\mathcal{E}_s^*(\vec{\kappa}_\perp') \rangle \qquad (11.5)$$

and denote the third cross-term of interest as

$$\mathcal{J}_{sr}(\vec{\kappa}_\perp, \vec{\kappa}_\perp') = \langle \mathcal{E}_s(\vec{\kappa}_\perp)\mathcal{E}_r^*(\vec{\kappa}_\perp') \rangle \qquad (11.6)$$

The intensity spectrum associated with this third cross-term is then

$$\widetilde{\mathcal{I}}_{sr}(\vec{\kappa}_{\perp d}) = \int d^2\vec{\kappa}_{\perp c}\, \mathcal{J}_{sr}(\vec{\kappa}_{\perp c}, \vec{\kappa}_{\perp d}) \qquad (11.7)$$

where we have adopted our usual coordinate transformation (see Eq. 4.13). Equations 11.4 and 11.7 are related by a Fourier transform. Because $\widetilde{I}_{sr}(\vec{\rho})$ need not be real, $\widetilde{\mathcal{I}}_{sr}(\vec{\kappa}_{\perp d})$ need not be Hermitian (hence the tilde here as well).

We note that Eq. 11.3 is similar to Eq. 10.7, though with its terms re-ordered. The main difference is that the interferogram in the previous chapter (terms T_2 and T_3) arose from the interference between the sample field and the unscattered (ballistic) incident field, rather than between the sample field and an external reference field as it does here. This difference will be crucial. Indeed, in the previous chapter, we had difficulty separating amplitude from phase information, and had to resort to idealized cases where the sample was assumed to be purely amplitude or phase in nature. We will not make such assumptions here. The next two sections will present an overview of various sample reconstruction strategies based on our assumption that we have isolated $\widetilde{I}_{sr}(\vec{\rho})$, or equivalently $\widetilde{\mathcal{I}}_{sr}\left(\vec{\kappa}_{\perp d}\right)$. A discussion of the critical step of how exactly to isolate $\widetilde{I}_{sr}(\vec{\rho})$ or $\widetilde{\mathcal{I}}_{sr}\left(\vec{\kappa}_{\perp d}\right)$ will be deferred to Section 11.4.2.

11.2 COHERENT ILLUMINATION

In the simplest interference microscopy configurations, both the illumination field incident on the sample and the reference field are spatially coherent. Such configurations that make use of spatially coherent fields to record interferograms are generally referred to as holography configurations, of which there are many variations. We will review some of the more common here.

We begin with a description of the radiant field produced by a volumetric sample. As in Chapter 10, we assume the sample is illuminated by an incident field which, at the nominal sample plane $z_0 = 0$, is given by $\mathcal{E}_i(\vec{\kappa}_{\perp}; 0)$. In fact, we have already derived an expression for this sample radiant field in Eq. 10.49, which we reproduce here

$$\mathcal{E}_{0s}(\vec{\kappa}_{\perp}; 0) = i\pi\kappa \iint d^2\vec{\kappa}_{\perp}''\, dz_0\, \mathcal{D}_+(\vec{\kappa}_{\perp}; -z_0)\delta\hat{\varepsilon}\left(\vec{\kappa}_{\perp} - \vec{\kappa}_{\perp}''; z_0\right)\mathcal{D}_+(\vec{\kappa}_{\perp}''; z_0)\mathcal{E}_i(\vec{\kappa}_{\perp}''; 0)$$

(11.8)

where, for simplicity, we have made use of the paraxial approximation. As a reminder, this equation is best interpreted when read from right to left. The incident radiant field is propagated to a sample at location z_0, whereupon, after interacting with the sample, it is propagated back to the nominal sample plane $z_0 = 0$. Again, this interpretation neglects the possibility of multiple scattering and thus treats the sample-field interactions at every location as independent, in accord with the Born approximation.

Throughout this section we consider only coherent illumination in a transmission geometry with a plane wave directed normal to the sample plane. That is, we write

$$\mathcal{E}_i(\vec{\kappa}_{\perp}; 0) = E_i\, \delta^2\left(\vec{\kappa}_{\perp}\right)$$

(11.9)

The sample radiant field becomes then, making use of the Fresnel approximation (Eq. 2.54),

$$\mathcal{E}_{0s}(\vec{\kappa}_{\perp}; 0) = i\pi\kappa E_i \int dz_0\, e^{i\pi\frac{z_0}{\kappa}\kappa_{\perp}^2}\delta\hat{\varepsilon}\left(\vec{\kappa}_{\perp}; z_0\right)$$

(11.10)

leading to

$$\mathcal{E}_{0s}(\vec{\kappa}_{\perp}; 0) = i\pi\kappa E_i\, \delta\hat{\varepsilon}\left(\vec{\kappa}_{\perp}, -\frac{\kappa_{\perp}^2}{2\kappa}\right)$$

(11.11)

where we note that $\delta\hat{\varepsilon}\left(\vec{\kappa}_{\perp}; z_0\right)$ is a mixed representation of the sample whereas $\delta\hat{\varepsilon}\left(\vec{\kappa}_{\perp}, -\frac{\kappa_{\perp}^2}{2\kappa}\right)$ is a fully 3D spatial-frequency representation of the sample.

We observe here the correspondence between the sample radiant field and the sample spectrum itself. The 3D frequency components of the sample encoded in $\delta\hat{\varepsilon}\left(\vec{\kappa}_\perp, -\frac{\kappa_\perp^2}{2\kappa}\right)$ are thus confined to a parabolical cap, with transverse radius extending to the maximum 2D transverse frequency $\kappa_{\perp,\max}$ encompassed by $\mathcal{E}_{0s}(\vec{\kappa}_\perp; 0)$. We have encountered such a frequency support before when we considered the Fourier diffraction theorem (Section 9.2.1 and specifically Fig 9.31). The reason the support here is in the form of a parabolical cap rather than a more exact spherical cap is that we have adopted the Fresnel approximation.

This correspondence between sample and sample field becomes even more direct when the sample is thin, of axial extent smaller than $\kappa/\kappa_{\perp,\max}^2$, in which case the sample may be considered 2D rather than 3D, and we can write

$$\delta\hat{\varepsilon}\left(\vec{\kappa}_\perp, -\frac{\kappa_\perp^2}{2\kappa}\right) \approx \delta\hat{\varepsilon}\left(\vec{\kappa}_\perp, 0\right) \tag{11.12}$$

For simplicity, we will adopt this thin-sample approximation below.

We examine now a few common digital holography configurations. In general, these involve only small variations in the detection geometry and in the shape of the reference field.

11.2.1 Fresnel Holography

The simplest holography configuration is similar to the one used by Gabor in 1948, and is depicted in Fig. 11.2. For reasons that will become apparent, this configuration is called Fresnel holography.

Here, the sample field propagates unimpeded to the camera located a distance z from the nominal sample plane. That is,

$$\mathcal{E}_s(\vec{\kappa}_\perp) = \mathcal{D}_+(\vec{\kappa}_\perp; z)\mathcal{E}_{0s}(\vec{\kappa}_\perp; 0) \tag{11.13}$$

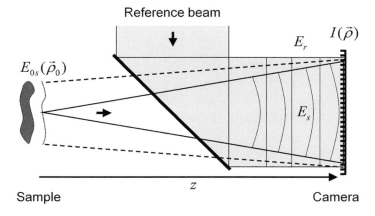

Figure 11.2. Fresnel digital holgraphy.

The reference field in this configuration is also a plane wave propagating normal to the camera plane. However, a defining feature of interference microscopy that separates it from standard widefield transmission microscopy (Section 10.1) is that the reference field follows a separate optical path than the sample field. For ease of notation, we assume that this path also originates a distance z from the camera plane, but is folded by a beamsplitter. At its origin, the plane-wave reference field is given by

$$\mathcal{E}_{0r}(\vec{\kappa}_\perp; 0) = E_r \, \delta^2(\vec{\kappa}_\perp) \tag{11.14}$$

such that at the camera it is given by

$$\mathcal{E}_r(\vec{\kappa}_\perp) = \mathcal{D}_+(\vec{\kappa}_\perp; z)\mathcal{E}_{0r}(\vec{\kappa}_\perp; 0) \tag{11.15}$$

We are interested here only in a single interference term between the sample and reference fields, namely $\langle \mathcal{E}_s(\vec{\kappa}_\perp)\mathcal{E}_r^*(\vec{\kappa}_\perp') \rangle$. From Eqs. 11.11, 11.13, and 11.15 this is given by

$$\mathcal{J}_{sr}(\vec{\kappa}_\perp, \vec{\kappa}_\perp') = i\pi\kappa E_i E_r^* \mathcal{D}_+(\vec{\kappa}_\perp; z)\mathcal{D}_+^*(\vec{\kappa}_\perp'; z) \, \delta\hat{\varepsilon}\left(\vec{\kappa}_\perp, -\frac{\kappa_\perp^2}{2\kappa}\right) \delta^2(\vec{\kappa}_\perp') \tag{11.16}$$

Following our usual coordinate transformation and integrating over $\vec{\kappa}_{\perp c}$ (Eq. 11.7) we arrive at, under the Fresnel approximation

$$\widetilde{\mathcal{I}}_{sr}(\vec{\kappa}_{\perp d}) = i\pi\kappa E_i E_r^* e^{-i\pi\frac{z}{\kappa}\kappa_{\perp d}^2} \, \delta\hat{\varepsilon}\left(\vec{\kappa}_{\perp d}, -\frac{\kappa_{\perp d}^2}{2\kappa}\right) \tag{11.17}$$

In other words, $\widetilde{\mathcal{I}}_{sr}(\vec{\kappa}_{\perp d})$ in Fresnel holography basically contains the same sample information as the radiant field $\mathcal{E}_{0s}(\vec{\kappa}_\perp; 0)$ in Eq. 11.11 (provided E_r^* is known). There is the difference, however, of an additional quadratic phase term. This difference, although seemingly small, leads to a computational difficulty, as becomes apparent when we express the transverse sample distribution $\delta\varepsilon(\vec{\rho}_0)$ in terms of $\widetilde{I}_{sr}(\vec{\rho})$. Making use of the thin-sample assumption (Eq. 11.12), and defining $\delta\varepsilon(\vec{\rho}_0)$ to be the 2D Fourier transform of $\delta\hat{\varepsilon}(\vec{\kappa}_{\perp d}, 0)$, we obtain

$$\delta\varepsilon(\vec{\rho}_0) = \frac{1}{i\pi\kappa E_i E_r^*} \iint d^2\rho \, d^2\kappa_{\perp d} \, e^{i2\pi(\vec{\rho}_0 - \vec{\rho})\cdot\vec{\kappa}_{\perp d}} e^{i\pi\frac{z}{\kappa}\kappa_{\perp d}^2} \widetilde{I}_{sr}(\vec{\rho}) \tag{11.18}$$

In other words, two integrals are required to calculate $\delta\varepsilon(\vec{\rho}_0)$, one a Fourier transform and the other a Fresnel transform (hence the name Fresnel holography). In addition to the computational costs of the two integrals, we will see that Fresnel holography imposes certain minimum requirements on the sampling capacity (or pixel size) of the camera used to measure $\widetilde{I}_{sr}(\vec{\rho})$. Both of these difficulties can be alleviated if we turn to slightly different configurations, called Fourier holography.

11.2.2 Lensless Fourier Holography

A simple modification can be made to Fresnel holography that streamlines the calculation of $\delta\varepsilon(\vec{\rho}_0)$ significantly. This modification involves replacing the plane-wave reference field with

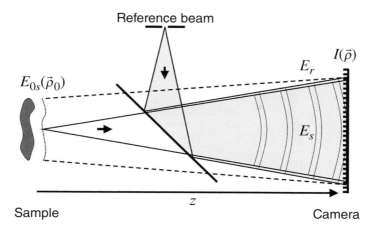

Figure 11.3. Lensless Fourier digital holography.

a diverging reference field produced by a point source (see Fig. 11.3). In other words, instead of Eq. 11.14 we use

$$\mathcal{E}_{0r}(\vec{\kappa}_\perp; 0) = \kappa^{-2} E_r \tag{11.19}$$

leading to

$$\mathcal{J}_{sr}(\vec{\kappa}_\perp, \vec{\kappa}'_\perp) = i\pi\kappa^{-1} E_i E_r^* \mathcal{D}_+(\vec{\kappa}_\perp; z) \mathcal{D}_+^*(\vec{\kappa}'_\perp; z)\, \delta\hat{\varepsilon}\left(\vec{\kappa}_\perp, -\frac{\kappa_\perp^2}{2\kappa}\right) \tag{11.20}$$

Making use of our thin-sample assumption (Eq. 11.12) and the Fresnel approximation, which gives

$$\mathcal{D}_+(\vec{\kappa}_\perp; z)\mathcal{D}_+^*(\vec{\kappa}'_\perp; z) = e^{-i\pi\frac{z}{\kappa}\left(\kappa_\perp^2 - \kappa_\perp'^2\right)} = e^{-i2\pi\frac{z}{\kappa}\vec{\kappa}_{\perp c}\cdot\vec{\kappa}_{\perp d}} \tag{11.21}$$

we arrive at

$$\widetilde{\mathcal{I}}_{sr}(\vec{\kappa}_{\perp d}) = i\pi\kappa^{-1} E_i E_r^* e^{i\pi\frac{z}{\kappa}\kappa_{\perp d}^2}\, \delta\varepsilon\left(-\frac{z}{\kappa}\vec{\kappa}_{\perp d}\right) \tag{11.22}$$

This looks similar to Eq. 11.17 with the notable difference that $\widetilde{\mathcal{I}}_{sr}(\vec{\kappa}_{\perp d})$ is now proportional to the sample itself ($\delta\varepsilon$) rather than to the sample spectrum ($\delta\hat{\varepsilon}$). In other words, we can now write

$$\delta\varepsilon\left(\vec{\rho}_0\right) = \frac{\kappa}{i\pi E_i E_r^*} e^{-i\pi\frac{\kappa}{z}\rho_0^2} \int d^2\vec{\rho}\, e^{i2\pi\frac{\kappa}{z}\vec{\rho}_0\cdot\vec{\rho}}\, \widetilde{I}_{sr}(\vec{\rho}) \tag{11.23}$$

The quadratic phase factor is now outside the integral, and only a single integral is required to calculate $\delta\varepsilon\left(\vec{\rho}_0\right)$ from $\widetilde{I}_{sr}(\vec{\rho})$, namely a simple, computationally efficient, Fourier transform (hence the name Fourier holography [26]). The additional designation of "lenless" comes from the fact that there are no lenses in the configuration depicted in Fig. 11.3, distinguishing it from another more common Fourier holography configuration.

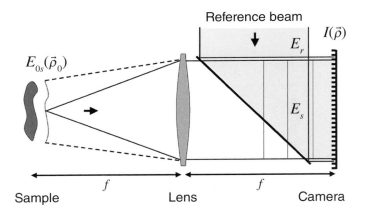

Figure 11.4. Fourier digital holography with a lens.

11.2.3 Fourier Holography with a Lens

A variant of Fourier holography that involves a lens is shown in Fig. 11.4 [10]. Here the reference field is a plane wave, as in Fresnel holography, but the sample field is projected onto the camera using a 2f lens configuration (see Section 3.2). It is simpler, in this case, to directly calculate the fields at the camera rather than the radiant fields. The reference field at the camera is given by

$$E_r(\vec{\rho}) = e^{i4\pi\kappa f}E_r \tag{11.24}$$

and the sample field at the camera, from Eqs. 11.11 and 3.10, is given by

$$E_s(\vec{\rho}) = -i\frac{\kappa}{f}e^{i4\pi\kappa f}\mathcal{E}_{0s}\left(\frac{\kappa}{f}\vec{\rho}; 0\right) = \pi\frac{\kappa^2}{f}E_i e^{i4\pi\kappa f}\delta\hat{\varepsilon}\left(\frac{\kappa}{f}\vec{\rho}, -\frac{\kappa}{2f^2}\rho^2\right) \tag{11.25}$$

leading to

$$\tilde{I}_{sr}(\vec{\rho}) = \pi\frac{\kappa^2}{f}E_i E_r^* \,\delta\hat{\varepsilon}\left(\frac{\kappa}{f}\vec{\rho}, -\frac{\kappa}{2f^2}\rho^2\right) \tag{11.26}$$

Making use of our thin-sample assumption, we arrive at

$$\delta\varepsilon(\vec{\rho}_0) = \frac{1}{\pi f E_i E_r^*}\int d^2\vec{\rho}\, e^{i2\pi\frac{\kappa}{f}\vec{\rho}_0\cdot\vec{\rho}}\,\tilde{I}_{sr}(\vec{\rho}) \tag{11.27}$$

which is essentially identical to Eq. 11.23, where f plays the role of z in the scaling of the Fourier transform.

11.2.4 Imaging Holography

Finally, we consider another example of interference microscopy where the nominal sample plane is imaged to the camera plane, typically with a 4f imaging configuration, as shown in Fig. 11.5.

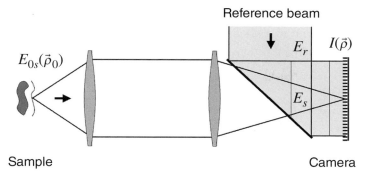

Figure 11.5. Imaging digital holography.

Transmission

We begin with a transmission geometry. In this example, let us assume that the illumination, sample and reference fields travel in the same direction and through similar imaging optics toward the camera plane. The calculation proceeds in much the same way as it did for Fresnel holography, though we follow a slightly different approach. Instead of starting with Eq. 11.11, we start with Eq. 11.10, leaving the integral over z_0 for later. We have then

$$\mathcal{J}_{sr}(\vec{\kappa}_\perp, \vec{\kappa}_\perp') = i\pi\kappa E_i E_r^* \mathcal{H}(\vec{\kappa}_\perp)\mathcal{H}^*(\vec{\kappa}_\perp')\delta^2\left(\vec{\kappa}_\perp'\right)\int dz_0\, e^{i\pi\frac{z_0}{\kappa}\kappa_\perp^2}\,\delta\hat{\varepsilon}\left(\vec{\kappa}_\perp; z_0\right) \quad (11.28)$$

which leads to

$$\widetilde{\mathcal{I}}_{sr}(\vec{\kappa}_{\perp d}) = i\pi\kappa E_i E_r^* \mathcal{H}(\vec{\kappa}_{\perp d})\mathcal{H}^*(0)\int dz_0\, e^{i\pi\frac{z_0}{\kappa}\kappa_{\perp d}^2}\,\delta\hat{\varepsilon}\left(\vec{\kappa}_{\perp d}; z_0\right) \quad (11.29)$$

In the case of an unobstructed pupil where $\mathcal{H}^*(0) = 1$, this may be recast in a simpler form as

$$\widetilde{\mathcal{I}}_{sr}(\vec{\kappa}_{\perp d}) = i\pi\kappa E_i E_r^* \mathcal{H}(\vec{\kappa}_{\perp d})\,\delta\hat{\varepsilon}\left(\vec{\kappa}_{\perp d}, -\frac{\kappa_{\perp d}^2}{2\kappa}\right) \quad (11.30)$$

where $\delta\hat{\varepsilon}$ is now expressed in a fully 3D spatial-frequency representation. We may compare this result with Eq. 11.17 and note that the quadratic phase factor occasioned by Fresnel holography is now replaced with the coherent transfer function $\mathcal{H}(\vec{\kappa}_{\perp d})$. For unobstructed pupils, $\mathcal{H}(\vec{\kappa}_{\perp d})$ represents a simple spatial-frequency cutoff that limits the bandwidth of $\vec{\kappa}_{\perp d}$. In other words, sample reconstruction in imaging holography is much more straightforward than it is in Fresnel holography.

We can also compare Eq. 11.30 with Eq. 3.28 and note that $\delta\hat{\varepsilon}\left(\vec{\kappa}_d\right)$ and $\widetilde{\mathcal{I}}_{sr}(\vec{\kappa}_{\perp d})$ are linked in a similar manner as input and output radiant fields in a coherent imaging system, even though they are not radiant fields at all. This notion that $\widetilde{\mathcal{I}}_{sr}(\vec{\kappa}_{\perp d})$ plays a role similar to a radiant field, or $\widetilde{I}_{sr}(\vec{\rho})$ a role similar to a field, can be taken a step further. Indeed, just as it is possible to numerically propagate fields using propagators such as $D_+\left(\vec{\rho}, z\right)$, so is it possible to numerically propagate $\widetilde{I}_{sr}(\vec{\rho})$. In other words, the camera need not even be placed in the nominal imaging plane to obtain an in-focus image, since the measurement of $\widetilde{I}_{sr}(\vec{\rho})$ can be numerically adjusted and brought back into focus a posteriori. Such numerical refocusing is entirely equivalent to a re-adjustment of the focus knob on a conventional microscope, even though, remarkably, it is done here with no moving parts.

Reflection

So far we have considered transmission geometries only. It is instructive at this point to consider what happens when holographic imaging is performed in a reflection geometry instead. For this we repeat our calculations, this time starting with

$$
\mathcal{E}_{0s}(\vec{\kappa}_\perp; 0) = i\pi\kappa \iint d^2\vec{\kappa}_\perp'' \, dz_0 \, \mathcal{D}_-(\vec{\kappa}_\perp; -z_0)\delta\hat{\varepsilon}\left(\vec{\kappa}_\perp - \vec{\kappa}_\perp''; z_0\right) \mathcal{D}_+(\vec{\kappa}_\perp''; z_0)\mathcal{E}_i(\vec{\kappa}_\perp''; 0)
$$

(11.31)

Note that the detected field is backward propagating here (forward is taken to be in the direction of the illumination beam). Making use of the Fresnel identity

$$
\mathcal{D}_-(\vec{\kappa}_\perp; -z_0)\mathcal{D}_+(\vec{\kappa}_\perp'; z_0) = e^{i4\pi\kappa z_0} e^{-i\pi z_0 \frac{\kappa_\perp^2 + \kappa_\perp'^2}{\kappa}}
$$

(11.32)

and considering again plane-wave illumination (Eq. 11.9), we find that Eq. 11.31 no longer simplifies to Eq. 11.10 but rather becomes

$$
\mathcal{E}_{0s}(\vec{\kappa}_\perp; 0) = i\pi\kappa E_i \int dz_0 e^{i4\pi\kappa z_0} e^{-i\pi \frac{z_0}{\kappa}\kappa_\perp^2}\delta\hat{\varepsilon}\left(\vec{\kappa}_\perp; z_0\right)
$$

(11.33)

which contains an additional rapidly varying phase factor $e^{i4\pi\kappa z_0}$. This rapidly varying phase along the axial direction is endemic to all reflection microscopies, as we will find again in this chapter and the next. Proceeding in the same manner as above, we arrive at finally

$$
\widetilde{\mathcal{I}}_{sr}(\vec{\kappa}_{\perp d}) = i\pi\kappa E_i E_r^* \mathcal{H}(\vec{\kappa}_{\perp d}) \, \delta\hat{\varepsilon}\left(\vec{\kappa}_{\perp d}, -2\kappa + \frac{\kappa_{\perp d}^2}{2\kappa}\right)
$$

(11.34)

This reflection result is similar to the transmission result given by Eq. 11.30, except that the rapidly varying axial phase has manifestly introduced an overall shift of the axial frequency support associated with $\delta\hat{\varepsilon}(\vec{\kappa})$ by an amount -2κ. The ramifications of this shift will be discussed in detail in Section 11.4.

As alluded to above, imaging holography has several advantages over non-imaging holography. For one, the sample is directly reproduced at the camera, without any requirement of Fourier or Fresnel transforms. Moreover, though we neglected it here, magnification can be introduced in the imaging optics, meaning that the configuration in Fig. 11.5 can perform holographic microscopy rather than simply holographic imaging. Holographic microscopy with a microscope objective was first proposed by VanLigten and Osterberg ([25]) and has since been routinely implemented with camera-based detection (e.g. [3, 18, 28]).

11.3 INCOHERENT ILLUMINATION

Interference microscopy is not restricted to coherent illumination, and can work equally well with incoherent illumination provided some extra care is taken in the alignment of the sample and reference fields, as will be explained below.

Transmission

For example, let us consider a transmission imaging geometry, as we did above. In this case, we write

$$\mathcal{E}_r\left(\vec{\kappa}_\perp\right) = \mathcal{H}(\vec{\kappa}_\perp)\mathcal{E}_i\left(\vec{\kappa}_\perp; 0\right) \tag{11.35}$$

$$\mathcal{E}_s(\vec{\kappa}_\perp) = i\pi\kappa\mathcal{H}(\vec{\kappa}_\perp)\iint d^2\vec{\kappa}_\perp'' \, dz_0 \, \mathcal{D}_+(\vec{\kappa}_\perp; -z_0)\delta\hat{\varepsilon}(\vec{\kappa}_\perp - \vec{\kappa}_\perp''; z_0)\mathcal{D}_+(\vec{\kappa}_\perp''; z_0)\mathcal{E}_i(\vec{\kappa}_\perp''; 0) \tag{11.36}$$

As before, our goal is to calculate $\mathcal{J}_{sr}(\vec{\kappa}_\perp, \vec{\kappa}_\perp')$ from which we can derive $\widetilde{\mathcal{I}}_{sr}(\vec{\kappa}_{\perp d})$. But now the illumination is not a plane wave, and Eq. 11.36 cannot be simplified in the same way as it was when we utilized Eq. 11.11. Instead, we must follow a similar approach as in Section 10.1.2, where we express $\mathcal{J}_{sr}(\vec{\kappa}_\perp, \vec{\kappa}_\perp')$ as a function of $\mathcal{J}_i\left(\vec{\kappa}_\perp, \vec{\kappa}_\perp'\right)$.

Making use of the fact that $\mathcal{J}_i\left(\vec{\kappa}_\perp, \vec{\kappa}_\perp'\right) = \kappa^{-2}I_\xi\left|\mathcal{H}_i\left(-\vec{\kappa}_\perp\right)\right|^2\delta^2\left(\vec{\kappa}_\perp - \vec{\kappa}_\perp'\right)$ for incoherent illumination (see Eq. 4.58), we obtain

$$\mathcal{J}_{sr}(\vec{\kappa}_\perp, \vec{\kappa}_\perp')$$
$$= i\pi\kappa^{-1}I_\xi\mathcal{H}(\vec{\kappa}_\perp)\mathcal{H}^*(\vec{\kappa}_\perp')\left|\mathcal{H}_i\left(-\vec{\kappa}_\perp'\right)\right|^2\int dz_0 \, \mathcal{D}_+(\vec{\kappa}_\perp; -z_0)\delta\hat{\varepsilon}\left(\vec{\kappa}_\perp - \vec{\kappa}_\perp'; z_0\right)\mathcal{D}_+(\vec{\kappa}_\perp'; z_0) \tag{11.37}$$

Making use also of Eq. 11.21, and assuming for simplicity that the illumination and detection pupils are the same, and that they are unobstructed and circular (i.e. binary and symmetric), we arrive at

$$\mathcal{J}_{sr}(\vec{\kappa}_{\perp c}, \vec{\kappa}_{\perp d}) = i\pi\kappa^{-1}I_\xi\mathcal{H}(\vec{\kappa}_{\perp c} + \tfrac{1}{2}\vec{\kappa}_{\perp d})\mathcal{H}(\vec{\kappa}_{\perp c} - \tfrac{1}{2}\vec{\kappa}_{\perp d})\int dz_0 \, e^{i2\pi\frac{z_0}{\kappa}\vec{\kappa}_{\perp c}\cdot\vec{\kappa}_{\perp d}}\delta\hat{\varepsilon}\left(\vec{\kappa}_{\perp d}; z_0\right) \tag{11.38}$$

At this point, we can follow two approaches. We can perform the integration over z_0 in Eq. 11.38, which is a Fourier transform, to obtain

$$\mathcal{J}_{sr}(\vec{\kappa}_{\perp c}, \vec{\kappa}_{\perp d}) = i\pi\kappa^{-1}I_\xi\mathcal{H}(\vec{\kappa}_{\perp c} + \tfrac{1}{2}\vec{\kappa}_{\perp d})\mathcal{H}(\vec{\kappa}_{\perp c} - \tfrac{1}{2}\vec{\kappa}_{\perp d})\delta\hat{\varepsilon}\left(\vec{\kappa}_{\perp d}, -\tfrac{1}{\kappa}\vec{\kappa}_{\perp c}\cdot\vec{\kappa}_{\perp d}\right) \tag{11.39}$$

from which we can then integrate over $\vec{\kappa}_{\perp c}$ to obtain $\widetilde{\mathcal{I}}_{sr}(\vec{\kappa}_{\perp d})$. Alternatively, we can first integrate Eq. 11.38 over $\vec{\kappa}_{\perp c}$. This last approach is more revealing, and leads to

$$\widetilde{\mathcal{I}}_{sr}(\vec{\kappa}_{\perp d}) = i\pi\kappa I_\xi\int dz_0 \, \text{OTF}\left(\vec{\kappa}_{\perp d}; z_0\right)\delta\hat{\varepsilon}(\vec{\kappa}_{\perp d}; z_0) \tag{11.40}$$

where we have made use of Eq. 5.37.

This equation warrants a closer look. For one, it can be compared with Eq. 11.29, which we obtained in the case of coherent illumination. Here, in the case of incoherent illumination, the transfer function linking object to image is the optical transfer function. Remarkably, this difference in transfer functions stems only from the difference in the state of the illumination light. In fact, we have seen Eq. 11.40 before in Chapter 5 (Eq. 5.51), suggesting that, in the case of incoherent illumination, the images produced by an interference microscope are the same as those obtained from a volumetric distribution of uncorrelated sources. Indeed, we will see a similar equation again in Chapter 13, when the sources become fluorescent.

Reflection

As in the case of coherent illumination, the situation is very different if we consider a reflection geometry. In Eq. 11.36 we must make the modification

$$\mathcal{E}_s(\vec{\kappa}_\perp) = i\pi\kappa\mathcal{H}(\vec{\kappa}_\perp) \iint d^2\vec{\kappa}_\perp'' \, dz_0 \, \mathcal{D}_-(\vec{\kappa}_\perp; -z_0)\delta\hat{\varepsilon}(\vec{\kappa}_\perp - \vec{\kappa}_\perp''; z_0)\mathcal{D}_+(\vec{\kappa}_\perp''; z_0)\mathcal{E}_i(\vec{\kappa}_\perp''; 0)$$

(11.41)

since in a reflection geometry the detected light travels in the backward direction.

Making use of Eq. 11.32 and following the same procedure we used to arrive at Eq. 11.39, we obtain finally

$$\mathcal{J}_{sr}(\vec{\kappa}_{\perp c}, \vec{\kappa}_{\perp d})$$
$$= i\pi\kappa^{-1}I_\xi\mathcal{H}(\vec{\kappa}_{\perp c} + \tfrac{1}{2}\vec{\kappa}_{\perp d})\mathcal{H}(\vec{\kappa}_{\perp c} - \tfrac{1}{2}\vec{\kappa}_{\perp d}) \, \delta\hat{\varepsilon}\big(\vec{\kappa}_{\perp d}; -2\kappa + \tfrac{1}{\kappa}\left(\kappa_{\perp c}^2 + \tfrac{1}{4}\kappa_{\perp d}^2\right)\big)$$

(11.42)

The frequency components of $\delta\hat{\varepsilon}(\vec{\kappa})$ that are captured in $\mathcal{J}_{sr}(\vec{\kappa}_{\perp c}, \vec{\kappa}_{\perp d})$ are manifestly quite different in a reflection geometry than they are in a transmission geometry, and again feature an axial shift by an amount -2κ.

11.4 FREQUENCY SUPPORT

By themselves, Eqs. 11.39 and 11.42 are not particularly intuitive. Moreover, these must be integrated over $\vec{\kappa}_{\perp c}$ to reveal the frequency components of $\delta\hat{\varepsilon}(\vec{\kappa})$ that are supported by $\widetilde{\mathcal{I}}_{sr}(\vec{\kappa}_{\perp d})$. A simpler procedure to gain an intuitive picture of this frequency support is the one described in Section 10.1.4, where the sample spatial frequencies captured by the camera are those within the range encompassed by $\vec{\kappa} - \vec{\kappa}_i$. Here $\vec{\kappa}_i$ corresponds to the input spatial frequencies incident on the sample, and $\vec{\kappa}$ corresponds to the output spatial frequencies emanating from the sample and ultimately detected by the camera. Note that in interference microscopy, the input spatial frequencies are encoded in the reference field, which is assumed to be a copy of the illumination field incident on the sample. The minus sign in $\vec{\kappa} - \vec{\kappa}_i$ comes from the fact that the camera detects the interference of the sample and reference fields, and in this interference one field is complex-conjugated while the other is not. Finally, the ranges of

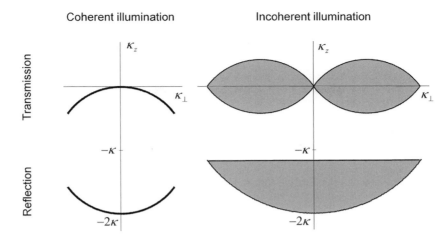

Figure 11.6. Frequency support of $\delta\hat{\varepsilon}(\vec{\kappa})$ for various interferometric microscopy configurations.

$\vec{\kappa}_i$ and $\vec{\kappa}$ are not arbitrary. The transverse components of the former ($\vec{\kappa}_{\perp i}$) are defined by the input pupil in the illumination optics while those of the latter ($\vec{\kappa}_{\perp}$) are defined by the output pupil of the detection optics. The axial components κ_{zi} and κ_z, in turn, are defined by the fact that the wavevectors must $\vec{\kappa}_i$ and $\vec{\kappa}$ must lie on their respective Ewald spheres. That is, they are constrained by the conditions $\kappa^2 = \kappa_{\perp i}^2 + \kappa_{zi}^2 = \kappa_{\perp}^2 + \kappa_z^2$. Bearing all these constraints in mind, a simple picture for the frequency support for $\delta\hat{\varepsilon}\left(\vec{\kappa}\right)$ ensues, as illustrated in Fig. 11.6.

In the case of coherent illumination with an on-axis illumination (and reference) plane wave, $\vec{\kappa}_i$ comprises only a single wavevector, namely $(\vec{0}, \kappa)$. In the case of incoherent illumination, $\vec{\kappa}_i$ comprises a range of wavevectors. In both cases, we observe that a simple change in orientation of the camera (forward or backward) leads to a very fundamental change in the nature of the frequency support. For example, the frequency support in transmission interference microscopy features a missing cone, meaning it cannot provide axial resolution for a uniform planar sample (as discussed in Section 5.4.1). In contrast, reflection interference microscopy does not feature a missing cone, meaning it can axially resolve a uniform planar sample. But one must be careful. Clearly, a reflection geometry cannot reveal sample structure that is slowly varying in the axial direction. Inherently, it can only reveal rapidly varying structure, such as occasioned from sharp interfaces or punctate objects. This stems from the fact that, inherently, illumination light must undergo a u-turn in a reflection geometry in order to be detected. That is, it must be subjected to a large change in axial momentum of order -2κ in the direction opposing the illumination. Such a large momentum transfer can only arise from sample structure that is axially rapidly varying.

11.4.1 Origin of Reference Beam

By definition, interference microscopy is based on interference with an external reference field. So far, the only constraint we have imposed on this reference field is that it be coherent

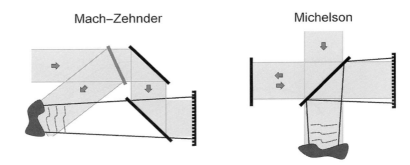

Figure 11.7. Reflection configurations.

with the illumination field. In practice, such coherence can be assured if the reference and illumination fields are derived from the same source, and are copies of one another. In the case of coherent fields, these could be derived from a common laser. In the case of incoherent fields, they could be derived from a laser passing through a common rotating diffuser (recall that we are considering only monochromatic, or near-monochromatic light, in this chapter – see also discussion in Section 7.4).

In general, the reference field can be extracted upstream or downstream from the sample. For example, Fig. 11.7 shows two reflection-geometry variants, where the reference field is extracted upstream. The left variant, where the beams follow widely separated optical paths and are recombined downstream, is called a Mach–Zehnder interferometer. The right variant, where the beams retrace their optical paths is called a Michelson interferometer. These classic configurations are discussed in detail in any textbook involving interferometry (e.g. [11]). The advantage of extracting the reference field upstream from the sample is that the power of this field is easily controllable and can be chosen to be much larger than that of the sample field (an advantage that will be exploited in the next chapter). However, the wide separation between reference and sample fields makes these variants susceptible to technical noises such as vibrations or air fluctuations that can cause random variations in relative optical pathlength difference between the reference and sample arms, leading to smearing of the interference pattern recorded at the camera.

Alternatively, Fig. 11.8 shows two transmission-geometry variants where the reference field is extracted downstream from the sample. In these cases, only the only the unscattered light traveling through the sample can serve as the reference, since only this light possesses a well-defined phase front that is unaffected by the sample (see Chapter 9). A standard technique for isolating the unscattered light is with the use of a pinhole, as shown in Fig. 11.8, though it should also be mentioned that the unscattered field is subject to extinction as it propagates through the sample, meaning it becomes attenuated with increasing sample thickness. As such, these variants are practical for thin samples only.

While the variants shown in Fig. 11.8 seem similar, there is a key difference between the two. To begin, we note that in both cases the reference and sample field are not collinear when they impinge upon the camera. As a result, their phase fronts, which are perpendicular to

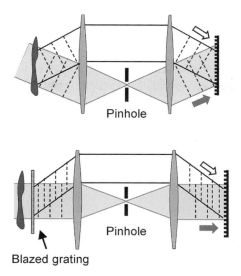

Figure 11.8. Transmission configurations (vertical dashed lines indicate coherence fronts).

their propagation directions (denoted by arrows), are not parallel to one another, leading to an overlying fringe pattern at the camera. Previously we had not made allowances for such a fringe pattern, since we had taken the sample and reference beams to be collinear at the camera. As it happens, such a fringe pattern can be exploited to advantage, as we will see in the next section, but this is not the key difference between the variants mentioned above. The key difference is that a grating is used to separate the fields in one variant and not the other. A grating has the peculiar property that while it changes the direction of the phase front of an incident beam, it does not change the direction of what can be referred to as the "coherence front" (denoted by dashed lines). That is, in the top variant, both the phase and coherence fronts are tilted relative to one another at the camera plane, while in the bottom variant only the phase fronts are tilted and not the coherence fronts, which remain parallel to one another and to the camera plane. This difference plays an important role when the illumination is spatially incoherent, in which case the coherence fronts are quite thin (see, for example, Section 7.2.3). Without a grating (top variant), the coherence fronts of the reference and sample beams overlap only over a limited intersection region at the camera, and $\widetilde{\mathcal{I}}_{sr}(\vec{\kappa}_{\perp d})$ can be recovered only in this region. In contrast, when beamsplitting is performed with a grating (bottom variant), the coherence fronts of the two beams overlap over the entire camera surface, and $\widetilde{\mathcal{I}}_{sr}(\vec{\kappa}_{\perp d})$ can be recovered throughout. This effect is discussed in [2], and is exploited in a technique called Spatial Light Interference Microscopy [5]. Indeed, this effect of a grating-induced separation of phase and coherence fronts occurs not only with spatially incoherent light, but also with temporally incoherent light, as is well-known in the ultrafast optics community. We will revisit this effect when we consider temporal focusing in Section 16.6. Note that the grating shown in Fig. 11.8 is blazed, meaning that it provides a beam separation that is dominantly one-sided, as opposed to the sinusoidal grating shown in Fig. 9.2. Blazed gratings are discussed in most optics textbooks (e.g. [12]).

11.4.2 Extraction of Complex Interference

Our entire discussion so far has been based on the crucial premise that we are able to isolate the third term in Eq. 11.3 (or Eq. 11.5), thereby allowing us to consider only Eq. 11.4 (or Eq. 11.7). So far, we have overlooked exactly how to do this. As emphasized above, it is important that we isolate the third term alone and not the sum of the third and fourth terms, since the latter would be real and hence would not contain sufficient information for a reconstruction of $\delta\varepsilon(\vec{\rho}_0)$, which in general is complex.

We refer to this third term as a complex interference and discuss some common strategies for its isolation.

Phase Stepping

In Section 11.1, the arbitrary phase between the reference and sample fields was systematically omitted because it was assumed fixed. While this phase indeed plays no role if it is fixed, it can play a quite significant role when it is varied in a controlled manner. Let us now consider this possibility by explicitly allowing control of the relative phase between the reference and sample fields, denoted by ϕ. That is, let us write instead of Eq. 11.2,

$$E^{(\phi)}(\vec{\rho}) = E_r(\vec{\rho})e^{i\phi} + E_s(\vec{\rho}) \tag{11.43}$$

In practice, ϕ can be controlled by adjusting the optical path length of either the reference or sample arm. For example, this can be readily accomplished in a Mach–Zehnder configuration by mounting a mirror, such as the mirror at the top right in Fig. 11.7, onto a piezoelectric transducer, allowing electrical control of the reference-beam path length. Alternatively, the fold mirror in the reference arm of a Michelson interferometer could be mounted on a piezoelectric transducer. Regardless of how the phase shift ϕ is applied, Eq. 11.43 leads to

$$I^{(\phi)}(\vec{\rho}) = I_{rr}(\vec{\rho}) + I_{ss}(\vec{\rho}) + e^{-i\phi}\left\langle E_r^*(\vec{\rho})E_s(\vec{\rho})\right\rangle + e^{i\phi}\left\langle E_r(\vec{\rho})E_s^*(\vec{\rho})\right\rangle \tag{11.44}$$

which differs from Eq. 11.3 in that it features a controllable parameter ϕ. This parameter will ultimately enable us to isolate the third term of interest in Eq. 11.44.

The technique of adjusting ϕ to isolate $\widetilde{I}_{sr}(\vec{\rho})$ (i.e. Eq. 11.4) is known as phase stepping, and was first applied to digital holography by Yamaguchi [27]. In fact, several implementations of phase stepping can achieve the same result. As an example, we consider a general K-step technique that involves the application of a circular phase sequence $\phi_k = \frac{2\pi k}{K}$, where k is incremented from 0 to $K-1$. The third term in Eq. 11.44 can then be isolated by the algorithm

$$\widetilde{I}_{sr}(\vec{\rho}) = \frac{1}{K}\sum_{k=0}^{K-1} e^{i\phi_k}I^{(\phi_k)}(\vec{\rho}) \tag{11.45}$$

In most applications, K is typically in the range 3 to 5. For example, for $K = 4$ we have

$$\widetilde{I}_{sr}(\vec{\rho}) = \frac{1}{4}\left[\left(I^{(0)}(\vec{\rho}) - I^{(\pi)}(\vec{\rho})\right) + i\left(I^{(\pi/2)}(\vec{\rho}) - I^{(3\pi/2)}(\vec{\rho})\right)\right] \tag{11.46}$$

As a reminder, $\tilde{I}_{sr}(\vec{\rho})$ is a complex function and therefore not directly measurable. Instead, it is synthesized here by a series of real functions $I^{(\phi_k)}(\vec{\rho})$ that are directly measurable.

As suggested above, a variety of alternative phase-stepping algorithms can be applied to isolate $\tilde{I}_{sr}(\vec{\rho})$. For example, a five-step algorithm has been shown to be more immune to artifacts [21]. Still other algorithms involve the application of continuous phase steps by sinusoidal phase modulation [14]. Regardless of which algorithm is adopted, the basic principle of phase stepping remains the same, the only requirement being that the steps are controllable and well defined.

Off-Axis Holography

A drawback of phase stepping is that multiple holograms are required to isolate $\tilde{I}_{sr}(\vec{\rho})$, limiting the speed of the reconstruction of $\delta\varepsilon(\vec{\rho}_0)$. This drawback can be circumvented by the entirely different strategy of off-axis holography, originally demonstrated by Leith and Upatnieks [15, 16].

Let us examine what happens when a slight tilt is introduced in the reference beam, as shown in Fig. 11.9 (or in Fig. 11.8). In particular, let us consider coherent illumination with a plane-wave reference beam of amplitude E_r and tilted wavevector $\vec{\kappa}_r$. We can thus write

$$E(\vec{\rho}) = E_r e^{-i2\pi\vec{\kappa}_{\perp r}\cdot\vec{\rho}} + E_s(\vec{\rho}) \tag{11.47}$$

where $|\vec{\kappa}_{\perp r}| = \kappa\sin\theta_r$, and θ_r is the tilt angle relative to the sample beam axis. Accordingly, the hologram intensity can be written as

$$I(\vec{\rho}) = I_{rr} + I_{ss}(\vec{\rho}) + e^{i2\pi\vec{\kappa}_{\perp r}\cdot\vec{\rho}}E_r^*E_s(\vec{\rho}) + e^{-i2\pi\vec{\kappa}_{\perp r}\cdot\vec{\rho}}E_rE_s^*(\vec{\rho}) \tag{11.48}$$

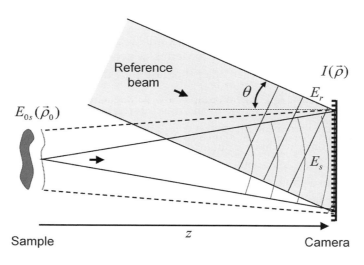

Figure 11.9. Fresnel holography in an off-axis configuration.

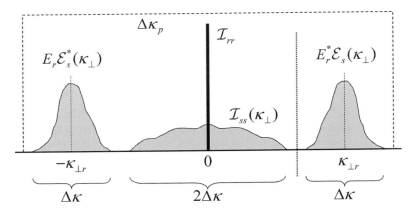

Figure 11.10. Amplitude spectrum of a reconstructed off-axis hologram. The vertical dotted line depicts a one-sided high-pass filter cutoff. The dashed box depicts the bandwidth of the detection electronics.

Equation 11.48 is similar to Eq. 11.44 except that now the phase shift between the reference and sample beams varies linearly across the camera plane. In other words, the phase shifts that were previously distributed in time in the phase-stepping approach are now distributed in space. In effect, we have introduced a spatial carrier frequency to the phase of $E_s(\vec{\rho})$. This becomes apparent when $I(\vec{\rho})$ is recast as an intensity spectrum, obtaining

$$\mathcal{I}(\vec{\kappa}_\perp) = I_{rr}\delta^2\left(\vec{\kappa}_\perp\right) + \mathcal{I}_{ss}(\vec{\kappa}_\perp) + E_r^*\mathcal{E}_s(\vec{\kappa}_\perp - \vec{\kappa}_{\perp r}) + E_r\mathcal{E}_s^*(\vec{\kappa}_\perp + \vec{\kappa}_{\perp r}) \qquad (11.49)$$

A schematic of this intensity spectrum is shown in Fig. 11.10. Note in particular that $\mathcal{I}(\vec{\kappa}_\perp)$ contains four terms. The first is the intensity spectrum of the reference beam alone. Since the reference beam intensity is presumed to be uniformly distributed over the camera, this first term is peaked about the frequency origin. The second term is the intensity spectrum of the sample beam alone. The bandwidth of this term depends on the sample in question. Finally, the third and fourth terms are the interference terms, which, aside from being shifted in frequency, are Hermitian conjugates of each other (because $I(\vec{\rho})$ must be real). Again, we will be interested in the third term only.

To understand the advantage of an off-axis reference beam configuration, let us first consider the case where there is no tilt at all (i.e. $\vec{\kappa}_{\perp r} = 0$). That is, let us revert to on-axis configurations such as the ones we have examined throughout this chapter. In this case, all four terms in Eq. 11.49 overlap one another and the isolation of the interference terms is impossible unless one resorts to a strategy such as phase stepping. On the other hand, by introducing a tilt in the reference beam, the interference terms become displaced in frequency both from each other and from the central dc terms. When this displacement is large enough, the isolation of the third term in Eq. 11.49 becomes trivial by simple numerical application of a one-sided high-pass filter [4, 24], as illustrated in Fig. 11.10. This isolation can be performed in a single shot and does not require multiple holograms as did the phase-stepping technique. Once the first, second, and fourth terms in Eq. 11.49 are removed by high-pass filtration, only the third term remains in the intensity spectrum, given by

$$\widetilde{\mathcal{I}}_{sr}(\vec{\kappa}_\perp) = E_r^* \mathcal{E}_s(\vec{\kappa}_\perp - \vec{\kappa}_{\perp r}) \tag{11.50}$$

This is almost what we are looking for, except that it is laterally shifted by the carrier frequency $\vec{\kappa}_{\perp r}$. Indeed, if we took an inverse Fourier transform of this to recover $\widetilde{I}_{sr}(\vec{\rho})$, and ultimately to reconstruct $\delta\varepsilon\left(\vec{\rho}_0\right)$, we would find that the result oscillates rapidly with spatial frequency $\vec{\kappa}_{\perp r}$. But the value of $\vec{\kappa}_{\perp r}$ is known, presumably, and hence it is an easy matter to remove this carrier frequency by simply shifting Eq. 11.50 numerically so that it is re-centered about the spatial frequency origin. That is, we write

$$\widetilde{\mathcal{I}}_{sr}^{\text{centered}}(\vec{\kappa}_\perp) = \widetilde{\mathcal{I}}_{sr}(\vec{\kappa}_\perp + \vec{\kappa}_{\perp r}) \tag{11.51}$$

from which $\delta\varepsilon\left(\vec{\rho}_0\right)$ can be reconstructed using the same techniques outlined in Section 11.2. An example of off-axis digital holographic imaging of red blood cells is provided in Fig. 11.11.

Because off-axis digital holography can be performed in a single shot, it is highly attractive when speed is a concern. However, several constraints must be kept in mind. As noted above, the third term in Eq. 11.49 can only be unambiguously isolated if it is shifted with a large enough $\kappa_{\perp r}$ (i.e. a large enough tilt angle). In practice, the minimum value for $\kappa_{\perp r}$ is prescribed by the bandwidth of the sample field at the camera [8]. For example, if the bandwidth of $\mathcal{E}_s(\vec{\kappa}_\perp)$ is $\Delta\kappa_\perp$, then the bandwidth of $\mathcal{I}_{ss}(\vec{\kappa}_\perp)$ is $2\Delta\kappa_\perp$ (see Chapters 3 and 4), and accordingly, the tilt angle must satisfy the constraint

$$\sin\theta_r \geq \frac{3\Delta\kappa_\perp}{2\kappa} \tag{11.52}$$

This result is valid for both coherent and incoherent illumination; however, it should be pointed out that $\Delta\kappa_\perp$ is larger with incoherent illumination than with coherent illumination because of the additional spatial-frequency diversity introduced by the illumination itself. The constraint imposed by Eq. 11.52 is thus more severe with incoherent illumination (more on this below).

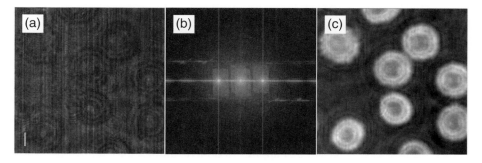

Figure 11.11. Red blood cells images with off-axis digital holography. (a) Raw hologram, (b) Fourier transfrom of raw hologram, and (c) reconstructed image. Courtesy of G. Popescu, University of Illinois at Urbana-Champaign.

11.5 RESOLUTION AND SAMPLING

Our treatment of digital holography has been idealized so far in the sense that we have not taken into account limitations resulting from the camera itself. One such limitation comes from the fact that the camera is finite in size. For example, when considering Fresnel and Fourier holography, we arrived at the expressions 11.18, 11.23, and 11.27. Each of these involves an integration of $\widetilde{I}_{sr}(\vec{\rho})$ over $\vec{\rho}$. Clearly, since the camera is finite in size, these integrals cannot span all of space, but rather span only the area over which $\widetilde{I}_{sr}(\vec{\rho})$ is detected, namely the surface area of the camera. Since each of these integrals is a Fourier transform, the restriction on the integration area ultimately leads to a restriction on the spatial resolution obtainable for $\delta\varepsilon\left(\vec{\rho}_0\right)$. But this problem is not new, and indeed it is the most basic problem associated with any imaging technique. In effect, the limited size of the camera leads to a limited range of the wavevector angles that can be collected from the sample, which, in turn, corresponds to a limited NA (see Section 3.3.4). For example, in the case of Fresnel or lensless Fourier holography, the NA is roughly given by $L/2z$, where L is the cross-sectional size of the camera. In the case of Fourier holography with a lens, it is given by $L/2f$. Each of these NAs is associated with a spatial resolution (Eq. 3.36). On the other hand, the case of imaging holography must be treated somewhat differently, since spatial resolution does not depend on camera size in this case.

Another fundamental limitation comes from the fact that a digital camera is pixellated. As such, a camera cannot provide a complete representation of the intensity distribution at the hologram plane, but rather only a sampled representation where the sampling resolution is defined by the camera pixel size. The importance of sampling resolution cannot be underestimated in holography since it was one of the main hurdles impeding the feasibility of digital holography in the first place. As explained below, different digital holography configurations require different sampling resolutions, some more stringent than others.

For example, let us return to the case of imaging holography. In this case, we established various correspondences between $\widetilde{\mathcal{I}}_{sr}(\vec{\kappa}_{\perp d})$ and $\delta\hat{\varepsilon}(\vec{\kappa}_{\perp d})$, meaning that the spatial resolution of the reconstructed $\delta\varepsilon\left(\vec{\rho}_0\right)$ is inversely related to the spatial-frequency bandwidth captured in $\widetilde{\mathcal{I}}_{sr}(\vec{\kappa}_{\perp d})$. For coherent illumination, we found this bandwidth to be restricted by $\mathcal{H}(\vec{\kappa}_{\perp d})$ (see Eqs. 11.30 or 11.34). For incoherent illumination, the restriction is twice this (as limited, for example, by the OTF in Eq. 11.40). But these results are based on the presupposition that the camera can accommodate such spatial frequency bandwidths. In the event it cannot, then the spatial resolution of the reconstructed $\delta\varepsilon\left(\vec{\rho}_0\right)$ becomes limited by the camera itself.

From sampling theory (e.g. see [8]), the spatial-frequency bandwidth that can be accommodated by a camera is $\Delta\kappa_p = 1/p$, where p is the camera pixel size. Denoting, as usual, the bandwidth accommodated by $\mathcal{H}(\vec{\kappa}_\perp)$ as $\Delta\kappa_\perp$, we find that the camera pixel size should be no larger than $1/\Delta\kappa_\perp$ for coherent illumination, and $1/2\Delta\kappa_\perp$ for incoherent illumination. But this is a best-case scenario that applies to on-axis digital holography only, such as holography based on phase stepping. In the case of off-axis digital holography the constraints on pixel size become more severe. This is apparent if we re-examine Fig. 11.10. The dashed box in this figure depicts the bandwidth allowed by the camera. Spatial frequencies beyond this bandwidth are suppressed, or, worse still, they are aliased into lower frequencies resulting in artifacts in the sample reconstruction. To ensure that the sideband of interest (e.g. the third term in Eq. 11.49) can fit inside this box, and assuming we are abiding by the constraint imposed by Eq. 11.52, we find that the camera pixel size should be no larger than $1/4\Delta\kappa_\perp$ for coherent illumination, and $1/8\Delta\kappa_\perp$ for incoherent illumination. Such small pixel sizes are required not only to resolve the fringe pattern on the camera that results from the non-collinear interference of the sample and reference fields, but also the increased bandwidth of the sample field occasioned by interaction with the sample and, when applicable, the incoherence of the illumination. In the event that the pixels are too large, it may be necessary to reduce the bandwidth of $\mathcal{H}(\vec{\kappa}_\perp)$ (or $\mathrm{OTF}(\vec{\kappa}_\perp)$) to avoid the introduction of reconstruction artifacts. This can be done, for example, by reducing the size of the detection pupil.

But the situation is not as dire as it appears. The reader is reminded that throughout this chapter we have assumed unit magnification for our imaging configurations, largely for ease of notation. In the more realistic case where the magnification $|M|$ is greater than unity, the bandwidth $\Delta\kappa_\perp$ impinging the camera is smaller than the bandwidth $\Delta\kappa_{\perp 0}$ associated with the sample reconstruction by a factor $|M|$. In other words, $\Delta\kappa_\perp$ can be made smaller simply by increasing magnification (for a particular desired sample reconstruction resolution, and at the cost of field of view), thus alleviating the pixel size constraint. Nevertheless, the fact remains that off-axis holography is more demanding on camera resolution than on-axis holography, so much so that it is generally reserved only for coherent illumination.

11.6 APPLICATIONS

Interference microscopy relies on the basic premise that if one knows the complex field distribution at one axial plane, then one can infer the complex field distribution at any other axial plane provided one also knows the field propagation function linking the two planes. Thus, by isolating the interference term in Eq. 11.3, one can calculate the complex field $E_s\left(\vec{\rho}\right)$ at the camera plane, from which one can infer the complex field $E_{0s}\left(\vec{\rho}_0\right)$ at the sample plane, ultimately enabling us to perform a holographic reconstruction of the sample, also complex (e.g. see Fig. 11.11).

Quantification of Index of Refraction

An advantage of interference microscopy that separates it from the widefield microscopy techniques outlined in Chapter 10 is its ability to distinguish phase from amplitude contrast, allowing it to perform quantitative phase imaging (the reader is referred to [19] for a thorough treatise). This advantage is important in particular when one is interested in accurately characterizing the index of refraction of a sample. For example, let us consider the case where the scattering from a sample is weak and highly forward directed, meaning the phase fronts of a transilluminating plane wave are only slightly perturbed. In this limit, the wave propagating through the sample can be thought of as a bundle of parallel rays that, while remaining roughly

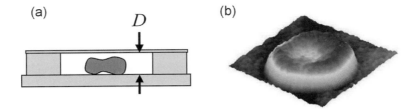

Figure 11.12. (a) Chamber for thin sample (e.g. blood cell). (b) Reconstructed phase profile.

parallel throughout the sample, nevertheless incur slight phase shifts (if the phase shifts are not slight, the rays do not remain parallel). The relation between the measured phase of $E_{0s}(\vec{\rho}_0)$ emerging from the sample and the sample index of refraction then becomes

$$\phi(\vec{\rho}_0) = \kappa \left[\int_0^{L(\vec{\rho}_0)} dz_0 \, \delta n(\vec{\rho}_0, z_0) + D \right] \qquad (11.53)$$

where $\delta n(\vec{\rho}_0, z_0)$ are the index of refraction variations in the sample relative to the surrounding medium of index n (see Eq. 9.3), and $L(\vec{\rho}_0)$ and D are the thicknesses of the sample and sample chamber, respectively (see Fig. 11.12). Note that $\phi(\vec{\rho}_0)$ can vary as a result of changes in either sample thickness or index of refraction, the latter being tantamount to a change in sample density (see, for example, [18]). A distinction between these two types of variation in $\phi(\vec{\rho}_0)$ is not always obvious and can require restraining the cell in a chamber to eliminate the possibility of changes in cell thickness.

Effect of Sample Dynamics

Another application of interference microscopy is the monitoring of temporal dynamics in a sample resulting, for example, from motion or flow [9, 23]. A straightforward method to monitor such dynamics is to perform two sequential measurements of $E_{0s}(\vec{\rho}_0)$. The sample dynamics can then be inferred from the changes observed in either the amplitude or phase of $E_{0s}(\vec{\rho}_0)$, provided a model exists to relate the observed changes to the dynamics of interest. For very fast dynamics, the sequential measurements can even be made within the same camera exposure, in which case the superposed fields interfere with one another in the hologram (even though they were acquired at different times), and changes in $E_{0s}(\vec{\rho}_0)$ become manifest as interference fringes in the reconstructed image. When applied to the study of sample deformations, this latter technique is generally called digital holographic interferometry.

A particularly effective implementation of digital holography for the measurement of sample dynamics stems directly from the application of phase stepping. For example, let us imagine applying phase stepping in combination with off-axis reference illumination. As discussed above, either method is suitable for the isolation of the interference term in Eq. 11.3, and so the application of both might appear to be redundant. However, in certain cases, their combination can be advantageous.

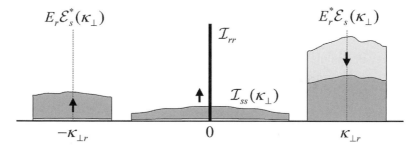

Figure 11.13. Effect of decorrelations on frequency bands in phase-stepping hologram reconstruction.

An obvious case where phase stepping is warranted is when the spatial frequency bandwidth $\Delta\kappa_\perp$ is too large to allow a proper spectral isolation of the sideband of interest (see Fig. 11.10). For example, the application of a phase-stepping algorithm such as the one given by Eq. 11.46 (or more generally 11.45) effectively cancels all but the right-most sideband in Fig. 11.10, to within the limitations of noise. This cancellation does not require the application of a one-sided high-pass filter, and hence is effective even if there is spectral overlap of the bands.

However, let us now imagine that there is motion in the sample that occurs on a time scale comparable to the time required to acquire the holograms for phase stepping. In this case, the holograms become "decorrelated" and the cancellation of the central and left bands by phase stepping becomes imperfect (see Fig. 11.13). The spectral separation afforded by off-axis reference illumination then becomes indispensable. Moreover, the residual components that remain in the central and left bands now contain useful information. Inasmuch as they arise from decorrelations in the holograms resulting from the sample dynamics, they can serve as signatures of these dynamics.

A more sophisticated version of this technique can be implemented with the application of a frequency shift to either the sample or the reference beam. In this manner, high-frequency dynamics of the sample can be mixed with this frequency shift, causing a demodulation of the sample dynamics to lower frequencies. Such a technique falls under the general rubric of heterodyne detection, and will be revisited in Sections 12.2.1 and 13.3.2. In the context of digital holography, heterodyne detection has been used to characterize the dynamics of blood flow in tissue, even though these dynamics extend far beyond the bandwidth of the digital camera used for their recording [1].

11.7 PROBLEMS

Problem 11.1

Equations 11.23 and 11.27 are idealized in that they consider the integration over $\vec{\rho}$ to be infinite. In practice, the integration can only be performed over the area of the camera, which has a finite size $L_x \times L_y$. Derive the effect of this finite size on the transverse spatial resolution of the reconstructed sample $\delta\varepsilon(\vec{\rho}_0)$. In particular...

(a) Show that in the case of lensless Fourier holography (Fig. 11.3), the resolution is given by $\delta x_0 = \frac{\lambda}{2n\theta_x}$ and $\delta y_0 = \frac{\lambda}{2n\theta_y}$, where $\theta_x = \frac{L_x}{2z}$ and $\theta_y = \frac{L_y}{2z}$. (Assume δx_0 and δy_0 are small.)

(b) Show that in the case of Fourier holography with a lens (Fig. 11.4), the resolution is given by $\delta x_0 = \frac{\lambda}{2n\theta_x}$ and $\delta y_0 = \frac{\lambda}{2n\theta_y}$, where $\theta_x = \frac{L_x}{2f}$ and $\theta_y = \frac{L_y}{2f}$.

Hint: derive expressions for how the actual sample reconstructions are related to the idealized sample reconstructions.

Note the analogy of these results with the standard resolution criterion given by Eq. 3.36. In performing these calculations, you will run into sinc functions. Define the width of $\mathrm{sinc}(ax)$ to be $\delta x = \frac{1}{a}$.

Problem 11.2

Consider the three lensless digital holographic microscopy configurations shown in Figs. 11.2, 11.3, and 11.9, and assume all parameters in these configurations are the same, and that the camera has sensor size L and pixel size p (both square). Assume also coherent illumination.

(a) What camera parameter defines the transverse spatial resolution δx_0 and δy_0 of the reconstructed sample in all three configurations?

(b) Provide estimates for the transverse fields of view Δx_0 and Δy_0 of the reconstructed sample in all three configurations, assuming these are smaller than L and much smaller than z (roughly as depicted in the the figures), and assuming the reference-beam tilt angle θ in the case of off-axis Fresnel holography is in the x direction only.

Hint: the sample and reference beams interfere at the camera, and produce fringes. Estimate the largest fringe frequencies for each configuration. These must be properly sampled according the Nyquist sampling criterion.

Problem 11.3

(a) On-axis digital holography is performed with circular phase stepping. Consider an arbitrary camera pixel and assume a camera gain of 1 (i.e. the camera directly reports the number of detected photoelectrons). The phase-stepping algorithm applied to this pixel may be written as

$$\widetilde{N} = \frac{1}{K}\sum_{k=0}^{K-1} e^{i\phi_k} N^{(\phi_k)}$$

where $N^{(\phi_k)}$ is the pixel value recorded at reference phase ϕ_k (for a given integration time). Neglect all noise contributions except shot noise. Show that the variances of the real and imaginary components of $\widetilde{N}$ are given by

$$\mathrm{Var}\left[\widetilde{N}_{\mathrm{Re}}\right] = \mathrm{Var}\left[\widetilde{N}_{\mathrm{Im}}\right] = \frac{1}{2K^2}\langle N_{\mathrm{total}}\rangle$$

where $\langle N_{\mathrm{total}}\rangle$ it the *total* number of pixel values accumulated over all phase steps.

Hint: start by writing $N^{(\phi_k)} = \langle N \rangle + \delta N^{(\phi_k)}$, where $\delta N^{(\phi_k)}$ corresponds to shot noise variations in the number of detected photoelectrons. Use your knowledge of the statistics of these variations.

(b) What happens to the above result if the camera gain is G?

Problem 11.4

Consider the technique of phase-stepping, as described in Section 11.4.2.

(a) Why are a minimum of three phase steps required to determine $\widetilde{I}_{sr}(\vec{\rho})$?

(b) The phase steps need not be circular. For example, show how one can recover the amplitude and phase of $\widetilde{I}_{sr}(\vec{\rho})$ using the phase sequence $\phi_k = \left\{0, \frac{\pi}{2}, \pi\right\}$.

(c) Consider a reference beam that is frequency shifted (as opposed to phase shifted) relative to the sample beam, such that $E_r(\vec{\rho}) \rightarrow E_r(\vec{\rho})e^{i2\pi\delta\nu t}$. Assume that the camera frame rate is $3\delta\nu$, and the camera exposure time is $(3\delta\nu)^{-1}$. Write an algorithm to recover $\widetilde{I}_{sr}(\vec{\rho})$ from a sequence of three camera exposures.

Problem 11.5

Most widefield microscopes are based on 4f configurations. Here we consider a 6f configuration. In particular, a 2f system projects an in-focus sample field $E_0(\vec{\rho}_0)$ onto a Fourier plane, where it is denoted by $E_\xi\left(\vec{\xi}\right)$. The Fourier field is then re-imaged with a unit magnification 4f system that separates and re-combines the field through two paths, one of which is inverting, as shown in the figure. That is, the output field is given by

$$E^{(\phi_k)}\left(\vec{\rho}\right) = \frac{1}{2}\left[E_\xi\left(\vec{\rho}\right) + E_\xi\left(-\vec{\rho}\right)e^{i\phi_k}\right]$$

where ϕ_k is a controllable phase shift that can be applied to the inverting path. Phase-stepping interferometry then allows one to synthesize the complex intensity

$$\widetilde{I}\left(\vec{\rho}\right) = \left\langle E_\xi\left(\vec{\rho}\right) E_\xi^*\left(-\vec{\rho}\right)\right\rangle$$

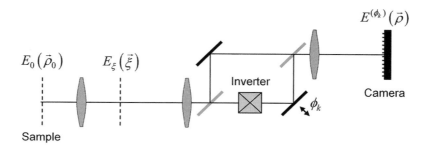

(a) Derive an expression for $\widetilde{I}\left(\vec{\rho}\right)$ in terms of the radiant mutual intensity at the sample plane (adopting the usual coordinate transformation of Eq. 4.17). Hint: use results from Chapter 4.

(b) Assume that the intensity distribution at the sample plane is spatially incoherent, meaning that the mutual intensity at this plane can be expressed in the form of Eq. 4.43. Show that the sample intensity $I_0 \left(\vec{\rho}_0 \right)$ can be numerically calculated from the above radiant mutual intensity, and hence $\widetilde{I} \left(\vec{\rho} \right)$.

(c) Show that even if the sample is displaced away from the focal plane by a distance z_0, the calculated sample intensity $I_{z_0} \left(\vec{\rho}_0 \right)$ remains equal to $I_0 \left(\vec{\rho}_0 \right)$, independently of z_0. That is, the 6f system described here provides extended-depth-of-field imaging, where the recovered image remains in focus independently of the axial location of the (spatially incoherent) sample.

References

[1] Atlan, M., Gross, M., Vitalis, T., Rancillac, A., Rossier, J. and Boccara, A. C. "High-speed wave-mixing laser Doppler imaging in vivo," *Opt. Lett.* 33, 842–844 (2008).

[2] Choi, Y., Yang, T. D., Lee, K. J. and Choi, W. "Full-field and single-shot quantitative phase microscopy using dynamic speckle illumination," *Opt. Lett.* 36, 2465–2467 (2011).

[3] Cuche, E., Marquet, P. and Depeursinge, C. "Simultaneous amplitude-contrast and quantitative phase-contrast microscopy by numerical reconstruction of Fresnel off-axis holograms," *Appl. Opt.* 38, 6994–7001 (1999).

[4] Cuche, E., Marquet, P. and Depeursinge, C. "Spatial filtering for zero-order and twin-image elimination in digital off-axis holography," *Appl. Opt.* 39, 4070–4075 (2000).

[5] Ding, H. and Popescu, G. "Instantaneous spatial light interference microscopy," *Opt. Express* 18, 1569–1575 (2010).

[6] Gabor, D. "A new microscope principle," *Nature* 161, 777–778 (1948).

[7] Goodman, J. W. and Lawrence, R. W. "Digital image formation from electronically detected holograms," *Appl. Phys. Lett.* 11, 77–79 (1967).

[8] Goodman, J. W. *Introduction to Fourier Optics*, 4th edn, W. H. Freeman (2017).

[9] Gross, M., Goy, P., Forget, B. C., Atlan, M., Ramaz, F., Boccara, A. C. and Dunn, A. K. "Heterodyne detection of multiply scattered monochromatic light with a multipixel detector," *Opt. Lett.* 30, 1357–1359 (2005).

[10] Haddad, W. S., Cullen, D., Solem, J. C., Longworth, J. M., McPherson, A., Boyer, K. and Rhodes, C. K. "Fourier-transform holographic microscope," *Appl. Opt.* 31, 4973–4978 (1992).

[11] Hariharan, P. *Optical Holography: Principles, Techniques and Applications*, Cambridge University Press (1984).

[12] Hecht, E. *Optics*, Addison Wesley (2002).

[13] Kronrod, M. A., Merzlyakov, N. S. and Yaroslavski, L. P. "Reconstruction of holograms with a computer," *Sov. Phys.-Tech. Phys.* 17, 333–334 (1972).

[14] Le Clerc, F., Collot, L. and Gross, M. "Numerical heterodyne holography with two-dimensional photodetector arrays," *Opt. Lett.* 25, 716–718 (2000).

[15] Leith, E. N. and Upatnieks, J. "Wavefront reconstruction with continuous-tone objects," *J. Opt. Soc. Am.* 53, 1377 (1963)

[16] Leith, E. N. and Upatnieks, J. "Wavefront reconstruction with diffused illumination and three-dimensional objects," *J. Opt. Soc. Am.* 54, 1295–1301 (1964).

[17] Lohmann, A. W. and Paris, D. P. "Binary Fraunhofer holograms, generated by computer," *Appl. Opt.* 6, 1739 (1967).

[18] Marquet, P., Rappaz, B., Magistretti, P. J., Cuche, E., Emery, Y., Colomb, T. and Depeursinge, C. "Digital holographic microscopy: a noninvasive contrast imaging technique allowing quantitative visualization of living cells with subwavelength axial accuracy," *Opt. Lett.* 30, 468–470 (2005).

[19] Popescu, G. *Quantitative Phase Imaging of Cells and Tissues*, McGraw-Hill (2011).

[20] Schnars, U. and Jüptner, W. P. O. "Digital recording and numerical reconstruction of holograms," *Meas. Sci. Technol.* 13, R85–R101 (2002).

[21] Schwinder, J., Burow, R., Elssner, K.-E., Grzanna, J., Spolaczyk, R. and Merkel, K. "Digital wave-front measuring interferometry: some systematic error sources," *Appl. Opt.* 22, 3421–3432 (1983).

[22] Sentenac, A. and Mertz, J. "A unified description of three-dimensional optical diffraction microscopy: from transmission microscopy to optical coherence tomography," *J. Opt. Soc. Am.* A 35, 748–754 (2018).

[23] Skarman, B., Becker, J. and Wozniak, K. "Simultaneous 3D-PIV and temperature measurements using a new CCD-based holographic interferometer," *Flow. Meas. Instrum.* 7, 1–6 (1996).

[24] Takeda, M., Ina, H. and Kobayashi, S. "Fourier-transform method of fringe-pattern analysis for computer-based topography and interferometry," *J. Opt. Soc. Am.* 72, 156–160 (1982).

[25] VanLigten, R. F. and Osterberg, H. "Holographic microscopy," *Nature* 211, 282–283 (1966).

[26] Wagner, C., Seebacher, S., Osten, W., and Jüptner, W. "Digital recording and numerical reconstruction of lensless Fourier holograms in optical metrology," *Appl. Opt.* 38, 4812–4820 (1999).

[27] Yamaguchi, I. and Zhang, T. "Phase-shifting digital holography," *Opt. Lett.* 22, 1268–1270 (1997).

[28] Zhang, T. and Yamaguchi, I. "Three-dimensional microscopy with phase-shifting digital holography," *Opt. Lett.* 23, 1221–1223 (1998).

12 | Optical Coherence Tomography

In Chapters 10 and 11, we concentrated on imaging with monochromatic, or near monochromatic, illumination. Spatial resolution in these cases came from the diversity of wavevector angles present in the detection path, and, if the illumination was spatially incoherent or partially coherent, also in the illumination path. In this chapter, we will explore a different imaging technique where spatial resolution comes instead from the diversity of wavevector magnitudes (i.e. wavenumbers). The technique is called optical coherence tomography, or OCT, which was introduced around 1990 by Fercher [9] and Fujimoto [13].

OCT is an interference microscopy. It relies on interference with an external reference beam, as did holographic microscopy in the previous chapter, except that the illumination now spans a broad range of wavelengths. As discussed in Chapter 7, a broad span of wavelengths is generally associated with a low temporal coherence length, and OCT is thus often referred to as low-coherence interferometry (LCI) microscopy. While low temporal coherence is a defining characteristic of OCT, this does not exclude the possibility of combining it with low spatial coherence, as we will see later in the chapter.

OCT is also a reflection microscopy. As we saw in Chapters 10 and 11, reflection microscopes have the advantage that they do not feature a missing cone, meaning they are capable of axial resolution even of sample structure that is slowly varying in the transverse direction. Such axial resolution, in turn, means that out-of-focus sample structure can be rejected by the imaging process. Reflection microscopes are thus capable of performing optical sectioning (see Chapter 5). We have not previously dwelled on optical sectioning, since this was not a key advantage of most of the microscope techniques we have seen so far. However, we will dwell on it here. Indeed, OCT features an extraordinarily high capacity for out-of-focus background rejection, making it one of the highest performing microscope techniques to date in terms of depth penetration.

There are many variants of OCT. At its origin, OCT was based on beam scanning, and only more recently have non-scanning camera-based approaches become available. This chapter more or less follows the development of OCT in chronological order. In other words, we begin with variants of OCT based on beam scanning. We will only consider scanning in its most rudimentary form, where the beam is treated basically as a planar wave that is localized to some extent. That is, the beam is considered only weakly focused. A more detailed discussion of scanning with strongly focused beams will be relegated to Chapter 14.

Because of its many advantages, particularly in depth penetration in tissue, considerable effort has gone into the development of OCT for biomedical applications. Extensive treatments of OCT can be found in books [2, 4] as well as in review articles [11, 18], the first of these being particularly thorough.

12.1 COHERENCE GATING

The principle of optical sectioning in OCT is based on coherence gating. To understand such gating, let us imagine sending a short pulse of light into a thick sample and detecting the return pulse backreflected from the sample. This pulse will be broadened in time because reflections from increasing depths become increasingly delayed. In principle, reflections from different depths in the sample can be identified by adding a simple temporal gate to the detection. In such a manner, different depths in the sample can be probed by applying different time delays to the gate. The narrower the width of the gate, the finer the axial resolution.

Unfortunately, this simple strategy cannot be realized in practice with an electronic gating mechanism because the gating times required for microscopic optical sectioning are far too short for current detector technology. However, this strategy can be realized with an alternative gating mechanism based on optics (see Fig. 12.1). In particular, if a small part of the initial light pulse is split off from the main pulse prior to entering the sample, then it retains its initial short temporal duration. By recording only the interference (overlap) between this short reference pulse and the broadened sample pulse, then the reference pulse can itself serve as the temporal gate. Different depths in the sample are thus probed by adjusting the time delay of the reference pulse, which can be achieved by simply adjusting its optical path length. This is the principle of coherence gating.

The layout shown in Fig. 12.1 is that of a Michelson interferometer, as we saw in Section 11.4.1. The separation of the incident beam into sample and reference beams is performed by a beamsplitter. The sample beam is directed into the sample, whereas the reference beam is directed onto a mirror, the location of which defines the path length traveled by the reference beam. The backreflected sample and reference beams are then recombined by the same beamsplitter and directed onto a detector. When the sample and reference beams overlap in time they interfere at the detector, and it is precisely this interference that constitutes the signal of interest in OCT. Various techniques to isolate this interference are described below.

Since we are concerned here with relative time delays between backreflected sample and reference beams, we explicitly write these as functions of time. Thus, we define $E_i(t)$ to be the field incident on the sample and $\zeta E_i(t)$ to be the field incident on the reference mirror (derived from the same source). From the Born approximation, the corresponding field incident on the detector is then

$$E(z_r, t) = \zeta E_i \left(t - \tfrac{2}{v_r} z_r \right) + \int dz_0 \, r_z(z_0) E_i \left(t - \tfrac{2}{v_0} z_0 \right) \tag{12.1}$$

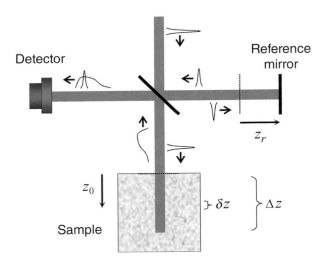

Figure 12.1. Temporal gating with a Michelson interferometer.

where $r_z(z_0)$ is the sample reflectivity (reflectance per unit depth). The axial position of the reference mirror z_r and the axial coordinate inside the sample z_0 are defined relative to a common origin. This common origin can arbitrarily be chosen to be at the beamsplitter or, as depicted in Fig. 12.1, at the location of the sample surface projected into both sample and reference paths. The velocities of the fields in the sample and reference arms are denoted by v_0 and v_r. Because these are associated with time delays, they correspond to the respective group velocities of the fields in both arms, which, in general may be different if the media in both arms are different (see Section 7.2.3 for a definition of group velocity).

From Eq. 12.1, the intensity incident at the detector is

$$I(z_r) = I_{rr} + I_{ss} + \widetilde{I}_{sr}(z_r) + \widetilde{I}^*_{sr}(z_r) \tag{12.2}$$

where I_{rr} and I_{ss} are the powers originating, respectively, from the reference mirror alone and the sample alone, and $\widetilde{I}_{sr}(z_r) + \widetilde{I}^*_{sr}(z_r)$ is the (real-valued) interferogram. We observe an immediate similarity with Eq. 11.3, though with the difference that $I(z_r)$ depends on an axial coordinate z_r rather than on a transverse coordinate $\vec{\rho}$. Indeed, the detector here is only a single element detector and not a camera. Any lateral dependence of $I(z_r)$ comes from the transverse scanning of the sample beam, which, for ease of notation, we disregard for now.

Assuming constant incident intensity I_i, and making use of the temporal coherence function (see Section 7.1.1)

$$\gamma(\tau) = \frac{\langle E_i(t+\tau) E^*_i(t) \rangle}{I_i} \tag{12.3}$$

we have

$$I_{rr} = \zeta^2 I_i \tag{12.4}$$

$$I_{ss} = I_i \iint dz_0 \, dz_0' \, r_z(z_0) r_z^*(z_0') \, \gamma\left(\frac{2}{v_g}(z_0' - z_0)\right) \tag{12.5}$$

$$\widetilde{I}_{sr}(z_r) = \zeta I_i \int dz_0 \, r_z(z_0) \, \gamma\left(\frac{2}{v_g}(z_r - z_0)\right) \tag{12.6}$$

where, for simplicity, we have assumed that the media in the reference and sample arms are matched, meaning that their group velocities are the same (i.e. $v_r = v_s \equiv v_g$). As in the previous chapter, the tilde in Eq. 12.6 indicates that $\widetilde{I}_{sr}(z_r)$ is a complex function and hence does not represent a directly measurable intensity. It should be noted that while Eqs. 12.5 and 12.6 involve the temporal coherence function of the illumination beam, they do not require in any way that this beam be pulsed, as suggested in Fig. 12.1. In fact, many OCT configurations make use of continuous light beams rather than pulsed light beams. We will consider both cases below.

As in all interference microscopies, it is the interference term Eq. 12.6 that is of most interest. This equation resembles a convolution, and we observe that $\gamma(\tau)$ plays a role equivalent to an amplitude point spread function acting on $r_z(z_0)$. From Section 7.1.1 we saw that the envelope of $\gamma(\tau)$ is peaked about $\tau = 0$, and thus serves as a temporal gate. In other words, $\widetilde{I}_{sr}(z_r)$ captures sample information that is dominantly centered about $z_0 = z_r$. By adjusting the location z_r of the reference mirror, different sample locations become probed.

To gain an idea of what axial sample frequencies are supported by this mechanism, we can recast Eq. 12.6 as

$$\widetilde{I}_{sr}(z_r) = \zeta I_i \int d\nu \, e^{-i4\pi\nu z_r/v_g} \hat{r}_z\left(-\frac{2}{v_g}\nu\right) S(\nu) \tag{12.7}$$

where $\hat{r}_z(\kappa_z)$ is the Fourier transform of $r_z(z_0)$, and $S(\nu)$ is the spectral density of the illumination beam (Eq. 7.6). Taking the Fourier transform of both sides, we obtain

$$\widetilde{\mathcal{I}}_{sr}(\kappa_z) = \tfrac{1}{2} v_g \zeta I_i \hat{r}_z(\kappa_z) S\left(-\tfrac{1}{2} v_g \kappa_z\right) \tag{12.8}$$

enabling the reconstruction of $\hat{r}_z(\kappa_z)$ from $\widetilde{\mathcal{I}}_{sr}(\kappa_z)$, which is measured, and $S(\nu)$, which is presumably known a priori.

We observe here that $S(\nu)$ plays a role equivalent to a coherent transfer function acting on $\hat{r}_z(\kappa_z)$. But this CTF is somewhat unusual in that it is one-sided, since $S(\nu)$ spans only positive frequencies (see Section 1.1). Denoting the mean frequency of this span by $\bar{\nu}$ and making allowances for a frequency-dependent index of refraction $n(\nu)$ (see Section 7.2.3), we find that the frequency support for $\hat{r}_z(\kappa_z)$ is centered about $\bar{\kappa}_z = -2\bar{\kappa}\frac{v_p}{v_g}$, where $v_p = \frac{\bar{\nu}}{\bar{\kappa}}$ is the phase velocity of light. In most cases $\frac{v_p}{v_g}$ is generally only slightly greater than one and the frequency support for $\hat{r}_z(\kappa_z)$ is thus centered only slightly more negative than $-2\bar{\kappa}$. We have encountered such an offset axial frequency support before, since it is intrinsic to all microscope techniques based on reflection. OCT is no exception.

But we must remember that, as we have described OCT so far, $\widetilde{I}_{sr}(z_r)$ (and hence $\widetilde{\mathcal{I}}_{sr}(\kappa_z)$) is not measured directly. Instead, what is measured is $I(z_r)$ (Eq. 12.2). Two problems are associated with this. The first is that the component I_{ss} (Eq. 12.5), which depends nonlinearly on sample reflectivity, can contaminate our results and impair the reconstruction of $r_z(z_0)$. Fortunately, when $r_z(z_0)$ is small, this term can usually be neglected. A second problem comes from the fact that $\widetilde{I}_{sr}(z_r)$ is systematically accompanied by $\widetilde{I}^*_{sr}(z_r)$, meaning that a measure of $I(z_r)$ provides information only about the real part of $\widetilde{I}_{sr}(z_r)$. Bearing in mind that $\mathcal{S}(\nu)$ is itself real-valued, we infer from Eq. 12.7 that the interferogram $\widetilde{I}_{sr}(z_r) + \widetilde{I}^*_{sr}(z_r)$ provides information only about the real part of $r_z(z_0)$. While it is possible to numerically recover information about the imaginary part of $r_z(z_0)$, this requires assumptions about sample analyticity (cf. [11]). More robust strategies to recover a fully complex-valued $r_z(z_0)$ are described below.

Finally, it should be noted that we have tacitly assumed that the media in question are dispersionless. In other words, we have assumed that the dependence of $n(\nu)$ on ν is strictly linear (or, equivalently, that the group velocity v_g is independent of frequency). In the event that the media exhibit dispersion *and* that the two arms are not properly matched, then the interpretation of the interferogram becomes more complicated. We will not consider this complication here, and instead refer the reader to [11].

12.2 EXTRACTION OF COMPLEX INTERFERENCE

Our goal, as stated above, is to extract $\widetilde{I}_{sr}(z_r)$ from Eq. 12.2. It should come as no surprise that, given the parallels between OCT and interference microscopy, this section will be quite similar in content to Section 11.4.2. We consider the two cases of a static sample and a dynamic sample.

12.2.1 Static Sample

In the case of a static sample, the sample properties, such as its reflectivity $r_z(z_0)$, are assumed to be time invariant. We tacitly made this same assumption of time invariance in Chapter 11, and, accordingly, the same strategies for the extraction of complex interference directly carry over.

Phase Stepping

The first strategy we (re)consider is that of phase stepping. As a reminder, this strategy involves applying a user-controllable phase shift to the reference beam, as can be obtained, for example, by mounting the reference mirror onto a piezoelectric transducer. A sequence of intensities $I^{(\phi)}(z_r)$ is then recorded with different phase shifts, and Eq. 12.2 becomes

$$I^{(\phi)}(z_r) = I_{rr} + I_{ss} + e^{-i\phi}\,\widetilde{I}_{sr}(z_r) + e^{i\phi}\,\widetilde{I}^*_{sr}(z_r) \tag{12.9}$$

Since the unknowns in this equation are $I_{rr} + I_{ss}$ and the real and imaginary components of $\widetilde{I}_{sr}(z_r)$, the application of at least three phase shifts is required to fully isolate $\widetilde{I}_{sr}(z_r)$. Typically, these are applied in a circular phase sequence $\phi_k = \frac{2\pi k}{K}$ where k is incremented from 0 to $K-1$, and K is generally chosen between 3 and 5. The construction of $\widetilde{I}_{sr}(z_r)$ is then performed numerically using the algorithm

$$\widetilde{I}_{sr}(z_r) = \frac{1}{K} \sum_{k=0}^{K-1} e^{i\phi_k} I^{(\phi_k)}(z_r) \tag{12.10}$$

Phase stepping is particularly adapted to widefield OCT configurations, as will be described below.

Frequency Shifting

As noted in Section 11.4.2, alternative phase sequences involving any number of phase steps could also do the job. In particular, variants of phase stepping can involve the use of a continuously shifting phase. For example, a widely employed technique for the extraction of $\widetilde{I}_{sr}(z_r)$ is known as heterodyne interference. The basic idea is to impart a frequency shift to the reference beam, rather than a sequence of phase steps. This can be achieved by scanning the reference beam mirror with a constant velocity V_r, whereupon the reference beam frequency ν_r incurs a Doppler shift defined by

$$\delta\nu_r = 2\frac{V_r}{v_p}\bar{\nu} = 2\bar{\kappa}V_r \tag{12.11}$$

The detected intensity becomes then

$$I(z_r(t)) = I_{rr} + I_{ss} + e^{-i2\pi\,\delta\nu_r t}\widetilde{I}_{sr}(z_r(t)) + e^{i2\pi\,\delta\nu_r t}\widetilde{I}^*_{sr}(z_r(t)) \tag{12.12}$$

where we note that $I(z_r(t))$ is now time dependent because of the explicit time dependence of $z_r(t) = V_r t$.

Our goal, again, is to extract $\widetilde{I}_{sr}(z_r(t))$. We can do this in several ways, but we focus here on a technique known as demodulation. This technique involves first multiplying both sides of Eq. 12.12 by the same harmonic modulation as imparted by the Doppler shift, but this time complex conjugated. In other words, the first step in the demodulation procedure is to obtain

$$e^{i2\pi\,\delta\nu_r t}I(z_r(t)) = e^{i2\pi\,\delta\nu_r t}(I_{rr} + I_{ss}) + \widetilde{I}_{sr}(z_r(t)) + e^{i4\pi\,\delta\nu_r t}\widetilde{I}^*_{sr}(z_r(t)) \tag{12.13}$$

either numerically or electronically.

We observe that all the terms here are rapidly oscillating because of the exponential prefactors, except one, the term of interest. To extract this term of interest, it thus suffices to apply a lowpass filter to both sides of Eq. 12.13. This is the second step in the demodulation procedure. The filter should have a bandwidth wide enough to encompass the time variations in $\widetilde{I}_{sr}(z_r(t))$, and narrow enough to suppress the faster time variations of all the other components in Eq. 12.13 that are rapidly oscillating. The demodulation procedure is then summarized by

$$\widetilde{I}_{sr}(z_r(t)) = \mathrm{LF}\left[e^{i2\pi\,\delta\nu_r t}I(z_r(t))\right] \tag{12.14}$$

where $\mathrm{LF}[...]$ denotes the lowpass filter operation.

To gain a better understanding of this procedure, let us clarify the bandwidth constraints imposed on the lowpass filter. For starters, the only reason $\widetilde{I}_{sr}\left(z_r(t)\right)$ varies in time is because $z_r(t)$ varies in time. That is, by sweeping the location of the reference mirror with velocity V_r, we are also sweeping the coherence gate through the sample with velocity V_r. The passage of sample structures through this coherence gate thus causes $\widetilde{I}_{sr}\left(z_r(t)\right)$ to vary in time. It is a simple matter to show that the maximum bandwidth $\Delta\nu_r$ associated with these time variations is given by

$$\Delta\nu_r = 2\frac{V_r}{\nu_g}\Delta\nu \qquad (12.15)$$

where $\Delta\nu$ is the bandwidth of the illumination light, as defined by the span of $\mathcal{S}\left(\nu\right)$. The subscript r is used here to emphasize that this bandwidth arises from the motion of the reference mirror alone (assuming the sample is fixed in time).

To capture the time variations in $\widetilde{I}_{sr}\left(z_r(t)\right)$, the bandwidth of the lowpass filter in Eq. 12.14 should be no smaller than $\Delta\nu_r$ (Eq. 12.15). On the other hand, to avoid capturing the time variations in the other components of Eq. 12.13, it should be no larger than $2\delta\nu_r$ (Eq. 12.11). Provided $\Delta\nu < 2\frac{v_g}{v_p}\bar{\nu}$, we arrive at the conclusion that it is always possible to find a lowpass filter bandwidth that separates $\widetilde{I}_{sr}\left(z_r(t)\right)$ from the other components in Eq. 12.14 no matter what the value of V_r. And indeed, as V_r approaches zero the frequency shifting strategy begins to resemble the phase-stepping strategy, as we can see by the direct correspondence between Eqs. 12.10 and 12.14 (the averaging in the former plays the role of the lowpass filter in the latter).

12.2.2 Dynamic Sample

But samples, particularly biological samples, are rarely perfectly static. There are at least two reasons the detected intensity may exhibit temporal variations in addition to those caused by the motion of the reference mirror. One reason is that the sample possesses intrinsic motion, for example due to blood flow, heart pulsations, tissue mechanics, etc. But another reason is more prosaic. We have neglected so far the fact that, to create an image, we must scan the illumination beam in the transverse direction. Such transverse scanning is often performed simultaneously with the axial scanning of the reference mirror, leading to additional variations in the detected intensity which can be quite rapid. To take these variations into account, we generalize Eq. 12.12 and write

$$I\left(t\right) = I_{rr} + I_{ss}\left(t\right) + e^{-i2\pi\,\delta\nu_r t}\,\widetilde{I}_{sr}\left(t\right) + e^{i2\pi\,\delta\nu_r t}\,\widetilde{I}_{sr}^{*}\left(t\right) \qquad (12.16)$$

While a static sample imposed no constraints on the minimum required velocity V_r, this is no longer true for a dynamic sample. We can understand this by noting that Eq. 12.16 bears a resemblance to Eq. 11.48, where the spatial variables are replaced by temporal variables. In other words, frequency shifting, in the context of OCT, can be thought of as a temporal analog of off-axis digital holography. Given this connection between the two techniques, we

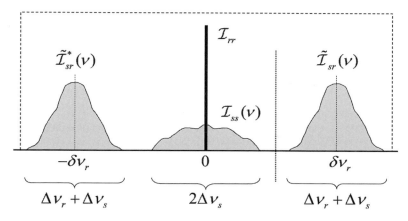

Figure 12.2. Frequency components of $\widetilde{\mathcal{I}}_{sr}(\nu)$ when using heterodyne detection (shown in amplitude).

can make use of a similar pictorial interpretation as Fig. 11.7. Denoting $\Delta\nu_s$ as the temporal bandwidth associated with the sample dynamics, we observe from Fig. 12.2 that the minimum $\delta\nu_r$ required for the isolation of $\widetilde{\mathcal{I}}_{sr}(\nu)$ is $\delta\nu_r \geq \frac{3}{2}\Delta\nu_s + \frac{1}{2}\Delta\nu_r$. Accordingly, the minimum reference velocity must satisfy

$$V_r \geq \frac{3}{2}\frac{v_g\Delta\nu_s}{2\frac{v_g}{v_p}\bar{\nu} - \Delta\nu} \qquad (12.17)$$

But V_r cannot be too large either, for there is another constraint to consider. Just as the rightmost complex interference in Fig. 11.7 had to fit within the spatial bandwidth allowed by the camera, so too in Fig. 12.2 it must fit within the temporal bandwidth allowed by the detector (dashed box). Previously, the spatial bandwidth was defined by the size of the camera pixels; here the temporal bandwidth is defined by the speed of the detector electronics. In general, OCT systems are operated with the fastest mirror velocities possible so as to minimize overall image acquisition time. However, the translation of a mirror at high speed over distances typically in the range of millimeters can be technically challenging. Several techniques have been developed to address this challenge. For example, a time delay can be imparted to the reference beam by use of a clever scan mechanism that applies a linear phase ramp to the beam's spectrum [19].

As depicted in Fig. 12.2, the basic operations for the isolation of $\widetilde{\mathcal{I}}_{sr}(\nu)$ are (1) filtering (i.e. setting to zero all components to the left of the vertical dashed line), and (2) frequency shifting (i.e. re-centering $\widetilde{\mathcal{I}}_{sr}(\nu)$ about zero). But it should be pointed out that the ordering of these operations can, equivalently, be reversed, as in Eq. 12.14.

Doppler OCT

Let us examine more carefully the case where the object possesses intrinsic motion, such as that arising from flow. Such motion can vary as a function of position within the sample. That is, to each position in the sample z_0 we ascribe a local sample velocity $V_s(z_0)$ projected along the optical axis (which may or may not be zero). The sample beam reflected from position z_0 then incurs its own Doppler shift given by $\delta\nu_s(z_0) = 2\bar{\kappa}V_s(z_0)$. Upon interference of the sample and reference beams, the respective Doppler shifts of each beam subtract. Following the same demodulation procedure as outlined in Section 12.2.1, we obtain

$$\widetilde{I}_{sr}(t) \approx e^{i2\pi \, \delta\nu_s(z_r)\, t} \, \widetilde{I}_{sr}(z_r) \tag{12.18}$$

where $\widetilde{I}_{sr}(z_r)$ is the demodulated intensity that would be obtained if the sample were static (i.e. 12.14 – for ease of notation, we have suppressed the explicit time dependence of $z_r(t)$). In other words, the sample-induced Doppler shift has introduced a residual oscillation in the demodulated intensity.

The principle of Doppler OCT is to derive $V_s(z_r)$ from Eq. 12.18. While seemingly straightforward, this derivation is subject to several constraints. A standard technique involves calculating the instantaneous spectra of $\widetilde{I}_{sr}(t)$ (denoted in Fig. 12.3 by $\widetilde{\mathcal{I}}_{sr}(\Omega; t)$) associated with each time t, and then simply inferring the frequency shifts of these instantaneous spectra (see, for example, [14]). Bearing in mind that each time t corresponds to an axial position $z_r = V_r t$, and that each frequency shift is given by $\delta\nu_s(t) \to \delta\nu_s(z_r) = 2\bar{\kappa}V_s(z_r)$, this procedure provides a map of the sample velocities $V_s(z_r)$ as a function of axial position. However, one must be careful, since the reference arm is in constant motion during the acquisition of $\widetilde{I}_{sr}(t)$. To properly calculate the instantaneous spectrum $\widetilde{\mathcal{I}}_{sr}(\Omega; t)$ associated with a particular position z_r, $\widetilde{I}_{sr}(t)$ must be recorded over a certain duration, say Δt. The shorter this duration Δt, the shorter the distance traveled by the reference arm during the recording process, and hence the more localized $\widetilde{\mathcal{I}}_{sr}(\Omega; t)$ is about $z_r = V_r t$. However, from the Fourier transform uncertainty relation (see [3]), the spectral width of $\widetilde{\mathcal{I}}_{sr}(\Omega; t)$ is at least as broad as $1/\Delta t$. That is, if Δt is short the spectral width $\Delta\Omega(t)$ must inevitably be large, meaning that the calculation of the frequency shift becomes prone to uncertainty. In effect, the technique

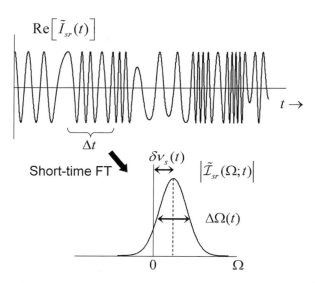

Figure 12.3. Principle of Doppler OCT based on short-time Fourier transforms of $\widetilde{I}_{sr}(t)$.

of calculating $V_s(z_r)$ based on short-time spectra of $\widetilde{I}_{sr}(t)$ suffers from an inherent tradeoff between velocity resolution (denoted by δV_s) and spatial resolution (denoted by δz_r), roughly characterized by

$$\delta V_s \delta z_r \approx \frac{V_r}{\bar{\kappa}} \tag{12.19}$$

with the consequence that, for a given reference mirror velocity V_r, spatial resolution must be compromised when measuring small sample velocities.

The above difficulty can be readily circumvented by adopting an alternative strategy that decouples δV_s from δz_r [22]. In particular, if two axial scans are performed separated by an arbitrary time interval $\Delta\tau$, these can provide phase information pertaining to a same location z_r. That is, a first scan supplies $\widetilde{I}_{sr}(t)$, given by Eq. 12.18, and a second scan a time $\Delta\tau$ later supplies

$$\widetilde{I}_{sr}(t + \Delta\tau) \approx e^{i2\pi\,\delta\nu_s(z_r)\,(t+\Delta\tau)}\,\widetilde{I}_{sr}(z_r) \tag{12.20}$$

The axial velocity of the sample at position z_r is then derived from

$$\angle\widetilde{I}_{sr}(t + \Delta\tau) - \angle\widetilde{I}_{sr}(t) = 2\pi\,\delta\nu_s(z_r)\,\Delta\tau = 4\pi\bar{\kappa}V_s(z_r)\,\Delta\tau \tag{12.21}$$

where $\angle$ denotes phase. Such a sequential scanning approach is not constrained by short-time Fourier transforms and thus can be used to record arbitrarily small velocities, to within the limits of detection noise. To date, blood flow imaging (or angiography) is by far the most important application of Doppler OCT [2] and has been used to reveal flow velocities as small as a few microns per second.

12.3 OCT IMPLEMENTATIONS

Before discussing different strategies for the implementation of OCT, we briefly digress with a bit of nomenclature. To begin, we recall that $\bar{\nu}$ and $\Delta\nu$ denote the mean frequency and spectral bandwidth of the illumination beam respectively, as defined by $\mathcal{S}(\nu)$. We adopt the following notation:

- Δz is the total axial scan range.
- δz is the width of an axial slice (i.e. the axial resolution).
- $N = \frac{\Delta z}{\delta z}$ is the number of resolution slices within an axial scan.

We define these parameters for two limiting cases, one where the illumination bandwidth is broad ($\Delta\nu \lesssim \bar{\nu}$), and the other where it is narrow ($\Delta\nu \ll \bar{\nu}$). The first case corresponds to temporally incoherent illumination, the extreme case being white-light illumination; the second corresponds to temporally coherent illumination, such as that produced by a continuous laser. Both cases will be treated in the context of very different OCT implementations.

12.3.1 Time-Domain OCT

Time-domain OCT, or TD-OCT, is based on the use of temporally incoherent illumination. Associated with the broad spectral bandwidth of this illumination is a short coherence time $\tau_\gamma = 1/\Delta\nu$ (see Section 7.1.1). Ultimately it is this coherence time that serves as the temporal gate in TD-OCT and hence provides axial resolution.

We can define axial resolution for TD-OCT based on the illumination spectral bandwidth. From Eq. 12.8, we find that the axial information contained in $\hat{r}_z(\kappa_z)$ is confined to a frequency support defined by the span of the $\mathcal{S}(\nu)$. Specifically, we find $\Delta\kappa_z = \frac{2}{v_g}\Delta\nu$, meaning that the larger the spectral bandwidth of the illumination, the greater the span of the axial frequency support, and correspondingly the finer the axial resolution. Conventionally, the axial resolution of TD-OCT is defined by

$$\delta z = \frac{1}{\Delta\kappa_z} = \frac{1}{2}\frac{v_g}{\Delta\nu} \tag{12.22}$$

But we must be careful with this definition. In particular, we must recall that the frequency support is centered about $\bar{\kappa}_z = -\frac{2}{v_g}\bar{\nu} = -2\bar{\kappa}\frac{v_p}{v_g}$. In other words, even if it provides high axial resolution, TD-OCT is not able to reveal slowly varying axial structure in the sample characterized by small values of κ_z. Instead, it can only reveal rapidly varying structure of spatial frequencies centered about $\bar{\kappa}_z \approx -2\bar{\kappa}$. As discussed previously (Section 11.4), this inability to reveal slowly varying axial structure is endemic to all reflection microscopes. But it should not be considered wholly a disadvantage. Though offset, the frequency support also exhibits no missing cone, meaning it can provide optical sectioning. In particular, samples possessing no transverse features are generally quite difficult to axially resolve with transmission microscopes. This is not the case with reflection microscopes, such as TD-OCT.

As an example, let us consider a sample that possesses no lateral features, but that also possesses rapidly varying axial features, such as a thin plane of uniform reflectance r located at axial position z_s. That is, we write

$$r_z(z_0) = r\,\delta(z_0 - z_s) \tag{12.23}$$

The interference signal produced by such a sample is most easily interpreted by Eq. 12.6, obtaining

$$\widetilde{I}_{sr}(z_r) = \zeta I_i r\,\gamma\left(\frac{2}{v_g}(z_r - z_s)\right) \tag{12.24}$$

We recall from Section 7.1.1 that the coherence function $\gamma(\tau)$ is rapidly oscillating, but it is also localized about $\tau = 0$. In other words, $\widetilde{I}_{sr}(z_r)$ is localized about $z_r = z_s$. The greater the spectral bandwidth of the illumination, the tighter this localization and the finer the axial resolution of the sample. We also recall from Section 7.1.1 that $\gamma(\tau)$ is a complex function. The reflectivity r can also be complex, in general. But since $\widetilde{I}_{sr}(z_r)$ is itself complex (provided we have applied our measurement tricks such as phase stepping or frequency shifting), and since $\gamma(\tau)$ is presumably known from Eq. 7.6, based on an a priori measurement of $\mathcal{S}(\nu)$, a full reconstruction of a complex r becomes straightforward.

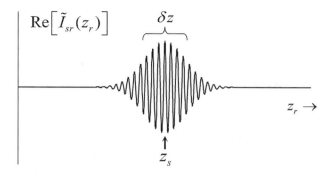

Figure 12.4. Real quadrature of TD-OCT signal from a reflecting planar sample.

We note that in many samples of interest, the reflectivity r is real-valued. In the simple example of a planar sample, the reconstruction of a real-valued r does not require the extraction of the complex interference $\widetilde{I}_{sr}(z_r)$, but instead can be obtained from the interferogram itself. The real-valued interferogram is given by

$$\widetilde{I}_{sr}(z_r) + \widetilde{I}_{sr}^{*}(z_r) = 2\zeta I_i r \, \mathrm{Re}\left[\gamma\left(\frac{2}{v_g}(z_r - z_s)\right)\right] \tag{12.25}$$

which is illustrated in Fig. 12.4. The recovery of a real-valued r thus requires a knowledge only of the real component of $\gamma(\tau)$.

In the above analysis, we have implicitly assumed that the axial resolution in TD-OCT is limited by the spectral bandwidth of the illumination beam. In practice, this assumption may not be fulfilled. For example, we have not taken into account the spectral bandwidth of the detector itself. If the detector exhibits a spectral bandwidth that is narrower than the illumination bandwidth, then the overall spectral bandwidth of a TD-OCT device is limited by the detector and not the illumination. While a restricted detector bandwidth may be problematic in TD-OCT, since it leads to a degradation of axial resolution, it can actually be used to significant advantage in an alternative OCT strategy based not on time gating but on frequency gating, which we will examine below.

12.3.2 Frequency-Domain OCT

Frequency-domain OCT, or FD-OCT, is based on the use of temporally coherent illumination. However, like TD-OCT, FD-OCT is also based on the use of a broad overall spectral bandwidth $\Delta\nu$ (or $\Delta\kappa$). At first glance, it may seem difficult to reconcile temporal coherence with a broad spectral bandwidth, and we consider two techniques to achieve this reconciliation, commonly referred to as swept source and spectral domain, as illustrated in Fig. 12.5.

The two techniques are defined by:

1. Swept-source FD-OCT makes use of narrowband illumination of frequency ν_n and bandwidth $\delta\nu$. However, the illumination frequency can be tuned over a broad bandwidth $\Delta\nu$

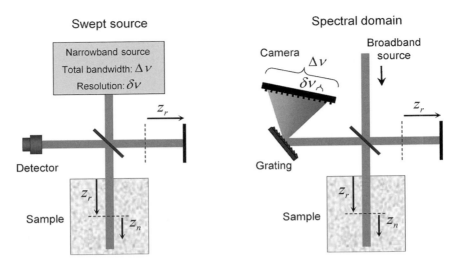

Figure 12.5. FD-OCT configurations.

centered about $\bar{\nu}$ (to facilitate a comparison between FD-OCT and TD-OCT, we take $\Delta\nu$ and $\bar{\nu}$ to be the same in both domains). In practice, swept-source OCT typically involves the use of a narrowband laser whose frequency can be rapidly scanned using an intracavity tilting grating or tunable wavelength filter.

2. Spectral-domain FD-OCT makes use of broadband illumination of spectral width $\Delta\nu$, as in TD-OCT, but instead makes use of an array of narrowband detector elements each sensitive to discrete frequencies ν_n of bandwidth $\delta\nu$, where ν_n spans the frequency range $\Delta\nu$ centered about $\bar{\nu}$ (again, we take $\Delta\nu$ and $\bar{\nu}$ to be the same in both FD-OCT and TD-OCT). In practice, this method typically involves the use of a broadband source, such as a superluminescent diode or supercontinuum laser, and a spectrograph.

For both methods, we have $\Delta\nu \gg \delta\nu$.

FD-OCT was first introduced by Fercher [10] and is characterized by the fact that axial scanning is performed without any need for the reference mirror to be translated. To see how this works, let us assume, as before, that we have managed to isolate the third term in Eq. 12.2. In practice, this can be accomplished, for example, by phase stepping. As in TD-OCT, the goal is to reconstruct $r_z(z_0)$ from Eq. 12.6, or, equivalently, $\hat{r}_z(\kappa_z)$ from Eq. 12.7. However, this time the spectrum $S(\nu)$ is narrow rather than broad. That is, its width is defined by $\delta\nu$ rather than $\Delta\nu$. For simplicity, let us assume $S(\nu)$ is so narrow that it may be approximated by a delta function, centered on ν_n. In FD-OCT, this center frequency ν_n is "scanned," either by the swept-source laser or effectively by the spectrograph. That is, we can write

$$S(\nu) = \delta\nu\, S(\nu_n)\, \delta(\nu - \nu_n) \tag{12.26}$$

where $S(\nu_n)$ is the envelope of the frequency scan, corresponding to the output profile of the swept-source laser or to the response profile of the spectrograph. In either case, $S(\nu_n)$ spans the range $\bar{\nu} - \frac{1}{2}\Delta\nu$ to $\bar{\nu} + \frac{1}{2}\Delta\nu$. We note here that the prefactor $\delta\nu$ is introduced for dimensional consistency, but also to characterize the narrowness of $S(\nu)$.

By inserting Eq. 12.26 into Eq. 12.7 we obtain the complex interference acquired at each frequency ν_n, given by

$$\widetilde{I}_{sr}(z_r; \nu_n) = \zeta I_i \delta\nu \, e^{-i4\pi\nu_n z_r/\upsilon_g} \, \hat{r}_z\left(-\frac{2}{\upsilon_g}\nu_n\right) S(\nu_n) \tag{12.27}$$

We recall that the parameter z_r is fixed here, since it corresponds to the location of the fixed reference mirror. The parameter that is scanning is instead ν_n. When we make the association $\nu_n = -\frac{1}{2}\upsilon_g\kappa_z$, we observe that Eq. 12.27 bears striking resemblance to Eq. 12.8 , aside from an additional phase shift $e^{i2\pi\kappa_z z_r}$, suggesting that FD-OCT and TD-OCT provide similar information. This is made more clear if we perform a scaled Fourier transform of both sides of Eq. 12.27, obtaining

$$\frac{1}{\delta\nu}\int \mathrm{d}\nu_n \, e^{i4\pi\nu_n z_n/\upsilon_g} \widetilde{I}_{sr}(z_r; \nu_n) = \zeta I_i \int \mathrm{d}\nu_n \, e^{-i4\pi\nu_n(z_r-z_n)/\upsilon_g} \, \hat{r}_z\left(-\frac{2}{\upsilon_g}\nu_n\right) S(\nu_n) \tag{12.28}$$

(we treat ν_n here as a continuous variable, though in practice it is discrete).

If we compare Eq. 12.28 with Eq. 12.7, we again observe a striking resemblance. The only difference, aside from an axial offset z_r, is that the illumination spectrum $S(\nu)$ in the latter has been replaced by the frequency-scan envelope $S(\nu_n)$ in the former. If the profiles $S(\nu)$ and $S(\nu_n)$ are the same, then Eqs. 12.28 and 12.7 are essentially identical, leading to

$$\widetilde{I}_{sr}(z_r - z_n) = \frac{1}{\delta\nu}\int \mathrm{d}\nu_n \, e^{i4\pi\nu_n z_n/\upsilon_g} \widetilde{I}_{sr}(z_r; \nu_n) \tag{12.29}$$

In other words, both FD-OCT and TD-OCT provide a measurement of the same axially dependent complex interference $\widetilde{I}_{sr}(z)$, and thus, ultimately, lead to the same sample reconstruction. In FD-OCT the axial variable is z_n, which is obtained by numerical calculation (namely the application of Eq. 12.29), while in TD-OCT it is z_r, which is obtained by mechanical translation of the reference mirror.

We have not yet considered the axial range of FD-OCT, which we denoted above by Δz. In our analysis, it was assumed that the spectrum $S(\nu)$ was infinitely narrow (see Eq. 12.26). This is, of course, an idealization since in practice $\delta\nu$ is defined by the spectral resolution of either the swept-source laser or the spectrograph, depending on which variation of FD-OCT is adopted. In either case, the scanning of ν_n effectively occurs in independent steps of width $\delta\nu$, as illustrated in Fig. 12.6. A straightforward extension of our calculations then reveals that the axial range of FD-OCT is in fact defined by the inverse of $\delta\nu$.

The reader is also reminded that in our derivation of Eq. 12.29 we made the assumption that the third term in Eq. 12.2 was isolated, for example by phase stepping (see, for example, [20]). Such phase stepping actually requires that the reference mirror be moved a small amount. But it also requires the acquisition of multiple real-valued intensities $I^{(\phi)}(z_r)$ (see Eq. 12.9), which limits some of the speed advantage of FD-OCT. Let us imagine what would happen if we did not isolate the complex interference and instead performed FD-OCT with the

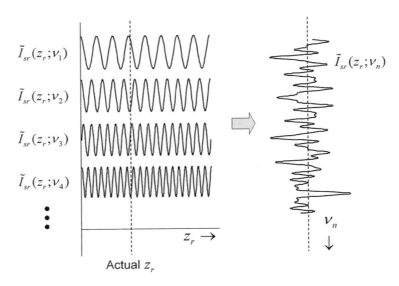

$\tilde{I}_{sr}(z_r;\nu_1)$

$\tilde{I}_{sr}(z_r;\nu_2)$

$\tilde{I}_{sr}(z_r;\nu_3)$

$\tilde{I}_{sr}(z_r;\nu_4)$

$z_r \rightarrow$

Actual z_r

$\tilde{I}_{sr}(z_r;\nu_n)$

ν_n

Figure 12.6. Synthesis of complex $\tilde{I}_{sr}(z_r;\nu_n)$.

real-valued interferogram obtained from only a single intensity measurement. From Eq. 12.6 we observe that $\widetilde{I}^*_{sr}(z_r)$ is the same as $\widetilde{I}_{sr}(z_r)$ when $r_z(z_r - z_0)$ is replaced by $r^*_z(z_r + z_0)$. As a consequence for FD-OCT, the sample reconstruction obtained from $\widetilde{I}^*_{sr}(z_r;\nu_n)$ is the same as that obtained from $\widetilde{I}_{sr}(z_r;\nu_n)$, except that it is complex conjugated and flipped about the axial position z_r. The reconstruction obtained from the interferogram $\widetilde{I}_{sr}(z_r;\nu_n) + \widetilde{I}^*_{sr}(z_r;\nu_n)$ is thus the superposition of $r_z(z_r - z_n)$ and the mirror image $r^*_z(z_r + z_n)$. In general, this superposition is messy and the separation of $r_z(z_r - z_n)$ from $r^*_z(z_r + z_n)$ within a single image is quite problematic. However, the position of the reference mirror z_r can be chosen arbitrarily. If it is chosen to correspond to the sample surface, or just above the sample surface (see Fig. 12.1), then $r_z(z_r - z_n)$ vanishes wherever $r^*_z(z_r + z_n)$ is non-zero, and vice versa. In this case $r_z(z_r - z_n)$ and $r^*_z(z_r + z_n)$ do not overlap and their superposition becomes readily separable. Sample reconstruction can thus be obtained in a single shot, without the complexity or requirement of multiple shots associated with phase stepping. Of course, the price paid for this speed advantage is that the overall axial range of single-shot FD-OCT is reduced by a factor of at least two. But inasmuch as speed is often of paramount concern, particularly when imaging highly dynamic samples, this price is often quite acceptable. An example of a single-shot retinal image acquired with spectral-domain FD-OCT is illustrated in Fig. 12.7.

12.3.3 Comparison

A summary of our results is presented in Fig. 12.8. In TD-OCT the axial resolution is given by $\delta z = \frac{1}{2}\frac{v_g}{\Delta\nu}$ and the total axial range is correspondingly $\Delta z = N\delta z$, where again N is the number of resolution elements acquired. In comparison, in FD-OCT, the total axial range is

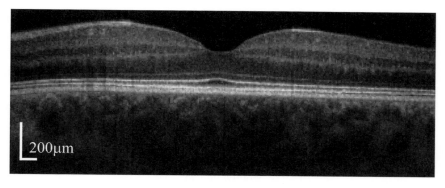

Figure 12.7. FD-OCT image of a retina (x–z scan). Note striations arising from nerve fiber and cell layers. Adapted from [21].

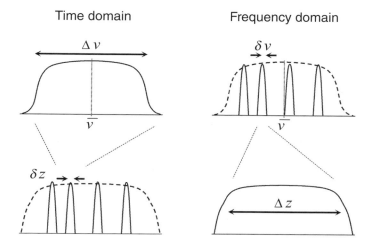

Figure 12.8. Relation between OCT bandwidth and resolution.

given by $\Delta z = \frac{1}{2}\frac{v_g}{\delta\nu}$ and the axial resolution is correspondingly $\delta z = \frac{\Delta z}{N}$. Since the number of resolution elements is given by $N = \frac{\Delta\nu}{\delta\nu}$, both TD and FD techniques provide the same axial range and resolution. The natural question to ask then is which is better? To properly address this question, we must turn to the issues of acquisition speed and SNR.

In TD-OCT, $\widetilde{I}_{sr}(z_r)$ is obtained by translating the reference mirror from one axial position z_r to another, either physically or by using a time-delay mechanism. In FD-OCT, $\widetilde{I}_{sr}(z_r - z_n)$ is instead synthesized numerically from $\widetilde{I}_{sr}(z_r;\nu_n)$. That is, the scan parameter in FD-OCT is ν_n rather than z_r. This can lead to a significant speed advantage for FD-OCT. For example, in the case of swept-source OCT, ν_n corresponds to the output frequency of the tunable laser. The acquisition speed is thus limited by the time required to perform a full sweep of the laser wavelength, which can be on the order of fractions of a millisecond when using an intracavity acousto-optical tunable filter (AOTF). Similarly, in the case of spectral-domain

OCT, ν_n corresponds to the detection frequency of a spectrograph pixel. The acquisition speed is thus limited by the frame rate of the spectrograph. This too can be can be on the order of fractions of a millisecond when using a line-array camera.

But FD-OCT provides yet a more fundamental advantage. To appreciate this advantage, let us compare the respective SNRs of TD-OCT and FD-OCT. We begin by evaluating the signal associated with an OCT measurement. For simplicity, let us consider the idealized case where the total spectral bandwidth $\Delta\nu$ of the illumination and detection are effectively infinite. In this case, the temporal coherence gate effectively reduces to a delta function, and we can make the replacements $\gamma(\tau) \to \Delta\nu^{-1}\delta(\tau)$ and $\mathcal{S}(\nu_n) \to \Delta\nu^{-1}$. Whether we consider TD-OCT (Eq. 12.6) or FD-OCT (Eq. 12.29), the signal is basically the same for equivalent scan ranges, given by

$$\widetilde{I}_{sr}(z_0) = \zeta I_i \delta z\, r_z(z_0) \tag{12.30}$$

Just as we divided the total axial scan range Δz into discrete axial resolution elements δz, we can divide the total axial scan duration Δt into discrete duration elements δt required for the acquisition of each axial resolution element, such that $\Delta t = N\delta t$. The signal energy associated with the measurement of a single axial resolution element in TD-OCT is given by $\zeta\Phi_i\delta z\, r_z\,\delta t$, where $\Phi_i = A_d I_i$ and A_d is the detector area. In contrast, the signal energy associated with the measurement of a full axial scan in FD-OCT is given by $\zeta\Phi_i\delta z\, r_z\,\Delta t$.

Turning now to an evaluation of noise, we note that in typical OCT applications the beamsplitter transmission ζ is adjusted to a fixed value so that the first term in Eq. 12.2 is dominant. The noise associated with an OCT measurement is then given by the shot noise associated with the reference beam alone, assuming all other noise sources negligible. Importantly, the reference beam power remains constant, independent of signal strength, meaning that the measurement noise in OCT is also independent of signal strength.

Based on the above considerations and on the results in Section 8.1, we derive the following amplitude SNRs:

$$\text{SNR}_{\text{td}} = \frac{\zeta\frac{\eta}{h\nu}\Phi_i\delta z\, r_z\,\delta t}{\sqrt{\zeta^2\frac{\eta}{h\nu}\Phi_i\,\delta t}} = \delta z\, r_z\sqrt{\frac{\eta}{h\nu}\Phi_i\,\delta t} \tag{12.31}$$

$$\text{SNR}_{\text{fd}} = \frac{\zeta\frac{\eta}{h\nu}\Phi_i\delta z\, r_z\,\Delta t}{\sqrt{\zeta^2\frac{\eta}{h\nu}\Phi_i\,\Delta t}} = \delta z\, r_z\sqrt{\frac{\eta}{h\nu}\Phi_i\,\Delta t} \tag{12.32}$$

where η is the detector quantum efficiency, assumed to the same over all wavelengths. In both cases, Eqs. 12.31 and 12.32 refer to amplitude SNRs of the reconstructed $r_z(z_0)$. We conclude that for equal illumination power and total acquisition time, then

$$\frac{\text{SNR}_{\text{fd}}}{\text{SNR}_{\text{td}}} = \sqrt{N} \tag{12.33}$$

Such an improvement in SNR is quite remarkable given that in practice N is typically two to three orders of magnitude! The origin of this improvement is clear. Whereas in TD-OCT most of the sample beam power is rejected by a narrow coherence function γ, in FD-OCT the coherence function is broad and the sample beam power is utilized much more efficiently. This advantage of FD over TD is well known in spectroscopy applications, particularly in cases where detector noise (i.e. noise that is independent of signal) can be dominant, such as in infrared wavelength ranges. For example, in the context of Fourier transform infrared (FTIR) spectroscopy, this advantage is referred to as the Fellgett advantage. For more information the reader can consult [7, 12, 16]. The reader should also bear in mind that the derivation of Eq. 12.32 was based on the assumption that the third term in Eq. 12.2 was fully isolated, for example by phase stepping. Had this not been the case, then half of the signal power would have been diverted into a spurious mirror image of $r_z^*(z_0)$ and the ratio of SNRs in Eq. 12.33 would have been reduced by a factor of $\sqrt{2}$. More detailed comparisons of TD and FD techniques are provided in [5, 6, 15].

Finally, it should be noted that while FD-OCT provides several benefits, it also has drawbacks. In particular, one cannot use a tightly focused laser beam when performing FD-OCT since this leads to a transverse resolution that is highly dependent on axial depth. To maintain a relatively uniform transverse resolution over an extended axial range, the laser beam must be approximately collimated within the sample, meaning that in general the transverse resolution of FD-OCT is poor. This is not the case with TD-OCT since only one axial depth is acquired at a time. By using a tight laser focus at this depth, the transverse resolution of TD-OCT can be quite high. Furthermore, by scanning the reference mirror and laser beam focus in unison, this high transverse resolution can be maintained at different depths. Various techniques have been devised to improve the transverse resolution of FD-OCT. For example, a hybrid technique that combines FD (numerical scanning of z_n) and TD (physical scanning of z_r) can be a good compromise [6]. Alternatively, recent advances in image processing based on the concept of a synthetic aperture have made it possible to operate FD-OCT with a tight laser focus while maintaining high resolution even away from the focal plane [17]. More will be said about synthetic aperture imaging in Chapter 19.

12.4 WIDEFIELD OCT

As described in Fig. 12.1, OCT retrieves information from a beam of light directed into a sample. This beam is more or less focused, depending on whether the implementation of OCT is in the time or frequency domain. In either case, we have only considered the axial information provided by these implementations. To obtain transverse sample information, additional degrees of freedom are required. So far we have assumed are that these additional degrees of freedom are provided by lateral translations of either the sample or the illumination beam. In other words, we have considered only variants of OCT that involve lateral scanning. For example, lateral beam scanning can be obtained with the use of tilting mirrors, or alternatively, if the light delivery is performed by way of an optical fiber, by translating the end of the illumination fiber itself. We will not dwell on this issue of lateral scanning here since it will be considered in more detail in Chapter 14. We note only that construction of an en-face image in scanning microscopes is performed in sequential manner, one x–y transverse coordinate after another. This differs fundamentally from the construction of en-face images in widefield microscopes, where x–y coordinates are obtained in parallel using a 2D array detector (i.e. a camera).

Historically, OCT was first implemented in scanning configurations, which prescribes the use of illumination sources that are spatially coherent. The illumination from such sources is coherent in the transverse direction, enabling the light to be focused or collimated, while being incoherent in the longitudinal direction, providing temporal coherence gating in the interference signal (in the case of swept-source FD-OCT, the longitudinal incoherence is synthesized from a wavelength sweep). Such temporal gating is a defining characteristic of OCT.

More recently, it was realized that OCT can also be performed in a widefield configuration using a camera [1, 8]. In this case, no lateral scanning is required, and sample reconstruction is performed in parallel at each camera pixel. Such widefield OCT is very similar to reflection-mode digital holographic microscopy, which we briefly examined in Section 11.3, but with a key difference. In Chapter 11, we considered only monochromatic, or narrowband, illumination. In other words, we relied on spatial incoherence of the illumination beam to obtain axial resolution. Here, instead, we consider broadband illumination, and rely on temporal incoherence. But a remarkable advantage of widefield OCT is that it can be operated with an illumination source that is *both* spatially and temporally incoherent. That is, it can be operated with sources as simple as a halogen light bulb! An example of a widefield OCT configuration is shown in Fig. 12.9. This is similar to the Michelson interferometer configuration shown in Fig. 12.1 except that it comprises a microscope objective in both the sample and reference arms. In practice, the objectives are generally chosen to be the same so as to balance any

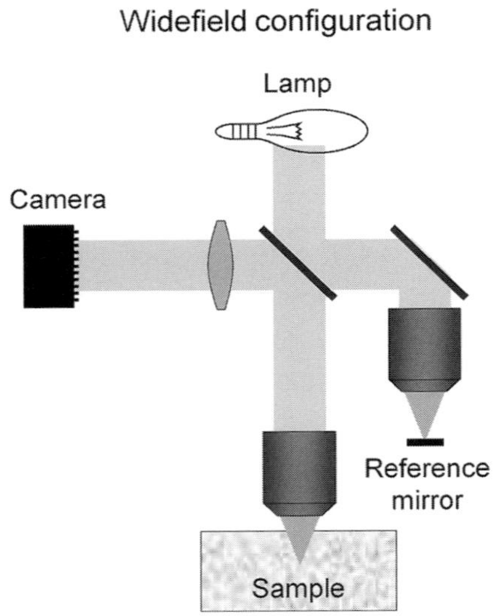

Figure 12.9. Widefield OCT configuration with white-light Linnik interferometry.

possible aberrations or dispersion introduced by the objectives themselves. Such a widefield Michelson interferometer configuration that makes use of microscope objectives is called a Linnik interferometer. Transverse image resolution is thus provided by the NA of the microscope objectives, as in any widefield reflection microscope, interferometric or otherwise. Longitudinal (i.e. axial) image resolution is provided instead by coherence gating. But this coherence gating is now no longer simply temporal – it is spatiotemporal.

12.4.1 Frequency Support

To better appreciate the resolution provided by spatiotemporal gating, let us examine the frequency support associated with the sample $\hat{r}_z\left(\vec{\kappa}\right)$ reconstruction. As always, this frequency support is given by the span of $\vec{\kappa} - \vec{\kappa}_i$, where $\vec{\kappa}_i$ corresponds to the illumination (input) wavevectors and $\vec{\kappa}$ corresponds to the detected (output) wavevectors arising from the interaction of the illumination light with the sample. For reflection microscopes, such as OCT, the output wavevectors propagate in an axial direction opposite to the input wavevectors, meaning that both input and output wavevectors contribute to the frequency support in the negative direction. But the difference between holographic microscopy and widefield OCT is that only the wavevector directions are diverse in the former, whereas both the wavevector directions and magnitudes are diverse in the latter. This increase in wavevector diversity leads to an increase in the span of the frequency support. Schematics of this frequency support are provided in Fig. 12.10. These may be compared with Fig. 11.6, which provides examples of reflection-mode frequency supports when the illumination is monochromatic (i.e. $\left|\vec{\kappa}_i\right| = \left|\vec{\kappa}\right| = \kappa$ is fixed). Again two scenarios are considered in Fig. 12.10, one where the illumination is spatially coherent (a plane wave), and another where it is spatially incoherent, spanning the same NA as the detection optics (corresponding to coherence parameters equal to 0 and 1, respectively – see Eq. 10.20). In both cases, the reference beam is assumed identical to the illumination beam.

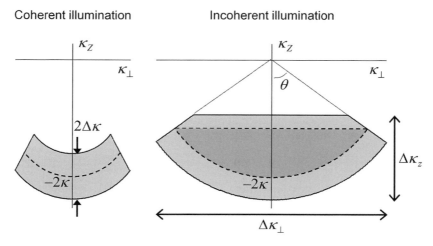

Figure 12.10. Frequency support of $\hat{r}_z(\vec{\kappa})$ for different illumination coherences.

In the case of incoherent illumination, from geometric considerations, we find that the frequency support associated with $\hat{r}_z\left(\vec{\kappa}\right)$ has transverse and axial spans given by

$$\Delta\kappa_\perp = 4\bar{\kappa}\sin\theta + 2\Delta\kappa\sin\theta \tag{12.34}$$

$$\Delta\kappa_z = 4\bar{\kappa}\sin^2\frac{\theta}{2} + 2\Delta\kappa\cos^2\frac{\theta}{2} \tag{12.35}$$

where $\bar{\kappa} = \bar{\nu}/v_p$ is the mean illumination wavenumber, $\Delta\kappa = \Delta\nu/v_g$ is the wavenumber bandwidth, and θ is the half-angular span of the illumination and detection optics, defined by NA$= n\sin\theta$.

We encountered similar results (albeit scaled by a factor of two) when we examined the spatiotemporal coherence of light itself in Section 7.2.3. As previously, the first terms in Eqs. 12.34 and 12.35 arise from the diversity of wavevector directions (associated here with spatial incoherence); the second terms arise from the diversity in wavevector magnitudes (associated here with temporal incoherence). For example, in the case of low-NA optics where θ is small, the axial frequency support $\Delta\kappa_z$ is defined primarily by temporal gating. In the case of high-NA optics, it is defined by both spatial and temporal gating. Which of these is dominant depends on the experimental configuration. Specifically, when $\tan^2\frac{\theta}{2} > \frac{\Delta\kappa}{2\bar{\kappa}}$, spatial gating becomes dominant.

From Eqs. 12.34 and 12.35 we can define associated transverse and axial resolutions, defined by $\delta\rho = 1/\Delta\kappa_\perp$ and $\delta z = 1/\Delta\kappa_z$. In both cases, the presence of a non-zero $\Delta\kappa$ leads to a broadening of frequency support and hence to an improvement in 3D resolution. Moreover, this new parameter supplies an additional degree of freedom that effectively allows the transverse and axial resolutions to be decoupled, and defined more or less independently. This differs in a fundamental way from other sectioning microscopes that we will encounter later, such as confocal or structured illumination microscopes, where transverse and axial resolutions are intimately coupled (see Chapters 14 and 15).

Since the invention of OCT, a multitude of configurations have been reported in the literature, many of which are available commercially. These generally have in common that they are based on light sources of short coherence lengths, though in practice even this is not required (cf. Section 12.3.2). Ultimately, the choice of configuration depends on a variety of considerations including acquisition speed, source brightness, technical complexity, and, of course, the specific application of interest, be it microscopy, opthalmology, endoscopy, etc. For more detailed information on such applications, the reader is referred again to [2, 4].

12.5 PROBLEMS

Problem 12.1

Derive Eq. 12.15. Hint: consider first what type of sample structure leads to the fastest possible signal variations.

Problem 12.2

Figure 12.6 provides a plot of $\widetilde{I}_{sr}\left(z_r; \nu_n\right)$ as a function of ν_n for a given value of z_r. What does this plot look like if the sample consists of a single reflecting plane located at depth z_s? What happens when $z_s = z_r$? When $z_s > z_r$? When $z_s < z_r$? Make sketches of the real and imaginary components assuming $S\left(\nu_n\right)$ is Gaussian in shape, centered on $\bar{\nu}_n$.

Problem 12.3

Show that the axial range of swept-source OCT scales inversely with $\delta\nu$. Specifically, consider a swept-source laser that produces a spectrum $S\left(\nu_n\right) e^{-(\nu-\nu_n)^2/\delta\nu^2}$, where $S\left(\nu_n\right)$ is the envelope of the frequency scan. As in the text, treat ν_n as a continuous variable and assume that $S\left(\nu_n\right)$ is slowly varying on the scale of $\delta\nu$. Show that $\widetilde{I}_{sr}\left(z_r - z_n\right)$, as defined by Eq. 12.29, is confined to within a range of z_ns defined by a Gaussian envelope. What is the extent Δz_n of this axial range? That is, what is the width (waist) of this Gaussian envelope?

Problem 12.4

Our goal in this problem is to compare direct versus multiplexed signal acquisition, under conditions of constant illumination intensity I_i and equal total measurement time Δt.

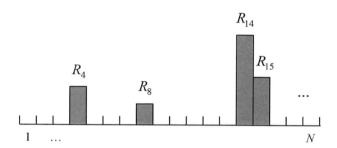

Consider objects of reflectance strength R_n distributed along a line at positions indexed by n, where $n = 1, ..., N$ (as illustrated in the figure), such that the power reflected from each object is given by $\Phi_n = R_n I_i$. These objects can be imaged directly, by illuminating each position one after the other in sequence, and recording the reflected power from each position with a single detector. Alternatively, the objects can be imaged in a multiplexed manner by illuminating the objects in parallel with a sequence of quasi-uniform intensity patterns (of average intensity I_i), and detecting the total reflected power for every illumination pattern with the same single detector. In both cases, N measurements are made, each of duration $\delta t = \Delta t/N$. In the case of multiplexed acquisition, the final image must be reconstructed numerically based on some kind of demultiplexing algorithm. Assume that each object signal adds "coherently" upon reconstruction (i.e. scales with the number of measurements N), and is given by $NR_n I_i \delta t$. Assume also that noise adds "incoherently" upon reconstruction (i.e. scales as $\sqrt{N}$ times the noise associated with each measurement).

(a) Consider a sparse distribution of objects, namely a single object of reflectance strength R located at arbitrary position m. Assume a noiseless detector of unity gain and take into account only shot noise. Compare the SNRs associated with this object for direct versus multiplexed acquisition. Which is best? Specifically, derive SNR_{dir}/SNR_{mul}.

(b) Consider a dense distribution of objects, each of equal reflectance strength R, such that an object is located at every position n. Derive SNR_{dir}/SNR_{mul} for any given object. You should find that the multiplex (or Fellgett) advantage has disappeared.

(c) Consider the same dense distribution of objects, except that one object located at arbitrary position m has reflectance strength $\frac{1}{10}R$. Derive SNR_{dir}/SNR_{mul} for this weaker object. You should find that you are better off using direct acquisition. This difficulty with multiplexed acquisition is generally referred to as the multiplex disadvantage.

(d) Repeat calculations **(a)**–**(c)** taking into account a detector readout noise of standard deviation σ_r. Assume σ_r is so large that shot noise can be neglected. Does the sample sparsity matter in this case? Note that this scenario is similar to the scenario of FD-OCT.

(e) Consider now the condition of constant illumination power Φ_i rather than constant illumination intensity. In other words, in the case of direct acquisition, the illumination intensity sequentially delivered to each location n is given by $I_i = \Phi_i/\delta A$, where δA is the focused illumination area. In the case of multiplexed acquisition, the illumination is spread over an area $\Delta A = N\delta A$ spanning all N locations, and delivers instead an average illumination intensity given by $I_i = \Phi_i/\Delta A$. Are there any cases where multiplexing is advantageous?

Problem 12.5

Widefield phase-sensitive OCT is performed with circular phase stepping (4 steps). Consider an arbitrary camera pixel and assume a camera gain of unity (i.e. the camera directly reports the number of detected photoelectrons). The phase stepping algorithm applied to this pixel may be written as

$$\widetilde{N} = \frac{1}{4}\sum_{k=0}^{3} e^{i\phi_k} N^{(\phi_k)}$$

where $N^{(\phi_k)}$ is the pixel value recorded at reference phase $\phi_k = \frac{2\pi k}{4}$ (for a given integration time T). Our goal is to determine the phase of r_z recorded by this pixel. To do this, we must determine the phase of $\widetilde{N}$, which we denote here by φ_N.

(a) Derive an expression for φ_N in terms of the four measured pixel values $N^{(\phi_k)}$.

(b) Consider two noise sources: shot noise and dark noise. The latter is modeled as producing background photoelectron counts obeying Poisson statistics. Let $\langle N_S \rangle$, $\langle N_R \rangle$, and $\langle N_D \rangle$ be the average pixel values obtained from separate measurements of the sample beam, the reference beam, and the dark current respectively, using a *total* integration time required for all four steps (i.e. $4T$).

Show that the error in the determination of φ_N has a standard deviation given by

$$\sigma_{\varphi_N} = \sqrt{\frac{1}{2\langle N_S \rangle}\left(1 + \frac{\langle N_S \rangle}{\langle N_R \rangle} + \frac{\langle N_D \rangle}{\langle N_R \rangle}\right)}$$

(Without loss of generality, you may set the actual φ_N to be any arbitrary value – in particular, you may assign it to be equal to zero.)

Hint: start by writing $N = \langle N \rangle + \delta N$, where δN corresponds to shot noise variations in the number of detected photoelectrons. Use your knowledge of the statistics of these variations.

Observe that when the reference beam power is increased to such a point that $\langle N_R \rangle \gg \langle N_S \rangle$ and $\langle N_R \rangle \gg \langle N_D \rangle$, then $\sigma_{\varphi_N} \to \sqrt{\frac{1}{2\langle N_S \rangle}}$, meaning that the phase measurement accuracy becomes limited by sample-beam shot noise alone (i.e. dark noise becomes negligible). This is one of the main advantages of interferometric detection with a reference beam.

References

[1] Akiba, M., Chan, K. P. and Tanno, N. "Full-field optical coherence tomography by two-dimensional heterodyne detection with a pair of CCD cameras," *Opt. Lett.* 28, 816–818 (2003).

[2] Bouma, B. E. and Tearney, G. J. Eds., *Handbook of Optical Coherence Tomography*, Marcel Dekker, Inc. (2002).

[3] Bracewell, R. M. *The Fourier Transform and its Applications*, McGaw-Hill (1965).

[4] Brezinski, M. E. *Optical Coherence Tomography: Principles and Applications*, Academic Press (2006).

[5] Choma, M. A., Sarunic, M. V., Yang, C. and Izatt, J. A. "Sensitivity advantage of swept source and Fourier domain optical coherence tomography," *Opt. Express* 11, 2183–2189 (2003).

[6] de Boer, J. F., Cense, B., Park, B. H., Pierce, M. C., Tearney, G. J. and Bouma, B. E. "Improved signal-to-noise ratio in spectral-domain compared with time-domain optical coherence tomography," *Opt. Lett.* 28, 2067–2069 (2003).

[7] Dorrer, C., Belabas, N., Likforman, J. P. and Joffre, M. "Spectral resolution and sampling issues in Fourier transform spectral interferometry," *J. Opt. Soc. Am.* B 17, 1795–1802 (2000).

[8] Dubois, A., Vabre, L., Boccara, A. -C. and Beaurepaire, E. "High-resolution full-field optical coherence tomography with a Linnik microscope," *Appl. Opt.* 41, 805–812 (2002).

[9] Fercher, A. F., Mengedoht, K. and Werner, W. "Eye length measurement by interferometry with partially coherent light," *Opt. Lett.* 13, 1867–1869 (1988).

[10] Fercher, A. F., Hitzenberger, C. K., Kamp, G. and El-Zaiat, S. Y. "Measurement of intraocular distances by backscattering spectral interferometry," *Opt. Commun.* 117, 43–48 (1995).

[11] Fercher, A. F., Drexler, W., Hitzenberger, C. K. and Lasser, T. "Optical coherence tomography – principles and applications," *Rep. Prog. Phys.* 66, 239–303 (2003).

[12] Griffiths, P. R. and de Haseth, J. A. *Fourier Transform Infrared Spectroscopy*, 2nd edn, Wiley-Interscience (2007).

[13] Huang, D., Swanson, E. A., Lin, C. P., Schuman, J. S., Stinson, W. G., Chang, W., Hee, M. R., Flotte, T., Gregory, K., Puliafito, C. A. and Fujimoto, J. G. "Optical coherence tomography," *Science* 254, 1178–1181 (1991).

[14] Kulkarni, M. D., van Leeuwen, T. G., Yazdanfar, S. and Izatt, J. "Velocity-estimation accuracy and frame-rate limitations in color Doppler optical coherence tomography," *Opt. Lett.* 23, 1057–1059 (1998).

[15] Leitgeb, R., Hitzenberger, C. K. and Fercher, A. F. "Performance of Fourier-domain vs. time domain optical coherence tomography," *Opt. Express* 11, 889 (2003).

[16] Mertz, L. *Transformations in Optics*, Wiley (1965).

[17] Ralston, T. S., Marks, D. L., Boppart, S. A. and Carney, P. S. "Inverse scattering for high-resolution interferometric microscopy," *Opt. Lett.* 31, 3585–3587 (2006).

[18] Schmitt, J. M. "Optical coherence tomography (OCT): A review," *IEEE J. Sel. Top. Quant. Elec.* 5, 1205–1215 (1999).

[19] Tearney, G. J., Bouma, B. E. and Fujimoto, J. G. "High-speed phase- and group-delay scanning with a grating-based phase control delay line," *Opt. Lett.* 22, 1811–1813 (1997).

[20] Wotjkowski, M., Kowalczyk, A., Leitgeb, R. and Fercher, A. F. "Full range complex spectral optical coherence tomography technique in eye imaging," *Opt. Lett.* 27, 1415–1417 (2002).

[21] Zawadski, R. J., Jones, S. M., Olivier, S. S., Zhao, M., Bower, B. A., Izatt, J. A., Choi, S., Laut, S., and Werner, J. S. "Adaptive-optics optical coherence tomography for high-resolution and high-speed 3D retinal in vivo imaging," *Opt. Express* 13, 8532–8546 (2005).

[22] Zhao, Y., Chen, Z., Saxer, C., Xiang, S., de Boer, J. F. and Stuart Nelson, J. "Phase-resolved optical coherence tomography and optical Doppler tomography for imaging blood flow in human skin with fast scanning speed and high velocity sensitivity," *Opt. Lett.* 25, 114–116 (2000).

13 Fluorescence

In Chapters 9 through 12, we dealt with contrast mechanisms based on light scattering. In these mechanisms, the sample was irradiated with an illumination beam, and an image of the sample was constructed based on the resulting scattered light or, more generally, on the interference between scattered and non-scattered light. Inasmuch as interference was involved, the phase of the scattered light was found to play a key role in the establishment of contrast.

We turn now to a conceptually simpler contrast mechanism where phase no longer plays such a significant role. In this mechanism, the sample is irradiated with an illumination beam. Instead of being scattered, however, the illumination beam is absorbed, thereby depositing energy into the sample. In Chapter 9 we discussed the possibility of absorption, though we did not consider the subsequent fate of the deposited energy in the sample. Indeed, we tacitly assumed this energy simply disappeared, lost perhaps as heat or vibrations, etc. In this chapter, we consider the specific case where the energy, or at least part of the energy, is re-released as light. Such re-released light is called fluorescence (or, in rare cases, phosphorescence).

Ultimately, in most fluorescence imaging microscopes, it is the intensity of the fluorescence that is of interest. As we saw in Chapter 7, intensity variations produced by multiple radiators can arise from phase variations in their fields. These phase variations in turn can be driven by an incident illumination beam, in the case of scattering. In contrast, in the case of fluorescence, the phase variations are essentially random. That is, fluorescent molecules can be thought of as random-phase radiators of the type described in Section 7.2.1. The coherence time associated with fluorescence is usually extremely short, on the order of picoseconds or less for molecules in solution (cf. [28]). Intensity fluctuations arising from such short coherence times are too rapid to be resolved by detectors, and are subject to such severe temporal filtering as to be immeasurable (see Chapter 7). As a result, when evaluating the imaged intensity obtained from a population of fluorescent molecules, the cross terms in Eq. 7.16 can be neglected, meaning that the total intensity can be derived simply from the sum of the intensities produced by each individual molecule. That is, fluorescent sources are effectively spatially incoherent. Such incoherence leads to enormous simplifications when analyzing fluorescence microscopes since, in general, these need be characterized only by their PSF (or OTF) rather than by H (or $\mathcal{H}$).

Imaging systems characterized by a simple PSF have already been briefly considered in Section 5.3, where the treatment for self-luminous samples pertains equally well to fluorescent samples. More complicated fluorescence imaging systems will be introduced in this and the next chapters. However, before examining these, we first review some basic properties of fluorescence generation at the molecular level, and how these properties can be translated into mechanisms of contrast. Needless to say, fluorescence microscopy has had an enormous impact in biological imaging. For reference, extensive discussions on fluorescence and its applications to biological imaging can be found in several textbooks, including [16, 18, 28, 29].

13.1 RATE EQUATIONS

For a molecule to generate fluorescence, it must absorb energy and then subsequently release this energy as fluorescence. In the simplest description of fluorescence generation, the rate of energy absorption, corresponding to the power absorbed from the illumination beam, is characterized by an absorption cross-section, as it was in Section 9.3. Thus,

$$\Phi_a = \sigma_a I_i \tag{13.1}$$

where I_i is the illumination intensity (assumed to be constant), and σ_a is implicitly dependent on illumination wavelength.

The absorbed power is released as fluorescence power with an efficiency q_r, called the radiative quantum yield. The net fluorescence power generated by the molecule is then

$$\Phi_f = q_r \sigma_a I_i = \sigma_f I_i \tag{13.2}$$

where σ_f is called a fluorescence cross-section.

Equations 13.1 and 13.2 provide a generally adequate description of fluorescence generation for most microscopy applications. However, experimental conditions can easily arise where this description becomes flawed. For example, Eq. 13.2 suggests that the fluorescence power scales linearly with illumination intensity. While this is true at low intensities, it is no longer true at higher intensities. We turn, therefore, to a more accurate description of fluorescence that involves the use of rate equations to characterize the internal dynamics of fluorescent molecules.

13.1.1 Two-Level Fluorescent Molecule

In its simplest form, a fluorescent molecule can be thought of as a two-level system comprising a ground state denoted by $|g\rangle$ and an excited state denoted by $|e\rangle$, separated by an electronic transition energy ΔE. In the absence of illumination, ΔE is large enough that the molecule resides with quasi-certainty in its ground state at room temperature. In the presence of illumination of frequency ν, such that $\Delta E = h\nu$ where h is Planck's constant, the molecule can be promoted to its excited state by absorption of an illumination photon. Once excited, the molecule can then relax back to its ground state by release of an emission photon, also of frequency ν.

This idealized picture of a molecule as a perfect two-level system turns out to be deceptively oversimplified. In particular, let us consider the instantaneous probabilities of finding a

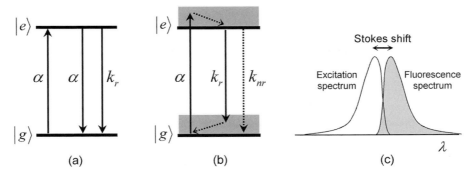

Figure 13.1. (a) Two-level system featuring both stimulated and spontaneous emission. (b) Molecular two-level system featuring non-radiative decay and Stokes shift. (c) Molecular excitation and fluorescence spectra.

two-level system in its ground or excited state, denoted by $g(t)$ and $e(t)$ respectively. By conservation of probability, we must have

$$g(t) + e(t) = 1 \tag{13.3}$$

The rate equations governing these probabilities were derived by Einstein based purely on thermodynamic considerations [5], and are given by

$$\frac{dg(t)}{dt} = -\frac{de(t)}{dt} = -\alpha\left(g(t) - e(t)\right) + k_r e(t) \tag{13.4}$$

where $\alpha = \sigma_a\left(\frac{I_i}{h\nu}\right)$ corresponds to the rate of energy absorption *provided* the molecule is definitely in its ground state. The parameter k_r is called the radiative decay rate constant. These rate equations for a perfect two-level system are schematically depicted in Fig. 13.1a. The rate constant for the promotion of the system from the ground state to the excited state is α, while the rate constant for the decay of the excited state back to the ground state involves two decay channels with rate constants α and k_r. The first of these depends on the illumination intensity and is referred to as stimulated; the second, since it is independent of illumination intensity, is referred to as spontaneous. Both decay channels release energy in the form of light, called stimulated and spontaneous emission. By far the most useful of these in microscopy applications is the latter, better known as fluorescence.

The reason stimulated emission plays little role in microscopy applications (though we will encounter an exception in Chapter 19) is because the perfect two-level system depicted in Fig. 13.1a is, as indicated above, an oversimplification. A real molecule in solution possesses many vibrational and rotational sublevels which, in effect, spread the electronic energy levels into multilevel bands. The promotion of a molecule from the ground-state band to the excited-state band is thus spread over an extended wavelength range, with varying efficiency characterized by a wavelength-dependent absorption cross-section $\sigma_a(\lambda)$. Once promoted to somewhere within the excited-state band, the molecule quickly relaxes to the bottom of this band by collisional equilibration with its solvent environment. Under normal illumination conditions, the rate of this relaxation is so rapid (on the order of picoseconds) that it overwhelms the possibility of stimulated emission, making the latter negligible. As for stimulated emission from the bottom of the excited state band, this is not possible without the presence of illumination light at the frequency commensurate with the energy gap (now reduced compared to the excitation energy) separating the bottom of the excited-state band from the ground-state band.

The presence of the molecular environment has yet the additional effect of allowing another form of energy release through collisions. In particular, it introduces another decay channel from the excited-state band, this time non-radiative, with decay rate constant k_{nr}. Modifying our definitions of $g(t)$ and $e(t)$ such that they correspond now to the instantaneous probabilities of finding the molecule in its ground or excited state *bands*, the rate equations governing these probabilities become

$$\frac{dg(t)}{dt} = -\frac{de(t)}{dt} = -\alpha g(t) + (k_r + k_{nr}) e(t) \tag{13.5}$$

again satisfying the conservation of probability prescribed by Eq. 13.3. In effect, we have recovered our original simple picture of a two-level system with the difference now that the excitation energy is no longer the same as the emission energy (see Fig. 13.1b). The balance in energy is dissipated as heat into the molecular environment, leading to a red shift in the fluorescence emission spectrum relative to the absorption spectrum, known as a Stokes shift (Fig. 13.1c). This Stokes shift turns out to be quite convenient for imaging applications, since it enables an easy separation of excitation from emission light by simple spectral filtering.

Returning to the rate equations in Eq. 13.5, the radiative quantum yield is now defined by

$$q_r = \frac{k_r}{k_r + k_{nr}} \tag{13.6}$$

corresponding to the probability that the excited-state energy is released as fluorescence. As expected, this probability equals unity in the absence of non-radiative decay channels.

The radiative lifetime is defined by

$$\tau_e = \frac{1}{k_r + k_{nr}} \tag{13.7}$$

corresponding to the average lifetime of the excited state. From Eq. 13.5 we find that if the molecule is definitely in its excited state at time $t = 0$, the probability of its still being in the excited state at time $t \geq 0$ is given by $e(t) = e^{-t/\tau_e}$. The time delay between excitation and fluorescence emission therefore obeys a negative-exponential probability distribution (see Chapter 7), and the radiative lifetime can be interpreted as the average of this time delay.

If the instantaneous excitation rate constant α does not change with time, then the steady-state solution of Eq. 13.5 becomes

$$\langle e \rangle = \frac{\alpha}{\alpha + k_r + k_{nr}} = \frac{\alpha \tau_e}{1 + \alpha \tau_e} \tag{13.8}$$

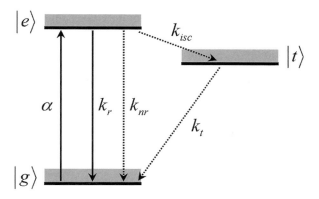

Figure 13.2. Molecular three-level system.

from which we derive the steady-state fluorescence emission rate of the molecule to be

$$\frac{\Phi_f}{h\nu_f} = k_r \langle e \rangle \tag{13.9}$$

where ν_f is the fluorescence frequency. Both sides of Eq. 13.9 correspond to a rate of emission of fluorescence photons per second. Equation 13.9 is more exact than Eq. 13.2 since it now accounts for a nonlinearity in Φ_f as a function of I_i. In particular, at very high illumination intensity, then $\Phi_f \rightarrow h\nu_f k_r$, which becomes saturated.

13.1.2 Three-Level Molecule: Triplet State

In many cases, an intermediate third-state band, called the triplet state, is involved in the internal dynamics of a fluorescent molecule. This state is rarely luminous itself, but has the effect of providing the molecule with an alternative pathway for the release of energy from the excited state (see Fig. 13.2). The transition from the excited state $|e\rangle$ to the triplet state $|t\rangle$ is called intersystem crossing (ISC), suggesting that the triplet state belongs to a different "system" than the excited state. In standard nomenclature, $|t\rangle$ belongs to a triplet manifold whereas both $|e\rangle$ and $|g\rangle$ belong to a singlet manifold, the manifolds being defined by the spins of the electrons involved in the molecular excitation (cf. [28]). The quantum mechanics necessary to properly define these manifolds is beyond the scope of our discussion, and we remark here only that the probability of ISC is typically small, usually less than a percent. Nevertheless, despite such a low ISC probability, the triplet state can be quite deleterious to fluorescence, as we will see below.

To take into account the triplet state, we include it in the rate equations governing the molecular dynamics. In matrix form, these rate equations become

$$\frac{d}{dt} \begin{pmatrix} g \\ e \\ t \end{pmatrix} = \begin{pmatrix} -\alpha & (k_r + k_{nr}) & k_t \\ \alpha & -(k_r + k_{nr} + k_{isc}) & 0 \\ 0 & k_{isc} & -k_t \end{pmatrix} \begin{pmatrix} g \\ e \\ t \end{pmatrix} = \mathbf{M} \begin{pmatrix} g \\ e \\ t \end{pmatrix} \tag{13.10}$$

where k_{isc} is the intersystem-crossing (ISC) rate constant, and k_t is the triplet-state decay rate constant. We note that $\mathrm{Det}[\mathbf{M}] = 0$ because we are again dealing with a closed system, meaning that the molecule can only reside in one of the bands $|g\rangle$, $|e\rangle$, or $|t\rangle$.

Accordingly, we define a new radiative quantum yield

$$q_r = \frac{k_r}{k_r + k_{nr} + k_{isc}} \tag{13.11}$$

and a new radiative lifetime

$$\tau_e = \frac{1}{k_r + k_{nr} + k_{isc}} \tag{13.12}$$

We also introduce an ISC quantum yield

$$q_{isc} = \frac{k_{isc}}{k_r + k_{nr} + k_{isc}} \tag{13.13}$$

and triplet-state lifetime

$$\tau_t = \frac{1}{k_t} \tag{13.14}$$

Note that while the addition of the triplet state leads to modifications in both the radiative quantum yield and lifetime, it has no effect on k_r and k_{nr}. As such, the rate constants are more fundamental in characterizing molecular photodynamics than are quantum yields or lifetimes.

As an example, let us consider typical values for a molecule in solution: $k_r \approx 10^9$ Hz and $k_{isc} \approx k_t \approx 10^6$ Hz. Assuming there are no other non-radiative decay channels, we find $\tau_e \approx 1$ ns, $\tau_t \approx 1$ μs, and $q_{isc} \approx 0.1\%$. In steady state, then, we have $\frac{\langle t \rangle}{\langle e \rangle} = q_{isc} \frac{\tau_t}{\tau_e} \approx 1$. This last result may seem surprising since it suggests that despite a low probability of ISC the molecule spends an equal amount of time in $|e\rangle$ and $|t\rangle$. However, a low probability of ISC goes both ways. The more infrequent the transition from $|e\rangle$ to $|t\rangle$, the correspondingly longer the molecule remains trapped in $|t\rangle$. At low excitation rates (e.g. $\alpha \ll k_t$) the fluorescence reduction caused by ISC is quite negligible. On the other hand, at high excitation rates the maximum steady-state fluorescence is reduced by a factor $\bar{e}/(\bar{e} + \bar{t})$, and becomes essentially halved in our example.

The decay of the triplet state $|t\rangle$ to the ground state $|g\rangle$ is usually non-radiative. In the rare instances where it is radiative, the light emitted from $|t\rangle$ is called phosphorescence (as opposed to fluorescence, which is emitted from $|e\rangle$).

13.1.3 Photobleaching

Intersystem-crossing removes a molecule from its singlet manifold and hence effectively causes a fluorescent molecule to turn off. However, this removal is transient since, on average, it lasts only the duration of the triplet-state lifetime. A much more pernicious problem in fluorescence imaging is that of photobleaching, which permanently removes the molecule from its singlet manifold. The physical mechanisms of photobleaching vary from molecule to molecule and are usually dependent on molecular environment. We will not discuss these mechanisms here

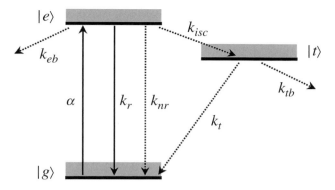

Figure 13.3. Molecular three-level system with photobleaching.

and describe only how they can be accounted for in the rate equations governing molecular dynamics. Specifically, we allow for the possibility of photobleaching from either $|e\rangle$ or $|t\rangle$ by introducing phenomenological rate constants k_{eb} and k_{tb}, characterizing the respective decay channels to a photobleached state $|b\rangle$ (see Fig. 13.3). The rate equations become then

$$\frac{d}{dt}\begin{pmatrix} g \\ e \\ t \end{pmatrix} = \begin{pmatrix} -\alpha & (k_r + k_{nr}) & k_t \\ \alpha & -(k_r + k_{nr} + k_{isc} + k_{eb}) & 0 \\ 0 & k_{isc} & -(k_t + k_{tb}) \end{pmatrix} \begin{pmatrix} g \\ e \\ t \end{pmatrix} = \mathbf{M} \begin{pmatrix} g \\ e \\ t \end{pmatrix}$$

(13.15)

We observe that $\mathrm{Det}[\mathbf{M}] \neq 0$ because we are now dealing with an open system. For more information on the mechanisms of photobleaching and on the photochemistry of fluorescence in general the reader is referred to [16] and [28].

13.2 FÖRSTER RESONANCE ENERGY TRANSFER (FRET)

So far, we have dealt with rate equations only for a single fluorescent molecule in solution. While it was posited early in this chapter that fluorescence is incoherent and that the fluorescence from multiple molecules can be treated independently, this is in fact not the case when fluorescent molecules are in very close proximity, so close as to be within the range of each other's near field. We have already seen some effects of near-field proximity in the case of two radiating dipoles (see Section 9.4). We now examine similar proximity effects in the case of two fluorescent molecules. For simplicity, the molecules are taken to be two-level systems, as defined in Section 13.1.1, but with unequal energy gaps separating their ground and excited state bands. If the illumination frequency is so high as to excite only the higher energy molecule, one might naturally expect fluorescence to be generated only from this higher energy molecule and not from the lower energy molecule. This is indeed the case if the two

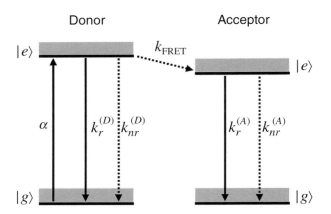

Figure 13.4. FRET between donor and acceptor molecules.

molecules are far apart. However, if the molecules are so close as to be within each other's near field, then a new decay channel becomes available to the higher energy molecule, this time by way of the lower energy molecule. This additional channel enables the excited-state energy from the higher energy molecule to transfer to the lower energy molecule, a process known as Förster resonance energy transfer, or FRET. Accordingly, the higher energy molecule is called the FRET donor (D), and the lower energy molecule is called the FRET acceptor (A), as illustrated in Fig. 13.4. The rate constant for FRET is denoted by k_{FRET}. An expression for this rate constant for an intermolecular separation R was derived by Förster [9], and is given by

$$k_{FRET} = \frac{1}{\tau_e^{(D)}} \left(\frac{R_0}{R} \right)^6 \tag{13.16}$$

where R_0 is known, befittingly, as the Förster distance, typically on the order of a few nanometers, and $\tau_e^{(D)}$ is the radiative lifetime of the donor molecule in the absence of FRET. That is, $\tau_e^{(D)}$ is given by Eq. 13.7. The FRET efficiency, or FRET quantum yield, can thus be written as

$$q_{FRET} = \frac{k_{FRET}}{k_r^{(D)} + k_{nr}^{(D)} + k_{FRET}} = \frac{1}{1 + (R/R_0)^6} \tag{13.17}$$

The dependence of k_{FRET} on R^{-6} leads to a highly sensitive dependence of FRET efficiency on intermolecular separation. On the one hand, this separation has only to be somewhat larger than R_0 for $q_{FRET} \ll 1$ and the FRET efficiency to be negligible, in which case the donor and acceptor molecules become effectively decoupled. On the other hand, when the intermolecular separation is within the range R_0 then $q_{FRET} \simeq 1$ and the donor excited-state energy becomes very efficiently transferred to the acceptor molecule. In turn, this energy transfer leads to the excitation of the acceptor molecule which, provided $q_r^{(A)} \neq 0$, induces a sudden appearance of acceptor fluorescence that would otherwise have been absent. It should be pointed out that

FRET can occur equally well if $q_r^{(A)} = 0$; however, the transferred energy in this case becomes dissipated entirely non-radiatively.

From the above considerations, it is clear that the Förster distance plays a critical role in governing FRET efficiency. We will not go into details in the derivation of R_0, which we defer to various references such as [4, 12, 16, 28, 30]. We note here only that $(R_0)^6$ scales with three parameters of importance. The first is $q_r^{(D)}$, the radiative quantum yield of the donor molecule in the absence of FRET. The second is a geometric factor that depends on the orientation of the donor and acceptor molecular axes relative to themselves and to their separation axis. The third parameter is usually denoted by J and corresponds to the overlap integral between the donor emission spectrum and the acceptor excitation spectrum. This last parameter ensures that FRET occurs from a higher energy molecule to a lower energy molecule, and not the other way around. It also tells us that FRET can occur between molecules of the same species. Such homo-FRET, as it is called, can lead to a nonlinear dependence of fluorescence on molecular concentration.

The use of FRET in optical microscopy applications has now become routine. Perhaps the most common use was proposed by Stryer [26] and is based on its exquisitely sensitive dependence on intermolecular separation, allowing it to serve as a spectroscopic ruler. As noted above, R_0 is typically in the range of a few nanometers, which is about two orders of magnitude smaller than an optical wavelength. While intermolecular separations in this range are quite beyond the resolution capabilities of standard diffraction-limited microscopes, they become readily measurable by FRET. Several measurement techniques are available for this, the most common of which rely on power measurements of either the donor or acceptor fluorescence, or both, to infer FRET efficiency. As an example, if a sample is illuminated with a wavelength low enough to excite only donor molecules, then acceptor fluorescence is produced only in regions where the donor and acceptor molecules are co-localized (i.e. fall within the range R_0).

Alternatively, the donor fluorescence can be recorded. The donor quantum yield in the presence of the acceptor is given by

$$q_r^{(DA)} = \frac{k_r^{(D)}}{k_r^{(D)} + k_{nr}^{(D)} + k_{\text{FRET}}} \qquad (13.18)$$

which becomes vanishingly small for large k_{FRET}. The donor fluorescence is then said to be quenched by FRET, and an assessment of the FRET efficiency can be inferred by monitoring the degree of quenching. Thus, if we assume that only a fraction f of the donor molecules become co-localized with acceptor molecules, the recorded donor fluorescence power in the presence of the acceptor is given by $\Phi^{(DA)} \propto \alpha \left(q_r^{(DA)} f + q_r^{(D)} (1 - f) \right)$, from which we derive the corresponding FRET efficiency

$$q_{\text{FRET}} = 1 - \frac{q_r^{(DA)}}{q_r^{(D)}} = \frac{1}{f} \left(1 - \frac{\Phi^{(DA)}}{\Phi^{(D)}} \right) \qquad (13.19)$$

It should be noted that the technique of measuring q_{FRET} based on Eq. 13.19 requires two separate recordings of the donor fluorescence power, namely with and without the presence of the acceptor. These can be obtained by first recording the donor fluorescence power and then introducing the acceptor, or alternatively by first recording the donor–acceptor fluorescence power and then removing the acceptor, for example by photobleaching. In either case, measurements of $\Phi^{(D)}$ and $\Phi^{(DA)}$ alone are not sufficient to fully determine q_{FRET}, since a knowledge of f is also required. In general, f is difficult to infer experimentally and this requirement can pose significant problems (see [11] for examples).

An alternative and more robust technique for measuring q_{FRET} is based not on the second expression in Eq. 13.19 but rather on the first. Rewriting this as

$$q_{\text{FRET}} = 1 - \frac{\tau_e^{(DA)}}{\tau_e^{(D)}} \tag{13.20}$$

we observe that q_{FRET} can be fully determined from measurements of $\tau_e^{(D)}$ and $\tau_e^{(DA)}$ alone, without any need for an additional knowledge of f. Such measurements are called fluorescence lifetime measurements, to which we turn our attention.

13.3 FLUORESCENCE LIFETIME IMAGING MICROSCOPY (FLIM)

In Section 13.1, we concentrated on steady-state solutions of the radiative rate equations. That is, the excitation rate constant α, or equivalently the illumination intensity I_i, was assumed to be constant, or at least not to vary significantly on the time scale of the molecular photodynamics. Such was the case, in particular, when we wrote Eq. 13.1.

Let us consider the possibility that $I_i(t)$ does vary with time, so rapidly that the variations are on the time scale of the molecular photodynamics of interest. Based on their rate equations, the radiating molecules cannot respond to these variations instantaneously, and the emitted fluorescence power becomes effectively temporally filtered. That is, we replace Eq. 13.1 with

$$\Phi_f(t) = \overline{\sigma}_f \int dt' \, \overline{R}(t - t') I_i(t') \tag{13.21}$$

where $\overline{R}(t)$ represents the response function of the fluorescing sample. As in Section 7.3.1, $\overline{R}(t)$ is real and normalized so that

$$\int dt \, \overline{R}(t) = 1 \tag{13.22}$$

and also obeys the causality condition $\overline{R}(t < 0) = 0$.

The overbars in Eq. 13.21 indicate that both $\overline{\sigma}_f$ and $\overline{R}(t)$ are sample averaged, meaning they represent the averaged cross-section and response associated with an arbitrary collection of different fluorescing species. More precisely, defining f_n to be the concentration fraction of each fluorescent species n, we write

$$\overline{\sigma}_f = \sum_n f_n \sigma_f^{(n)} \tag{13.23}$$

$$\overline{R}(t) = \frac{1}{\overline{\sigma}_f} \sum_n f_n \sigma_f^{(n)} R_n(t) \tag{13.24}$$

where $\sigma_f^{(n)}$ and $R_n(t)$ are, respectively, the cross-sections and responses associated with each fluorescent species. As defined, $R_n(t)$ obeys the same normalization and causality conditions as $\overline{R}(t)$. For example, in the case of simple two-state fluorescing molecules, $R_n(t)$ is given by

$$R_n(t) = \frac{1}{\tau_e^{(n)}} e^{-t/\tau_e^{(n)}} \qquad (t > 0) \tag{13.25}$$

which depends only on the fluorescence lifetime $\tau_e^{(n)}$ of the nth fluorescent species.

The goal of fluorescence lifetime imaging microscopy (FLIM) is to obtain a map of $\overline{R}(t)$ with microscopic spatial resolution and with sufficient temporal resolution to infer the underlying f_n and $\tau_e^{(n)}$ on which it depends. Such an inference, of course, is based on a specific model for $\overline{R}(t)$, such as that provided by Eqs. 13.24 and 13.25, which is assumed to be known a priori. In general, the greater the number of fluorescent species included in the model, the higher the required temporal resolution. We will not go into the fitting procedures that have been developed to infer f_n and $\tau_e^{(n)}$ from $\overline{R}(t)$, as these are beyond the scope of this book (see [16] for more information). Instead, we concentrate specifically on techniques to retrieve $\overline{R}(t)$ itself. These techniques largely originate from the pioneering work of Weber [32], and, as usual, may be separated into two categories based on time domain and frequency domain.

13.3.1 Time Domain

Conceptually, the most straightforward approach for measuring $\overline{R}(t)$ is with a short illumination pulse. If this pulse is much shorter than the time scale of the dynamics of $\overline{R}(t)$, then $I_i(t)$ may be replaced by a delta function in Eq. 13.21, as illustrated in Fig. 13.5, leading to

$$\Phi_f(t) \propto \overline{R}(t) \tag{13.26}$$

That is, a measure of the temporal response of $\Phi_f(t)$ provides a direct measure of $\overline{R}(t)$.

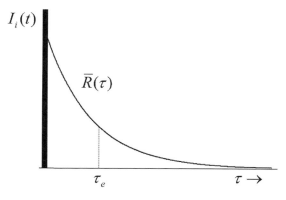

Figure 13.5. Excitation pulse and corresponding fluorescence response function for single molecular species.

Fluorescence lifetimes are typically on the order of nanoseconds for molecules in solution, meaning that to obtain an accurate representation of $\overline{R}(t)$ the temporal resolution of the detector must be on the order of fractions of a nanosecond. Such resolutions are just attainable with current detector technologies. For example, time-domain FLIM can be achieved in a widefield imaging configuration by abutting a multichannel plate (MCP) to the active area of a camera. Such a detector configuration is called an intensified camera, as we have seen before in Section 8.3.2. Fluorescence lifetimes can then be determined by applying gated gains to the MCP at different time delays relative to the illumination pulses [31]. A significant drawback of this gating technique, however, is that it is very wasteful of fluorescence photons.

An alternative time-domain technique involves the use of time-to-amplitude converters (TACs) [22]. These measure the time delay between the illumination pulse and the first detected photoelectron. An accumulation of such measurements from a series of illumination pulses allows one to progressively reconstruct $\overline{R}(t)$. While less wasteful of photons than a gating technique, the TAC technique is slow since a maximum of only one photoelectron can be detected per pulse, constraining the illumination intensity to be low and the pulse repetition rate to be no faster than a few tens of MHz. Moreover, while the TAC technique is well adapted to scanning microscopes, as described in the next chapter, it is less well adapted to widefield imaging unless one is equipped with fast detector arrays or a specialized imaging PMT (see, for example, [13]).

13.3.2 Frequency Domain

A formally equivalent strategy for measuring $\overline{R}(t)$ involves instead measuring its Fourier transform $\overline{\mathcal{R}}(\nu)$. For example, the Fourier equivalent of the simple model given by Eqs. 13.24 and 13.25 becomes

$$\overline{\mathcal{R}}(\nu) = \frac{1}{\overline{\sigma}_f} \sum_n \frac{f_n \sigma_f^{(n)}}{1 - i2\pi\nu\tau_e^{(n)}} \tag{13.27}$$

and a fit of $\overline{\mathcal{R}}(\nu)$ again reveals the underlying parameters of interest, namely f_n and $\tau_e^{(n)}$.

To experimentally reconstruct $\overline{\mathcal{R}}(\nu)$, its value must be measured at various frequencies. This can be achieved by applying sinusoidal modulations to the illumination intensity $I_i(t)$ and recording the resulting modulations in the fluorescence power $\Phi_f(t)$ (see Fig. 13.6). The value of $\overline{\mathcal{R}}(\nu)$ is then inferred from the relation

$$\hat{\Phi}_f(\nu) = \overline{\sigma}_f \overline{\mathcal{R}}(\nu)\mathcal{I}_i(\nu) \tag{13.28}$$

which is the Fourier equivalent of Eq. 13.21 ($\hat{\Phi}_f(\nu)$ is the Fourier transform of $\Phi_f(t)$).

To begin, let us consider a single modulation frequency ν_m. A slight complication arises here because $I_i(t)$ is an intensity rather than a field, meaning it is positive definite. Any modulation in $I_i(t)$ must therefore be accompanied by a dc offset, and we write

$$I_i(t) = \langle I_i \rangle \left[1 + M_i \cos(2\pi\nu_m t) \right] \tag{13.29}$$

where M_i is the modulation depth.

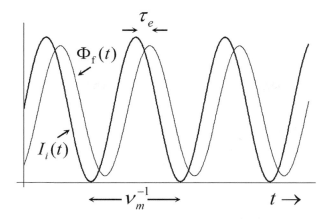

Figure 13.6. Modulated excitation and correspondingly modulated fluorescence for single molecular species.

In the frequency domain, Eq. 13.29 translates to

$$\mathcal{I}_i(\nu) = \langle I_i \rangle \, \delta(\nu) + \tfrac{1}{2} M_i \, \langle I_i \rangle \left[\delta(\nu - \nu_m) + \delta(\nu + \nu_m) \right] \tag{13.30}$$

In principle, by measuring the amplitude and phase of $\hat{\Phi}_f(\nu_m)$ relative to the amplitude and phase of $\mathcal{I}_i(\nu_m)$, it is possible to determine the complex value of $\overline{\mathcal{R}}(\nu_m)$ (no extra information is gained at $-\nu_m$ because $\overline{\mathcal{R}}(\nu)$ is Hermitian). However, we recall that typical values for $\tau_e^{(n)}$ are on the order of nanoseconds. According to Eqs. 13.28 and 13.30, this means that the amplitude or phase changes in $\hat{\Phi}_f(\nu_m)$ only become significant for modulation frequencies on the order of several tens of MHz. While such high frequencies are easily within the bandwidth range of single photodiodes or PMTs, they are very far from the range of standard cameras. Fortunately, we have a trick to get around this last difficulty, which we have seen before (Section 12.2.1). This trick is called demodulation.

As a reminder, the principle of demodulation is to modulate again a modulated signal. By frequency mixing, part of the signal is then transferred to a lower more manageable frequency. Thus, with homodyne demodulation, part of the signal is transferred to dc frequency, which is patently within the range of even the slowest cameras. In FLIM, demodulation can be achieved by multiplying the fluorescence power $\Phi_f(t)$ with a demodulation function $Q(t)$ prior to its being detected by the camera. This can be performed, for example, by applying a sinusoidal modulation to the MCP gain of an intensified camera. In effect, the power incident on the camera active area becomes

$$\Phi_{dm}(t) = Q(t) \Phi_f(t) \tag{13.31}$$

or, equivalently, in the frequency domain

$$\hat{\Phi}_{dm}(\nu) = \overline{\sigma}_f \, Q(\nu) * \left(\overline{\mathcal{R}}(\nu) \mathcal{I}_i(\nu) \right) \tag{13.32}$$

Here again, we encounter the same slight difficulty that the MCP gain is positive definite, meaning that any gain modulation must be accompanied by a dc offset. We write then

$$Q^{(\phi)}(t) = 1 + M_Q \cos(2\pi\nu_m t + \phi) \tag{13.33}$$

where ν_m is the gain modulation frequency (taken to be the same as the illumination modulation frequency since we are considering homodyne demodulation), M_Q is the associated gain modulation depth, and ϕ is an arbitrary phase shift relative to the illumination modulation. As we will see below, ϕ will play an important role.

In the frequency domain, Eq. 13.33 translates to

$$Q^{(\phi)}(\nu) = \delta(\nu) + \tfrac{1}{2}M_Q\left[e^{i\phi}\delta(\nu - \nu_m) + e^{-i\phi}\delta(\nu + \nu_m)\right] \tag{13.34}$$

Making use of Eqs. 13.34 and 13.30, and performing the convolution prescribed by Eq. 13.32, we find that $\hat{\Phi}_{dm}(\nu)$ reduces to a set of five peaks located at the frequencies $\nu = \{0, \pm\nu_m, \pm 2\nu_m\}$. Because of the slow response of the camera or of the phosphor screen interposed between the MCP and the camera, only the peak around zero frequency survives the detection process, and we are left with

$$\hat{\Phi}_{dm}^{(\phi)}(\nu) \to \overline{\sigma}_f \langle I_i \rangle \left[\mathcal{R}(\nu) + \tfrac{1}{4}e^{i\phi}M_i M_Q \overline{\mathcal{R}}(\nu - \nu_m) + \tfrac{1}{4}e^{-i\phi}M_i M_Q \overline{\mathcal{R}}(\nu + \nu_m)\right]\delta(\nu) \tag{13.35}$$

Recalling that a dc component ($\nu = 0$) in the frequency domain corresponds to a time average in the time domain, we obtain finally

$$\left\langle \Phi_{dm}^{(\phi)} \right\rangle = \overline{\sigma}_f \langle I_i \rangle \left[1 + \tfrac{1}{4}e^{i\phi}M_i M_Q \overline{\mathcal{R}}(-\nu_m) + \tfrac{1}{4}e^{-i\phi}M_i M_Q \overline{\mathcal{R}}(\nu_m)\right] \tag{13.36}$$

where we have made use of the fact that $\overline{\mathcal{R}}(0) = 1$ (see Eq. 13.22).

Equation 13.36 is the power ultimately reported by the camera. This power contains three terms. The first term ($\overline{\sigma}_f \langle I_i \rangle$) corresponds to the average fluorescence power incident on the camera. This could have been obtained by direct detection without any need for modulation or demodulation, and, as such, constitutes a standard fluorescence image that contains no lifetime information. The last two terms are more interesting since they contain information on $\overline{\mathcal{R}}(\nu_m)$, which we have set out to retrieve. However, these last two terms, when summed together as in Eq. 13.36, form a real function ($\overline{\mathcal{R}}(\nu)$ is Hermitian). To fully retrieve $\overline{\mathcal{R}}(\nu_m)$, which is a complex function, we must isolate these terms. By now, a technique to perform this should be familiar. This is the technique of phase stepping, which we have seen before in Sections 11.4.2 and 12.2.1.

In practice, then, FLIM in the frequency domain requires only minor modifications to a standard widefield microscope, at least conceptually. In particular, both the illumination and the detection are modulated at a high frequency ν_m (several tens of MHz). A sequence of images is then acquired where the phase shift between the illumination and detection modulations is stepped through various phases. Finally, the fluorescence response function at the frequency ν_m is reconstructed at every pixel. To reconstruct the response function at different values of ν_m requires changing the modulation frequency. Alternatively, multiple values of ν_m can be applied in parallel with the use of non-sinusoidal modulation, such as pulsed modulation; however, the retrieval of $\overline{\mathcal{R}}(\nu)$ becomes somewhat more complicated in this case (cf. [25]). For a description of various experimental FLIM configurations, the reader is referred to, for example, references [8, 10, 17, 24]. For theoretical comparisons of the performance of time-domain versus frequency-domain techniques, the reader is referred to [14, 23].

13.4 FLUORESCENCE CORRELATION SPECTROSCOPY (FCS)

In Chapter 7, we saw how a measurement of the power fluctuations from a collection of externally driven radiators can provide information about their temporal dynamics, a technique known as dynamic light scattering (DLS). We close this chapter by examining yet another technique to monitor the dynamics of radiators called fluorescence correlation spectroscopy (FCS), originally developed by Webb [19] and Rigler [6]. The principal difference between DLS and FCS, as their names imply, is that the former is based on scattering whereas the latter is based on fluorescence. That is, whereas in DLS power fluctuations stem largely from the interference of fields from different radiators, in FCS interference plays no role and fluorescence powers from different molecules simply add. This leads to considerable simplifications in the interpretation of power fluctuations. In brief, the technique of FCS consists of analyzing these power fluctuations to extract information about the internal or external dynamics of fluorescent molecules. Particularly thorough reviews of FCS can be found in [15] and [27], and the reader is also referred to [7].

We begin by evaluating the conditions where FCS is feasible. Since FCS is based on measurements of power fluctuations rather than power averages, let us compare how these scale with the number of detected molecules N. Because the fluorescence emitted by different molecules can be treated as independent, the power average scales as N whereas the power fluctuations scale as $\sqrt{N}$ (see Section 7.2.4). To avoid problems with detection dynamic range, the latter must be significant compared to the former, and hence N should be small. This is a general requirement for FCS, which is most often ensured experimentally by restricting the detection of molecules to as small a volume as possible, called a probe volume. Chapters 14 and 16 specifically discuss microscope techniques that achieve very small probe volumes. In the present chapter, however, it will be assumed that such a small probe volume already exists and is well defined. In particular, a unitless spatial profile is ascribed to this probe volume, denoted by $\Psi(\vec{r})$. The fluorescence power generated from the volume is then given by

$$\Phi_f(t) = \phi_f \int \mathrm{d}^3\vec{r}\ \Psi(\vec{r})C(\vec{r},t) \tag{13.37}$$

where $C(\vec{r},t)$ is the density of molecules (units: 1/volume) and $\phi_f = \sigma_f \langle I_i(\vec{r}=0) \rangle$ is the time-averaged fluorescence power generated by a single molecule located exactly at the probe-volume center (we consider only a single molecular species here). Equation 13.37 will be the starting point of our analysis.

Following our usual convention, we normalize $\Psi(\vec{r})$ such that $\Psi(0) = 1$ and define the total probe volume to be

$$V_\psi = \frac{\left(\int d^3\vec{r}\ \Psi(\vec{r})\right)^2}{\int d^3\vec{r}\ \Psi^2(\vec{r})} \tag{13.38}$$

A condition for this volume to be well defined is that $\int d^3\vec{r}\ \Psi(\vec{r})$ be bounded; however, aside from this condition, the probe volume is allowed to be arbitrary. In particular, its boundary can be soft or hard. A convenient parameter to characterize the volume boundary is the volume contrast, defined by

$$\gamma_\psi = \frac{\int d^3\vec{r}\ \Psi^2(\vec{r})}{\int d^3\vec{r}\ \Psi(\vec{r})} \tag{13.39}$$

For a hard boundary $\gamma_\psi = 1$; for a soft boundary $\gamma_\psi < 1$. As we will see in Chapters 14 and 16, the volume contrasts γ_ψ typically obtained in practice are in the range 0.2–0.3. In terms of volume contrast, the probe volume takes the form

$$V_\psi = \frac{1}{\gamma_\psi} \int d^3\vec{r}\ \Psi(\vec{r}) \tag{13.40}$$

The main difference between Eq. 13.37 and Eq. 13.21 is that $I_i(\vec{r})$ is now taken to be fixed in Eq. 13.37, and the temporal variations in $\Phi_f(t)$ are caused instead by temporal fluctuations in $C(\vec{r}, t)$. These fluctuations can arise from many mechanisms. For example, molecules in solution may freely diffuse in and out of the probe volume by thermally driven Brownian motion, or may be actively transported through this volume by directed motion. Alternatively, the molecules may simply vanish, transiently or permanently, as a result, respectively, of ISC or photobleaching. Regardless of the mechanism responsible for concentration fluctuations, we write

$$C(\vec{r}, t) = \langle C \rangle + \delta C(\vec{r}, t) \tag{13.41}$$

The corresponding fluorescence power generated from the probe volume is then

$$\Phi_f(t) = \langle \Phi_f \rangle + \delta\Phi_f(t) \tag{13.42}$$

where, from Eq. 13.37, we have

$$\langle \Phi_f \rangle = \phi_f \langle C \rangle \int d^3\vec{r}\ \Psi(\vec{r}) = \phi_f \langle C \rangle\, \gamma_\psi V_\psi \tag{13.43}$$

$$\delta\Phi_f(t) = \phi_f \int d^3\vec{r}\ \Psi(\vec{r})\, \delta C(\vec{r}, t) \tag{13.44}$$

The core idea of FCS is to characterize the concentration fluctuations $\delta C(\vec{r}, t)$ based on a measurement of the fluorescence power fluctuations $\delta\Phi_f(t)$. To this end, we introduce the normalized autocorrelation function of $\delta\Phi_f(t)$, defined by

$$\Gamma_f(\tau) = \frac{\langle \delta\Phi_f(t + \tau)\delta\Phi_f(t) \rangle}{\langle \Phi_f \rangle^2} \tag{13.45}$$

which leads directly to

$$\Gamma_f(\tau) = \frac{1}{\left(\langle C \rangle \, \gamma_\psi V_\psi\right)^2} \iint d^3 \vec{r} \, d^3 \vec{r}' \, \Psi(\vec{r}) \Psi(\vec{r}') \left\langle \delta C(\vec{r}, t + \tau) \delta C(\vec{r}', t) \right\rangle \qquad (13.46)$$

Equation 13.46 serves as the basis of FCS. In particular, it provides a direct link between a measurable parameter $\Gamma_f(\tau)$ and the autocorrelation function $\left\langle \delta C(\vec{r}, t + \tau) \delta C(\vec{r}', t) \right\rangle$, which in turn provides key information about the underlying mechanisms governing molecular dynamics. As noted above, several mechanisms can influence these dynamics. We begin our discussion by considering the most straightforward of these mechanisms, namely free diffusion.

13.4.1 Molecular Diffusion

Molecules that undergo free diffusion are, by definition, non-interacting. Their motion dynamics are governed by two basic laws, known as Fick's laws [1], that describe Brownian motion:

$$\begin{array}{ll}
\text{Fick's 1st law:} & \vec{J}(\vec{r}, t) = -D\vec{\nabla}C(\vec{r}, t) \\
\text{Fick's 2nd law:} & \frac{\partial C(\vec{r}, t)}{\partial t} = D\nabla^2 C(\vec{r}, t)
\end{array} \qquad (13.47)$$

where $\vec{J}(\vec{r}, t)$ is the molecular flux (i.e. the number of molecules traversing a unit area in a unit time), and D is the diffusion constant (units of area/time). From Fick's laws, we can derive the autocorrelation function for the concentration fluctuations (cf. [2]), obtaining

$$\left\langle \delta C(\vec{r}, t + \tau) \delta C(\vec{r}', t) \right\rangle = \langle C \rangle \int d^m \vec{\kappa} \, e^{i2\pi \vec{\kappa} \cdot (\vec{r} - \vec{r}')} e^{-4\pi^2 \kappa^2 D\tau} \qquad (13.48)$$

or, equivalently,

$$\left\langle \delta C(\vec{r}, t + \tau) \delta C(\vec{r}', t) \right\rangle = \frac{\langle C \rangle}{(4\pi D\tau)^{m/2}} \exp\left(-\frac{|\vec{r} - \vec{r}'|^2}{4D\tau} \right) \qquad (13.49)$$

where m is the number of dimensions through which the molecules can diffuse. We note that, in three dimensions and in the limit $\tau \to 0$, we have $\left\langle \delta C(\vec{r}, t) \delta C(\vec{r}', t) \right\rangle \to \langle C \rangle \, \delta^3 \left(\vec{r} - \vec{r}' \right)$, and hence

$$\Gamma_f(\tau \to 0) = \frac{1}{\left(\langle C \rangle \, \gamma_\psi V_\psi\right)^2} \langle C \rangle \int d^3 \vec{r} \, \Psi^2(\vec{r}) = \frac{1}{\langle N \rangle} \qquad (13.50)$$

where $\langle N \rangle = \langle C \rangle \, V_\psi$ is the average number of molecules residing inside the probe volume at any given time. A measurement of $\Gamma_f(0)^{-1}$ therefore provides a convenient indicator of the number of active molecules contributing to the fluorescence signal (see Fig. 13.7). The corresponding average fluorescence power per active molecule is then

$$\frac{\langle \Phi_f \rangle}{\langle N \rangle} = \gamma_\psi \phi_f \qquad (13.51)$$

As expected, this average power is in general smaller than the peak power ϕ_f produced by a molecule located exactly at the volume center.

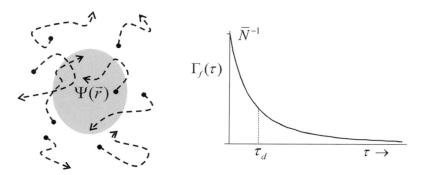

Figure 13.7. Molecules freely diffusing through a probe volume and corresponding fluorescence-fluctuation autocorrelation function.

When a molecule freely diffuses into the probe volume, it produces a detectable fluorescence burst. The average duration of this burst corresponds to the average dwell time of the molecule inside the volume, defined by

$$\tau_d = \frac{1}{\Gamma_f(0)} \int_0^\infty d\tau \, \Gamma_f(\tau) \tag{13.52}$$

This dwell time includes the possibility that the molecule can revisit the volume several times, and as a result τ_d converges when the diffusion is in a 3D volume, whereas it diverges when the diffusion is constrained to a 2D plane or a 1D line. This last result ultimately stems from the theory of random walks, which tells us that in one or two dimensions a random walker always eventually returns to its point of origin, whereas in three dimensions it can wander away never to return again (see [1, 21] for more information on this fascinating topic).

Finally, it is a simple matter to show that when τ_d is well defined, the average rate at which fluorescence bursts occur is given by $\langle N \rangle / \tau_d$.

13.4.2 Example: Hard-Sphere Volume with Photobleaching

As an example of information that can be obtained with FCS, let us consider molecules undergoing free diffusion in solution. Moreover let us assume that, while fluorescing in the probe volume, these molecules can undergo photobleaching. In this example, fluorescence fluctuations arise from both external (diffusion) and internal (photobleaching) molecular dynamics.

To properly take photobleaching into account, Fick's second law must be modified to include a diffusive term (first term) and a decay term (second term) such that

$$\frac{\partial C(\vec{r}, t)}{\partial t} = D\nabla^2 C(\vec{r}, t) - k_b(\vec{r}) C(\vec{r}, t) \tag{13.53}$$

where $k_b(\vec{r})$ is the local photobleaching rate constant, which depends on the local illumination intensity. Standard boundary conditions that can be applied to this modified Fick's second law are $C(\vec{r} \to \infty, t) = C_\infty$ and $C(\vec{r}, 0) = C_\infty$, where C_∞ corresponds to the steady-state concentration in the absence of photobleaching. In general, Eq. 13.53 cannot be solved analytically for arbitrary $k_b(\vec{r})$, and for purposes of discussion we limit ourselves only to an approximate solution valid for an idealized hard-sphere probe volume (see [3]).

To this end, we introduce a volume-averaged molecular concentration given by

$$C_{\text{in}}(t) = \frac{1}{V_\psi} \int_{V_\psi} d^3\vec{r}\, C(\vec{r}, t) \tag{13.54}$$

and a volume-averaged photobleaching rate constant given by

$$k_b = \frac{1}{V_\psi} \int_{V_\psi} d^3\vec{r}\, k_b(\vec{r}) \tag{13.55}$$

obtaining then, from Fick's first law,

$$J_{\text{in}} = 4\pi r_\psi D C_\infty \tag{13.56}$$

where J_{in} is the molecular flux into the spherical probe volume of radius r_ψ. The molecular dwell time inside the volume is accordingly

$$\tau_d = \frac{V}{J_{\text{in}}} C_\infty = \frac{r_\psi^2}{3D} \tag{13.57}$$

Finally, the modified Fick's second law (Eq. 13.53) can be approximated by

$$\frac{dC_{\text{in}}(t)}{dt} \approx \frac{1}{\tau_d}\left(C_\infty - C_{\text{in}}(t)\right) - k_b C_{\text{in}}(t) \tag{13.58}$$

which yields the steady-state solution

$$\langle C_{\text{in}} \rangle = C_\infty \left(\frac{1}{1 + k_b \tau_d}\right) \tag{13.59}$$

The interpretation of Eq. 13.59 is clear. The average dwell time of the molecules in the probe volume is curtailed because of photobleaching and is given by

$$\tau_d^{(\text{eff})} = \tau_d \left(\frac{1}{1 + k_b \tau_d}\right) \tag{13.60}$$

A plot of $\Gamma_f(\tau)$ without and with photobleaching is illustrated in Fig. 13.8. When there is no photobleaching, $\Gamma_f(0)$ provides a direct measure of C_∞ (or rather its inverse), whereas the elbow in $\Gamma_f(\tau)$ provides a measure of τ_d, and hence of D (provided r_ψ is known a priori). In contrast, in the presence of photobleaching $\Gamma_f(0)$ becomes larger and τ_d becomes shorter, both providing measures of k_b.

It should be noted that the solution provided by Eq. 13.58 is only an approximation. A full analytical solution can be found in [20]. Moreover, a hard-sphere illumination profile is of somewhat academic interest since it cannot be realized in practice. We defer a consideration of more realistic probe volumes to Chapters 14 and 16.

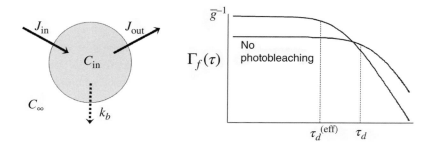

Figure 13.8. Hard-sphere probe volume with photobleaching. Note: fluorescence-fluctuation autocorrelation functions are plotted in a loglog scale.

13.5 PROBLEMS

Problem 13.1

Consider a solution of two-level fluorescent molecules such as the one depicted in Fig. 13.1b. The fluorescence from this solution is decreased by the addition of a quencher Q. The effect of this quencher is to induce an additional non-radiative decay of the excited state such that

$$|e\rangle + Q \xrightarrow{k_q} |g\rangle$$

where k_q is the quenching rate constant, in units $s^{-1}M^{-1}$ (M = molar concentration).

(a) Show that

$$\frac{\tau_e}{\tau_e^{(Q)}} = 1 + \tau_e k_q [Q]$$

where $\tau_e^{(Q)}$ and τ_e are the excited state lifetimes with and without the presence of the quencher, and $[Q]$ is the molar concentration of the quencher.

Such quenching is said to obey a Stern–Volmer relation.

(b) Show that, based on our simple model,

$$\frac{\Phi_f}{\Phi_f^{(Q)}} \leq \frac{\tau_e}{\tau_e^{(Q)}}$$

where the equality holds only in a particular limit. What is this limit?

Problem 13.2

Molecules in solution undergo both translational and rotational diffusion. A method for characterizing rotational diffusion is by measuring fluorescence anisotropy. This can be done using the standard configuration shown in the figure.

An illumination beam of intensity I_i is vertically polarized (x direction). The resultant fluorescence emission power is measured in the y direction within a small solid angle Ω.

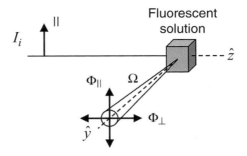

A polarizer is used to distinguish the measured vertical and horizontal powers, denoted by $\Phi_\parallel$ and $\Phi_\perp$ respectively. It can be shown that these powers are given by

$$\Phi_\parallel(t) = \Omega\sigma_f \int dt'\, R_\parallel(t-t')I_i(t')$$

$$\Phi_\perp(t) = \Omega\sigma_f \int dt'\, R_\perp(t-t')I_i(t')$$

where

$$R_\parallel(t) = \tfrac{1}{3}\left[1 + 2K(t)\right] R(t)$$
$$R_\perp(t) = \tfrac{1}{3}\left[1 - K(t)\right] R(t)$$

and where σ_f is the fluorescence cross-section, $R(t)$ is given by Eq. 13.25 (assume a single two-level fluorescent species), and $K(t)$ comes from rotational diffusion. In particular, if the rotational diffusion is isotropic, then

$$K(t) = r_0 e^{-6D_\theta t} = r_0 e^{-t/\tau_\theta}$$

where D_θ is a rotational diffusion constant and, concomitantly, τ_θ is a rotational diffusion time.

The measured fluorescence anisotropy is defined by

$$r(t) = \frac{\Phi_\parallel(t) - \Phi_\perp(t)}{\Phi_\parallel(t) + 2\Phi_\perp(t)}$$

(a) Show that if the illumination intensity is constant, then the steady-state fluorescence anisotropy is given by

$$\langle r \rangle = \frac{r_0}{1 + \tau_e/\tau_\theta}$$

This is known as the Perrin relation. In deriving this relationship, bear in mind that the denominator of $r(t)$ remains constant over time.

(b) Denote Φ_f as the total *emitted* fluorescence power in all solid angles. Derive an expression for the total *measured* fluorescence power $\Phi_\parallel(t)+\Phi_\perp(t)$ when $\tau_e/\tau_\theta \to 0$ (i.e. the rotation is slow compared to the excited state lifetime). When is this measured fluorescence power equal to $\Omega\Phi_f$?

(c) Derive an expression for the total *measured* fluorescence power when $\tau_e/\tau_\theta \to \infty$ (i.e. the rotation is fast compared to the excited state lifetime). In the case, the molecule orientation is essentially randomized before fluorescence emission can occur. Explain why the measured fluorescence power in this case is smaller than $\Omega\Phi_f$.

Problem 13.3

Consider using a microscope to image a sample containing N different fluorescent species of unknown concentrations $C_1, ..., C_N$. Each of these species emits fluorescence with a different spectrum. Accordingly, the microscope is equipped with N different spectral channels. The microscope has been pre-calibrated so that a unit concentration of species j is known to produce a signal M_{ij} in channel i. When the unknown sample is measured, signals $S_1, ..., S_N$ are obtained in each channel.

(a) Derive a general "spectral unmixing" formula to deduce $C_1, ..., C_N$ from the measured signals. Hint: your formula should be in terms of the matrix of cofactors of $\mathbf{M}$.

(b) Consider using a two-channel microscope to look at a fluorescent molecule [1]. The introduction of a non-fluorescent binding agent $[x]$ leads to an interaction $[1] + [x] \to [2]$, where the species $[2]$ produces fluorescence that is spectrally different from $[1]$. The efficiency of the interaction is defined by

$$q_{int} = 1 - \frac{C_1'}{C_1}$$

where C_1 and C_1' are concentrations of species $[1]$ before and after the response.
    Show that

$$q_{int} = \frac{-M_{22}\Delta S_1 + M_{21}\Delta S_2}{M_{22}S_1 - M_{21}S_2}$$

where $\Delta S = S' - S$ and $\mathbf{M}$ is assumed to be non-singular.

Problem 13.4

Frequency-domain FLIM provides a measurement of $\overline{\mathcal{R}}(\nu)$, as defined by Eq. 13.27, for a given modulation frequency ν.

(a) Consider a solution containing only a single fluorescent species. Draw a plot of $\mathcal{R}(\nu)$ in the complex plane as a function of increasing ν. This is called a "polar" or "phasor" plot of the fluorescence response, where $\mathrm{Re}[\mathcal{R}(\nu)]$ and $\mathrm{Im}[\mathcal{R}(\nu)]$ are the in-phase and quadrature components respectively. At what value of ν is the quadrature component peaked? For any given ν, where does a small fluorescence lifetime place $\mathcal{R}(\nu)$ in your plot? Where does a large fluorescence lifetime place $\mathcal{R}(\nu)$?

(b) Consider two fluorescent species of known cross-sections and lifetimes. Arbitrarily choose locations for $\mathcal{R}^{(1)}(\nu)$ and $\mathcal{R}^{(2)}(\nu)$ in your plot (for a given modulation frequency). Now consider making a measurement of $\overline{\mathcal{R}}(\nu)$ for a mixture of these two species of unknown concentration fractions. Where must $\overline{\mathcal{R}}(\nu)$ be located relative to $\mathcal{R}^{(1)}(\nu)$ and $\mathcal{R}^{(2(}(\nu)$? From a single measurement, can you infer the concentration fractions of the two species?

(c) Consider a solution containing an unknown number of fluorescent species of responses characterized by Eq. 13.25. What can you say about the resulting $\overline{\mathcal{R}}(\nu)$? Specifically, where must $\overline{\mathcal{R}}(\nu)$ be located for low, high, and mid-range modulation frequencies?

Problem 13.5

Consider performing FCS with a solution of freely diffusing fluorescent molecules and a 3D Gaussian probe volume defined by $\Psi(\vec{r}) = \exp\left(-r^2/w_0^2\right)$. The average concentration of molecules is $\langle C \rangle$. Their diffusion constant is D.

(a) Define a corresponding probe volume V_ψ.

(b) Show that $\Gamma_f(\tau) = \frac{1}{\langle N \rangle}\left(1 + \frac{2D\tau}{w_0^2}\right)^{-3/2}$, where $\langle N \rangle$ is the average number of molecules in the probe volume.

References

[1] Berg, H. *Random Walks in Biology*, Princeton University Press (1993).

[2] Berne, B. J. and Pecora, R. *Dynamic Light Scattering: with Applications to Chemistry, Biology, and Physics*, Dover (2000).

[3] Carslaw, H. S. and Jaeger, J. C. *Conduction of Heat in Solids*, 2nd edn, Oxford University Press (1986).

[4] Clegg, R. M. "Fluorescence resonance energy transfer and nucleic acids," *Meth. Enzymol.* 211, 353–388 (1992).

[5] Einstein, A. "On the quantum theory of radiation," *Phys. Z.* 18, 121–128 (1917).

[6] Ehrenberg, M. and Rigler, R. "Rotational Brownian motion and. fluorescence intensity fluctuations," *Chem. Phys.* 4, 390–401 (1974).

[7] Elson, E. L. and Magde, D. "Fluorescence correlation spectroscopy. I. Conceptual basis and theory," *Biopolymers* 13, 1–27 (1974).

[8] Esposito, A., Oggier, T., Gerritsen, H. C., Lustenberger, F. and Wouters, F. S. "All solid-state lock-in imaging for widefield fluorescence lifetime sensing," *Opt. Express* 13, 9812–9821 (2005).

[9] Förster, T. "Transfer mechanisms of electronic excitation," *Discuss. Faraday Soc.* 27, 7–17 (1959).

[10] Gadella, T. W., Jovin, T. M. and Clegg, R. M. "Fluorescence lifetime imaging microscopy (FLIM) – spatial resolutions of microstructures on the nanosecond time scale," *Biophys. Chem.* 48, 221–239 (1993).

[11] Gordon, G. W., Berry, G., Liang, X. H., Levine, B. and Herman, B. "Quantitative fluorescence resonance energy transfer measurements using fluorescence microscopy," *Biophys. J.* 74, 2702–2713 (1998).

[12] Jares-Erijman, E. A. and Jovin, T. M. "FRET imaging," *Nat. Biotech.* 21, 1387–1395 (2003).

[13] Kemnitz, K., Pfeifer, L., Paul, R. and Coppey-Moisan, M. "Novel detectors for fluorescence lifetime imaging on the picosecond time scale," *J. Fluoresc.* 7, 93–98 (1997).

[14] Köllner, M. and Wolfrum, J. "How many photons are necessary for fluorescence lifetime measurements?" *Chem. Phys. Lett.* 200, 199–204 (1992).

[15] Krichevsky, O. and Bonnet, G. "Fluorescence correlation spectroscopy: the technique and its applications," *Rev. Prog. Phys.* 65, 251–297 (2002).

[16] Lakowicz, J. R. *Principles of Fluorescence Spectroscopy*, 3rd edn, Springer (2006).

[17] Lakowicz, J. R., Laczko, G. and Cherek, H. "Analysis of fluorescence decay kinetics from variable-frequency phase shift and modulation data," *Biophys. J.* 46, 463–477 (1984).

[18] Loudon, R. *The Quantum Theory of Light*, 3rd edn, Oxford University Press (2000).

[19] Magde, D., Elson, E. and Webb, W. W. "Thermodynamic fluctuations in a reacting system – Measurement by fluorescence correlation spectroscopy," *Phys. Rev. Lett.* 29, 705–708 (1972).

[20] Mertz, J. "Molecular photodynamics involved in multiphoton excitation fluorescence microscopy," *Eur. Phys. J.* D 3, 53–66 (1998).

[21] Montroll, E. W. and West, B. J. "On an enriched collection of stochastic processes," in *Fluctuation Phenomena*, Eds. E. W. Montroll and J. L. Lebowitz, North-Holland Publishing (1979).

[22] O'Connor, D. V. and Phillips, D. *Time-Correlated Single Photon Counting*, Academic Press (1984).

[23] Philip, J. and Carlsson, K. "Theoretical investigation of the signal-to-noise ratio in fluorescence lifetime imaging," *J. Opt. Soc. Am.* A 20, 368–379 (2003).

[24] Schneider, P. C. and Clegg, R. M. "Rapid acquisition, analysis, and display of fluorescence lifetime-resolved images for real-time applications," *Rev. Sci. Instrum.* 68, 4107–4119 (1997).

[25] Squire, A. and Bastiaens, P. I. H. "Three dimensional image restoration in fluorescence lifetime imaging microscopy," *J. Microsc.* 193, 36–49 (1999).

[26] Stryer, L. "Fluorescence energy transfer as a spectroscopic ruler," *Ann. Rev. Biochem.* 47, 819–846 (1978).

[27] Thompson, N. L. "Fluorescence correlation spectroscopy" in *Topics in Fluorescence Spectroscopy, Vol. 1: Techniques*, Ed. J. R. Lakowicz, Springer (2005).

[28] Turro, N. J. *Modern Molecular Photochemistry*, University Science Books (1991).

[29] Valeur, B. *Molecular Fluorescence: Principles and Applications*, Wiley-VCH (2002).

[30] Van der Meer, B. W., Coker, I. G. and Chen, S.-Y. S. *Resonance Energy Transfer: Theory and Data*, VCH (1994).

[31] Wang, X. F., Uchida, T., Coleman, D. M. and Minami, S. "A 2-dimensional fluorescence lifetime imaging-system using a gated image intensifier," *Appl. Spectrosc.* 45, 360–366 (1991).

[32] Weber, G. "Resolution of the fluorescence lifetimes in a heterogeneous system by phase and modulations measurements," *J. Phys. Chem.* 85, 949–953 (1981).

14 Confocal Microscopy

As we have seen throughout this book, the issue of out-of-focus background rejection in microscopy is a problematic one. The difficulty stems from the inability of standard (i.e. wide-field) microscopes to perform optical sectioning, and, in particular, to discriminate blurred background from signal. For such discrimination to be possible, the microscope must be able to axially resolve even low spatial-frequency distributions of light, including distributions that are laterally uniform, corresponding to zero spatial frequency. But, as we have found, transmission microscopes are systematically bedevilled by the problem of a missing cone in their frequency support, which prohibits the possibility of axial resolution of uniform light distributions. It was only in the case of reflection microscopies that axial resolution was found to be possible, though at the cost of an insensitivity to low-frequency axial structure owing to an axial offset of the frequency support. Such axial resolution in reflection microscopes relied on some kind of gating mechanism, either spatial, as in the case of holographic microscopy with spatially incoherent illumination (Section 11.3), or temporal, as in the case of scanning OCT (Section 12.1), or both, as in the case of widefield OCT (Section 12.4). In all cases, whether spatial or temporal, the gating relied on the interference of the sample light with an external reference beam.

We turn now to an alternative spatial gating mechanism that does not rely on interference with an external reference beam. Remarkably, the background rejection in this mechanism is performed physically, by preventing background light from impinging on the detector, rather than computationally as was required in interference-based mechanisms (recall that the interferograms or complex interference terms had to be extracted post-detection by, at the very least, subtracting away the dc bias inherent in the detected intensities). This technique is called confocal microscopy.

In fact, confocal microscopy, which was initially devised by Minsky [10], predates even the invention of the laser. As we will see below, the gating mechanism in confocal microscopy relies on the spatial overlap of a focused illumination distribution and a "focused" detection distribution that comes from signal collection through a pinhole. Confocal microscopy represents one of the major advances in microscopy of the last century. Not only does it provide out-of-focus background rejection, but it does so while offering higher spatial resolution than widefield microscopy, and can work with either non-fluorescent or fluorescent samples. Since many of the more recent developments in optical microscopy have been motivated by precisely these issues, they are all systematically compared with confocal microscopy, which to this day still serves as the standard of reference.

This chapter will present various confocal microscope configurations and will re-evaluate the concept of out-of-focus background rejection in terms of an optical sectioning strength. Once again, this chapter serves only as a cursory introduction. For more thorough discussions, the reader is referred to the seminal work by Wilson and Sheppard [21] as well as several reference books [1, 2, 5, 13].

14.1 SCANNING CONFIGURATIONS

The layout of a confocal microscope is similar to that of any microscope in that it involves illumination and detection optics. However, instead of illuminating the sample with an extended uniform light beam (widefield illumination), a confocal microscope illuminates the sample with a light beam that is tightly focused. Correspondingly, instead of detecting signal from an extended area of the sample (widefield detection), a confocal microscope detects signal only from the same tightly focused region where the sample is illuminated. As we will see, such an illumination and detection configuration effectively confines the signal to a small 3D probe volume. To acquire a volumetric image of the sample, either the probe volume is scanned in three dimensions within a stationary sample, or instead it is held fixed and the sample is scanned. Both techniques are formally equivalent. In both cases the final 3D image must be synthesized from the time sequence of detected signals obtained from the predefined scan pattern of the probe volume.

By definition, focused illumination in a confocal microscope is achieved by critical illumination. From Section 6.5.2, we recall that a basic requirement for critical illumination is that the source beam be spatially coherent. When the source beam is derived from a laser, this spatial coherence is innate; however; when it is derived from a lamp, an extra step must be taken, namely the lamp light must be sent through a pinhole. Because the invention of the confocal microscope predates the laser, scanning was initially performed in the latter manner, with an array of pinholes mounted on a spinning disk (see [13] and [14] for descriptions).

Most modern confocal microscopes, however, are based on laser-beam scanning. A standard technique to perform laser-beam scanning involves moving mirrors that tilt the angle of the beam entering the condenser back focal plane. From the Fourier shift theorem (Eq. A.3), such an angular tilt leads to a lateral translation of the beam focus inside the sample. Beam tilting along orthogonal axes thus allows probe-volume translation in the sample x–y plane, enabling a synthesis of a transverse 2D image. To extend the scanning to the axial z dimension, a series of images can be acquired at different focal depths by axial translation of the condenser lens, finally enabling the synthesis of a 3D image stack.

Confocal microscopes can be configured in either reflection or transmission modes (see Fig. 14.1). A technical advantage of the reflection mode is that most of the microscope optics is common to the illumination and detection paths. Thus, a single lens can play the role of both the illumination condenser and detection objective. Moreover, the same mechanism of

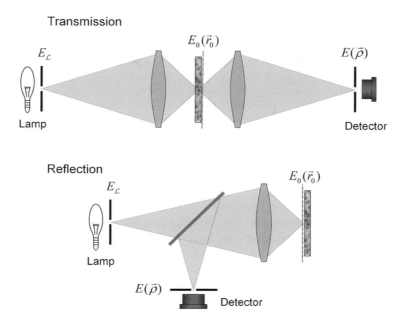

Figure 14.1. Transmission and reflection geometries for a confocal microscope.

moving mirrors can be used to simultaneously scan the illumination beam in the forward path and the signal beam in the backward path. Such backward scanning automatically re-aligns the signal beam onto the optical axis independently of the scanner position, allowing relatively easy positioning of the detection pinhole. This re-alignment of the signal beam on the optical axis is called de-scanning. While de-scanning is straightforward and automatic in a reflection geometry, it is generally quite cumbersome in a transmission geometry, often imposing the use of a second precisely aligned and synchronized scanning mechanism downstream from the sample (see [13]). Because of this technical complexity, and also because of its limited applicability only to thin samples, transmission confocal microscopes are rarely found in practice.

As pointed out above, beam scanning and sample scanning are formally equivalent. For mathematical convenience, we will take the point of view of the latter.

14.2 NON-FLUORESCENCE CONFOCAL MICROSCOPY

We begin by considering confocal microscopy with non-fluorescent samples. In Chapters 9 we saw that non-fluorescent samples can be characterized by a scattering potential $4\pi^2 \kappa^2 \delta \varepsilon \left(\vec{r} \right)$ (see Eq. 9.21), We return to this characterization here. In fact, the analysis of non-confocal fluorescence microscopy is very similar to that of widefield microscopy, and we could proceed along similar lines as in Chapter 10; however, we adopt here a slightly different viewpoint by remaining in real-space rather than working in Fourier space. This is done largely for the sake of variety, but also because it will turn out to be somewhat simpler.

Without loss of generality, we take our illumination source to be point-like and therefore spatially coherent. That is, we write $E_{\mathcal{L}}(\vec{\rho}_{\mathcal{L}}) = \kappa^{-2}E_{\mathcal{L}}\,\delta^2(\vec{\rho}_{\mathcal{L}})$, where we have adopted our usual convention of defining the minimum area of a radiative point to be κ^{-2} (see Chapter 2). By definition of critical illumination, the fields at the source and sample planes are conjugate to one another, meaning they are related by a 3D amplitude point spread function H_i. From Eq. 5.17 (assuming $M = 1$), we then write

$$E_i(\vec{r}_0) = \kappa^{-2}E_{\mathcal{L}}\,H_i(\vec{r}_0) \tag{14.1}$$

where $E_i(\vec{r}_0)$ denotes the illumination field incident on the location $\vec{r}_0$ inside the sample. Equation 14.1 highlights one of the key differences between confocal and widefield microscopy. In the latter case, the illumination is generally partially coherent, and can be described only statistically by a mutual intensity. Here it is described deterministically by a coherent field.

At this point, we can distinguish two different microscopy configurations.

14.2.1 Transmission Confocal

In a transmission configuration, both the ballistic and sample fields propagate through the detection optics. We have, then, in the region of the detector,

$$E(\vec{r}) = E_b(\vec{r}) + E_s(\vec{r}) \tag{14.2}$$

where

$$E_b(\vec{r}) = \int d^2\rho_0\,H_+(\vec{\rho} - \vec{\rho}_0, z)\,E_i(\vec{\rho}_0, 0) \tag{14.3}$$

and

$$E_s(\vec{r}) = i\pi\kappa \int d^3r_0\,H_+(\vec{r} - \vec{r}_0)\,\delta\varepsilon(\vec{r}_0)E_i(\vec{r}_0) \tag{14.4}$$

The reader is reminded here of the fundamental difference between Eqs. 14.3 and 14.4. The former involves the propagation of a field from the focal plane to the detector region as if the sample were not present. The latter involves the propagation of fields arising from the sample itself, hence the additional phase shift of $\frac{\pi}{2}$ associated with primary sources (see Section 2.3.1). More exactly, this latter propagation is mediated by the inverse Fourier transform of $\mathcal{H}(\vec{\kappa}_\perp)\,\mathcal{G}_+(\vec{\kappa}_\perp; z)$, but as usual we have invoked the Fresnel approximation $\mathcal{G}_+(\vec{\kappa}_\perp; z) \rightarrow \frac{1}{i4\pi\kappa}D_+(\vec{\kappa}_\perp; z)$, enabling us to cast Eq. 14.4 in terms of the 3D amplitude point spread function (see Section 5.2.2). Finally, the additional integration over z_0 in Eq. 14.4 comes from our use of the Born approximation, which allows us to treat the primary-source fields produced at different sample locations as independent (see Section 9.2.1).

Equations 14.3 and 14.4 are general and could be applied to any transmission microscope. Here, we are considering specifically a confocal transmission microscope. This involves placing a small pinhole in the detector region at $\vec{r} = 0$, meaning that the ballistic and sample fields incident on the detector reduce to

$$E_b(0) = \kappa^{-2} E_{\mathcal{L}} \int d^2 \vec{\rho}_0 \, H(-\vec{\rho}_0) \, H_i(\vec{\rho}_0) \tag{14.5}$$

$$E_s(0) = i\pi\kappa^{-1} E_{\mathcal{L}} \int d^3 \vec{r}_0 \, H_+(-\vec{r}_0) \, H_i(\vec{r}_0) \delta\varepsilon(\vec{r}_0) \tag{14.6}$$

Confocal microscopy also involves scanning. As noted above, we adopt the point of view of sample scanning. That is, we translate the sample by a 3D scan vector $\vec{r}_s$ such that $\delta\varepsilon(\vec{r}_0) \to \delta\varepsilon(\vec{r}_0 + \vec{r}_s)$. Assuming that the pupils in the illumination and detection optics are the same, we have $H_i(\vec{r}_0) = H_+(\vec{r}_0)$. Assuming further the pupils are real (i.e. $\mathcal{H}(\vec{\kappa}_\perp) = \mathcal{H}^*(\vec{\kappa}_\perp)$), we have $H_+(-\vec{r}_0) = H_+^*(\vec{r}_0)$, and write

$$E(\vec{r}_s) = E_{\mathcal{L}} \int d^2 \vec{\rho}_0 \, \mathrm{PSF}(\vec{\rho}_0) + i\pi\kappa E_{\mathcal{L}} \int d^3 \vec{r}_0 \, \mathrm{PSF}(\vec{r}_0) \, \delta\varepsilon(\vec{r}_0 + \vec{r}_s) \tag{14.7}$$

where we have made use of Eq. 5.28.

Finally, bearing in mind that detectors measure intensities rather than fields, we arrive at

$$I(\vec{r}_s) = \Omega_0^2 I_{\mathcal{L}} - 2\pi\kappa\Omega_0 I_{\mathcal{L}} \int d^3 \vec{r}_0 \, \mathrm{PSF}(\vec{r}_0) \, \mathrm{Im}\left[\delta\varepsilon(\vec{r}_0 + \vec{r}_s)\right] \tag{14.8}$$

where $I_{\mathcal{L}} = |E_{\mathcal{L}}|^2$ and, by virtue of the Born approximation, we have neglected the term second order in $\delta\varepsilon$.

Several conclusions can be drawn from this. For one, we observe that, to first order in $\delta\varepsilon$, a transmission confocal microscope cannot reveal a pure phase object. However, it should be noted that the various solutions to this problem identified in Chapter 10, such as Schlieren or oblique-field microscopy, apply equally well to transmission confocal microscopy.

Another conclusion we can draw from Eq. 14.8 is that transmission confocal microscopy provides the same object frequency support as shown in Fig. 5.6. In other words, it does not provide optical sectioning.

14.2.2 Reflection Confocal

Much more prevalent in practice are reflection confocal microscopes. We can derive the imaging properties of these in a similar way as above, with the difference that the pinhole is now located on the same side of the sample as the illumination source. As a result, $E_b(\vec{r}) = 0$ and the only light detected is that reflected from the sample, namely

$$E_s(\vec{r}) = i\pi\kappa \int d^3 \vec{r}_0 \, H_-(\vec{r} - \vec{r}_0) \, \delta\varepsilon(\vec{r}_0) E_i(\vec{r}_0) \tag{14.9}$$

Note the change in sign of the propagator direction compared to Eq. 14.4.

Making use of Eq. 14.1 and introducing both the pinhole and sample scanning, as above, we obtain

$$E_s\left(\vec{r}_s\right) = i\pi\kappa^{-1}E_{\mathcal{L}} \int d^3\vec{r}_0\, H_-\left(-\vec{r}_0\right) H_i(\vec{r}_0)\delta\varepsilon(\vec{r}_0 + \vec{r}_s) \qquad (14.10)$$

In most cases the illumination and detection share a common pupil in reflection confocal microscopy, meaning we can write $H_i(\vec{r}_0) = H_+(\vec{r}_0)$. Assuming this pupil is symmetric (i.e. $\mathcal{H}\left(\vec{\kappa}_\perp\right) = \mathcal{H}\left(-\vec{\kappa}_\perp\right)$), we have $H_-(-\vec{r}_0) = H_+(\vec{r}_0)$, and obtain

$$E_s\left(\vec{r}_s\right) = i\pi\kappa^{-1}E_{\mathcal{L}} \int d^3\vec{r}_0\, H_+^2\left(\vec{r}_0\right) \delta\varepsilon(\vec{r}_0 + \vec{r}_s) \qquad (14.11)$$

A comparison of Eqs. 14.7 and 14.11 reveals that while the propagation of the sample field to the detector is governed by $\left|H_+\left(\vec{r}_0\right)\right|^2$ in transmission confocal microscopy, it is governed by $H_+^2\left(\vec{r}_0\right)$ in reflection confocal microscopy (i.e. the former is real while the latter can be complex). Again this difference, seemingly minor, has important ramifications. In frequency space, the former corresponds to the auto-correlation of $\mathcal{H}_+\left(\vec{\kappa}\right)$; the latter to the auto-convolution of $\mathcal{H}_+\left(\vec{\kappa}\right)$. The frequency support of the latter is thus the same as illustrated in the bottom right of Fig. 11.6, indicating that, again, as is common to all reflection microscopes, sample information can only be gleaned from its high axial spatial-frequency components.

But one must be careful here. Equation 14.11 refers to the field incident on the detector. The actual measured intensity is given by

$$I_s\left(\vec{r}_s\right) = \pi^2\kappa^{-2}I_{\mathcal{L}} \iint d^3\vec{r}_0\, d^3\vec{r}_0'\, H_+\left(\vec{r}_0\right)^2 H_+^*\left(\vec{r}_0'\right)^2 \delta\varepsilon(\vec{r}_0 + \vec{r}_s)\delta\varepsilon^*(\vec{r}_0' + \vec{r}_s) \quad (14.12)$$

This expression is second order in $\delta\varepsilon$, meaning that the image obtained by a reflection confocal microscope is a nonlinear function of the sample. In other words, we cannot describe this type of microscope in terms of a conventional point spread function. Nevertheless, some limiting examples are instructive:

Point Object
Consider, for example, a sample consisting of an isolated point object located at $\vec{r}_\sigma$. That is, we write $\delta\varepsilon(\vec{r}_0) = \delta\varepsilon_\sigma\delta^3(\vec{r}_0 - \vec{r}_\sigma)$. Inserting this into Eq. 14.12 we obtain,

$$I_s\left(\vec{r}_s\right) = \pi^2\kappa^2 I_{\mathcal{L}}\left|\delta\varepsilon_\sigma\right|^2 \text{PSF}^2\left(\vec{r}_\sigma - \vec{r}_s\right) \qquad (14.13)$$

As expected, $I_s(\vec{r}_s)$ is peaked at $\vec{r}_s = \vec{r}_\sigma$, providing a 3D image of the point object. The resolution of this image is defined by the transverse and axial widths of $\text{PSF}^2(\vec{r})$.

Uniform Plane Object
Consider now a sample consisting of a uniform planar object located at $z_0 = z_\rho$. That is, we write $\delta\varepsilon(\vec{r}_0) = \delta\varepsilon_\rho\delta\left(z_0 - z_\rho\right)$. Inserting this into Eq. 14.11 we obtain

$$E_s\left(z_s\right) = i\pi\kappa^{-1}E_{\mathcal{L}}\delta\varepsilon_\rho \int d^2\vec{\rho}_0\, H_+\left(\vec{\rho}_0, z_s - z_\rho\right)^2 \qquad (14.14)$$

Making use of the integration property given by Eq. 5.24, this simplifies to

$$E_s\left(z_s\right) = i\pi\kappa^{-1}E_\mathcal{L}\delta\varepsilon_\rho H_+\left(0, 2z_s - 2z_\rho\right) \tag{14.15}$$

leading to

$$I_s\left(z_s\right) = \pi^2 I_\mathcal{L}\left|\delta\varepsilon_\rho\right|^2 \mathrm{PSF}\left(0, 2z_s - 2z_\rho\right) \tag{14.16}$$

We conclude that $I_s(z_s)$ is peaked at $z_s = z_\rho$ and decays to zero with increasing defocus of the object (i.e. with increasing $|z_s|$). Remarkably, this decay is governed by a PSF, which in standard widefield microscopy is normally associated with a point object rather than a plane object. In other words, reflection confocal microscopy does what reflection widefield microscopy cannot: it provides axial resolution even for a uniformly reflecting planar object. The decay profiles of $H_+\left(0, z\right)$ and $\mathrm{PSF}(0, z)$ for circular pupils are given by Eqs. 5.21 and 5.33, respectively.

14.2.3 Comparison with Widefield Microscopy

As an aside, a well-known principle in optics is the Helmholtz principle of reciprocity [6], which states that a field produced by a source at point $\vec{r}_0$ and detected at point $\vec{r}$ is identical to the field produced by the same source at point $\vec{r}$ and detected at point $\vec{r}_0$. This principle is valid independent of any intervening sample between the points, provided only that scattering or attenuation caused by the sample is linear (i.e. does not modify the field frequency). The consequences of this principle of reciprocity are far-reaching. For example, in its most general form, a microscope can be conceptualized as comprised of a source, a sample, and a detector. The source can be spatially coherent, in which case it is a point (for plane-wave illumination, this source is displaced to infinity, or a lens is interposed). Alternatively, the source can be spatially incoherent, in which case it is comprised of a multiplicity of uncorrelated point emitters. Similarly, a detector can be coherent (a point detector) or incoherent (a multiplicity of point detectors whose outputs are summed, equivalent to a large detector).

Remarkably, as a consequence of reciprocity, scanning and widefield microscopes are formally identical [20]. The transformation from one to the other simply involves exchanging sources with detectors. For example, in an incoherent, widefield transmission microscope, a large source (typically a lamp) illuminates a sample and the transmitted field is detected by a camera (an array of point detectors). If each detector in the camera is replaced by a point source, sequentially turned on, and the source is replaced by a large detector of equal size, the result is a scanning microscope that produces an identical image of the sample. Note, this is not the same as a confocal transmission microscope since the detector is large here, meaning the pinhole is absent. And herein lies the difference between confocal and camera-based widefield microscopy. In a confocal microscope, both the source and detector are effectively point-like, whereas in a widefield microscope only the detector is point-like (or, rather, each detector element).

We will re-examine the principle of Helmholtz reciprocity in Chapter 20. Suffice it to say here that it is an extremely powerful tool that can provide a unified description of essentially all scattering-based microscopes (for example, see [16], which extends this description to vector fields). But, as noted above, a failing of this principle comes when considering interactions where wavelengths into and out of the sample differ, such as in the case of fluorescence, to which we presently turn our attention.

14.3 CONFOCAL FLUORESCENCE MICROSCOPY

By far, the most common application of confocal microscopy is not transmission or reflection imaging, as described above, but rather fluorescence imaging. This is particularly true in the bioimaging community, since fluorescent markers can be designed to label well-defined targets in cells or tissue samples, thus providing a degree of molecular specificity that is unattainable with transmission or reflection alone. As noted in Chapter 13, some characteristics of molecular fluorescence are that its wavelength is generally Stokes-shifted compared to that of the illumination beam, allowing easy spectral separation of the two. Moreover, also as noted in Chapter 13, the phase of fluorescence is essentially random upon generation, and therefore uncorrelated with the phase of the illumination beam. Thus we may neglect phase when treating confocal fluorescence microscopy, and describe both the illumination and detection optics in terms of point spread functions (PSF) rather than amplitude point spread functions (H).

From Chapter 13, the efficiency of fluorescence generation inside a sample can be characterized by a fluorescence excitation cross-section σ_f (units: area). Defining I_i to be the illumination intensity incident upon the object, and F_{0z} to be the fluorescence intensity (or, more precisely, the emittance) generated by the object per unit depth, we have from Eq. 5.43

$$F_{0z}(\vec{r}_0) = \sigma_f I_i(\vec{r}_0) C(\vec{r}_0) \tag{14.17}$$

where $C(\vec{r}_0)$ is the concentration of fluorescent molecules (units: 1/volume). Note that we have slightly modified our notation compared to Eq. 5.43 and use the variable F_{0z} to explicitly refer to fluorescence.

From Eqs. 14.1 and 5.28 the illumination intensity is given by

$$I_i(\vec{r}_0) = \left\langle \left| E_i(\vec{r}_0) \right|^2 \right\rangle = \Phi_i \, \Omega_i^{-1} \text{PSF}_i(\vec{r}_0) \tag{14.18}$$

where $\Phi_i = \kappa^{-2} \Omega_i I_{\mathcal{L}}$ denotes the source power incident on the sample, and $\Omega_i = \int d^2\vec{\rho}_0 \, \text{PSF}_i(\vec{r}_0)$.

Hence, the fluorescence intensity per unit depth emitted from the sample is

$$F_{0z}(\vec{r}_0) = \sigma_f \Phi_i \, \Omega_i^{-1} \text{PSF}_i(\vec{r}_0) C(\vec{r}_0) \tag{14.19}$$

From Eq. 5.41 the fluorescence intensity in the vicinity of the detector is given by

$$I\left(\vec{\rho}, z\right) = \iint d^2\vec{\rho}_0 \, dz_0 \, \text{PSF}(\vec{\rho} - \vec{\rho}_0, -z + z_0) F_{0z}(\vec{\rho}_0, z_0) \tag{14.20}$$

where we have assumed a reflection-mode geometry, meaning that we have utilized the point spread function $\text{PSF}(\vec{\rho}, -z)$ rather than $\text{PSF}(\vec{\rho}, z)$. This geometry is generally referred to as an epifluorescence detection geometry. Note that in the case of imaging with a real and symmetric pupil, the two PSFs are equivalent.

Our treatment of confocal fluorescence microscopy will be a little different than our treatment of non-fluorescence confocal microscopy in that we will allow for the possibility of a pinhole of arbitrary size in the detector plane, defined by the unitless aperture function $A_p\left(\vec{\rho}\right)$. That is, the fluorescence power incident on the detector is given by $\Phi = \int d^2\vec{\rho} \, A_p\left(\vec{\rho}\right) I\left(\vec{\rho}, 0\right)$, or

$$\Phi = \sigma_f \Phi_i \Omega_i^{-1} \iiint d^2\vec{\rho} \, d^2\vec{\rho}_0 \, dz_0 \, A_p\left(\vec{\rho}\right) \text{PSF}(\vec{\rho} - \vec{\rho}_0, z_0) \text{PSF}_i(\vec{\rho}_0, z_0) C(\vec{\rho}_0, z_0) \tag{14.21}$$

Making allowances finally for sample scanning, we arrive at

$$\Phi\left(\vec{\rho}_s, z_s\right) = \sigma_f \Phi_i \iint d^2\vec{\rho}_0 \, dz_0 \, \text{PSF}_{\text{conf}}\left(\vec{\rho}_0, z_0\right) C(\vec{\rho}_0 + \vec{\rho}_s, z_0 + z_s) \tag{14.22}$$

where we have introduced an effective confocal point spread function given by

$$\text{PSF}_{\text{conf}}\left(\vec{\rho}, z\right) = \Omega_i^{-1} \int d^2\vec{\beta} \, A_p\left(\vec{\beta}\right) \text{PSF}(\vec{\beta} - \vec{\rho}, z) \text{PSF}_i(\vec{\rho}, z) \tag{14.23}$$

Equations 14.22 and 14.23 will serve as the basis of our analysis of confocal fluorescence microscopy. To begin, let us consider the extreme case where the pinhole is large, or removed altogether, and the detector is infinite in size, that is $A_p\left(\vec{\rho}\right) \rightarrow 1$. We find then

$$\text{PSF}_{\text{conf}}\left(\vec{\rho}, z\right) \rightarrow \Omega_0 \Omega_i^{-1} \text{PSF}_i(\vec{\rho}, z) \qquad \text{(large pinhole)} \tag{14.24}$$

where $\Omega_0 = \int d^2\vec{\rho} \, \text{PSF}(\vec{\rho}, z)$. That is, confocal microscopy provides similar imaging as standard widefield microscopy (see Eq. 5.41), with the only difference being that the resolution is governed by the illumination PSF rather than the detection PSF (in accord with Helmholtz reciprocity – see above).

In the opposite extreme where the pinhole is present, of infinitesimally small area A_p (i.e. $A_p\left(\vec{\rho}\right) \rightarrow A_p \delta^2\left(\vec{\rho}\right)$), we find instead

$$\text{PSF}_{\text{conf}}\left(\vec{\rho}, z\right) \rightarrow A_p \Omega_i^{-1} \text{PSF}(-\vec{\rho}, z) \text{PSF}_i(\vec{\rho}, z) \qquad \text{(small pinhole)} \tag{14.25}$$

The effective confocal point spread function is thus given by the product of two PSFs, rather than a single PSF as in widefield microscopy. As we will see below, the consequences of this difference are fundamental.

A further simplification comes from noting that in an epifluorescence geometry the illumination and fluorescence generally travel through the same microscope objective, though in opposite directions. The illumination and detection PSFs are then quite similar (notwithstanding differences due to the Stokes shift between fluorescence and illumination wavelengths). That is, in a similar manner as in Section 14.2, we can typically write $\mathrm{PSF}(\vec{\rho}, z) \simeq \mathrm{PSF}_i(\vec{\rho}, z)$, where, for simplicity, we assume $\mathrm{PSF}(\vec{\rho}, z)$ to be symmetric in $\vec{\rho}$. Equation 14.25 then reduces to

$$\mathrm{PSF}_{\mathrm{conf}}(\vec{\rho}, z) \to A_p \Omega_i^{-1} \mathrm{PSF}_i^2(\vec{\rho}, z) \qquad \text{(small pinhole)} \qquad (14.26)$$

14.3.1 Optical Sectioning

As noted at the beginning of this chapter, one of the defining characteristics of confocal fluorescence microscopy is that it can perform optical sectioning. To verify this, we consider a thin uniform fluorescent-plane sample, which we scan through the focus. That is, we write $C(\vec{\rho}_0, z_0) = C_\rho \delta(z_0 - z_s)$, where C_ρ is a surface concentration.

To compare the relative performances of confocal versus widefield fluorescence microscopy, let us consider the usual example of a PSF derived from a circular pupil function. We recall from Chapter 5 two basic properties of such a PSF:

- $\mathrm{PSF}(0, z)$ scales as z^{-2} for large z.
- $\int \mathrm{d}^2\vec{\rho}\, \mathrm{PSF}(\vec{\rho}, z) = \Omega_0$, independent of z.

Hence, in widefield fluorescence microscopy (governed by Eq. 5.41) the signal produced by a uniform fluorescent plane is given by

$$I(z_s) = \sigma_f I_i C_\rho \int \mathrm{d}^2\vec{\rho}_0\, \mathrm{PSF}(-\vec{\rho}_0, -z_s) = \Omega_0 \sigma_f I_i C_\rho \qquad (14.27)$$

which does not depend on the axial location z_s of the plane. A widefield fluorescence microscope therefore does not confer optical sectioning, as we concluded earlier in Section 5.4.1.

The basic properties listed above can be compared with those when the PSF is squared, as in Eq. 14.26:

- $\mathrm{PSF}^2(0, z)$ scales as z^{-4} for large z.
- $\int \mathrm{d}^2\vec{\rho}\, \mathrm{PSF}^2(\vec{\rho}, z)$ scales as z^{-2} for large z.

(Note that this last result is not obvious for the case of a circular pupil, but can be inferred from Parseval's theorem and a numerical integration of Eq. 5.60 – on the other hand, it becomes obvious if the PSF is Gaussian–Lorentzian.)

Hence, in confocal microscopy with a pinhole, where $\mathrm{PSF}_{\mathrm{conf}}(\vec{\rho}_0, -z_s)$ is governed by Eq. 14.23, or in the extreme case of a vanishingly small pinhole by Eq. 14.25, the signal produced by a uniform fluorescent plane is given by

$$\Phi(z_s) = \sigma_f \Phi_i C_\rho \int \mathrm{d}^2\vec{\rho}_0\, \mathrm{PSF}_{\mathrm{conf}}(\vec{\rho}_0, -z_s) \propto \frac{1}{z_s^2} \Omega_i^{-1} \sigma_f \Phi_i C_\rho \qquad (14.28)$$

which manifestly does depend on z_s. A confocal fluorescence microscope thus indeed provides axial resolution even for a uniform plane. The sectioning strength of this axial resolution can be characterized by the scaling law of signal decay as the plane becomes defocused (i.e. as z_s becomes displaced from 0). From the PSF properties listed above, we find that uniform-plane signal for a confocal microscope decays as z_s^{-2}, whereas for a widefield microscope it does not decay at all (no optical sectioning). Inasmuch as defocused signal can be interpreted as out-of-focus background, we conclude that a confocal microscope confers out-of-focus background rejection, whereas a widefield microscope does not. This conclusion represents one of the defining advantages of confocal fluorescence microscopy and is the principal reason for its enormous success.

14.3.2 Confocal Probe Volume

An optical sectioning strength that obeys a z_s^{-2} scaling law leads to another important result, namely intrinsic 3D confinement of signal. This is demonstrated by deriving the effective probe volume of a confocal fluorescence microscope. The derivation follows the identical steps outlined in Section 13.4 to derive the probe volume in fluorescence correlation spectroscopy. Thus, referring to Eq. 14.23, the signal detected in confocal fluorescence microscopy can be thought of as originating from a volume inside the sample whose spatial profile is defined by

$$\Psi(\vec{r}) = \frac{\text{PSF}_{\text{conf}}(\vec{r})}{\text{PSF}_{\text{conf}}(\vec{0})} \tag{14.29}$$

which obeys the normalization condition $\Psi(\vec{0}) = 1$, as prescribed in Section 13.4. Equation 14.29 applies for arbitrary illumination and detection PSFs.

From Section 13.4, we find then that the confocal probe volume is given by

$$V_{\text{conf}} = \frac{\left(\int \text{d}^3\vec{r}\, \Psi(\vec{r})\right)^2}{\int \text{d}^3\vec{r}\, \Psi^2(\vec{r})} = \frac{\left(\int \text{d}^3\vec{r}\, \text{PSF}_{\text{conf}}(\vec{r})\right)^2}{\int \text{d}^3\vec{r}\, \text{PSF}_{\text{conf}}^2(\vec{r})} \tag{14.30}$$

Based on the PSF properties outlined above, we conclude in particular that the numerator in Eq. 14.30 is finite when using a small pinhole, meaning that the probe volume V_{conf} is well defined independently of the volume of the sample itself (assuming this is bigger than the probe volume). In other words, the signal detected in confocal fluorescence microscopy is intrinsically confined in three dimensions. Remarkably, such 3D confinement arises from the spatial gating afforded by pinhole detection. No recourse has been made to temporal gating, as was the basis, for example, of standard OCT (see Chapter 12).

14.3.3 Frequency Support

From Chapter 5, we inferred some key properties of widefield microscopy by examining its 3D OTF. In particular, this OTF was found to exhibit a missing cone in its frequency support, which ultimately led to the inability of a widefield microscope to provide optical sectioning. Figure 14.2 provides an illustration (dashed lines) of the frequency support of this widefield OTF for a circular pupil function, as reproduced from Figs. 5.5 or 5.6. According to Section 5.4, this frequency support is bounded in the transverse direction by $\kappa_{\perp\text{max}} = \Delta\kappa_\perp$ and in the

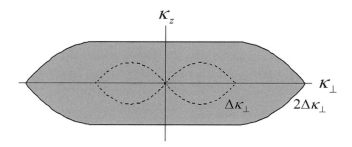

Figure 14.2. Frequency support of a widefield (dashed) and confocal (shaded) fluorescence microscope with a circular pupil.

axial direction by $|\kappa_z|_{max} = \frac{\kappa_\perp}{2\kappa}(\Delta\kappa_\perp - \kappa_\perp)$. The maximum range of $|\kappa_z|_{max}$ occurs when $\kappa_\perp = \frac{1}{2}\Delta\kappa_\perp$, at which point $|\kappa_z|_{max} = \frac{\Delta\kappa_\perp^2}{8\kappa}$.

We can compare this to the frequency support associated with a confocal fluorescence microscope. For simplicity, we consider the ideal case of a small pinhole and equal illumination and detection PSFs. That is, we consider $\text{PSF}_{conf}(\vec{r})$ to be given by Eq. 14.26. The effective confocal 3D OTF is then

$$\text{OTF}_{conf}(\vec{\kappa}) = \text{FT}_{3D}\left\{\text{PSF}_{conf}(\vec{r})\right\} \propto \left[\text{OTF}(\vec{\kappa}) * \text{OTF}(\vec{\kappa})\right]_{3D} \quad (14.31)$$

where the Fourier transform and auto-convolution are three-dimensional (see [12, 17, 18] for a more detailed discussion).

The frequency support of $\text{OTF}_{conf}(\vec{\kappa})$ is also shown in Fig. 14.2. Several conclusions can be drawn. First, as expected, the confocal frequency support exhibits no missing cone, indicating that $\text{OTF}_{conf}(\vec{\kappa})$ indeed provides axial resolution even at $\kappa_\perp = 0$ (i.e. for a laterally uniform object). Remarkably, the axial span of this confocal frequency support is roughly constant over quite a wide range of transverse spatial frequencies before it eventually tapers off to zero at $\kappa_\perp = 2\Delta\kappa_\perp$. Recalling from Section 5.4.1 that this axial span is roughly associated with the inverse of the axial resolution at spatial frequency $\kappa_\perp$, we conclude that the axial resolution is fairly uniform over a broad transverse bandwidth.

Based on the axial and transverse extents of this frequency support, we can also infer a relation between axial and transverse resolutions of a confocal fluorescence microscope, given by

$$\frac{\delta z_{conf}}{\delta\rho_{conf}} \approx \frac{4\kappa_{\perp max}}{4|\kappa_z|_{max}} = \frac{8\kappa}{\Delta\kappa_\perp} = \frac{4n}{\text{NA}} \quad (14.32)$$

Again, this expression is very rough, and applies under the Fresnel approximation only. Inasmuch as $\text{NA} < n$ always, the confocal resolution volume is more elongated axially than transversely, an elongation that becomes exacerbated with decreasing NA. Equation 14.32 indicates that the axial and transverse resolutions of a confocal fluorescence microscope are manifestly not independent, as they are in OCT, but instead are intimately related.

14.3.4 Finite-Size Pinhole

The reader is reminded that the conclusions drawn above were based on the assumption that the pinhole in the detection optics was infinitely small. Clearly, this cannot be the case in practice, since an infinitely small pinhole would imply an infinitely small detected signal. We turn now to a more realistic evaluation of resolution in the case of a finite pinhole size. For this, it is most useful to invoke the mixed representation of the 3D OTF, given by

$$\text{OTF}_{\text{conf}}(\vec{\kappa}_{\perp}; z) = \text{FT}_{2D}\left\{\text{PSF}_{\text{conf}}(\vec{\rho}, z)\right\} \tag{14.33}$$

where the Fourier transform is over the 2D coordinate $\vec{\rho}$ only. From Eq. 14.23, we arrive at

$$\text{OTF}_{\text{conf}}\left(\vec{\kappa}_{\perp}; z\right) = \Omega_i^{-1} \int d^2\vec{\kappa}_{\perp}' \, \mathcal{A}_p\left(-\vec{\kappa}_{\perp}'\right) \text{OTF}(\vec{\kappa}_{\perp}'; z)\text{OTF}_i\left(\vec{\kappa}_{\perp} + \vec{\kappa}_{\perp}'; z\right) \tag{14.34}$$

where $\mathcal{A}_p\left(\vec{\kappa}_{\perp}\right)$ is the Fourier transform of the pinhole function. For example, if the pinhole is a circular aperture of radius a_p, then $\mathcal{A}_p\left(\vec{\kappa}_{\perp}\right) = A_p \text{jinc}(2\pi\kappa_{\perp}a_p)$, where $A_p = \pi a_p^2$. When a_p becomes vanishingly small, then $\mathcal{A}_p\left(\vec{\kappa}_{\perp}\right) \rightarrow A_p$.

Figure 14.3 shows plots of the in-focus transverse PSF and OTF associated with widefield microscopy ($\text{PSF}(\rho, 0)$ and $\text{OTF}(\kappa_{\perp}; 0)$) and idealized confocal microscopy with zero pinhole size ($\text{PSF}_{\text{conf}}^{(0)}(\rho, 0) \propto \text{PSF}^2(\rho, 0)$ and $\text{OTF}_{\text{conf}}^{(0)}(\kappa_{\perp}; 0) \propto \left[\text{OTF}(\vec{\kappa}_{\perp}; 0) * \text{OTF}(\vec{\kappa}_{\perp}; 0)\right]_{2D}$—assuming the illumination and detection PSFs are the same). From these plots, it is clear that while the 3D span of the confocal frequency support has theoretically doubled compared to

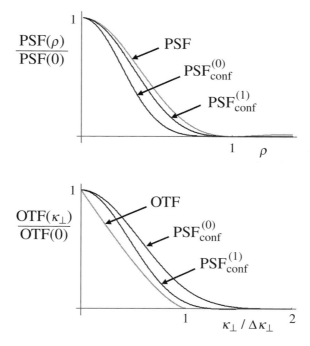

Figure 14.3. Widefield and confocal imaging functions for different pinhole sizes (ρ is in Airy units).

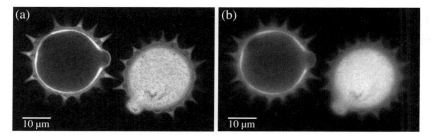

Figure 14.4. Confocal images of fluorescent pollen grains with pinhole diameter set to PSF width (a) and $6\times$ PSF width (b). Note the reduction in resolution and sectioning strength when the pinhole diameter becomes larger.

the widefield frequency support, one would be hard-pressed to arrive at a similar conclusion regarding resolution. The actual magnitude of the confocal $\mathrm{OTF}_{\mathrm{conf}}^{(0)}(\kappa_\perp; 0)$ is manifestly quite small beyond $\kappa_\perp \gtrsim 1.6\Delta\kappa_\perp$, so small as to be indiscernible from zero in Fig. 14.3. The resolution provided by a confocal microscope, even one with vanishingly small pinhole size, is thus manifestly not twice as good as that of a widefield microscope, but rather something less (the ratio $\sqrt{2}$ is often cited in the literature).

But again, such an improvement in resolution cannot be obtained in practice, since it would imply vanishingly small signal strength. A more practical analysis based on an optimization of signal contrast is summarized, for example, in [22], suggesting that a pinhole radius of ≈ 1 AU is more appropriate (1 Airy unit $= 0.61\lambda/\mathrm{NA} = 1.22/\Delta\kappa_\perp$, corresponding to the radius of the first zero in the Airy pattern given by Eq. 4.81). Plots of $\mathrm{PSF}_{\mathrm{conf}}^{(1)}(\rho, 0)$ and $\mathrm{OTF}_{\mathrm{conf}}^{(1)}(\kappa_\perp; 0)$ with this more realistic pinhole radius are also shown in Fig. 14.3, suggesting that the resolution of a confocal microscope, while still better than that of a widefield microscope, is not substantially better in practice. Examples of confocal images taken with different pinhole sizes are shown in Fig. 14.4

Image Scanning Microscopy

But there is a strategy to circumvent the above dilemma and recover a transverse resolution that is arguably better than that of a zero-pinhole-size confocal microscope *without* sacrificing signal strength. The strategy is called image scanning microscopy (variously known as optical photon reassignment microscopy) [11, 19]. The idea is as follows:

Let us consider using not one pinhole in the detection plane, but an array of many pinholes (such as provided, for example, by a pixellated camera). Let us further consider that each pinhole is smaller than an Airy unit, meaning that each pinhole, by itself, transmits very little signal. On the other hand, the overall signal obtained from the summation of all the pinhole signals is quite substantial. This is the basic idea. What is required now is a reconstruction algorithm to process these signals so as to recover an optimized resolution.

To begin, we must take into account the positions of each pinhole. Let us imagine that pinhole n is displaced from the optical axis by a transverse distance $\vec{\rho}_n$. From the Fourier shift theorem (see Eq. A.3) we have then $\mathcal{A}_p\left(\vec{\kappa}_\perp\right)_n = \mathcal{A}_p\left(\vec{\kappa}_\perp\right) e^{-i2\pi\vec{\kappa}_\perp \cdot \vec{\rho}_n}$. Each of these pinholes produces an image of the sample governed by its respective OTF, namely

$$\text{OTF}_{\text{conf}}^{(n)}\left(\vec{\kappa}_\perp; z\right) = \Omega_i^{-1} \int d^2\vec{\kappa}_\perp' \, \mathcal{A}_p\left(-\vec{\kappa}_\perp'\right) e^{i2\pi\vec{\kappa}_\perp'\cdot\vec{\rho}_n}\text{OTF}(\vec{\kappa}_\perp'; z)\text{OTF}_i\left(\vec{\kappa}_\perp + \vec{\kappa}_\perp'; z\right)$$

$$(14.35)$$

Let us imagine that each of the images is then numerically displaced a posteriori by a distance $-m\vec{\rho}_n$, where m is a user-defined scaling factor yet to be determined. Finally, let us perform a summation of all these numerically displaced images. The effective overall OTF that results from this photon reassignment procedure is given by

$$\text{OTF}_{\text{ism}}\left(\vec{\kappa}_\perp; z\right) = \int d^2\vec{\rho}_n \, e^{i2\pi\vec{\kappa}_\perp \cdot m\vec{\rho}_n}\text{OTF}_{\text{conf}}^{(n)}\left(\vec{\kappa}_\perp; z\right) \qquad (14.36)$$

(in practice, the summation is over discrete pinholes, which for simplicity we approximate here as an integration).

Equation 14.35 can be inserted into Eq. 14.36 and the integration over $\vec{\rho}_n$ is straightforward, obtaining

$$\text{OTF}_{\text{ism}}\left(\vec{\kappa}_\perp; z\right) = \Omega_i^{-1}\mathcal{A}_p\left(m\vec{\kappa}_\perp\right)\text{OTF}(-m\vec{\kappa}_\perp; z)\text{OTF}_i\left(\vec{\kappa}_\perp - m\vec{\kappa}_\perp; z\right) \qquad (14.37)$$

An optimal choice of m leads to an $\text{OTF}_{\text{ism}}\left(\vec{\kappa}_\perp; z\right)$ that is as broadly extended as possible. Assuming, as usual, that the illumination and detection PSFs are the same and symmetric, this optimal choice is given by $m = \frac{1}{2}$, leading to

$$\text{OTF}_{\text{ism}}\left(\vec{\kappa}_\perp; z\right) = \Omega_i^{-1}\mathcal{A}_p\left(\tfrac{1}{2}\vec{\kappa}_\perp\right)\text{OTF}^2\left(\tfrac{1}{2}\vec{\kappa}_\perp; z\right) \qquad (14.38)$$

We recall that the individual pinholes were expressly chosen to be smaller than 1 AU, meaning that the prefactor $\mathcal{A}_p\left(\tfrac{1}{2}\vec{\kappa}_\perp\right)$ is so broad it plays little role in determining the shape of $\text{OTF}_{\text{ism}}\left(\vec{\kappa}_\perp; z\right)$. A plot of $\text{OTF}_{\text{ism}}\left(\kappa_\perp; 0\right)$ is illustrated in Fig. 14.5. We observe that $\text{OTF}_{\text{ism}}\left(\kappa_\perp; 0\right)$ features the same transverse frequency support as $\text{OTF}_{\text{conf}}\left(\kappa_\perp; 0\right)$ with a vanishingly small pinhole. Moreover, it provides slightly better frequency response in the high-frequency range where the performance of $\text{OTF}_{\text{conf}}\left(\kappa_\perp; 0\right)$ is quite poor (albeit at the expense of a worse low-frequency response). But the key advantage of photon re-assigment is

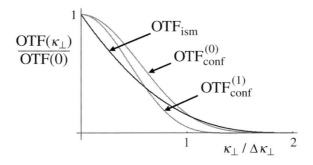

Figure 14.5. Comparison of OTFs for confocal and image scanning micrsocopy.

that essentially all the fluorescence signal is detected, assuming the pinholes are tightly packed, meaning it is remarkably light efficient.

14.3.5 Offset Illumination and Detection Pupils

Most of our discussion on confocal fluorescence microscopy has been based on the assumption that it is configured in an epifluorescence geometry. That is, the illumination and detection light are assumed to travel through the same pupil, albeit in different directions. However, this need not always be the case, and variants of confocal microscopy have been developed where the illumination and detection pupils are purposefully set apart. One such variant is called confocal theta microscopy [8]. In particular, let us consider what happens when a pupil is shifted to one side or the other. This leads to a tilting of its associated PSF, given by $\text{PSF}(\vec{\rho}, z) \rightarrow \text{PSF}(x \pm z \sin \Theta, y, z)$, where Θ is the resultant tilt angle, assumed here to be in the x direction. The associated OTF is then given by $\text{OTF}(\vec{\kappa}_\perp; z) \rightarrow \text{OTF}(\vec{\kappa}_\perp; z) e^{\pm i2\pi \kappa_{xz} \sin \Theta}$. If the illumination and detection pupils are increasingly shifted in opposing directions, the overlap region of their respective PSFs becomes increasingly smaller, as illustrated in Fig. 14.6. From purely geometrical arguments (i.e. neglecting diffraction), we observe that when this shift is so large that $n \sin \Theta > \text{NA}$, then the overlap region becomes axially confined about the focal plane (assuming, for simplicity, that the illumination and detection NAs are the same – we will relax this assumption below). The net result is that, for a given NA, confocal theta microscopy provides better axial resolution and optical sectioning capacity than standard confocal microscopy. This advantage is particularly beneficial in situations that require the use of small NAs, where in standard epifluorescence confocal microscopy the axial resolution is quite poor (see Eq. 14.32). An example of such a situation is when the distance between the objective and the sample, called the working distance, is required to be long.

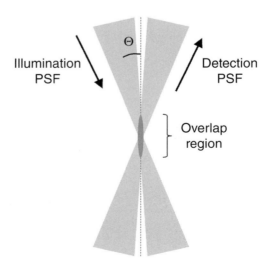

Figure 14.6. Geometry for theta microscopy.

Light-Sheet Microscopy

The extreme case where the illumination and detection PSFs are orthogonal to one another deserves special consideration. We have seen this case before in Section 10.2.2. Here we revisit it in the context of fluorescence imaging [3, 7]. In fact, this new context has led to a tremendous resurgence of interest in this technique for reasons that will be made clear below. Many names have been ascribed to this technique, the most common being selective plane illumination microscopy (SPIM) and light-sheet microscopy. We will use the latter.

To begin, we note that a layout where the illumination and detection PSFs are orthogonal requires more than a simple shifting of their respective pupils. Here, the pupils themselves must be oriented orthogonal to one another. In other words, we can no longer consider a common focal plane. Instead, the focal plane is defined to be the one associated with the detection pupil. The transverse resolution of the light-sheet microscope is thus defined by the detection PSF, and the axial resolution by the illumination PSF. The confocal probe volume is defined, as always, by the overlap between these two PSFs. And herein lies the key advantage of light-sheet microscopy. Because of the orthogonality of the illumination and detection PSFs, there is essentially no overlap in these PSFs outside the confocal probe volume. As a result, the probe volume is confined in three dimensions *even without the use of a pinhole*. All the benefits of confocal microscopy, such as axial resolution and optical sectioning, can thus be retained without the requirement of scanning, meaning that detection can be performed in a simple widefield geometry with the use of a camera, as illustrated in Fig. 14.7.

In most cases, light-sheet microscopes are configured such that the illumination NA is small and the detection NA is much larger (see Fig. 14.7). This asymmetric configuration produces a very elongated illumination PSF, which, in turn, provides a large field of illumination over which the axial resolution remains roughly uniform. The focusing of the illumination light can be achieved in one of two ways. It can be focused in the z direction, to provide axial

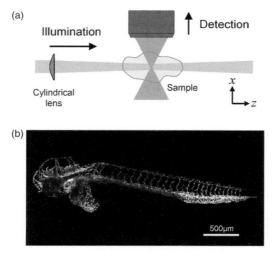

Figure 14.7. (a) Geometry for light-sheet microscope. (b) Fluorescently labeled endothelial and red blood cells in vasculature of zebrafish. Courtesy of Jan Huisken, Morgridge Institute.

resolution, while not focused in the y direction, to provide a large 2D field of illumination in the x–y plane (see Fig. 14.7). Such uni-dimensional focusing can be achieved, for example, with the use of a cylindrical lens, causing the illumination light to resemble a sheet of light (hence the appellation). Alternatively, the illumination can be focused in both the z and y directions, forming a narrow beam of light that is then rapidly swept along the y direction in a time shorter than the camera exposure time. Both techniques are essentially equivalent, though the latter tends to provide improved illumination homogeneity (at the cost of increased mechanical complexity). Both techniques provide rapid imaging of the focal plane. To obtain volumetric imaging, additional scanning in the z direction is required, either of the sample or of the light-sheet and detection optics in unison. A volume image is then created from a stack of 2D planar images.

But there is another advantage of light-sheet microscopy over standard confocal microscopy that is particularly relevant in biological imaging applications, where the samples are generally susceptible to light-induced damage. Because light-sheet microscopy dispenses with the use of a pinhole, all the fluorescence that propagates through the detection optics is captured by the camera and interpreted as in-focus signal. This is in distinct contrast with standard confocal microscopy, where only the fraction of the fluorescence that propagates through the pinhole is interpreted as in-focus signal. The rest that is rejected by the pinhole is considered to be out-of-focus background. As a result, much (if not most) of the fluorescence generated in standard confocal microscopy goes undetected. Such an inefficient usage of fluorescence means that energy is deposited into the sample, and hence damage is inflicted on the sample, needlessly. In other words, light-sheet microscopy has the notable advantage of rejecting out-of-focus background by not producing it in the first place. We will consider another type of microscopy that enjoys this same advantage in Chapter 16.

The above benefits of simplicity, speed, and reduced sample damage, along with its ability to provide good axial resolution even at long working distances, have made light-sheet microscopy highly attractive to biologists. The main impediments to its usage are that the sample must be largely transparent such that the light sheet remains thin throughout the entire field of view. The sample must also be amenable to side-on illumination (though this last impediment can circumvented with the use of tilted PSFs [4]). For more detailed reviews on light-sheet microscopy, the reader is referred to [9, 15].

14.4 PROBLEMS

Problem 14.1

From the result in Eq. 14.8 it is clear that a purely phase-shifting point object produces no discernible change in detected intensity in a transmission confocal microscope. That is, if $\delta\varepsilon$ is real then $I(\vec{\rho}_s, z_s)$ is independent of $\delta\varepsilon$ to first order. This result is based on the assumption that the microscope is well aligned and focused.

Consider now a transmission confocal microscope that is defocused. In particular, consider displacing the pinhole out of focus by a distance Δz_p. Show that this defocused transmission confocal microscope now becomes sensitive to a phase-shifting point object. For simplicity, assume that the illumination and detection amplitude-PSFs are identical and Gaussian (Eq. 5.23). Follow these steps:

(a) Calculate E_b.

(b) Calculate $E_s(\vec{\rho}_s, z_s)$. For simplicity, neglect scanning and set $\vec{\rho}_s$ and z_s to zero.

(c) From the resulting $E(0, 0) = E_b + E_s(0, 0)$, derive the detected intensity $I(0, 0)$ and show that this depends on (real) $\delta\varepsilon$ to first order (neglect any higher-order dependence on $\delta\varepsilon$).

Problem 14.2

Consider a fluorescence confocal microscope equipped with a reflecting pinhole, that is a pinhole of radius a surrounded by a reflecting annulus of outer radius b and inner radius a (assume that the beam is blocked beyond the annulus). A transmission detector records the power Φ_T transmitted through the pinhole. A reflection detector records the power Φ_R reflected from the annulus. The confocal signal is then given by the difference of these recorded powers, namely $\Delta\Phi = \Phi_T - \Phi_R$.

(a) Calculate $\Delta\Phi(z_s)$ if the sample is a thin uniform fluorescent plane located at a defocus position z_s. For simplicity, assume that PSF= PSF$_i$ (and hence OTF= OTF$_i$). Express your result in terms of OTF and omit extraneous prefactors.

(b) Show that for a particular ratio b/a, the optical sectioning strength of this microscope is greater than that of a standard confocal microscope. In particular, show that $\Delta\Phi(z_s) \propto |z_s|^{-3}$ when $|z_s|$ is large, for a particular ratio b/a. What is this ratio?

Hint: to solve this problem recall that OTF$(\vec{\kappa}_\perp; z_s)$ scales as $|z_s|^{-3/2}$ when $\kappa_\perp \neq 0$ and $|z_s|$ is large.

Problem 14.3

Consider a fluorescence confocal microscope where the illumination and detection PSFs are the same and Gaussian, as defined by Eq. 5.34, and the pinhole is small.

(a) Imagine dithering the pinhole of this microscope in the transverse direction by small amounts $\pm\frac{1}{2}\delta\vec{\rho}_p$, and demodulating the acquired image at the dither frequency. This is essentially equivalent to acquiring two images I_+ and I_- with the pinhole located at $+\frac{1}{2}\delta\vec{\rho}_p$ and $-\frac{1}{2}\delta\vec{\rho}_p$, respectively, and then subtracting, obtaining a final image given by $\Delta I = I_+ - I_-$. Derive the effective confocal PSF, or ΔPSF$_{conf}(\vec{\rho}, z)$, of this instrument to first order in $\delta\vec{\rho}_p$. Express your answer in terms of the conventional (undithered) PSF, or PSF$_{conf}(\vec{\rho}, z)$. Discuss some features of ΔPSF$_{conf}(\vec{\rho}, z)$, such as its axial profile and its response to a laterally uniform sample.

(b) Now consider acquiring a conventional image with this instrument, and simply calculating the gradient of this image along the direction $\delta\vec{\rho}_p$ (or, more precisely, the difference image using the same transverse shift $\delta\vec{\rho}_p$). Your answer should be the same to within a scaling factor. What is this scaling factor?

Problem 14.4

Consider the same as Problem 14.3, except that now the pinhole is dithered by small amounts $\pm\delta z_p$ in the axial direction. Derive the effective confocal PSF, or $\Delta\mathrm{PSF}_{\mathrm{conf}}\left(\vec{\rho},z\right)$, of this instrument to first order in δz_p. Express your answer in terms of the conventional (undithered) $\mathrm{PSF}_{\mathrm{conf}}\left(\vec{\rho},z\right)$. Qualitatively describe what sample features this instrument is sensitive to. Show that $\Delta\mathrm{PSF}_{\mathrm{conf}}\left(0,z\right)$ decays more rapidly than $\mathrm{PSF}_{\mathrm{conf}}\left(\vec{\rho},z\right)$ by a factor of $|z|^{-1}$ for large $|z|$.

Problem 14.5

Consider a fluorescence confocal microscope where the illumination and detection PSFs are the same and Gaussian, as defined by Eq. 5.34. Let the pinhole also be a Gaussian, defined by $A_p\left(\vec{\rho}\right) = e^{-\rho^2/w_p^2}$. Derive an expression for the normalized axial profile of the confocal PSF, namely $\mathrm{PSF}_{\mathrm{conf}}\left(0,z\right)/\mathrm{PSF}_{\mathrm{conf}}\left(0,0\right)$. Qualitatively describe the z dependence of this profile. Plot this profile for $w_0 = \lambda = 1\ \mu\mathrm{m}$, and for the different ratios $w_p/w_0 \approx 0, 1$, and 2.

References

[1] Corle, T. R. and Kino, G. S. *Confocal Scanning Optical Microscopy and Related Systems*, Academic Press (1996).

[2] Diaspro, A. Ed., *Confocal and Two-Photon Microscopy*, Wiley-Liss (2002).

[3] Dodt, H.-U., Leischner, U., Schierloh, A., Jährling, N., Mauch, C. P., Deininger, K., Deussing, J. M., Eder, M., Zieglgänsberger, W. and Becker, K. "Ultramicroscopy: three-dimensional visualization of neuronal networks in the whole mouse brain," *Nat. Meth.* 4, 331–336 (2007).

[4] Dunsby, C. "Optically sectioned imaging by oblique plane microscopy," *Opt. Express* 16, 20306–20316 (2008).

[5] Gu, M. *Principles of Three-Dimensional Imaging in Confocal Microscopes*, World Scientific (1996).

[6] von Helmholtz, H. *Handbuch der physiologischen Optik*, 1st edn, Leopold Voss (1856).

[7] Huisken, J., Swoger, J., Del Bene, F., Wittbrodt, J. and Stelzer, E. H. K. "Optical sectioning deep inside live embryos by selective plane illumination microscopy," *Science* 305, 1007–1009 (2004).

[8] Lindek S. and Stelzer, E. H. K. "Optical transfer functions for confocal theta fluorescence microscopy," *J. Opt. Soc. Am. A* 13, 479–482 (1996).

[9] Mertz, J. "Optical sectioning microscopy with planar or structured illumination," *Nat. Meth.* 8, 811–819 (2011).

[10] Minsky, M. *Microscopy Apparatus*, U.S. Patent 3,013,467, Dec. 19, 1961 (filed Nov. 7, 1957).

[11] Müller, C. B. and Enderlein, J. "Image scanning microscopy," *Phys. Rev. Lett.* 104, 198101 (2010).

[12] Nakamura, O. and Kawata, S. "Three dimensional transfer-function analysis of the tomographic capability of a confocal fluorescence microscope," *J. Opt. Soc. Am. A* 7, 522–526 (1990).

[13] Pawley, J. B. Ed., *Handbook of Biological Confocal Microscopy*, 3rd edn, Springer (2006).

[14] Petráň, M., Hadravský, M., Egger, M. D. and Galambos, R. "Tandem-scanning reflected-light microscope," *J. Opt. Soc. Am.* 58, 661–664 (1968).

[15] Reynaud, E. G. Kržič, U., Greger, K. and Stelzer, E. H. K. "Light sheet-based fluorescence microscopy: more dimensions, more photons, and less photodamage," *HFSP J.* 2, 266–275 (2008).

[16] Sentenac, A. and Mertz, J. "A unified description of three-dimensional optical diffraction microscopy: from transmission microscopy to optical coherence tomography," *J. Opt. Soc. Am.* A 35, 748–754 (2018).

[17] Sheppard, C. J. R. "The spatial frequency cutoff in three dimensional imaging," *Optik* 72, 131–133 (1986).

[18] Sheppard, C. J. R. "The spatial frequency cutoff in three dimensional imaging II," *Optik* 74, 128–129 (1986).

[19] Sheppard, C. J. R. "Super-resolution confocal microscopy," *Optik* 80, 53–54 (1988).

[20] Welford, W. T. "On the relationship between the modes of image formation in scanning microscopy and conventional microscopy," *J. Microsc.* 96, 105–07 (1972).

[21] Wilson, T. and Sheppard, C. J. R. *Theory and Practice of Scanning Optical Microscopy*, Academic Press (1984).

[22] Wilson, T. "Resolution and optical sectioning in the confocal microscope," *J. Microsc.* 244, 113–121 (2011).

15 Structured Illumination Microscopy

As we saw in the previous chapter, the main advantage of confocal microscopy is that it provides optical sectioning, which it achieves by physically rejecting out-of-focus background with a pinhole, or by not generating background in the first place with the use of crossed illumination and detection PSFs. In either case, confocal images are background-free because the detector is not subjected to background illumination.

In this chapter, we will examine alternative strategies to remove out-of-focus background that do not require the use of a pinhole and can be implemented even when the illumination and detection PSFs are collinear. These strategies fall under the general rubric of structured illumination microscopy, or SIM. In these strategies, the rejection of out-of-focus background is performed numerically, post-detection, rather than physically. Though SIM can be applied to non-fluorescent samples, we will consider it here only in the context of fluorescent samples.

The idea of exploiting structured illumination to obtain optical sectioning can be attributed to Lanni [1], who made use of axial fringes formed by the interference of counter-propagating laser beams. In more recent years, however, largely propelled by the seminal work of Wilson *et al.* [9, 10], implementations of SIM have been based instead on laterally structured illumination. As we will see in Chapter 19, this work was later expanded upon to circumvent the resolution limit of conventional widefield microscopes. But we are getting ahead of ourselves. In this chapter, we will consider SIM only as it applies to optical sectioning and out-of-focus background rejection.

15.1 PRINCIPLES

As the name suggests, SIM is based on the principle of illuminating a sample not with uniform light but rather with light that exhibits some sort of spatial structure. Aside from this simple principle, however, there is little difference between SIM and conventional widefield microscopy. That is, the detection optics in SIM consists of standard imaging lenses and a camera, and we can write

$$I^{(k)}(\vec{\rho}) = \sigma_f \iint d^2\vec{\rho}_0 \, dz_0 \, \text{PSF}(\vec{\rho} - \vec{\rho}_0, z_0) C(\vec{\rho}_0, z_0) I_i^{(k)}(\vec{\rho}_0, z_0) \tag{15.1}$$

where $\mathrm{PSF}(\vec{\rho}, -z)$ is the conventional widefield detection PSF (assumed to be in an epifluorescence configuration – thus the change of sign in z), σ_f and $C(\vec{r}_0)$ denote, respectively, the cross-section and concentration distribution of the fluorescent molecules, and $I_i^{(k)}(\vec{r}_0)$ is the illumination structure.

But SIM is based on a few other principles. In particular, SIM requires the acquisition of a sequence of images, each with different illumination structures (hence the superscripts k in Eq. 15.1). Moreover, even though each illumination structure in the sequence is non-uniform, their summation is assumed to be uniform. That is

$$\bar{I}_i = \frac{1}{K}\sum_{k=0}^{K-1} I_i^{(k)}(\vec{r}_0) \qquad (15.2)$$

where K is at least two (though typically it is more). In other words, each sample point is subjected to the same illumination intensity *on average*. If we perform an average of the resultant sequence of K images we arrive at

$$\bar{I}(\vec{\rho}) = \frac{1}{K}\sum_{k=0}^{K-1} I^{(k)}(\vec{\rho}) = \sigma_f \iint \mathrm{d}^2\vec{\rho}_0\,\mathrm{d}z_0\,\mathrm{PSF}(\vec{\rho} - \vec{\rho}_0, z_0)C(\vec{\rho}_0, z_0)\bar{I}_i \qquad (15.3)$$

corresponding to a standard widefield epifluorescence image with a uniform illumination $\bar{I}_i$. Manifestly, simple averaging has not brought us any closer to our goal of optical sectioning, and we must be more clever in how we numerically process our images.

Several versions of SIM have emerged over the years, differing mainly in the types of illumination structures they exploit. We will examine the most common of these below.

15.2 STRUCTURED ILLUMINATION WITH FRINGES

The most common illumination structures used in SIM are periodic patterns along one axis, such as fringe or grating patterns [6]. Different strategies can be used to generate such structures, but for now we assume these to be simple cosine patterns along the x axis, defined by a transverse spatial frequency q_x and a modulation depth $M(q_x; z_0)$. That is, we write

$$I_i^{(k)}(x_0, y_0, z_0) = \bar{I}_i\left[1 + M(q_x; z_0)\cos(2\pi q_x x_0 + \phi_k)\right] \qquad (15.4)$$

For generality, we have made allowances for a modulation depth that is dependent both on transverse frequency and defocus position z_0. We have also introduced an arbitrary phase ϕ_k. One of the key premises of fringe-based SIM is that this phase is user-defined and controllable, and it is precisely this phase that is varied between image acquisitions.

In particular, let us adopt our standard phase-stepping protocol where ϕ_k is varied according to the circular phase sequence $\phi_k = \frac{2\pi k}{K}$ (see Section 11.4.2). We readily observe that the sequence of illumination structures indeed obeys the condition prescribed by Eq. 15.2, meaning that an average of the resultant images captured by the camera yields a conventional image obtained with uniform illumination.

Instead of an average prescribed by Eq. 15.3, let us adopt the processing algorithm given by

$$\widetilde{I}(\vec{\rho}) = \frac{1}{K}\sum_{k=0}^{K-1} e^{i\phi_k} I^{(k)}(\vec{\rho}) \tag{15.5}$$

where each $I^{(k)}(\vec{\rho})$ is numerically multiplied by the factor $e^{i\phi_k}$ before performing the average. We have seen this algorithm many times before (Sections 11.4.2, 12.2.1, and 13.3.2), and at the risk of repeating ourselves we note that even though the measured intensities $I^{(k)}(\vec{\rho})$ are of course positive and real, the numerically synthesized "intensity" $\widetilde{I}(\vec{\rho})$ is complex, in general.

Making use of Eqs. 15.1, 15.4, and 15.5, we arrive at finally

$$\widetilde{I}(\vec{\rho}) = \sigma_f \iint \mathrm{d}^2\vec{\rho}_0\, \mathrm{d}z_0\, \mathrm{PSF}(\vec{\rho}-\vec{\rho}_0,z_0)C(\vec{\rho}_0,z_0)\widetilde{I}_i(\vec{\rho}_0,z_0) \tag{15.6}$$

where

$$\widetilde{I}_i(\vec{\rho}_0,z_0) = \frac{1}{K}\sum_{k=0}^{K-1} e^{i\phi_k} I_i^{(k)}(\vec{\rho}_0,z_0) \tag{15.7}$$

Equation 15.6 is remarkable. It suggests that the SIM procedure of acquiring K raw images and processing these with Eq. 15.5 is equivalent to acquiring a single image where the sample is illuminated by an effective complex illumination distribution $\widetilde{I}_i(\vec{\rho}_0,z_0)$. Indeed, Eqs. 15.7 and 15.4 readily simplify to

$$\widetilde{I}_i(\vec{\rho}_0,z_0) = \tfrac{1}{2}\bar{I}_i M(q_x;z_0)e^{-i2\pi q_x x_0} \tag{15.8}$$

which is manifestly complex.

But this interpretation of a complex illumination intensity, however attractive mathematically, leads to a practical difficulty in how to interpret the resultant complex image $\widetilde{I}(\vec{\rho})$. Ultimately, we would like to retrieve the concentration distribution of fluorescent molecules, particularly those that reside in focus, namely $C(\vec{\rho}_0,0)$. And certainly this distribution is real-valued, not complex. The question of how to retrieve $C(\vec{\rho}_0,0)$ from $\widetilde{I}(\vec{\rho})$ thus remains unresolved. Different strategies can be invoked here (and will be invoked later in this book). We begin by considering the simplest, which is to apply an absolute value, and write

$$I_{\mathrm{SIM}}(\vec{\rho}) = \left|\widetilde{I}(\vec{\rho})\right| \tag{15.9}$$

We must be careful, though. By taking an absolute value, we are performing a nonlinear operation, meaning that SIM imaging, as defined by Eq. 15.9, also becomes nonlinear. More precisely, $I_{\mathrm{SIM}}(\vec{\rho})$ can no longer be expressed as the convolution of a sample distribution $C(\vec{\rho}_0,z_0)$ with an effective SIM point spread function. In turn, this implies that the spatial resolution of SIM partially depends on sample structure. Nevertheless, Eq. 15.9 has the undeniable advantage of being simple and straightforward. Moreover, as we will see below, it provides a quite reasonable approximation to linear imaging.

15.2.1 Optical Sectioning

To evaluate the capacity of SIM to reject out-of-focus background, let us consider scanning a thin 2D sample through the focal plane. That is, we write $C(\vec{\rho}_0, z_0) = C_\rho(\vec{\rho}_0)\delta(z_0 - z_s)$. Equation 15.6 then becomes

$$\widetilde{I}(\vec{\rho}) = \tfrac{1}{2}\sigma_f \bar{I}_i M(q_x; z_s) \int d^2\vec{\rho}_0 \, \mathrm{PSF}(\vec{\rho} - \vec{\rho}_0, z_s)e^{-i2\pi q_x x_0} C_\rho(\vec{\rho}_0) \qquad (15.10)$$

We concentrate here on two cases, one where the sample is in focus ($z_s = 0$), and another where it is far out of focus.

In-Focus Signal

For the in-focus case, we have

$$\widetilde{I}(\vec{\rho}) = \tfrac{1}{2}\sigma_f \bar{I}_i M(q_x; 0) \int d^2\vec{\rho}_0 \, \mathrm{PSF}(\vec{\rho} - \vec{\rho}_0, 0)e^{-i2\pi q_x x_0} C_\rho(\vec{\rho}_0) \qquad (15.11)$$

If we assume that the in-focus detection PSF is tightly peaked over spatial dimensions that are small compared to the variations in $e^{-i2\pi q_x x_0}$ and $C_\rho(\vec{\rho}_0)$, we can make the rough approximation $\mathrm{PSF}(\vec{\rho} - \vec{\rho}_0, 0) \approx \delta^2(\vec{\rho} - \vec{\rho}_0)$ which, when substituted into Eq. 15.11, leads to

$$\widetilde{I}(\vec{\rho}) \approx \tfrac{1}{2}\sigma_f \bar{I}_i M(q_x; 0)e^{-i2\pi q_x x} C_\rho(\vec{\rho}) \qquad (15.12)$$

Once again, $\widetilde{I}(\vec{\rho})$ is complex because of the exponential term $e^{-i2\pi q_x x}$. However, this term disappears upon application of an absolute value to $\widetilde{I}(\vec{\rho})$, obtaining finally $I_{\mathrm{SIM}}(\vec{\rho}) \propto C_\rho(\vec{\rho})$. In other words, provided the imaging optics exhibits a high enough resolution compared to the fringe pattern and to the sample variations, we conclude that $I_{\mathrm{SIM}}(\vec{\rho})$ provides a reasonably faithful rendition of an in-focus sample. This rendition is directly proportional to the sample concentration.

Out-of-Focus Background

We turn now to the second case where the sample is far out of focus. To simplify this problem, let us further assume that the sample is a uniform plane. Such a sample represents an important test case since, as we have seen in previous chapters, for any imaging system to provide optical sectioning it must, by definition, be able to axially resolve a uniform plane. Substituting $C_\rho(\vec{\rho}_0) \to C_\rho$ into Eq. 15.10, we obtain

$$\widetilde{I}(\vec{\rho}) = \tfrac{1}{2}\sigma_f \bar{I}_i M(q_x; z_s)C_\rho \int d^2\vec{\rho}_0 \, \mathrm{PSF}(\vec{\rho} - \vec{\rho}_0, z_s)e^{-i2\pi q_x x_0} \qquad (15.13)$$

From Eq. 5.35, this may be recast in terms of the detection OTF, obtaining

$$\widetilde{I}(\vec{\rho}) = \tfrac{1}{2}\sigma_f \bar{I}_i M(q_x; z_s)C_\rho \mathrm{OTF}(-q_x, 0; z_s)e^{-i2\pi q_x x} \qquad (15.14)$$

Finally, upon taking the absolute value of $\widetilde{I}(\vec{\rho})$ we arrive at

$$I_{\mathrm{SIM}}(\vec{\rho}) = \tfrac{1}{2}\sigma_f \bar{I}_i C_\rho \, |M(q_x; z_s)| \, |\mathrm{OTF}(-q_x, 0; z_s)| \qquad (15.15)$$

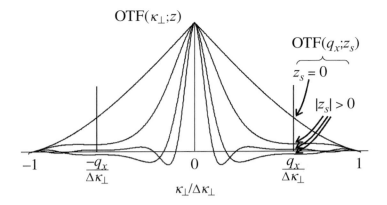

Figure 15.1. Incoherent modulation depth $OTF(q_x; z_0)$ depends on fringe frequency q_x and on defocus z_0.

As expected, $I_{SIM}(\vec{\rho})$ is independent of $\vec{\rho}$ since the sample is a uniform plane. On the other hand, $I_{SIM}(\vec{\rho})$ depends on z_s through the functions $M(q_x; z_s)$ and $OTF(-q_x, 0; z_s)$. If either of these functions decays to zero with increasing $|z_s|$, then $I_{SIM}(\vec{\rho})$ also decays to zero, and the planar sample becomes axially resolved. In other words, we have achieved our goal of optical sectioning. But we already know that $OTF(-q_x, 0; z_s)$ decays to zero from Section 5.2.4. In particular, assuming the detection pupil of the microscope to be circular, the Stokseth approximation tells us that $OTF(-q_x, 0; z_s)$ decays as $|z_s|^{-3/2}$ (see Fig. 15.1). It remains now to evaluate whether $M(q_x; z_s)$ also decays with defocus, in which case the modulation contrast provides additional sectioning capacity. As we will see below, this evaluation depends on the nature of the illumination itself.

15.2.2 Coherent Versus Incoherent Illumination

To obtain an illumination distribution in the sample given by Eq. 15.4 several options are available. We consider two such options where the illumination source producing the structured illumination is spatially coherent or incoherent.

Coherent Illumination

Perhaps the simplest way to generate a cosinusoidal fringe pattern is with two oblique plane waves. These can be obtained by sending a laser beam of amplitude $E_{\mathcal{L}}$ though a cosinusoidal phase grating, as described in Section 9.1.2. When the central non-diffracted beam is blocked, as illustrated in Fig. 15.2, the resultant radiant field emerging from the grating is given by

$$\mathcal{E}_{\mathcal{L}}^{(k)}(\kappa_x, \kappa_y) = \tfrac{1}{2}E_{\mathcal{L}}\left[e^{i\phi_k/2}\delta(\kappa_x - \tfrac{1}{2}q_x) + e^{-i\phi_k/2}\delta(\kappa_x + \tfrac{1}{2}q_x)\right]\delta(\kappa_y) \tag{15.16}$$

where $\tfrac{1}{2}q_x$ and $\tfrac{1}{2}\phi_k$ denote the grating frequency and phase, respectively. The latter can be adjusted by simply displacing the grating in the transverse direction.

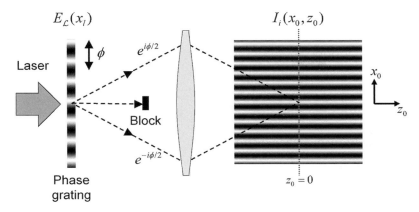

Figure 15.2. Projection of a coherent source fringe pattern into the sample.

This radiant field, when projected (imaged) into the sample, becomes

$$\mathcal{E}_i^{(k)}(\kappa_x, \kappa_y; z_0) = \mathcal{H}_i(\kappa_x, \kappa_y; z_0)\mathcal{E}_\mathcal{L}^{(k)}(\kappa_x, \kappa_y) \tag{15.17}$$

where $\mathcal{H}_i(\kappa_x, \kappa_y; z_0)$ is the amplitude point spread function associated with the illumination optics, assumed to be unit magnification for simplicity. The corresponding illumination field in the sample is then

$$E_i^{(k)}(x_0, y_0, z_0) = \tfrac{1}{2}E_\mathcal{L}\left[\mathcal{H}_i(\tfrac{1}{2}q_x, 0; z_0)e^{i\pi q_x x_0 + i\phi_k/2} + \mathcal{H}_i(-\tfrac{1}{2}q_x, 0; z_0)e^{-i\pi q_x x_0 - i\phi_k/2}\right] \tag{15.18}$$

Making use of Eq. 5.13 for $\mathcal{H}_i$, and provided the grating frequency $\tfrac{1}{2}q_x$ falls within the bandwidth of $\mathcal{H}_i$, we arrive at finally

$$I_i^{(k)}(\vec{\rho}_0, z_0) = \left|E_i^{(k)}(\vec{\rho}_0, z_0)\right|^2 = \tfrac{1}{2}|E_\mathcal{L}|^2\left[1 + \cos(2\pi q_x x_0 + \phi_k)\right] \tag{15.19}$$

In other words, we have arrived at an illumination intensity $I_i^{(k)}(\vec{\rho}_0, z_0)$ that has the same structure as Eq. 15.4, provided we make the associations $\bar{I}_i = \tfrac{1}{2}|E_\mathcal{L}|^2$ and $M(q_x; z_0) = 1$. This last association indicates that the contrast of the cosinusoidal illumination pattern remains unity independent of the defocus position z_0 (see Fig. 15.2). We conclude from Eq. 15.15 that the optical sectioning capacity of fringe-based SIM is governed by $\text{OTF}(-q_x, 0; z_s)$ alone when using coherent illumination. That is, $I_{\text{SIM}}(\vec{\rho})$ decays with defocus distance as $|z_s|^{-3/2}$ for a uniform planar sample. A different conclusion will be reached when using incoherent illumination.

Incoherent Illumination

To obtain a fringe-like pattern with an incoherent source, one can use the configuration shown in Fig. 15.3. Here, a spatially incoherent beam with uniform intensity distribution is transmitted through an amplitude grating (as opposed to a phase grating above). The amplitude grating is

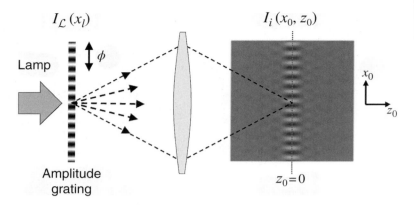

Figure 15.3. Projection of an incoherent source fringe pattern into the sample.

assumed to exhibit unity contrast, meaning the intensity distribution emerging from the grating is given by

$$I_{\mathcal{L}}^{(k)}(x_l, y_l) = I_{\mathcal{L}} \left[1 + \cos(2\pi q_x x_l + \phi_k) \right] \tag{15.20}$$

where q_x and ϕ_k denote the grating frequency and phase respectively, and, again, the latter can be adjusted by simply displacing the grating in the transverse direction.

Since the source $I_{\mathcal{L}}^{(k)}(x_l, y_l)$ is incoherent, we need not consider fields but can work directly with intensities. Converting, as before, to the frequency domain, the intensity spectrum of $I_{\mathcal{L}}^{(k)}(x_l, y_l)$ defined by Eq. 15.20 becomes

$$\mathcal{I}_{\mathcal{L}}^{(k)}(\kappa_x, \kappa_y) = I_{\mathcal{L}} \delta(\kappa_x) \delta(\kappa_y) + \tfrac{1}{2} I_{\mathcal{L}} \left[e^{i\phi_k} \delta(\kappa_x - q_x) + e^{-i\phi_k} \delta(\kappa_x + q_x) \right] \delta(\kappa_y) \tag{15.21}$$

Upon projecting (imaging) this intensity spectrum into the sample, we obtain

$$\mathcal{I}^{(k)}(\kappa_x, \kappa_y; z_0) = \mathrm{OTF}_i(\kappa_x, \kappa_y; z_0) \mathcal{I}_{\mathcal{L}}^{(k)}(\kappa_x, \kappa_y) \tag{15.22}$$

where $\mathrm{OTF}_i(\kappa_x, \kappa_y; z_0)$ is the optical transfer function associated with the illumination optics. Converting back to the spatial domain, the modulation pattern inside the sample is found to be

$$I_i^{(k)}(x_0, y_0, z_0) = \Omega_i I_{\mathcal{L}} + \tfrac{1}{2} I_{\mathcal{L}} \left[\mathrm{OTF}_i(q_x, 0; z_0) e^{i2\pi q_x x_0 + i\phi_k} + \mathrm{OTF}_i(-q_x, 0; z_0) e^{-i2\pi q_x x_0 - i\phi_k} \right] \tag{15.23}$$

where $\Omega_i = \mathrm{OTF}_i(\vec{0}; 0)$ is a constant (see Section 5.2.4).

In the case where the illumination pupil is circular, OTF_i is real and symmetric, and Eq. 15.23 reduces to

$$I_i^{(k)}(\vec{\rho}_0, z_0) = \Omega_i I_{\mathcal{L}} \left[1 + \Omega_i^{-1} \mathrm{OTF}_i(q_x, 0; z_0) \cos(2\pi q_x x_0 + \phi_k) \right] \tag{15.24}$$

We observe that $I_i^{(k)}(\vec{\rho}_0, z_0)$ again has the same structure given by Eq. 15.4, provided we make the new associations $\bar{I}_i = \Omega_i I_{\mathcal{L}}$ and $M(q_x; z_0) = \Omega_i^{-1} \mathrm{OTF}_i(q_x, 0; z_0)$.

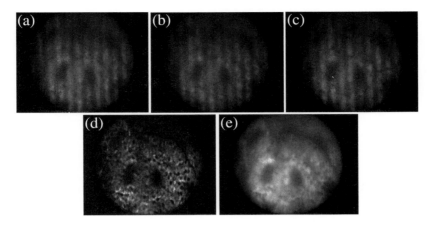

Figure 15.4. Images of fluorescently labelled rat colon mucosa acquired with SIM through a flexible fiber-optic bundle. Raw images (a–c), processed SIM image (d), and conventional image (e). Field of view diameter is 240 μm. Adapted from [2].

A comparison of Eqs. 15.24 and 15.19 reveals a fundamental difference between coherent and incoherent illumination, which is apparent in Figs. 15.2 and 15.3. In both cases, the projected patterns inside the sample are cosinusoidal modulations of equal frequency and phase. However, while the modulation contrast is independent of q_x and z_0 in the case of coherent illumination (since it is equal to 1), it is highly dependent on q_x and z_0 in the case of incoherent illumination (since it is equal to $\Omega_i^{-1}\mathrm{OTF}_i(q_x, 0; z_0)$).

In particular, we observe from Eq. 15.15 that in the case of incoherent illumination the net decrease in $I_{\mathrm{SIM}}(\vec{\rho})$ decays as the product $|\mathrm{OTF}_i(q_x, 0; z_s)|\,|\mathrm{OTF}(-q_x, 0; z_s)|$. That is, it decays as $|z_s|^{-3}$.

Both of these results with coherent and incoherent illumination are notable. Recalling that the corresponding sectioning strength for a confocal fluorescence microscope scales as z_s^{-2}, we conclude that fringe-based SIM confers optical sectioning that is almost as strong as confocal in the case of coherent illumination, and even stronger in the case of incoherent illumination. Note, however, this does not necessarily mean that incoherent is preferable to coherent illumination. Incoherent illumination provides weaker in-focus illumination contrast, particularly for large values of q_x. Inasmuch as in-focus illumination contrast is an essential requirement for SIM, the question of which illumination strategy is advantageous depends largely on the sample in question, and, in particular, on the ratio of in-focus signal power to out-of-focus background power.

15.3   STRUCTURED ILLUMINATION WITH SPECKLE

Thus far, we have concentrated on fringe-pattern illumination, coherent or incoherent, but SIM is far more general than this. As an example, let us turn to an altogether different type of

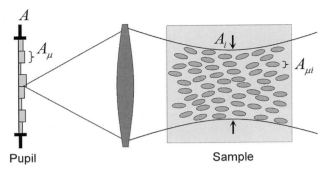

Figure 15.5. Schematic of speckle illumination where the pupil is comprised of multiple sub-pupils conferring random phase shifts.

structured illumination: speckle. As we have seen in Section 7.2.3, speckle arises from the superposition of multiple waves incident from various directions with random phases. In the case of Köhler illumination, each of these waves can be thought of as having propagated through its own individual sub-pupil, as illustrated in Fig. 15.5, where the relative phases of the sub-pupils are random, and the collection of all sub-pupils can be thought of as comprising the total illumination pupil. From the outset, it should be emphasized that speckle is generally regarded to be a bane in microscopy applications since it tends to corrupt images with highly granular artifactual noise. Though the strategies we will consider below are often more pedagogical than practical, the very idea that speckle can be used in a constructive manner for imaging applications, particularly given the ubiquitous nature of speckle, is worth investigating.

We begin with a review of some basic properties of speckle. Referring to Fig. 15.5 (or Fig. 6.1), the area of each sub-pupil is defined by a coherence area A_μ of the illumination beam within the full pupil, itself of area A. Correspondingly, at the sample, the illumination coherence area $A_{\mu i}$ is defined by the diffraction-limited spot size associated with A (see Eqs. 6.60 and 6.62), while the total illumination area A_i is defined by the diffraction-limited spot size associated with A_μ (see Eqs. 6.60 and 6.61), as illustrated in Fig. 15.5. That is, $A_{\mu i}$ is the characteristic area of a speckle-grain and A_i is the area of the field of illumination.

More rigorously, fully-developed speckle obeys well-defined circular Gaussian statistics [5]. Denoting a sequence of uncorrelated fully-developed speckle patterns by $I_i^{(k)}(\vec{r}_0)$, we thus have

$$\left\langle I_i^{(k)}(\vec{r}_0) \right\rangle_k = \bar{I}_i \tag{15.25}$$

$$\left\langle I_i^{(k)}(\vec{r}_0) I_i^{(k)}(\vec{r}_0') \right\rangle_k = \bar{I}_i^2 \left[1 + \left| \mu_i(\vec{r}_0 - \vec{r}_0') \right|^2 \right] \tag{15.26}$$

where $\langle ... \rangle_k$ indicates an average over the sequence. The first equation indicates that, on average, the illumination intensity of speckle is spatially uniform and given by $\bar{I}_i$ (provided k is large – more on this later). In effect, this is tantamount to the SIM condition given in Eq. 15.2. The second equation, in turn, is an re-statement of the spatial Siegert relation, which we saw in Section 7.2 (specifically Eq. 7.32).

In the case of Köhler illumination, which we assume here, we have from Section 4.3

$$\mu_i(\vec{r}) = \frac{H_i(\vec{r})}{H_i(\vec{0})} \tag{15.27}$$

and Eq. 15.26 can be recast as

$$\left\langle I_i^{(k)}(\vec{r}_0) I_i^{(k)}(\vec{r}_0') \right\rangle_k = \bar{I}_i^2 \left[1 + \frac{\mathrm{PSF}_i(\vec{r}_0 - \vec{r}_0')}{\mathrm{PSF}_i(\vec{0})} \right] \tag{15.28}$$

Equations 15.25 and 15.28 will serve as the basis of our analysis of SIM with speckle illumination. By now, it should be clear that there are key differences between fringe and speckle illumination. A fringe pattern is, by definition, well defined and periodic, whereas a speckle pattern is quite random. Moreover, as we saw above, fringe-based SIM requires a sequence of fringe patterns that are displaced in a well-defined step-by-step manner, presumably using circular phase shifts. In contrast, the sequence of speckle patterns here is obtained by a wholesale randomization of these patterns, thus ensuring that the speckle patterns are uncorrelated with one another.

15.3.1 Decoding with Known Patterns

A possible strategy to palliate some of the difficulties associated with random illumination patterns is to measure these patterns, somehow, prior to launching them into the sample of interest. Let us imagine that such measurements are possible, meaning that the sequence of speckle patterns used to illuminate the sample is actually known a priori. In this case, we can consider the following strategy for pupil synthesis, described in [3, 15]:

As before, SIM involves the acquisition of a series of K fluorescence images $(I^{(k)}(\vec{\rho}))$ resulting from the illumination of the sample with K illumination patterns $(I_i^{(k)}(\vec{r}_0))$. While in the previous section dealing with fringe-pattern illumination, K was small, typically 3 or 4, we will presume here that K is much larger. Moreover, our image synthesis strategy will be slightly different from that presented in Section 15.2. In particular, we will use the algorithm

$$\tilde{I}(\vec{\rho}) = \frac{1}{K} \sum_{k=0}^{K-1} Q^{(k)}(\vec{\rho}) I^{(k)}(\vec{\rho}) = \left\langle Q^{(k)}(\vec{\rho}) I^{(k)}(\vec{\rho}) \right\rangle_k \tag{15.29}$$

This is the same algorithm as given in Eq. 15.5 except that the phase factors $e^{i\phi_k}$ have been replaced by decoding masks, defined here to be

$$Q^{(k)}(\vec{\rho}) = \frac{I_i^{(k)}(\vec{\rho}, 0)}{\bar{I}_i} \tag{15.30}$$

where, again, we have made the assumption that the illumination patterns $I_i^{(k)}(\vec{\rho}_0, z_0)$ are known beforehand, at least at the focal plane $z_0 = 0$, meaning that the decoding masks $Q^{(k)}(\vec{\rho})$ are predefined.

Equation 15.29 can be rewritten as

$$\widetilde{I}(\vec{\rho}) = \sigma_f \iint d^2\vec{\rho}_0 \, dz_0 \, \text{PSF}(\vec{\rho} - \vec{\rho}_0, z_0) C(\vec{\rho}_0, z_0) \left\langle Q^{(k)}(\vec{\rho}) I_i^{(k)}(\vec{\rho}_0, z_0) \right\rangle_k \quad (15.31)$$

which may be compared with Eq. 15.6, the difference being that the term in brackets here depends on $\vec{\rho}$ whereas $\widetilde{I}_i(\vec{\rho}_0, z_0)$ in Eq. 15.6 does not. This seeming complication will lead to an advantage.

Finally, we define our speckle-based SIM image to be

$$I_{\text{SIM}}(\vec{\rho}) = \widetilde{I}(\vec{\rho}) - \bar{I}(\vec{\rho}) \quad (15.32)$$

where $\bar{I}(\vec{\rho})$ is the standard widefield image obtained with uniform illumination. This widefield image can be estimated from the image sequence average, as prescribed by 15.3, or alternatively it can be obtained by yet an additional image where the sample is illuminated by a uniform intensity $\bar{I}_i$.

An evaluation of Eq. 15.32 requires an evaluation of the term in brackets in Eq. 15.31. Making use of Eqs. 15.28 and 15.3, we arrive at

$$I_{\text{SIM}}(\vec{\rho}) = \frac{1}{\text{PSF}_i(\vec{0})} \sigma_f \bar{I}_i \iint d^2\vec{\rho}_0 \, dz_0 \, \text{PSF}(\vec{\rho} - \vec{\rho}_0, z_0) \text{PSF}_i(\vec{\rho} - \vec{\rho}_0, -z_0) C(\vec{\rho}_0, z_0) \quad (15.33)$$

where, for a circular pupil, $\text{PSF}_i(\vec{0})$ is given by $\kappa^2 \Omega_i^2$ (see Eq. 5.32).

In conclusion, SIM imaging with speckle illumination leads to identical results as confocal fluorescence microscopy with point illumination (see Section 14.3). That is, SIM imaging provides high-resolution optical sectioning, and moreover provides a linear, positive definite mapping between the object concentration and image intensity. However, these advantages should not belie the difficulty of the procedure. The SIM technique described here requires an a priori knowledge of each speckle illumination pattern. This last requirement, as the reader may well imagine, is quite difficult to realize in practice, particularly given that the speckle patterns must be known *inside* the sample of interest. In the next section, we will turn to a more robust strategy based on speckle illumination that, while sacrificing the benefits of linearity, fully dispenses with this last requirement that the speckle patterns be known.

15.3.2 Decoding with Unknown Patterns

Returning to our speckle illumination configuration depicted in Fig. 15.5, we examine an image processing strategy that requires a knowledge only of the speckle statistics. To understand this strategy, we return to the basic equation underlying any widefield imaging technique, namely Eq. 15.6. As in the previous strategy, a sequence of K fluorescence images $I^{(k)}(\vec{\rho})$ is acquired from K uncorrelated speckle illumination patterns $I_i^{(k)}(\vec{\rho}_0, z_0)$, K again being a large number.

In this new strategy, however, image processing is based on a calculation of the standard deviation of this image sequence, defined by

$$I_{\text{SIM}}(\vec{\rho}) = \sqrt{\text{Var}\left[I^{(k)}(\vec{\rho})\right]} = \sqrt{\left\langle I^{(k)}(\vec{\rho})^2\right\rangle_k - \left\langle I^{(k)}(\vec{\rho})\right\rangle_k^2} \tag{15.34}$$

where $\text{Var}[...]$ denotes variance.

The term $\left\langle I^{(k)}(\vec{\rho})\right\rangle_k$ in Eq. 15.34 is immediately recognized as an estimate of the standard widefield image, defined previously by Eq. 15.25. The term $\left\langle I^{(k)}(\vec{\rho})^2\right\rangle_k$, on the other hand, is new, and somewhat more involved. In effect, this term represents an application of decoding masks that are defined by the detected images themselves (compare with Eq. 15.29). Expanding this term out fully, we write

$$\left\langle I^{(k)}(\vec{\rho})^2\right\rangle_k = \sigma_f^2 \iiiint d^2\vec{\rho}_0\, d^2\vec{\rho}_0'\, dz_0\, dz_0'\, \text{PSF}(\vec{\rho} - \vec{\rho}_0, z_0)\text{PSF}(\vec{\rho} - \vec{\rho}_0', z_0')$$
$$\times C(\vec{\rho}_0, z_0)C(\vec{\rho}_0', z_0')\left\langle I_i^{(k)}(\vec{\rho}_0, z_0)I_i^{(k)}(\vec{\rho}_0', z_0')\right\rangle_k \tag{15.35}$$

From Eq. 15.28 and assuming a circular illumination pupil, we then obtain

$$\text{Var}\left[I^{(k)}(\vec{\rho})\right] = \frac{\sigma_f^2 \bar{I}_i^2}{\kappa^2 \Omega_i^2} \iiiint d^2\vec{\rho}_0\, d^2\vec{\rho}_0'\, dz_0\, dz_0'\, \text{PSF}(\vec{\rho} - \vec{\rho}_0, z_0)\text{PSF}(\vec{\rho} - \vec{\rho}_0', z_0')$$
$$\times C(\vec{\rho}_0, z_0)C(\vec{\rho}_0', z_0')\text{PSF}_i(\vec{\rho}_0 - \vec{\rho}_0', z_0 - z_0') \tag{15.36}$$

The interpretation of Eq. 15.36 is made clearer if we consider, as we did in Section 15.2.1, a thin sample defined by $C(\vec{\rho}_0, z_0) = C_\rho(\vec{\rho}_0)\delta(z_0 - z_s)$, and evaluate the resulting SIM image when the sample is in or out of focus.

When the sample is in focus, Eq. 15.36 reduces to

$$\text{Var}\left[I^{(k)}(\vec{\rho})\right] = \frac{\sigma_f^2 \bar{I}_i^2}{\kappa^2 \Omega_i^2} \iint d^2\vec{\rho}_0\, d^2\vec{\rho}_0'\, \text{PSF}(\vec{\rho} - \vec{\rho}_0, 0)\text{PSF}(\vec{\rho} - \vec{\rho}_0', 0)$$
$$\times C_\rho(\vec{\rho}_0)C_\rho(\vec{\rho}_0')\text{PSF}_i(\vec{\rho}_0 - \vec{\rho}_0', 0) \tag{15.37}$$

Let us assume, moreover, that the speckle grains are well resolved, meaning that the width of PSF is narrower than that of PSF_i. In this case, we can make the approximation

$$\text{Var}\left[I^{(k)}(\vec{\rho})\right] \approx \left(\sigma_f \bar{I}_i \int d^2\vec{\rho}_0\, \text{PSF}(\vec{\rho} - \vec{\rho}_0, 0)C_\rho(\vec{\rho}_0)\right)^2 \tag{15.38}$$

and, from Eq. 15.34, $I_{\text{SIM}}(\vec{\rho})$ corresponds to a standard widefield image.

For the situation when the sample is out of focus, we again adopt the simplification of a uniform sample, allowing us to write $C_\rho(\vec{\rho}_0) \to C_\rho$. In this case, Eq. 15.36 reduces to

$$\text{Var}\left[I^{(k)}(\vec{\rho})\right] = C_\rho^2 \frac{\sigma_f^2 \bar{I}_i^2}{\kappa^2 \Omega_i^2} \iint d^2\vec{\rho}_0 d^2\vec{\rho}_0'\, \text{PSF}(\vec{\rho} - \vec{\rho}_0, z_s)\text{PSF}(\vec{\rho} - \vec{\rho}_0', z_s)\text{PSF}_i(\vec{\rho}_0 - \vec{\rho}_0', 0)$$
$$\tag{15.39}$$

Now only PSF_i is in focus whereas both PSFs are out of focus. Effectively, this enables us to substitute $PSF_i(\vec{\rho}_0 - \vec{\rho}_0', 0) \rightarrow \delta^2(\vec{\rho}_0 - \vec{\rho}_0')$, obtaining

$$\text{Var}\left[I^{(k)}(\vec{\rho})\right] = C_\rho^2 \frac{\sigma_f^2 \bar{I}_i^2}{\kappa^2 \Omega_i^2} \int d^2\vec{\rho}_0 \, PSF^2(\vec{\rho} - \vec{\rho}_0, z_s) \tag{15.40}$$

A further simplification is occasioned by recalling that the cross-sectional area of $PSF(\vec{\rho}, z)$ can be defined as

$$A_0(z) = \frac{\left(\int d^2\vec{\rho} \, PSF(\vec{\rho}, z)\right)^2}{\int d^2\vec{\rho} \, PSF^2(\vec{\rho}, z)} = \frac{\Omega_0^2}{\int d^2\vec{\rho} \, PSF^2(\vec{\rho}, z)} \tag{15.41}$$

where we have made use of the PSF normalization given in Section 5.2.3.

In other words, Eq. 15.40 finally leads to

$$I_{\text{SIM}}(\vec{\rho}) \approx \sigma_f C_\rho \bar{I}_i \frac{\Omega_0}{\kappa \Omega_i \sqrt{A_0(z_s)}} \tag{15.42}$$

This is the out-of-focus SIM intensity produced by a uniform planar sample. Since the out-of-focus cross-sectional area of PSF obeys the scaling law $A_0(z_s) \propto z_s^2$, by power conservation, this out-of-focus SIM intensity obeys the scaling law $I_{\text{SIM}}(\vec{\rho}) \propto |z_s|^{-1}$. That is, $I_{\text{SIM}}(\vec{\rho})$ has the property that it is optically sectioned, though the sectioning strength is rather weak (compared with the z_s^{-2} sectioning strength of a confocal fluorescence microscope). Note that in both cases, whether in or out of focus, the final SIM image intensity is directly proportional to the sample concentration.

The imaging strategy described above is known as dynamic speckle illumination (DSI) microscopy [12]. The advantage of this strategy is that it is based only on Eq. 15.28 and does not require any a priori knowledge of the exact speckle illumination patterns. Since Eq. 15.28 is valid for any speckle patterns, provided they are fully developed, DSI microscopy is quite robust. However, it has the drawback that it requires several images to produce an accurate image. In particular, as we have seen in Section 7.3.1, the relative error in fulfilling Eq. 15.2 decreases only slowly as $1/\sqrt{K}$.

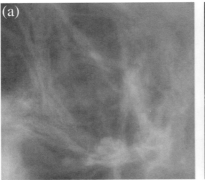 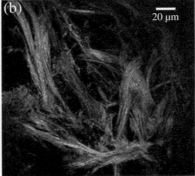

Figure 15.6. Fluorescently labelled axons in a mouse olfactory bulb imaged with conventional (a) and DSI (b) microscopy. Adapted from [13].

As a final note, we recall that speckle is a form of coherent illumination. As such, it might be expected that DSI microscopy should provide the same $|z_s|^{-3/2}$ optical sectioning strength as fringe-based SIM microscopy with coherent illumination. Indeed, this is the case with a little extra image prefiltering, as described in [13] (see Fig. 15.6).

15.4 HYBRID TECHNIQUES

A feature of SIM, at least of the variations we have considered so far, is that the microscope detection optics are exactly the same as those of a standard widefield fluorescence microscope. In particular, the rejection of out-of-focus background has so far been performed entirely numerically, post image acquisition. We will explore here some variations of SIM where the background rejection is not entirely numerical, but is aided by the presence of a physical mask in the detection optics. Confocal microscopy is, of course, an extreme example where background rejection is performed exclusively by a physical mask (namely a pinhole), and no additional numerical processing is required. Hybrid SIM approaches fall somewhere between these two extremes of no reliance on a physical mask and complete reliance on a physical mask. Some commonalities of these hybrid SIM approaches are that they still require numerical processing based on more than one acquired image. But they do not require beam scanning, at least not in the conventional sense of the term, as does, for example, a confocal microscope. In particular, they do not require beam focusing, and can be operated just as easily with incoherent illumination. Finally, the physical mask in these hybrid SIM techniques is shared in both the illumination and detection optics, as shown in Fig. 15.7, and this will be key to their ability to perform optical sectioning.

15.4.1 Aperture Correlation Microscopy

As shown in the previous sections, there are many ways to project an illumination structure into a sample. Let us return to one of the most straightforward, namely the projection of an amplitude mask into the sample using imaging optics and incoherent illumination, where the amplitude mask is located in a plane conjugate to the sample focal plane (see Fig. 15.7). Here we consider a sequence of masks denoted by $Q^{(k)}\left(\vec{\rho}\right)$. For simplicity, we also consider high-resolution, unit-magnification optics. That is, the in-focus illumination structures resulting from each mask can be approximated by

$$I_i^{(k)}(\vec{\rho}_0, z_0 = 0) \approx \bar{I}_i Q^{(k)}\left(\vec{\rho}_0\right) \tag{15.43}$$

Because the illumination is incoherent, the contrast of these structures fades with increasing defocus (as it does, for example in Fig. 15.3), meaning that the out-of-focus illumination is essentially uniform and given by

$$I_i^{(k)}(\vec{\rho}_0, |z_0| > 0) \approx \bar{I}_i \tag{15.44}$$

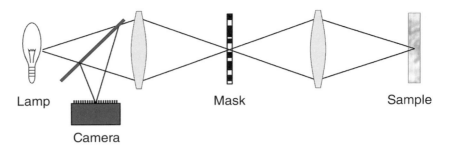

Figure 15.7. Aperture correlation microscopy with pseudo-random masks.

In turn, the fluorescence signal is imaged through the same mask pattern, and re-imaged onto the camera, leading to

$$I^{(k)}(\vec{\rho}) = \sigma_f Q^{(k)}(\vec{\rho}) \iint d^2\vec{\rho}_0 \, dz_0 \, \mathrm{PSF}(\vec{\rho} - \vec{\rho}_0, z_0) C(\vec{\rho}_0, z_0) I_i^{(k)}(\vec{\rho}_0, z_0) \qquad (15.45)$$

This may be compared with Eq. 15.1. The difference is that now the fluorescence image has been subjected to the mask pattern $Q^{(k)}(\vec{\rho})$. More importantly, there is an obvious correlation between $Q^{(k)}(\vec{\rho})$ and $I_i^{(k)}(\vec{\rho}_0, z_0)$, particularly at the focal plane where they are the same.

Finally, let us rapidly step through the sequence of K mask patterns, each time accumulating the resultant image $I^{(k)}(\vec{\rho})$ onto the camera sensor in a single camera exposure. The accumulated image is given by $I_Q(\vec{\rho}) = \sum_{k=0}^{K-1} I^{(k)}(\vec{\rho})$, or

$$I_Q(\vec{\rho}) = K\sigma_f \iint d^2\vec{\rho}_0 \, dz_0 \, \mathrm{PSF}(\vec{\rho} - \vec{\rho}_0, z_0) C(\vec{\rho}_0, z_0) \left\langle Q^{(k)}(\vec{\rho}) I_i^{(k)}(\vec{\rho}_0, z_0) \right\rangle_k \qquad (15.46)$$

This resembles 15.31, but now with some major differences. First, $Q^{(k)}(\vec{\rho})$ here is a physical mask rather than a numerical mask. Second, $I_Q(\vec{\rho})$ is a single measured image, rather than a numerical image synthesized from a sequence of multiple measured images. This leads to a significant boost in speed even if K is large (provided the mask patterns are shuffled rapidly).

But $I_Q(\vec{\rho})$ alone is not enough to provide optical sectioning. A difficulty is that $Q^{(k)}(\vec{\rho})$ corresponds to a physical intensity transmission function, and as such must be positive definite (as must $I_i^{(k)}(\vec{\rho}_0, 0)$). We can write then

$$Q^{(k)}(\vec{\rho}) = 1 + b^{(k)}(\vec{\rho}) \qquad (15.47)$$

where $b^{(k)}(\vec{\rho})$ is a function that is both positive or negative, such that

$$\left\langle b^{(k)}(\vec{\rho}) \right\rangle_k = 0 \qquad (15.48)$$

and $\left\langle Q^{(k)}(\vec{\rho}) \right\rangle_k$ is normalized to one.

Several choices can be made for the sequences $b^{(k)}(\vec{\rho})$. For example, $b^{(k)}(\vec{\rho})$ can be pseudo-random bi-modal functions equal to $+1$ or -1, chosen in such a way that the different sequences at different locations on the mask (or in the sample) are orthogonal. That is,

$$\left\langle b^{(k)}(\vec{\rho}) \, b^{(k)}(\vec{\rho}') \right\rangle_k = \delta_{\vec{\rho}, \vec{\rho}'} \qquad (15.49)$$

where $\delta_{\vec{\rho},\vec{\rho}'}$ is a Kronecker delta function. Classic examples of pseudo-random functions that satisfy both Eqs. 15.48 and 15.49 are Hadamard or Golay sequences [4]. Making use of Eqs. 15.43, 15.44, and 15.47, we arrive at

$$\left\langle Q^{(k)}(\vec{\rho})I_i^{(k)}(\vec{\rho}_0, z_0)\right\rangle_k \approx \bar{I}_i\left(1 + \delta_{\vec{\rho},\vec{\rho}_0}\delta_{0,z_0}\right) \tag{15.50}$$

Equation 15.50 bears striking resemblance to Eq. 15.28. We thus conclude that, in the same manner as Eq. 15.31 led to Eq. 15.33, Eq. 15.46 leads to optical confocal-like optical sectioning, provided we subtract from it a conventional widefield image as we did in Eq. 15.32. In other words, our final optically-sectioned SIM image is given here by

$$I_{\text{SIM}}(\vec{\rho}) = I_Q\left(\vec{\rho}\right) - K\bar{I}(\vec{\rho}) \tag{15.51}$$

Such a SIM technique relying on two images, one acquired with pseudo-random masks and another conventionally, is called aperture correlation microscopy [16].

15.4.2 Programmable Array Microscopy

The advantages of aperture correlation microscopy are speed, since only two images are required, and computational simplicity (as simple as Eq. 15.51). However, even the requirement of two images can be unattractive when these are acquired at different times. For example, the sample in question can be dynamic, changing on time scales faster than the camera frame rate, or the illumination conditions may be changing slightly with time, meaning that the precise balance of illumination intensities required for the cancellation of out-of-focus background in Eq. 15.51 becomes slightly compromised. A clever solution to this problem involves recording the two images simultaneously rather than sequentially, as shown in Fig. 15.8. In particular, imagine that a second camera is arranged such that it can record the fluorescence light not transmitted through the masks $Q^{(k)}\left(\vec{\rho}\right)$ but instead reflected from these masks. In this manner, the second camera records an image produced by the complementary mask sequence defined by

$$\bar{Q}^{(k)}(\vec{\rho}) = 1 - b^{(k)}\left(\vec{\rho}\right) \tag{15.52}$$

A resulting SIM image is then obtained from the modified algorithm

$$I_{\text{SIM}}(\vec{\rho}) = I_Q\left(\vec{\rho}\right) - I_{\bar{Q}}(\vec{\rho}) \tag{15.53}$$

It is straightforward to show that the in-focus contribution in this resulting SIM image is twice as strong as in the single-camera approach. But more importantly, the second image can be recorded at the same time as the first, allaying some of the difficulties associated with sequential image acquisition. This two-camera SIM approach is referred to as programmable array microscopy [14]. An even simpler approach comes from realizing that the orthogonality condition prescribed by Eq. 15.49 need not be enforced globally across the entire field of view, but rather only locally over regions somewhat larger than the resolution area of the microscope. As a result, reflecting masks as simple as a moving grid pattern are sufficient to provide optical sectioning. The grid pattern can be swept across the field of view by mounting it on a spinning

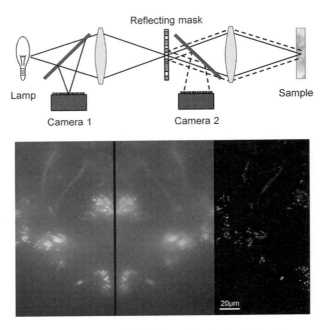

Figure 15.8. Fixed mold sample acquired with differential spinning disk microscopy. From left to right: Camera 1, Camera 2 (mirror image), and optically sectioned DSD image. Courtesy of Rimas Juskaitis, Aurox Ltd. © 2018.

disk (thus abiding by Eq. 15.2), a technique known as differential spinning disk microscopy ([11], see also review in [8]).

15.5 PUPIL SYNTHESIS

In general, SIM is founded on the basic principle that an optically-sectioned image can be calculated from a set of non-optically-sectioned images. This principle can be re-interpreted under the more basic framework of pupil synthesis, as introduced by Lohmann and Rhodes [7]. In this framework, a microscope PSF is no longer constrained to be positive definite, but instead can be effectively manipulated to be positive or negative, or even complex-valued, by manipulating the pupil functions associated with a set of real-valued PSFs and applying a little bit of math. For example, in the case of fringe-based SIM, it is the illumination PSF that is manipulated, thus leading to an effectively complex-valued illumination intensity (see Eq. 15.8). In the hybrid techniques, it is both the illumination and detection PSFs that are manipulated. In fact, we have already made use of this same concept of pupil synthesis in Chapter 11 when considering interferometric microscopes, though in the different context of coherent imaging.

We close this chapter, however, with a word of caution. In all cases, the rejection of out-of-focus background in SIM requires the application of math to a set of raw images (at least two). Each of these raw images contains both signal and background, and the purpose of the math is to preserve the signal while cancelling the background. But there are difficulties with this approach when the background level is high. First, out-of-focus background in the raw images can occupy so much of the camera dynamic range that only a small fraction of this range is left for the signal itself. Even more fundamental is the problem of shot noise arising from the background. While image math can nominally cancel the average background in a set of raw images, it cannot cancel the shot noise associated with this background, and indeed, with every application of image arithmetic (e.g. Eqs. 15.5, 15.32, 15.51, 15.53, etc.) the shot noise accumulates rather than cancels. In this regard, SIM does not fare well when the sample is thick or highly scattering, in which case the background level can be substantial. For such samples, strategies where the background is physically prevented from impinging upon the detector are preferable, such as confocal microscopy. But as we will see in the next chapter, even confocal microscopy has its limitations, and alternative nonlinear microscopy strategies may be preferable still. On the other hand, SIM is highly effective when the sample is largely transparent or sparse, in which case it can provide additional advantages, as we will discover in Chapter 19.

15.6   PROBLEMS

Problem 15.1

Consider performing coherent structured illumination microscopy with a modulated field source (as opposed to a modulated intensity source). That is, start with

$$E_{\mathcal{L}}(x_l, y_l) = E_{\mathcal{L}} \left(1 + \cos(2\pi q_x x_l + \phi)\right)$$

Such a field can be obtained, for example, by sending a plane wave through a sinusoidal amplitude grating. This field is imaged into the sample using an unobstructed circular aperture of sufficiently large bandwidth to transmit q_x.

(a) Derive an expression for the resulting intensity distribution $I_i(x_0, y_0, z_0)$ in the sample. You will note that this distribution exhibits different modulation frequencies at different defocus values z_0.

(b) At what values of z_0 does $I_i(x_0, y_0, z_0)$ correspond to an exact image of the source intensity $I_{\mathcal{L}}(x_l, y_l)$? These images are called Talbot images.

(c) At what values of z_0 does $I_i(x_0, y_0, z_0)$ correspond to the source intensity image, but with an inverted contrast? These images are called contrast-inverted Talbot images.

(d) At defocus planes situated halfway between the Talbot and the contrast-inverted Talbot images, $I_i(x_0, y_0, z_0)$ exhibits a new modulation frequency. What is this modulation frequency? What is the associated modulation contrast?

(e) Your solution for $I_i(x_0, y_0, z_0)$ should also exhibit a modulation in the z_0 direction. What is the spatial frequency of this modulation? Note: there is no control of the phase of the

z_0-direction modulation (i.e. there is no equivalent of ϕ in the z_0 direction). Devise an experimental strategy to obtain phase control in the z_0 direction.

Problem 15.2

Show that the absolute value of the complex intensity $\widetilde{I} = \frac{1}{K} \sum_{k=0}^{K-1} e^{i\phi_k} I_k$ obtained from phase stepping can be rewritten as

$$\left| \widetilde{I} \right| = \frac{1}{3\sqrt{2}} \sqrt{(I_0 - I_1)^2 + (I_1 - I_2)^2 + (I_2 - I_0)^2}$$

when $K = 3$.

Problem 15.3

Consider performing SIM with a coherent fringe pattern of arbitrary spatial frequency $\vec{q}$. Calculate the resulting sectioning strength when the detection aperture is square (as opposed to circular). That is, calculate how the signal from a uniform fluorescent plane decays as a function of defocus z_s (assumed to be large). Specifically, consider the fringe frequencies $\vec{q} = \{q_x, 0\}$ and $\{q_x, q_y\}$. Are the sectioning strengths for these two frequencies the same?

Problem 15.4

SIM strategies involving random illumination patterns, such as DSI, generally require the calculation of signal means and variances, defined respectively by

$$\mu = \frac{1}{N} \sum_{i=1}^{N} x_i$$

$$\sigma^2 = \frac{1}{N} \sum_{i=1}^{N} (x_i - \mu)^2$$

where x_i are independent signal realizations.

(a) Show that σ^2 can be written equivalently as $\sigma^2 = \frac{1}{N} \sum_{i=1}^{N} x_i^2 - \mu^2$

(b) ... and again as $\sigma^2 = \frac{1}{2N^2} \sum_{i,j=1}^{N} (x_i - x_j)^2$

(c) ... and yet again as $\sigma^2 = \frac{1}{2(N-1)} \sum_{i=1}^{N-1} (x_{i+1} - x_i)^2$, where N is assumed large. This last representation is particularly insensitive to slow signal fluctuations that may be due to extraneous instrumentation noise.

Problem 15.5

In Section 15.5, it was stated that SIM can interpreted as arising from pupil synthesis. The basic idea of pupil synthesis is to construct an effective (synthesized) pupil from a sequence of images obtained from multiple pupil configurations. For example, two-pupil synthesis can be performed either in the illumination or detection paths of a generic widefield fluorescence

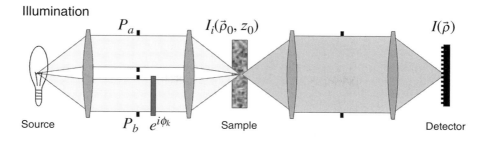

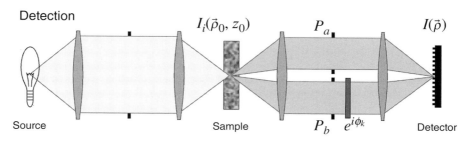

microscope, as shown in the figure above. In either case, it is assumed that one pupil, P_b, is phase-shifted relative to the other pupil P_a with a circular phase sequence ϕ_k.

(a) Consider illumination synthesis (figure, top). Light from any source point can equally travel through both illumination pupils, producing a field incident on the sample given by the coherent superposition

$$E_i^{(k)}\left(\vec{\rho}_0, z_0\right) = E_i^{(a)}\left(\vec{\rho}_0, z_0\right) + e^{i\phi_k} E_i^{(b)}\left(\vec{\rho}_0, z_0\right)$$

Show that conventional phase stepping leads to the synthesis of a complex image given by Eq. 15.6, with an effective complex illumination distribution given by Eq. 15.7. Express $\widetilde{I}_i(\vec{\rho}_0, z_0)$ in terms of $E_i^{(a)}(\vec{\rho}_0, z_0)$ and $E_i^{(b)}(\vec{\rho}_0, z_0)$.

(b) Consider detection synthesis (figure, bottom). Show that conventional phase-stepping leads to the synthesis of a complex image with an effective complex PSF. Derive corresponding expressions for $\widetilde{I}(\vec{\rho})$ and $\widetilde{\mathrm{PSF}}\left(\vec{\rho}, z\right)$, where the latter is in terms of amplitude point spread functions. Assume that the illumination intensity is spatially uniform.

References

[1] Bailey, B., Farkas, D. L., Taylor, D. L. and Lanni, F. "Enhancement of axial resolution in fluorescence microscopy by standing-wave excitation," *Nature* 366, 44–48 (1993).

[2] Bozinovic, N., Ventalon, C., Ford, T., and Mertz, J. "Fluorescence endomicroscopy with structured illumination," *Opt. Express* 16, 8016–8025 (2008).

[3] Garcia, J. Zalevsky, Z. and Fixler, D. "Synthetic aperture superresolution by speckle pattern projection," *Opt. Express* 13, 6073–6078 (2005).

[4] Golay, M. J. E. "Complementary series," *IRE Trans. Inform. Theory.* 7, 82–87 (1961).

[5] Goodman, J. W. *Speckle Phenomena in Optics: Theory and Applications*, Roberts & Co. (2007).

[6] Karadaglić, D. and Wilson, T. "Image formation in structured illumination wide-field fluorescence microscopy," *Micron* 39, 808–818 (2008).

[7] Lohmann, A. W. and Rhodes, W. T. "Two-pupil synthesis of optical transfer functions," *Appl. Opt.* 17, 1141–1151 (1978).

[8] Mertz, J. "Optical sectioning microscopy with planar or structured illumination," *Nat. Meth.* 8, 811–819 (2011).

[9] Neil, M. A. A., Juškaitis, R. and Wilson, T. "Method of obtaining optical sectioning by using structured light in a conventional microscope," *Opt. Lett.* 22, 1905–1907 (1997).

[10] Neil, M. A. A., Juškaitis, R. and Wilson, T. "Real time 3D fluorescence microscopy by two beam interference illumination," *Opt. Commun.* 153, 1–4 (1998).

[11] Neil, M. A. A., Wilson, T. and Juškaitis, R. "A light efficient optically sectioning microscope," *J. Microsc.* 189, 114–117 (1998).

[12] Ventalon, C. and Mertz, J. "Quasi-confocal fluorescence sectioning with dynamic speckle illumination," *Opt. Lett.* 30, 3350–3352 (2005).

[13] Ventalon, C., Heintzmann, R., Mertz, J. "Dynamic speckle illumination microscopy with wavelet prefiltering," *Opt. Lett.* 32, 1417–1419 (2007).

[14] Verveer, P. J., Hanley, Q. S., Verbeek, P. W., Van Vliet, L. J. and Jovin, T. M. "Theory of confocal fluorescence imaging in the programmable array microscope (PAM)," *J. Microsc.* 189, 192–198 (1997).

[15] Walker, J. G. "Non-scanning confocal fluorescence microscopy using speckle illumination," *Opt. Commun.* 189, 221–226 (2001).

[16] Wilson, T., Juškaitis, R., Neil, M. A. A. and Kozubek, M. "Confocal microscopy by aperture correlation," *Opt. Lett.* 21, 1879–1881 (1996).

16 Multiphoton Microscopy

Confocal microscopy represented a major breakthrough in that it was the first optical microscopy technique to provide true 3D resolution and optical sectioning. Since its invention in 1957, the success and popularity of confocal microscopy have established it as the gold standard in 3D imaging, unmatched in performance for more than two decades before alternative techniques could begin to compete. We turn now to a class of such alternative techniques called multiphoton microscopy. Like confocal fluorescence microscopy, multiphoton microscopy provides fluorescence contrast with 3D resolution and optical sectioning. It is also based on fluorescence excitation with a scanned, focused illumination beam, where 3D images are synthesized in the same way by sweeping a well-defined 3D probe volume, or voxel, throughout a fluorescent sample. However, the similarities end there. Multiphoton microscopy differs fundamentally from confocal fluorescence microscopy in that it is based on nonlinear rather than linear fluorescence excitation. As a consequence of this nonlinear excitation, the size of the 3D voxel in multiphoton microscopy is governed solely by the microscope illumination optics, as opposed to confocal microscopy where it is governed by both the illumination and detection optics.

Still another consequence of nonlinear excitation is that it prescribes the use of pulsed laser illumination. Indeed, were it not for the development of robust, user-friendly pulsed laser sources, multiphoton microscopy would not have gained its rapid popularity. For more details on pulsed lasers and their development, the reader is referred to [15] and [19]. Aside the requirement of a specialized laser source, however, the underlying hardware of a multiphoton microscope remains essentially identical to that of a laser scanning confocal microscope. If anything, a multiphoton microscope is technically simpler than a confocal microscope since, as we will see below, a multiphoton microscope requires no detection pinhole.

The term "multiphoton" microscopy is a generalization of the more specific variant called two-photon microscopy, originally invented by Denk and Webb in 1989 [4]. Most of this chapter will be focused on two-photon microscopy, since this is by far the most common multiphoton microscope. Recently, the extension of two-photon excitation to three-photon excitation has been shown to provide improved depth penetration in biological tissue, prompting renewed interest in higher-order nonlinearities for microscopy applications. General reviews of two-photon microscopy, which is now an established technique, can be found in [6, 8, 11, 20, 25]. For the more recent development of three-photon microscopy, the reader is referred to [9].

16.1 TWO-PHOTON EXCITED FLUORESCENCE (TPEF) CROSS-SECTION

Except for a brief digression in Sections 10.2.2 and 14.2.2, all the optical interactions we have dealt with so far, whether involving scattering or absorption, have been linear in nature. We turn now to a discussion nonlinear optical interactions between light and matter. In particular, we turn to the phenomenon of two-photon excitation. At least initially, the only characteristic of two-photon excitation relevant to our discussion will be that it scales quadratically with illumination intensity. That is, the two-photon excited fluorescence (TPEF) power emitted by a molecule initially in its ground state can be written as

$$\Phi = \sigma_{2f} I_i^2 \tag{16.1}$$

where I_i is the illumination intensity incident on the molecule, and σ_{2f} is the molecule TPEF cross-section (sometimes called an "action" cross-section [24]), in analogy with the linear one-photon fluorescence cross-section σ_f defined in Eq. 13.2. We note that the units of σ_{2f} are in $m^4/watts$ rather than m^2, meaning that σ_{2f} can no longer be interpreted as a simple area.

Conceptually, an interaction of the type described by Eq. 16.1 can be thought of as involving an absorption of two excitation photons of frequency ν_i followed by a subsequent emission of a single fluorescence photon of frequency ν_f, as illustrated in Fig. 16.1. By energy conservation, the frequency of the emitted fluorescence photon should nominally be twice that of the excitation photons; however, in practice this is rarely the case. Non-radiative decay mechanisms cause energy loss into the molecular environment and, in general, lead to $\nu_f \lesssim 2\nu_i$, where the balance in energy is lost as heat (see Section 13.1). Because of quantum mechanical selection rules, this balance of energy is often significantly larger than expected from the Stokes shift associated with one-photon fluorescence [24].

We note that, for historical reasons, the optical powers in Eq. 16.1 are usually expressed in units of photons/s, such that

$$\Phi_{2p} = \tfrac{1}{2}\sigma_{2p} I_\phi^2 \tag{16.2}$$

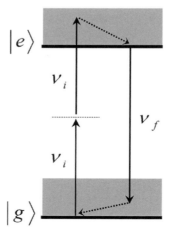

Figure 16.1. TPEF interaction. An absorption of two photons is followed by the emission of a fluorescence photon.

where $\Phi_{2p} = \frac{1}{h\nu_f}\Phi$ (units: photons/s), $I_\phi = \frac{10^{-4}}{h\nu_i}I_i$ (units: photons/s/cm^2), and σ_{2p} is expressed in units of Göppert-Mayer (1 GM = 10^{-50} cm^{4}s/photons) [7]. The conversion between σ_{2f} and σ_{2p} is given by

$$\sigma_{2f} = \frac{1}{2}\frac{10^{-58}}{h\nu_i}\left(\frac{\nu_f}{\nu_i}\right)\sigma_{2p} \approx \frac{10^{-58}}{h\nu_i}\sigma_{2p} \tag{16.3}$$

where h is Planck's constant and $h\nu_i$ is the energy of an excitation photon.

16.2 PULSED EXCITATION

Because of its nonlinear nature, TPEF is quite inefficient and requires high illumination power densities to compensate for the generally small cross-sections σ_{2f}. This requirement has advantages and disadvantages. An advantage, as we will see below, is that two-photon excitation becomes intrinsically confined to regions where the illumination power density is high. That is, when laser light is focused into a sample, TPEF is generated only from a small volume centered at the focal point [5]. Even when using focused beams, however, the power densities attainable by conventional lasers are generally insufficient to produce adequate TPEF for practical imaging applications. An additional mechanism for increasing power density is required. This additional mechanism comes from the "focusing" of light in time as well as in space. That is, it comes from the use of pulsed excitation.

Equation 16.1 is general and applies to both non-pulsed (i.e. continuous) and pulsed laser excitation. However, if the laser excitation is pulsed, the resulting TPEF must also be pulsed. In practice, any physical detector used to measure the TPEF must perform some degree of time averaging (see Section 7.3.1), and we assume here that the integration time of such averaging is much longer than the interval between pulses, or equivalently, that the detection bandwidth is much smaller than the pulse rate. Upon time averaging, we obtain then

$$\langle\Phi\rangle = \sigma_{2f}\langle I_i^2\rangle \tag{16.4}$$

where, again, we have made the tacit assumption that the molecule starts in its ground state at the onset of each pulse.

We must be careful in performing the above time average because if I_i is time varying, as certainly it is in the case of pulsed excitation, then $\langle I_i^2\rangle \neq \langle I_i\rangle^2$. It is usually more convenient to express $\langle\Phi\rangle$ directly in terms of $\langle I_i\rangle$, the average excitation intensity, since the latter can be readily measured in practice. To this end, we introduce what is known as the second-order temporal coherence factor of the illumination beam, defined by

$$g_2 = \frac{\langle I_i^2\rangle}{\langle I_i\rangle^2} \tag{16.5}$$

Equation 16.4 can then be rewritten as

$$\langle\Phi\rangle = \sigma_{2f}g_2\langle I_i\rangle^2 \tag{16.6}$$

The advantage of using pulsed rather than continuous laser excitation is now manifest. If the excitation is continuous then $g_2 = 1$; if the excitation is pulsed, then $g_2 > 1$. In particular, if the laser intensity arrives in short bursts of duration τ_p at regular intervals of period τ_l, we have

$$g_2 = \frac{\tau_l}{\tau_p} \tag{16.7}$$

In practice, typical lasers employed in two-photon microscopy possess second-order temporal coherence factors on the order of $g_2 \approx 10^4$–10^5, meaning that 10^4–10^5 times more fluorescence is generated from such a pulsed beam than from a continuous beam of equal average power. This gain in TPEF is quite dramatic and indispensable for a practical implementation of two-photon microscopy. For this reason, g_2 is often referred to as a pulsed-laser "advantage factor." An excellent discussion of the implications of g_2 in both classical and quantum optics is presented in [10].

A rearrangement of Eqs. 16.5 and 16.7 leads to

$$\tau_p = \tau_l \frac{\langle I_i \rangle^2}{\langle I_i^2 \rangle} \tag{16.8}$$

As defined here, τ_p corresponds to the coherence time associated with a single laser pulse (see Section 7.1). Accordingly, a characteristic peak pulse intensity can be defined as

$$\hat{I}_i = \frac{\tau_l}{\tau_p} \langle I_i \rangle \tag{16.9}$$

The parameters τ_p and $\hat{I}_i$, as illustrated in Fig. 16.2, are sufficient for our purposes to fully characterize the laser pulses for their two-photon excitation properties. We note that when τ_l is much longer than the excited-state lifetime of the fluorescent molecule, the molecule can safely be considered to be in its ground state at the onset of every illumination pulse, thus justifying our use of Eq. 16.4 to begin with.

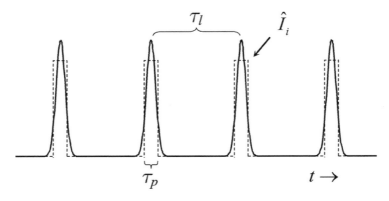

Figure 16.2. Pulsed intensity and square function approximation (dashed) as a function of time.

16.3   TWO-PHOTON EXCITATION VOLUME

In practice, the high power densities required for two-photon excitation generally impose the use of a tightly focused laser beam. That is, the time-averaged excitation intensity in TPEF microscopy can be written as

$$\left\langle I_i\left(\vec{r}\right)\right\rangle = \left\langle\Phi_i\right\rangle \Omega_i^{-1}\mathrm{PSF}_i\left(\vec{r}\right) \tag{16.10}$$

where $\left\langle\Phi_i\right\rangle$ denotes the laser power incident on the sample, $\mathrm{PSF}_i\left(\vec{r}\right)$ characterizes the laser focus, and $\Omega_i = \int d^2\vec{\rho}\,\mathrm{PSF}_i\left(\vec{r}\right)$.

The total fluorescence power obtained from a collection of molecules with concentration distribution $C\left(\vec{r}\right)$ (units: 1/volume) is then, from Eq. 16.6,

$$\left\langle\Phi\right\rangle = \sigma_{2f}g_2\int d^3\vec{r}\,\left\langle I_i\left(\vec{r}\right)\right\rangle^2 C(\vec{r}) \tag{16.11}$$

This may be recast in a more convenient form as

$$\left\langle\Phi\right\rangle = \phi_f\int d^3\vec{r}\,\Psi(\vec{r})C(\vec{r}) \tag{16.12}$$

where $\Psi(\vec{r})$ is a unitless function characterizing the TPE spatial profile, normalized so that $\Psi(\vec{0}) = 1$, and ϕ_f is the time-averaged fluorescence power collected from a single molecule located exactly at the focus center (units: watts), defined by, respectively

$$\Psi(\vec{r}) = \frac{\mathrm{PSF}_i^2\left(\vec{r}\right)}{\mathrm{PSF}_i^2\left(0\right)} \tag{16.13}$$

$$\phi_f = \sigma_{2f}g_2\left\langle\Phi_i\right\rangle^2\Omega_i^{-2}\mathrm{PSF}_i^2\left(0\right) \tag{16.14}$$

Equation 16.13 is similar to Eq. 14.29 which was used to define the probe volume in confocal fluorescence microscopy, and indeed, $\Psi(\vec{r})$ can be used in an identical manner to define a two-photon excitation (TPE) volume. As before, $\Psi(\vec{r})$ satisfies the properties listed in Section 13.4. In particular, $\Psi(\vec{r})$ is real and positive, and, as we will show below, $\int d^3\vec{r}\,\Psi(\vec{r})$ is bounded. This second condition is required for both optical sectioning and 3D image resolution, as illustrated in Fig. 16.3.

From Eq. 13.38, the TPE volume is thus defined by

$$V_{2f} = \frac{1}{\gamma_{2f}}\int d^3\vec{r}\,\Psi(\vec{r}) \tag{16.15}$$

where γ_{2f} is the TPE volume contrast given by Eq. 13.39. Once again, this volume contrast characterizes the sharpness of the volume profile. If the profile exhibits a hard step-function boundary then $\gamma_{2f} = 1$; if, as is the case for a physical TPE volume, the profile exhibits a soft boundary then $\gamma_{2f} < 1$.

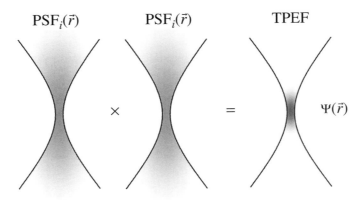

$\mathrm{PSF}_i(\vec{r})$ $\mathrm{PSF}_i(\vec{r})$ TPEF

$\times$ $=$ $\Psi(\vec{r})$

Figure 16.3. TPEF profile corresponds to product of two one-photon excitation profiles.

In the case where the average concentration of molecules is well defined and given by $\langle C \rangle$ we find that Eq. 16.11 simplifies to

$$\langle \Phi \rangle = \langle C \rangle \, V_{2f} \sigma_{2f} \gamma_{2f} g_2 \, \langle I_i(0) \rangle^2 \tag{16.16}$$

where $\langle I_i(0) \rangle$ is the time-averaged laser intensity exactly at the focus. Since the average number of molecules encompassed by the TPE volume is given by $N = \langle C \rangle V_{2f}$, we can rewrite Eq. 16.16 as

$$\langle \Phi \rangle = N \times \left(\sigma_{2f} \gamma_{2f} g_2 \, \langle I_i(0) \rangle^2 \right) \tag{16.17}$$

where the quantity in parentheses is interpreted as the average fluorescence power emitted per molecule residing within the TPE volume. This result is similar to that obtained when considering molecular fluorescence in FCS (cf. Eq. 13.51), and it comes as no surprise that two-photon microscopy, which provides TPE volumes on the order of femtoliters, is often used for FCS applications.

We turn now to two models that are routinely used to characterize the TPE volume.

16.3.1 Gaussian–Lorentzian Volume

The excitation laser in TPEF microscopy is generally focused with a microscope objective. In the common scenario where the width of the laser beam is smaller than the pupil of the objective, the objective is said to be underfilled. Assuming the shape of the laser beam is Gaussian at the pupil plane, then the 3D intensity profile of the focus is given by a Gaussian–Lorentzian PSF (Eq. 5.34). The corresponding TPE volume profile is

$$\Psi(\vec{r}) = \frac{w_0^4}{w^4(z)} e^{-4\rho^2/w^2(z)} \tag{16.18}$$

where $w^2(z) = w_0^2 \left(1 + \frac{z^2}{z_R^2}\right)$ and $z_R = \pi w_0^2 \kappa$ is the Rayleigh length, which we have seen before in Section 5.1.1.

From Eqs. 16.15 and 13.39, we obtain then for this profile

$$V_{2f} = \frac{4}{3}\pi^2 w_0^2 z_R \tag{16.19}$$

$$\gamma_{2f} = \frac{3}{16} \tag{16.20}$$

As expected, $\gamma_{2f} < 1$ because the TPE volume profile $\Psi(\vec{r})$ varies smoothly as a function of $\vec{r}$.

As an interesting aside, a feature of Gaussian–Lorentzian TPEF excitation is that the total fluorescence power $\langle \Phi \rangle$ obtained from a uniform concentration of fluorescence molecules is found to be independent of the laser focus waist w_0. This feature may appear surprising given that a larger focus waist leads to a reduced intensity at the focus, defined by $\langle I_i(0) \rangle = \frac{2\langle \Phi_i \rangle}{\pi w_0^2}$. However, in the case of Gaussian–Lorentzian excitation this reduction in intensity is exactly counterbalanced by an increase in TPE volume, yielding a net independence of TPEF on w_0, and hence on illumination NA, as pointed out in [25].

16.3.2 3D-Gaussian Volume

An alternative and simpler model to describe the TPE volume is that of a 3D Gaussian. This model is generally quite adequate for most TPEF calculations, particularly in the case where the illumination pupil is overfilled and the illumination NA is large (NA$_i \gtrsim 0.7$). We write in this case

$$\Psi(\vec{r}) \approx e^{-4\rho^2/w_0^2} e^{-4z^2/w_z^2} \tag{16.21}$$

where reasonably accurate dimensions for the 3D-Gaussian, found by numerical fitting [13], are given by

$$w_0 \approx \frac{0.52}{\kappa \sin\theta} \tag{16.22}$$

$$w_z \approx \frac{0.76}{\kappa(1 - \cos\theta)} \tag{16.23}$$

where θ corresponds to the half-angle spanned by the illumination pupil.

From Eqs. 14.29 and 13.39, we obtain then for this profile

$$V_{2f} = \left(\frac{\pi}{2}\right)^{3/2} w_0^2 w_z \tag{16.24}$$

$$\gamma_{2f} = \frac{1}{\sqrt{8}} \tag{16.25}$$

The volume contrast for a 3D-Gaussian volume is found to be almost a factor of two larger than for a Gaussian–Lorentzian volume, highlighting a certain deficiency of the above approximations, particularly in the transition region of moderate aperture size where the two models should merge. In fact, neither the Gaussian–Lorentzian nor 3D-Gaussian models can be exactly realized in practice. On the one hand, the illumination pupil always produces some degree of clipping of the Gaussian beam, and on the other hand, the excitation profile defined by Eq. 16.21 violates conservation of the illumination beam power along the axial direction. Nevertheless, the inaccuracies associated with these approximations are mostly limited to the wings of the illumination beam profile away from the focal center. These inaccuracies are abated by the fact that TPEF is based on nonlinear excitation, which confines the excitation to a region close to the focal center where the above approximations remain acceptable.

16.4   TWO-PHOTON SCANNING MICROSCOPY

So far, we have discussed the fluorescence power obtained from a TPE volume, but we have not yet considered how to create an image. In general, this is done the same way as with confocal fluorescence microscopy, by scanning a laser focus throughout a sample and creating an image voxel by voxel. Indeed, it is well worth comparing the two techniques.

Following the same notation as used in Section 14.3 and making use of Eq. 16.6, we can write the fluorescence intensity (or more precisely, the emittance) generated per unit depth inside the sample as

$$\left\langle F_{0z}(\vec{r}_0)\right\rangle = \sigma_{2f}g_2 \left\langle I_i(\vec{r}_0)\right\rangle^2 C(\vec{r}_0) \tag{16.26}$$

This equation is similar to Eq. 14.17, except that the generated fluorescence now scales quadratically with the illumination intensity. As before, we formally introduce beam scanning by applying a scan vector $(\vec{\rho}_s, z_s)$ to the sample, enabling us to write

$$\left\langle F_{0z}(\vec{\rho}_0, z_0; \vec{\rho}_s, z_s)\right\rangle = \sigma_{2f}g_2 \left\langle \Phi_i\right\rangle^2 \Omega_i^{-2}\mathrm{PSF}_i^2(\vec{\rho}_0, z_0)C(\vec{\rho}_0 + \vec{\rho}_s, z_0 + z_s) \tag{16.27}$$

where PSF_i is the point spread function associated with the illumination optics, and, for simplicity, the laser power is assumed to be relatively independent of depth z_0 within the sample. That is, $\left\langle \Phi_i(z_0)\right\rangle = \int \mathrm{d}^2\vec{\rho}_0 \left\langle I_i(\vec{\rho}_0, z_0)\right\rangle$ is taken to be a constant, equal to $\left\langle \Phi_i\right\rangle$.

Assuming an epifluorescence collection geometry, the total fluorescence intensity at the detector plane becomes, from Eq. 5.41,

$$\left\langle I(\vec{\rho}; \vec{\rho}_s, z_s)\right\rangle$$
$$= \sigma_{2f}g_2 \left\langle \Phi_i\right\rangle^2 \Omega_i^{-2} \iint \mathrm{d}^2\vec{\rho}_0\, \mathrm{d}z_0\, \mathrm{PSF}(\vec{\rho} - \vec{\rho}_0, z_0)\mathrm{PSF}_i^2(\vec{\rho}_0, z_0)C(\vec{\rho}_0 + \vec{\rho}_s, z_0 + z_s)$$
$$\tag{16.28}$$

where $\mathrm{PSF}(\vec{\rho}, -z)$ is the point spread function associated with the collection optics (note the change in sign of z owing to the collection geometry).

A characteristic feature of two-photon microscopy is that it obviates the need for a pinhole in the detector plane. That is, instead of recording the local fluorescence intensity only about $\vec{\rho} = 0$, as does a confocal microscope ($\vec{\rho} = 0$ being the pinhole location), a two-photon microscope ideally records the total fluorescence power across the entire detection plane. This total power is given by the integration of $\langle I(\vec{\rho}; \vec{\rho}_s, z_s) \rangle$ over $\vec{\rho}$, which, provided the detector area is large, can be taken to extend to infinity. Making use of Eq. 4.84, we have

$$\langle \Phi(\vec{\rho}_s, z_s) \rangle = \sigma_{2f} g_2 \langle \Phi_i \rangle^2 \Omega_0 \Omega_i^{-2} \iint d^2\vec{\rho}_0 \, dz_0 \, \text{PSF}_i^2(\vec{\rho}_0, z_0) C(\vec{\rho}_0 + \vec{\rho}_s, z_0 + z_s) \quad (16.29)$$

Equation 16.29 characterizes the imaging properties of a two-photon microscope and may be directly compared with its counterpart for confocal fluorescence microscopy, given by Eq. 14.22. In the same way that we defined an effective confocal PSF (Eq. 14.23), we can define here an effective TPEF PSF, given by $\text{PSF}_{\text{TPEF}}(\vec{r}) = \text{PSF}_i^2(\vec{r})$. The key difference between the two techniques is apparent. While 3D imaging resolution is governed by both the illumination and detection PSFs in confocal microscopy, it is governed by only the illumination PSF in TPEF microscopy. But in both cases, the effective PSF is governed by a product of two PSFs. As such, despite their differences, two-photon and confocal microscopes exhibit similar imaging properties. In particular, we saw in Section 14.3.1 that the product of two PSFs led to intrinsic optical sectioning in confocal microscopy. The same holds true for TPEF microscopy (see Fig. 16.3). That is, we have

- $\text{PSF}_i^2(0, z_0)$ scales as z_0^{-4} for large z_0
- $\int d^2\vec{\rho}_0 \, \text{PSF}_i^2(\rho_0, z_0)$ scales as z_0^{-2} for large z_0

meaning that the sectioning strength of a TPEF microscope is essentially identical to that of a confocal fluorescence microscope.

16.5 BROADBAND EXCITATION

In fact, Eq. 16.1, which has been the basis of our entire discussion so far, is not quite correct. As it happens, this equation relies on the implicit assumption that the excitation is monochromatic, or near monochromatic. But clearly, for a laser beam to be pulsed, it must exhibit a spectral bandwidth on the order $\Delta \nu_i = 1/\tau_p$ and therefore cannot be monochromatic. Indeed, we have completely neglected the issue of spectral bandwidth. A more exact description of two-photon excitation, which is only summarily presented here, takes this bandwidth into account (cf. [16]). In brief, the rate of energy absorbed by a two-photon interaction with a molecule is more correctly given by

$$\Phi_a = \int d\nu_2 \, \sigma_{2a}(\nu_2) \iint dt \, dt' \, e^{i2\pi\nu_2(t-t')} E_i(t)^2 E_i^*(t')^2 \quad (16.30)$$

where $\sigma_{2a}(\nu_2)$ is the spectral density of the two-photon absorption cross-section, and $E_i(t)$ is the incident excitation field. This may be recast in a more revealing form as

$$\Phi_a = \int d\nu_2 \, \sigma_{2a}(\nu_2) \left| \int d\nu' \, \mathcal{E}_i(\nu') \, \mathcal{E}_i(\nu_2 - \nu') \right|^2 \quad (16.31)$$

where the Fourier transform of the incident field is defined by

$$\mathcal{E}_i(\nu) = \int dt\, e^{i2\pi\nu t} E_i(t) \tag{16.32}$$

The interpretation of Eq. 16.31 is clear. If the incident field is sharply pulsed, then $\mathcal{E}_i(\nu)$ spans a broad range of frequencies. Two-photon absorption involves the product of pairwise components of this field for all combinations of frequencies ν' and $\nu_2 - \nu'$ such that the sum of the frequencies is ν_2. The absorbed power at frequency ν_2 is then weighted by $\sigma_{2a}(\nu_2)$.

As an example, let us consider an incident field that is relatively narrowband such that $\Delta\nu_i \ll \nu_{eg}$, where $h\nu_{eg}$ is energy gap between the molecular ground and excited states. In this case we are allowed the approximation $\mathcal{E}_i(\nu) \approx E_i\delta(\nu - \nu_i)$. The integral in Eq. 16.31 becomes non-zero only when $\nu_2 \approx 2\nu_i$, leading to

$$\Phi_a = \sigma_{2a}I_i^2 \tag{16.33}$$

where $\sigma_{2a} = \int d\nu_2\, \sigma_{2a}(\nu_2)$. We thus recover our initial premise of Eq. 16.1 (bearing in mind the link provided by the radiative quantum yield – see Eqs. 13.1 and 13.2).

For broadband pulses, however, Eq. 16.31 provides substantially more latitude. In particular, the Fourier transform $\mathcal{E}(\nu) = |\mathcal{E}(\nu)|\, e^{i\phi(\nu)}$ is a complex function. The spectral phase $\phi(\nu)$ associated with this function can be readily manipulated using techniques borrowed from the ultrafast optics community (for a review see [23]). This opens the remarkable possibility of controlling the rate of two-photon absorption by manipulating the excitation fields themselves, as first demonstrated by Silberberg [14]. More generally, such a technique involving the manipulation of spectral phases falls under the broad nomenclature of "coherent control." In addition to enabling the control of nonlinear light–matter interactions, coherent control is also routinely used for such applications as laser pulse shaping or pulse compression. And indeed, yet another example of coherent control is provided below.

16.6 TEMPORAL FOCUSING

In most cases, only a single laser focus is scanned throughout the sample, though variations are also possible where the excitation beam is divided into multiple beamlets (e.g. [1]). A clever such variation is based on coherent control, and, in particular, on the principle of temporal focusing [18, 27]. While a full description of temporal focusing is beyond the scope of this chapter, various interpretations can provide an intuitive understanding. We consider a first interpretation shown in Fig. 16.4a, where a blazed reflection grating is placed in an image plane of a standard widefield microscope configured in a 4f geometry. Imagine irradiating a sample with a pulsed laser beam that has been formed into a light sheet by a cylindrical lens (in a similar manner as in Section 14.3.5), and reflected from this grating. By virtue of the laser beam's being pulsed, it is non-monochromatic with spectral bandwidth given by $\Delta\nu_i$. The various frequency components $\mathcal{E}(\nu)$ comprising the beam are then spread by the grating into different directions about the optical axis. Each of these components, individually, is more monochromatic, and hence exhibits a longer pulse duration than the intact laser beam where the components are combined. In other words, when the components are spatially separated from one another, as they are upstream and downstream from the microscope focal plane, the illumination pulse duration is longer than it is at the focal plane where the components are brought together. In fact, at the focal plane, we expect to recover the initial pulse duration τ_p of the intact laser beam. Inasmuch as the resultant TPEF scales as τ_p^{-1} (see Eqs. 16.6 and 16.7) we expect it to decay with increasing distance from the focal plane. This decay turns out to be identical to the decay experienced in standard TPEF microscopy with a single point focus. But the difference here is that the illumination profile at the focal plane is an image (de-magnified) of the cross-sectional profile of the light sheet incident on the grating. In other words, the illumination profile is a line rather than a single point. As a result, we have gained a significant advantage in speed, since we have spared ourselves the need to perform scanning along the x direction.

While the interpretation provided in Fig. 16.4a provides a first-pass understanding of temporal focusing, it is incomplete in that it does not fully explain *why* the resultant sectioning strength is identical to that of a standard single-point TPEF. A more detailed interpretation is provided in Fig. 16.4b. In fact, what is imaged into the sample is not a cross-sectional line, as surmised above, but rather the point-like region where the laser pulse front and the grating surface intersect. This point-like region rapidly sweeps along the x axis of the grating as each pulse progresses, meaning the image of this region sweeps along the x axis of the focal plane in the opposite direction, with a speed given by $c\bar{\kappa}_i/Mq$ ($\bar{\kappa}_i$ being the mean wavenumber, q the grating frequency, and M the microscope magnification). Because, at any given time, the illumi-

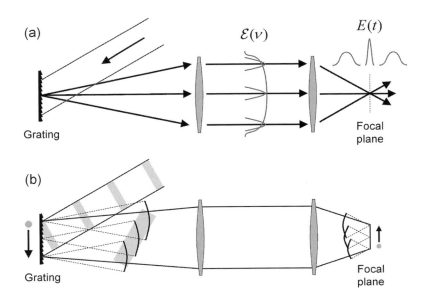

Figure 16.4. Interpretations of temporal focusing. Shaded rectangles in panel (b) indicate pulse fronts. Cylindrical lens at beam input not shown.

nation is point-like at the focal plane, the resulting sectioning corresponds to that of a point-like focus. What is important here is that the points arrive at the focal plane at different times (hence the sweep). This stems from the remarkable property of a grating that it can disassociate the phase fronts (i.e. direction) of a beam from the pulse fronts. This property is well known in the ultrafast optics community (e.g. [12]). We also saw an example of this property applied to non-pulsed light in Section 11.4.1, where the term "pulse front" was replaced by "coherence front." In effect, temporal focusing provides extremely fast scanning in the x direction, faster than can be resolved by a detector, and thus the illumination *appears* to be a line.

Alternatively, if the cylindrical lens is removed from the setup in Fig. 16.4, the illumination *appears* to be an area. In this case, actual lines of illumination along the y direction are projected into the sample, which rapidly sweep along the x direction. The sectioning capacity of such a line-scanning microscope is not as strong as that of the point-scanning microscope above, but the benefit is that physical scanning of the laser beam is now completely obviated.

But temporal focusing has its drawbacks. For one, it must project an image of the fluorescence signal onto a detector array, either a linear array in the case of point-scanning temporal focus, or a full-frame camera in the case of line-scanning temporal focus. This causes temporal-focusing to be sensitive to any blurring of the fluorescence signal induced by the sample, which, as we will see below, undermines one of the key advantages of standard two-photon microscopy. Another drawback of temporal focusing is that it does not utilize laser power efficiently. This is best seen in Fig. 16.4b. Each laser pulse is subdivided into smaller point-like regions. If N such regions produce N focal spots at the focal plane per laser pulse, then each of these spots yields a fluorescence energy reduced by a factor $1/N^2$ compared to the yield of a full pulse. Since N focal spots are occasioned per pulse, the net reduction in fluorescence is $1/N$. In brief, if A denotes the total illumination area provided by a temporal focusing microscope and A_0 denotes the area of a diffraction-limited focus, then the net reduction in fluorescence is $\sqrt{A_0/A}$ for a point-scanning temporal focus, and A_0/A for a line-scanning temporal focus. As a result, because of its inefficient use of laser power, two-photon microscopy based on temporal focusing has difficulty providing large fields of illumination.

16.7 MULTIPHOTON MICROSCOPY

The principle of two-photon fluorescence excitation can, of course, be extended to higher nonlinear orders. For example, in the case where the excitation laser is quasi-monochromatic compared to the molecular transition frequency ν_{eg}, Eq. 16.4 becomes generalized to

$$\langle \Phi \rangle = \sigma_{mf} \langle I_i^m \rangle \qquad (16.34)$$

where m is the multiphoton order, and the frequency of the excitation beam must nominally be reduced to ν_{eg}/m to excite the same molecular transition. In other words, higher nonlinear orders in multiphoton microscopy prescribe the use of longer-wavelength lasers.

Following the same arguments as in Section 16.2, Eq. 16.34 can be recast in terms of the average illumination intensity, obtaining

$$\langle \Phi \rangle = \sigma_{mf} \left(\frac{\tau_l}{\tau_p} \right)^{m-1} \langle I_i \rangle^m \tag{16.35}$$

Unfortunately a difficulty with multiphoton excitation is that the nonlinear cross-section σ_{mf} decreases precipitously with increasing nonlinear order. As a result, multiphoton excitation requires ever greater concentrations of illumination power to produce adequate fluorescence to be practical. To compensate for this reduction in cross-section, few options are available, from Eq. 16.35. One option is to increase the laser power so as to increase $\langle I_i \rangle$, but this rapidly becomes unwieldy, if not impossible. The remaining option is to decrease the pulse duty cycle τ_p/τ_l (for a given average intensity), by increasing τ_l or decreasing τ_p. But this, too, can only go so far. For example, when performing multiphoton microscopy, the minimum number of excitation pulses required per image voxel is obviously one. The maximum voxel acquisition rate is then determined by τ_l^{-1}, which is the laser pulse repetition rate. Typical acquisition rates in multiphoton microscopy are on the order of 1 MHz, thus setting an upper limit to τ_l. Alternatively, decreasing the pulse duty cycle by decreasing the pulse duration τ_p, in turn, requires the careful management of very broad laser bandwidths, which becomes technically challenging.

For all the above reasons, it was long assumed that multiphoton microscopy of any order higher than two would be hopelessly impractical. However, this assumption failed to consider a crucial element: the sample. As will be shown below, when sample-induced absorption and scattering are properly taken into account, higher-order nonlinearities can actually turn out to be advantageous.

16.7.1 Advantages of Multiphoton Microscopy

The benefits of multiphoton microscopy are particularly apparent when imaging in scattering media, such as biological tissue [3, 5, 22, 26]. Inevitably, when focusing laser light into tissue, part of the laser light becomes scattered before attaining the focus. In the case of standard one-photon confocal fluorescence microscopy, this scattered laser light leads to the excitation of out-of-focus fluorescence outside the focal volume of interest. While most of this out-of-focus fluorescence ultimately becomes rejected by the confocal pinhole, some nevertheless finds its way to the detector, also because of scattering, and thereby constitutes an increase in background. In the case of multiphoton microscopy, this problem of fluorescence background is largely resolved because the scattered excitation exhibits such a low power-density that it is generally too weak to generate significant out-of-focus fluorescence. The higher the order of the multiphoton excitation, the weaker this background.

But scattering does not simply lead to an increase in background. Once the signal fluorescence is generated inside the focal volume of interest, it too scatters as it propagates and exits

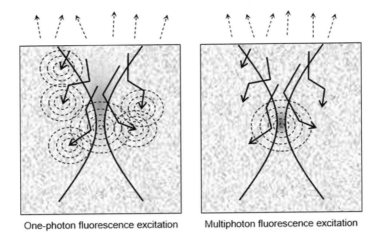

One-photon fluorescence excitation   Multiphoton fluorescence excitation

Figure 16.5. One-photon versus two-photon excited fluorescence (dashed lines) in a scattering sample.

the sample. In the case of standard confocal microscopy, such scattering poses a problem, since it deflects the signal fluorescence from its geometrical path toward the pinhole, and thus prevents it from being detected, causing a net reduction in signal. On the other hand, in the case of multiphoton microscopy, such scattering does not lead to a significant reduction in signal, because no pinhole is required in the detection path and the detector area can be quite large. It is again emphasized that the resolution of a multiphoton microscope is entirely defined by the excitation optics, as opposed to a standard confocal microscope, where it is defined by both the excitation and detection optics (unless temporal focusing is involved). Hence, because multiphoton fluorescence is dominantly generated by ballistic (i.e. non-scattered) excitation light, high-resolution images can be obtained that are essentially background free, even in scattering or turbid media. Figure 16.5 provides an illustration of this fundamental localization improvement.

But the most impactful advantage of multiphoton microscopy relates to penetration depth, which we consider now in detail.

Penetration Depth

To gain a more quantitative understanding of the penetration depth that can be achieved with a multiphoton microscope, let us estimate the detected fluorescence power as a function of focal depth [2, 17, 21]. As emphasized above, multiphoton fluorescence is almost exclusively generated by ballistic light. However, as a light beam penetrates into a scattering medium, its ballistic component decays with penetration distance. This decay is governed by the well-known Lambert–Beer law, which stipulates that the average ballistic power $\langle \Phi_i (z_0) \rangle$ decays as

$$\langle \Phi_i (z) \rangle = \langle \Phi_i \rangle \, e^{-\mu_e z_0} \tag{16.36}$$

where $\langle \Phi_i \rangle$ is the average excitation power incident at the sample surface, and μ_e is called the extinction coefficient. Just as in Section 9.3 where the extinction cross-section was defined as the sum of absorption and scattering cross-sections, here the extinction coefficient is defined as the sum of absorption and scattering coefficients. That is, $\mu_e = \mu_a + \mu_s$ (these will be discussed in more detail in Chapter 20).

We begin by generalizing Eq. 16.29 for multiphoton imaging, and write

$$\langle \Phi(\vec{\rho}_s, z_s) \rangle = \eta \Omega_i^{-m} \iint d^2\vec{\rho}_0 \, dz_0 \, \mathrm{PSF}_i^m(\vec{\rho}_0 - \vec{\rho}_s, z_0 - z_s) e^{-m\mu_e z_0} C(\vec{\rho}_{0s}, z_0) \qquad (16.37)$$

where $\eta = \Omega_0 \sigma_{mf} \left(\frac{\tau_l}{\tau_p} \right)^{m-1} \langle \Phi_i \rangle^m$, and we have taken into account the extinction of the illumination beam. Note that whereas for Eq. 16.29 we assumed the sample was scanned while PSF$_i$ remained static, here we assume the opposite. Moreover, the coordinate z_0 has been modified slightly. Previously $z_0 = 0$ corresponded to the center of PSF$_i$; here, it corresponds instead to the surface of the sample. That is, $C(\vec{\rho}_{0s}, z_0 < 0) = 0$. As a final note, the parameter Ω_0, which is associated with the detection PSF and may be interpreted as the fluorescence collection efficiency, is treated here as a constant. While this is correct for non-scattering samples, it is no longer strictly correct for scattering samples, where Ω_0 becomes dependent not only on depth but also on the microscope FOV (see [3]). Nevertheless, for depths no larger than the span of the FOV, Ω_0 may be regarded as reasonably constant.

To proceed, let us consider the specific case of a Gaussian–Lorentzian illumination PSF, as defined by Eq. 5.34. Let us also consider a dense distribution of fluorescent molecules, so dense that the concentration distribution may be regarded as essentially uniform. That is, $C(\vec{\rho}_0, z_0 > 0) \to C$. Equation 16.37 can then be integrated analytically.

Performing this integration first over $\vec{\rho}_0$, we obtain

$$\langle \Phi(z_s) \rangle_{z_0} = \left(2/\pi w_0^2 \right)^{m-1} \frac{\eta C}{m} \left[\frac{e^{-m\mu_e z_0}}{\left(1 + (z_0 - z_s)^2 / z_R^2 \right)^{m-1}} \right] \qquad (16.38)$$

This is the differential fluorescence power produced at depth z_0 within the sample, provided that the illumination beam is focused at depth z_s. Plots of $\langle \Phi(z_s) \rangle_{z_0}$ are shown in Fig. 16.6 for two- and three-photon excitation. Several conclusions can be drawn from these plots. First, we observe that the fluorescence signals arising from the focal volume are quite localized and pronounced. These are the signals of interest. But they are riding on top of sloping pedestals of background fluorescence that may not be as insignificant as we had previously assumed. The interpretation of these pedestals is clear. According to the Lambert–Beer law, with increasing penetration into the sample, exponentially less ballistic power is available to produce fluorescence. This is true for the signal, and it is also true for the background. But it is just as clear that with increasing penetration of the excitation focus z_s, there inevitably comes a point where the in-focus fluorescence signal becomes so diminished that it can no longer compete with the out-of-focus background, particularly the background produced near

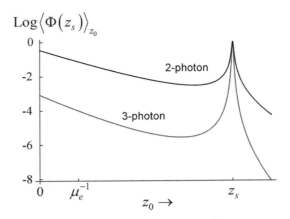

Figure 16.6. Representative $\langle \Phi (z_s) \rangle_{z_0}$ for two- and three-photon excitation inside a medium of extinction coefficient μ_e, normalized to peak at focus.

the sample surface. When this happens, the signal becomes lost and the multiphoton image becomes essentially featureless.

A more quantitative understanding of the depth limit imposed on z_s may be gained by explicitly separating the term in parentheses in Eq. 16.38 into signal (left) and background (right) components, by adopting the approximation (valid for $m\mu_e z_s \gg 1$ and $z_s^2/z_R^2 \gg 1$)

$$[...] \approx \frac{e^{-m\mu_e z_s}}{\left(1 + (z_0 - z_s)^2 / z_R^2\right)^{m-1}} + \frac{e^{-m\mu_e z_0}}{\left(1 + z_s^2/z_R^2\right)^{m-1}} \tag{16.39}$$

An integration of Eq. 16.38 over z_0 then leads to final expressions for the total signal and background produced by the sample when the illumination is focused at depth z_s. These are, respectively (where Γ is the Gamma function)

$$\langle \Phi(z_s) \rangle_{\text{signal}} = \left(2/\pi w_0^2\right)^{m-1} \frac{\eta C}{m} \left[\frac{\sqrt{\pi} \Gamma \left(m - \frac{3}{2}\right)}{\Gamma (m - 1)} z_R e^{-m\mu_e z_s} \right] \tag{16.40}$$

$$\langle \Phi(z_s) \rangle_{\text{background}} = \left(2/\pi w_0^2\right)^{m-1} \frac{\eta C}{m} \left[\frac{s}{m\mu_e \left(1 + z_s^2/z_R^2\right)^{m-1}} \right] \tag{16.41}$$

Note that, for more generality, we have introduced a sparsity factor s for the background, which ranges from 0 to 1. In other words, while the fluorescence concentration is assumed to be dense (i.e. uniform) in the vicinity of the excitation focus, we make allowances for the possibility that it is sparser ($s < 1$) elsewhere throughout the sample.

As expected, the total signal decreases exponentially with focus depth. The total background also decreases with focus depth, but more slowly. The point where the two become equal thus identifies the maximum focus depth allowed before the signal fades into the background and image contrast is lost. A derivation of this point is not quite analytical. Nevertheless, it can be reduced to simple equations that can be numerically solved self-consistently, which for two- and three-photon imaging become (for $z_s/z_R \gg 1$)

$$z_{s,\max} = \frac{1}{2\mu_e} \ln \left(\frac{2\pi\mu_e z_{s,\max}^2}{s z_R} \right) \qquad \text{(two-photon)} \qquad (16.42)$$

$$z_{s,\max} = \frac{1}{3\mu_e} \ln \left(\frac{3\pi\mu_e z_{s,\max}^4}{2 s z_R^3} \right) \qquad \text{(three-photon)} \qquad (16.43)$$

bearing in mind that the illumination wavelength, and hence both z_R and the extinction coefficient, is generally different for different excitation orders.

But one must be careful with the above results. Indeed, it would appear from Eq. 16.42 that the maximum depth penetration is governed by two parameters only, namely z_R and μ_e. This is quite remarkable, and seems to suggest that other more obvious parameters such as laser power, laser duty cycle, molecular cross-section, fluorescence collection efficiency, all encapsulated in the parameter η, play no role. But Eqs. 16.42 and 16.43 were derived from a comparison of signal with background. Noise was overlooked. In other words, it was assumed that both $\langle \Phi(z_s) \rangle_{\text{signal}}$ and $\langle \Phi(z_s) \rangle_{\text{background}}$ were strong enough that noise, such as shot noise or detector noise, could be neglected. Of course, to actually achieve these conditions, η plays a critical role.

It was noted above that several conclusions could be drawn from the plots in Fig. 16.6. Yet another conclusion relates to the qualitative difference between the background levels for different excitation orders. Manifestly, three-photon excitation provides significantly less background than two-photon microscopy, for equal signal levels. In other words, three-photon microscopy provides significantly improved contrast. This turns out to be a deciding advantage for depth penetration. One might even wonder if higher excitation orders could be more advantageous still. But herein arises a difficulty.

Clearly, the most important parameter that gives rise to background is μ_e. If the sample causes no illumination extinction, that is, if $\mu_e \rightarrow 0$, then all of Eq. 16.38 becomes signal and background becomes absent (note that the approximation in Eq. 16.39 becomes invalid in this case). Unfortunately, μ_e is difficult to control since it is a property of the sample itself. Nevertheless, it turns out that μ_e is highly dependent on the illumination wavelength. For example, in biological tissue, this wavelength dependence is governed by two competing trends. While the scattering coefficient μ_s tends to decrease, roughly inversely with wavelength for typical tissue heterogeneities, the absorption coefficient μ_a tends instead to increase with wavelength owing to the extensive presence of water. This increase becomes so rapid at longer infrared wavelengths ($\lambda \gtrsim 1800$ nm) that absorption-induced heating makes multiphoton imaging impossible at long wavelengths. Inasmuch as typical fluorescent molecules of interest exhibit transition frequencies ν_{eg} around 400 nm to 600 nm, this precludes the use of excitation orders greater than three. On the other hand, two- and three-photon excitation are quite feasible. And indeed, in the more specific case of brain tissue imaging, a fortuitous concurrence of increased contrast and decreased extinction coefficient is particularly favorable for three-photon imaging, as demonstrated by Xu [9] (see Fig. 16.7). A quick

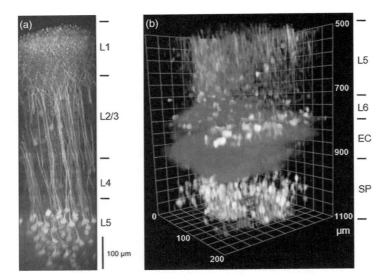

Figure 16.7. Two-photon (a) and three-photon (b) images of fluorescently labelled neurons in a mouse brain, illustrating depth penetration (L1-6 are neocortical layers, EC is external capsule, SP is stratum pyramidale). (a) adapted from [8]. (b) courtesy of C. Xu, Cornell University (note that darker gray structures correspond to third harmonic generation – see Chapter 17).

calculation based on Eq. 16.40 reveals that, for typical parameters associated with brain tissue imaging (e.g. $z_R^{(2)} \approx 2$ μm and $\mu_e^{(2)} = (150 \,\text{μm})^{-1}$; $z_R^{(3)} \approx 4$ μm and $\mu_e^{(3)} = (200 \,\text{μm})^{-1}$), the maximum depth penetrations are $z_{s,\text{max}}^{(2)} \approx 700$ μm and $z_{s,\text{max}}^{(3)} \approx 1400$ μm. That is, three-photon imaging provides effectively twice the depth penetration of two-photon (noise-related issues notwithstanding). This is quite an improvement over the depth penetrations achieved by more conventional one-photon approaches such as confocal microscopy, which, for the same samples, are typically limited to about 200 μm.

16.8 PROBLEMS

Problem 16.1

Consider a collimated beam that undergoes both one- and two-photon absorption as it propagates within a sample. That is, the beam intensity obeys the relation

$$\frac{dI}{dz} = -\alpha_1 I - \alpha_2 I^2$$

(a) Show that the resulting intensity at depth z is given by

$$I(z) = \frac{\alpha_1 I(0) e^{-\alpha_1 z}}{\alpha_1 + \alpha_2 \left(1 - e^{-\alpha_1 z}\right) I(0)}$$

(b) In the limit $a_2 \rightarrow 0$, the above result reduces to the familiar Lambert–Beer law. What does it reduce to in the limit $\alpha_1 \rightarrow 0$?

Problem 16.2

The fluorescence power emitted by a molecule under continuous two-photon excitation is given by Eq. 13.9. This equation is no longer valid in the case of pulsed illumination. In particular, consider pulsed illumination with pulse period τ_l and pulse width τ_p. Assume $\tau_p \ll \tau_e$ such that, at most, only one excitation can occur per pulse. Define g_p to be the probability of finding the molecule in the ground state at the *onset* of every pulse (in steady state). Moreover, define ξ to be the probability of excitation per pulse provided the molecule is in the ground state.

(a) Derive an expression for the average fluorescence power emitted by a molecule under pulsed illumination, in terms of g_p. (For simplicity, assume that the molecule is a simple two-level system with a radiative quantum yield equal to 1.)

(b) Derive an expression for g_p in steady state and show that

$$\frac{\langle \Phi_f \rangle}{h\nu_f} = \frac{\xi}{\tau_l} \left(\frac{1 - e^{-\tau_l/\tau_e}}{1 - e^{-\tau_l/\tau_e} + \xi e^{-\tau_l/\tau_e}} \right)$$

where τ_e is the excited state lifetime. Hint: to solve this problem, start by deriving the probability e_p of finding the molecule in the excited state at the *onset* of a pulse. To achieve steady state, this probability must be in balance with the residual probability from the previous pulse

(c) Let α be the excitation rate (two-photon or otherwise) during each pulse, and assume that the pulse width is so short that $\alpha\tau_p \ll 1$. Derive an expression for e_p when the repetition rate of the illumination becomes so high that the illumination becomes effectively a continuous wave (i.e. when $\tau_l \rightarrow \tau_p$). How does this expression compare with Eq. 13.8?

Problem 16.3

In the case of two-photon excitation, show that if the sample is a thin uniform plane at a defocus position z_s, with concentration defined by $C(\vec{r}) = C_\rho \delta(z - z_s)$, then the total generated fluorescence power is inversely proportional to the illumination beam cross-sectional area $A_i(z_s)$, independently of the shape of the illumination PSF_i. Is this also true of three-photon excitation?

Hint: define cross-sectional area in a similar manner as Eq. 6.24.

Problem 16.4

A Gaussian–Lorentzian focus is used to produce two- and three-photon excited fluorescence.

(a) Show that the three-photon excitation volume is given by $V_{3f} = \frac{32}{105}\pi^2 w_0^2 z_R$ and the volume contrast is given by $\gamma_{3f} = \frac{35}{128}$.

(b) Show that if the sample is a volume of uniform concentration C, then the total generated fluorescence is independent of the beam waist w_0 in the case of two-photon excitation, and it scales as w_0^{-2} in the case of three-photon excitation.

Problem 16.5

Consider a laser whose average power is a constant $\langle \Phi_i \rangle$ but whose repetition rate $R = \tau_l^{-1}$ can be varied. This laser is used to perform multiphoton excitation in a thick sample of extinction coefficient μ_e. But there is a problem. The maximum peak intensity that the sample can tolerate at the laser focus is $\hat{I}_{\max}$. Derive a strategy of adjusting the repetition rate to maximize the fluorescence produced at the beam focus, for arbitrary depth within the sample. That is, find an expression for the optimal $R(z_s)$.

References

[1] Andresen, V., Egner, A. and Hell, S. W. "Time-multiplexed multifocal multiphoton microscope," *Opt. Lett.* 26, 75–77 (2001).

[2] Beaurepaire, E., Oheim, M. and Mertz, J. "Ultra-deep two-photon fluorescence excitation in turbid media," *Opt. Comm.* 188, 25–29 (2001).

[3] Beaurepaire, E. and Mertz, J. "Epifluorescence collection in two-photon microscopy," *Appl. Opt.* 41, 5376–5382 (2002).

[4] Denk, W., Strickler, D. and Webb, W. W. "Two-photon laser scanning fluorescence microscopy," *Science* 248, 73–76 (1990).

[5] Denk, W. "Two-photon excitation in functional biological imaging," *J. Biomed. Opt.* 1, 296–304 (1996).

[6] Diaspro, A. Ed., *Confocal and Two-Photon Microscopy: Foundations, Applications and Advances*, Wiley-Liss (2002).

[7] Göppert-Mayer, M. "Über Elementarakte mit zwei Quantensprüngen (On elementary processes with two quantum steps)," *Ann. Phys.* 9, 273–294 (1931).

[8] Helmchen, F. and Denk, W. "Deep tissue two-photon microscopy," *Nat. Meth.* 2, 932–940 (2005).

[9] Horton, N. G., Wang, K., Kobat, D., Clark, C. G., Wise, F. W., Schaffer, C. B. and Xu, C. "In vivo three-photon microscopy of subcortical structures within an intact mouse brain," *Nat. Phot.* 7, 205–209 (2013).

[10] Loudon, R. *The Quantum Theory of Light*, 3rd edn, Oxford University Press (2000).

[11] Masters, B. R. and So, P. T. C. Eds., *Handbook of Biomedical Nonlinear Optical Microscopy*, Oxford University Press (2008).

[12] Maznev, A. A., Crimmins, T. F. and Nelson, K. A. "How to make femtosecond pulses overlap," *Opt. Lett.* 23, 1378–1380 (1998).

[13] Mertz, J. "Molecular photodynamics involved in multiphoton excitation fluorescence microscopy," *Eur. Phys. J.* D 3, 53–66 (1998).

[14] Meshulach, D. and Silberberg, Y. "Coherent quantum control of two-photon transitions by a femtosecond laser pulse," *Nature* 396, 239–242 (1998).

[15] Milonni, P. W. and Eberly, J. H. *Lasers*, Wiley-Interscience (1988).

[16] Ogilvie, J. P., Kubarych, K. J., Alexandrou, A. and Joffre, M. "Fourier transform measurement of two-photon excitation spectra: applications to microscopy and optimal control," *Opt. Lett.* 30, 911–913 (2005).

[17] Oheim, M., Beaurepaire, E., Chaigneau, E., Mertz, J. and Charpak, S. "Two-photon microscopy in brain tissue: parameters influencing the imaging depth," *J. Neurosci. Methods* 111, 29–37 (2001).

[18] Oron, D., Tal, E. and Silberberg, Y. "Scanningless depth-resolved microscopy," *Opt. Express* 13, 1468–1476 (2005).

[19] Siegman, A. E. *Lasers*, University Science Books (1986).

[20] Svoboda, K. "Principles of two-photon excitation microscopy and its applications to neuroscience," *Neuron* 50, 823–839 (2006).

[21] Theer, P., Hasan, M. T. and Denk, W. "Two-photon imaging at a depth of 1000 μm in living brain slices by use of a Ti:Al$_2$O$_3$ regenerative amplifier," *Opt. Lett.* 28, 1022–1024 (2003).

[22] Theer, P. and Denk, W. "On the fundamental imaging-depth limit in two-photon microscopy," *J. Opt. Soc. Am.* A 23, 3139–3149 (2006).

[23] Weiner, A. M. "Femtosecond pulse shaping using spatial light modulators," *Rev. Sci. Instrum.* 71, 1929–1960 (2000).

[24] Xu, C. and Webb, W. W. "Measurement of two-photon excitation cross-sections of molecular fluorophores with data from 690 to 1050nm," *J. Opt. Soc. Am.* B 13, 481–491 (1996).

[25] Xu, C. and Webb, W. W. "Multiphoton excitation of molecular fluorophores and nonlinear laser microscopy" in *Topics in Fluorescence Spectroscopy, Vol. 5*, Ed. J. R. Lakowicz, 471–450 Springer (1997).

[26] Ying, J., Liu, F. and Alfano, R. "Spatial distribution of two-photon-excited fluorescence in scattering media," *Appl. Opt.* 38, 224–229 (1999).

[27] Zhu, G., v-Howe, J., Durst, M., Zipfel, W. and Xu, C. "Simultaneous spatial and temporal focusing of femtosecond pulses," *Opt. Express* 13, 2153–2159 (2005).

17 | Multiharmonic Microscopy

The invention of two-photon excited fluorescence (TPEF) microscopy sparked a tremendous resurgence in the development of novel optical microscopy techniques based on nonlinear interactions. In particular, it was realized that TPEF is but one of a broad range of possible nonlinear interactions that can produce contrast. As we will see below, these interactions are generally divided into two categories: incoherent and coherent. In the first of these categories, there is no phase correlation between the incoming illumination and the outgoing signal, whereas in the second, this phase correlation is well defined. We recall that fluorescence excitation is an example of an incoherent interaction since the phase of fluorescence is random and uncorrelated with the phase of the illumination beam. TPEF, or more generally multiphoton excited fluorescence, is thus also an incoherent interaction. In contrast, scattering is an example of a coherent interaction, since the phase of scattered light is tightly correlated with that of the illumination beam.

In this chapter and the next we will examine microscopy techniques based on nonlinear versions of scattering. We begin with the simplest of these based on second- and third-harmonic generation, or more generally, multiharmonic generation. As we saw in Chapter 9, when the phase of the scattered field is well defined, the radiation pattern of the scattered intensity can be quite intricate depending on the distribution of the scattering centers within the sample. The same is true with nonlinear scattering.

With some exceptions, multiharmonic microscopes are operated in scanning configurations where a focused illumination beam is scanned within a sample and the resultant scattered signal is detected with a single-element detector, typically a photomultiplier tube. An important feature of multiharmonic microscopy is that it provides intrinsic 3D resolution and optical sectioning, in the same manner as multiphoton microscopy. The origin of this 3D resolution again resides in a supra-linear dependence of the signal power on illumination intensity, as was discussed in the previous chapter and will be briefly iterated here. It should be emphasized that the development of coherent nonlinear microscopy techniques is still ongoing. As such, this chapter and the next are meant only to outline basic fundamentals of these techniques. More detailed information on nonlinear optics in general can be found in a vast body of literature, including seminal work by Bloembergen [2] and classic textbooks by Boyd [4], Butcher and Cotter [5] and Shen [24]. For information specifically concerning molecular nonlinear interactions, the reader can also consult [19].

17.1 NONLINEAR SUSCEPTIBILITIES

In Chapter 9 scattering was treated as a process of re-radiation from field-induced primary sources. We proceed along the same lines for nonlinear scattering, the only difference being that the induced sources are now driven by nonlinear optical interactions. To better understand the nature of these sources, we adopt a classical picture of light–matter interactions mediated by polarizations resulting from the displacements of electrons. That is, light induces an oscillating polarization within matter, and concomitantly an oscillating polarization generates light. In the latter case, the oscillating polarization constitutes a primary source. More precisely, according to Maxwell's equations, it is the *acceleration* of this oscillating polarization that constitutes the primary source. That is, if $P(t)$ denotes a time-dependent polarization density, then the corresponding time-dependent primary source is $S(t) = \mu_0 \frac{d^2 P(t)}{dt^2}$, where μ_0 is a scaling factor (magnetic permeability). Equivalently, if $\mathcal{P}(\nu)$ denotes the Fourier transform of the time-dependent polarization density, then $\hat{S}(\nu) = -4\pi^2 \mu_0 \nu^2 \mathcal{P}(\nu)$ (we use the notation $\hat{S}(\nu)$ to avoid confusion with the spectral density $\mathcal{S}(\nu)$, which we will return to later).

In effect, scattering can be thought of as a two-step process. An incident field induces a polarization density, and this polarization density in turn produces a field at its oscillation frequency that may or may not interfere with the incident field. In the case of linear scattering, the polarization density and incident field oscillate at the same frequency, and interference occurs. Indeed, this interference has been manifest throughout this book as an interference between the illumination field E_i and the sample-induced signal field E_s, as in, for example, the transmission microscopes described in Chapter 10. In the case of nonlinear scattering, the illumination and polarization density frequencies may be the same or different. In this chapter, which deals with multiharmonic generation, they are different, and no interference occurs. The polarization density, once generated, thus radiates independently of illumination.

In general, the polarization density may be treated as a response to an illumination field, in a similar manner as fluorescence was treated as a response to an illumination intensity in Section 13.3, though on a much faster time scale. As we will see, the vectorial nature of light plays an important role here, and we explicitly include this in our notation. The simplest interaction is linear, written as

$$\vec{P}^{(1)}(t) = \epsilon_0 \int d\tau_1 \, \mathbf{R}^{(1)}(\tau_1) \, \vec{E}_1(t - \tau_1) \tag{17.1}$$

where ϵ_0 is a scaling factor (electric permittivity), and the boldface indicates that the sample response $\mathbf{R}^{(1)}(\tau_1)$ is a tensor, meaning it can couple a field component along one axis into a polarization density component along another (for isotropic media, $\mathbf{R}^{(1)}(\tau_1)$ is diagonal). Taking the Fourier transform of both sides, Eq. 17.1 can be recast as

$$\vec{\mathcal{P}}^{(1)}(\nu) = \epsilon_0 \chi^{(1)}(\nu) \, \vec{\mathcal{E}}_1(\nu) \tag{17.2}$$

where $\chi^{(1)}(\nu)$, called a linear susceptibility, denotes the Fourier transform of $\mathbf{R}^{(1)}(\tau)$. While these equations have not previously been made explicit, they have been the underlying origin of scattering in our discussions so far.

In general, light–matter interactions are more complicated than simple linear interactions, and can include higher-order nonlinear dependences. That is, in general

$$\vec{P}(t) = \vec{P}^{(1)}(t) + \vec{P}^{(2)}(t) + \vec{P}^{(3)}(t) + \cdots \tag{17.3}$$

where the polarization of order n is given by

$$\vec{P}^{(n)}(t) = \epsilon_0 \int d\tau_1 \cdots d\tau_n \, \mathbf{R}^{(n)}(\tau_1, \ldots, \tau_n) \, \vec{E}_1(t - \tau_1) \cdots \vec{E}_n(t - \tau_n) \tag{17.4}$$

and $\mathbf{R}^{(n)}(\tau_1, \ldots, \tau_n)$ is a generalized nonlinear response function. The term $\mathbf{R}^{(n)}(\tau_1, \ldots, \tau_n)$ vanishes when any of its arguments is negative, to ensure causality.

Equation 17.4 is rewritten as

$$\vec{P}^{(n)}(t) = \epsilon_0 \int d\nu_1 \cdots d\nu_n \, \chi^{(n)}(\nu_1, \ldots, \nu_n) \, \vec{\mathcal{E}}_1(\nu_1) \cdots \vec{\mathcal{E}}_n(\nu_n) \, e^{-i2\pi\nu_\Sigma t} \tag{17.5}$$

where $\nu_\Sigma = \nu_1 + \ldots + \nu_n$, and the nonlinear susceptibility and sample response of order n are related by the n-dimensional Fourier transform

$$\chi^{(n)}(\nu_1, \ldots, \nu_n) = \int d\tau_1 \ldots d\tau_n \, \mathbf{R}^{(n)}(\tau_1, \ldots, \tau_n) \, e^{i2\pi(\nu_1\tau_1 + \cdots + \nu_n\tau_n)} \tag{17.6}$$

Finally, taking the Fourier transform of both sides of Eq. 17.5, we arrive at

$$\vec{\mathcal{P}}^{(n)}(\nu) = \epsilon_0 \int d\nu_1 \cdots d\nu_n \, \chi^{(n)}(\nu_1, \ldots, \nu_n) \, \vec{\mathcal{E}}_1(\nu_1) \cdots \vec{\mathcal{E}}_n(\nu_n) \, \delta(\nu - \nu_\Sigma) \tag{17.7}$$

which is usually written in simplified notation as

$$\vec{\mathcal{P}}^{(n)}(\nu) = \epsilon_0 \chi^{(n)}(-\nu; \nu_1, \ldots, \nu_n) : \vec{\mathcal{E}}_1(\nu_1) \cdots \vec{\mathcal{E}}_n(\nu_n) \tag{17.8}$$

such that the sum of all the arguments of $\chi^{(n)}$ is implicitly equal to zero (i.e. $\nu = \nu_\Sigma$). This equality, which is ensured by the delta function in Eq. 17.7, is called frequency matching.

It should be noted that the susceptibilities $\chi^{(n)}$ are local properties of the sample that, in general, can vary in space. The driving fields $\vec{\mathcal{E}}_1, \ldots, \vec{\mathcal{E}}_n$ are also local and can vary in space. These comprise the illumination fields at frequencies $\nu_1, \ldots, \nu_n$, but also any re-radiated fields generated by the same polarization densities they have induced. In other words, Eq. 17.8 should, in principle, be solved self consistently. However, as was pointed out in Chapter 9, this difficulty can be sidestepped if we assume the interactions are weak, meaning the re-radiated fields play a negligible role. Effectively, this is equivalent to the Born approximation, which we again adopt. In other words, the driving fields $\vec{\mathcal{E}}_1, \ldots, \vec{\mathcal{E}}_n$ are assumed here to be the same as the unperturbed illumination fields themselves.

Finally, it should be noted that the susceptibilities $\chi^{(n)}$, as defined by Eq. 17.8, are not the same as those encountered in the literature. In particular, in keeping with our convention throughout this book, the complex fields $\vec{E}_1, \ldots, \vec{E}_n$ differ in units from physical electric fields (see Section 1.2). Similarly, the induced polarization densities $\vec{P}^{(n)}$ differ in units from physical induced polarization densities. As such, the susceptibilities $\chi^{(n)}$ in Eq. 17.8 do not correspond to electronic susceptibilities. Moreover, we have overlooked issues related to the permutation of field indices or possible degeneracies in field frequencies which, to properly keep track of, requires courageous efforts in bookkeeping. Several conventions exist in the literature for such bookkeeping, most notably those defined in [5, 20], or alternatively in [4, 24], or alternatively still in [28]. Since the purpose of this chapter and the next is to present an overview of nonlinear microscopy techniques rather than delve into the details of their underlying nonlinear optical interactions, we will content ourselves with our abbreviated notation.

17.1.1 Coherent Interactions

Figure 17.1 provides examples of two nonlinear interactions that we will consider in this chapter. The arrows schematically denote field-induced molecular transitions. The upward arrows indicate the multiharmonic generation of a polarization density at frequency ν_Σ while the downward arrow indicates the re-radiation of this polarization density at frequency ν, where frequency matching imposes the condition $\nu = \nu_\Sigma$. Figure 17.1 may be compared with Fig. 16.1, which depicts two-photon absorption followed by fluorescence emission. There are several fundamental differences between these phenomena.

For one, multiharmonic generation is a scattering process. No energy is deposited in the molecule, whose role is simply to re-package the incident energy into a different frequency. Such an interaction is said to be parametric. In contrast, multiphoton excitation is an absorption process. Energy is deposited into the molecule. This energy is eventually released into the environment, but whether it is released as light (fluorescence) or as heat (vibrations), or, more generally, as a combination of the two, depends on the specifics of the molecule in its environment.

Another difference between multiharmonic and multiphoton interactions relates to the time scales involved. Multiharmonic generation arises from a near instantaneous response of the sample, on the order of femtoseconds (or, more precisely, $\sim \nu_{eg}^{-1}$). In contrast, while the

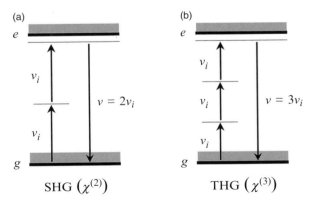

Figure 17.1. Second-harmonic generation (a) and third-harmonic generation (b).

multiphoton absorption process itself is fast, the subsequent release of energy from the excited state is in general several orders of magnitude slower, on the order of nanoseconds. In other words, the field-induced response function $\mathbf{R}^{(n)}$ in Eq. 17.4 is essentially a delta function in time, while the intensity-induced function $\overline{R}$ in Eq. 13.21 is quite spread in time, so spread that it can be readily measured with conventional electronics, as discussed in Section 13.3. Correspondingly, $\chi^{(n)}$ in Eq. 17.6 is quite flat and hardly depends on frequency at all. In particular, as we have already seen with two-photon absorption, not all light-induced transitions visit states that can be accommodated by a molecule. Such transitions to virtual states are called non-resonant, and are very short lived, again on the order of femtoseconds. Virtual states are depicted as thin lines in Fig. 17.1. On the other hand, when virtual states happen to lie in the vicinity of real states (i.e. states that can be accommodated by the molecule), the transitions are said to become resonantly enhanced (see [4]), with transition-state lifetimes extending to the order of picoseconds. We will not consider such resonantly enhanced transitions in this chapter, and defer these to our discussion of pump-probe microscopies in the next chapter.

Finally, in the same manner that multiharmonic generation does not deposit energy into the sample, it also does not deposit momentum. As such, a multiharmonic interaction must obey what is known as the phase-matching condition $\vec{\kappa} = \vec{\kappa}_{\Sigma}$, where, similar to the frequency-matching condition, $\vec{\kappa}$ is the wavevector of the outgoing field while $\vec{\kappa}_{\Sigma} = \vec{\kappa}_1 + \vec{\kappa}_2 + \cdots$ is the sum of wavevectors of the incoming fields that drive the nonlinear polarization density. As we will see below, the phase-matching condition plays a major role in determining the radiation patterns resulting from multiharmonic generation, and, indeed, whether multiharmonic generation is possible at all.

In effect, the phase-matching condition stems from the overarching difference between multiharmonic and multiphoton interactions. Multiharmonic interactions are coherent while multiphoton interactions are not. Said differently, in a coherent interaction there is tight phase correlation between the incoming driving field(s) and the outgoing radiated field, while in an incoherent interaction this phase correlation is lost. Said differently again, because the induced polarization densities have well-defined phases prescribed by their driving fields, the radiation produced by these polarization densities can exhibit interference, constructive or destructive (hence the repercussion on radiation patterns). In contrast, fluorescence is emitted with random phase. The fluorescence contributions produced by different molecules, even though driven by the same field(s), are uncorrelated and cannot interfere, or, more precisely, if they do interfere this interference is too rapid to observe.

But of course, as with any physical interaction, nothing is ever so clear cut. For example, we have completely neglected the possibility of such phenomena as superfluorescence (see, for example, [1]). Moreover, the distinction between coherent and incoherent processes is often confounded by subtleties related to correlations between transitions themselves, particularly when transition levels are degenerate, or when dealing with extremely short matter-field interaction times. This separation of coherent versus incoherent processes, as defined by a distinction between correlated and uncorrelated phases, should therefore be regarded as an approximate rule of thumb, which, for our purposes, will be adequate. For detailed discussions of some of these subtleties, the reader is referred to the nonlinear optics textbooks listed above, and also to [23].

17.1.2 Radiation Patterns

As noted above, and illustrated in Section 9.4, the radiation patterns produced by a collection of scattering centers can be quite complicated. An understanding of these patterns is of paramount importance in nonlinear microscopy applications, as it influences both the design of the detection optics and the interpretation of images. We begin here with a qualitative description of these radiation patterns.

As in Chapter 9, we consider scattering to be a process of re-radiation from an induced source distribution. If the source distribution is known, then the resultant radiated scattered field, in a fully vectorial treatment, is given by

$$\vec{E}_\nu(\vec{r}) = 4\pi^2 \mu_0 \nu^2 \int d^3\vec{r}_0 \, G_+(\vec{r} - \vec{r}_0) \left[\hat{r} \times \left(\hat{r} \times \vec{P}_\nu(\vec{r}_0) \right) \right] \tag{17.9}$$

where we have run into the slight notational difficulty that while $\vec{\mathcal{E}}$ was used to denote a temporal Fourier transform above, it was previously used to denote a spatial Fourier transform. We introduce here a different notation $\vec{E}_\nu$ to denote only the temporal Fourier transform. That is, we define

$$\vec{E}_\nu(\vec{r}) = \int dt \, e^{i2\pi\nu t} \vec{E}(\vec{r}, t) \tag{17.10}$$

with the same convention applying to $\vec{P}_\nu$.

Equation 17.9 differs from Eq. 9.56 only in that we are considering an induced source distribution $-4\pi^2 \mu_0 \nu^2 \vec{P}_\nu(\vec{r}_0)$ that arises from a nonlinear rather than a linear optical interaction. Specifically, $\vec{P}_\nu(\vec{r}_0)$ is taken here to be derived from Eq. 17.7. Because $\vec{P}_\nu(\vec{r}_0)$ oscillates with frequency ν, so too does the resultant scattered field $\vec{E}_\nu(\vec{r})$. The two are linked by the Green function of the Helmholtz equation, taken to be forward directed and defined by $G_+(\vec{r}) = -\frac{e^{i2\pi\kappa r}}{4\pi r}$, where $\kappa = \frac{n(\nu)}{c}\nu$ is the associated wavenumber of the scattered field (see Eqs. 2.17 and 2.13). For generality, we allow the sample medium to be dispersive. That is, we allow the index of refraction $n(\nu)$ to be frequency dependent.

Equation 17.9 provides a full description of the radiated scattered field, provided $\vec{P}_\nu(\vec{r}_0)$ is known. To gain further insight into this description, let us assume that the induced source distribution is limited in size, and the scattered radiation is observed far from the source, conditions for the Fraunhofer approximation (see Fig. 17.2). These conditions are generally adequate for most nonlinear microscopy applications. From Section 9.4, Eq. 17.9 then takes the familiar form

$$E_\nu^{(p,s)}(\vec{r}) = \pi \mu_0 \nu^2 \frac{e^{i2\pi\kappa r}}{r} \int d^3\vec{r}_0 \, e^{-i2\pi\kappa \hat{r} \cdot \vec{r}_0} P_\nu^{(p,s)}(\vec{r}_0) \tag{17.11}$$

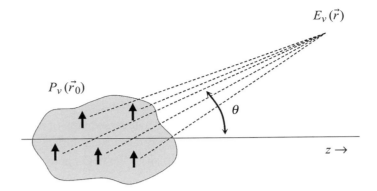

Figure 17.2. Nonlinear scattering arising from an induced polarization density.

where

$$\left(\begin{array}{c} E_\nu^{(p)}(\vec{r}) \\ E_\nu^{(s)}(\vec{r}) \end{array} \right) = \left(\begin{array}{c} \hat{\theta} \\ \hat{\varphi} \end{array} \right) \cdot \vec{E}_\nu(\vec{r}) \tag{17.12}$$

$$\left(\begin{array}{c} P_\nu^{(p)}(\vec{r}) \\ P_\nu^{(s)}(\vec{r}) \end{array} \right) = \left(\begin{array}{c} \hat{\theta} \\ \hat{\varphi} \end{array} \right) \cdot \vec{P}_\nu(\vec{r}) \tag{17.13}$$

and where $\hat{r}$ denotes the scattered field propagation direction, and $\hat{\theta}$ and $\hat{\varphi}$ denote the polarization directions, defined by

$$\left(\begin{array}{c} \hat{r} \\ \hat{\theta} \\ \hat{\varphi} \end{array} \right) = \left(\begin{array}{ccc} \sin\theta\cos\varphi & \sin\theta\sin\varphi & \cos\theta \\ \cos\theta\cos\varphi & \cos\theta\sin\varphi & -\sin\theta \\ -\sin\varphi & \cos\varphi & 0 \end{array} \right) \left(\begin{array}{c} \hat{x} \\ \hat{y} \\ \hat{z} \end{array} \right) \tag{17.14}$$

(see Fig. 9.6). Written more succinctly, Eq. 17.11 becomes

$$E_\nu^{(p,s)}(\vec{r}) = \pi\mu_0\nu^2 \frac{e^{i2\pi\kappa r}}{r} \mathcal{P}_\nu^{(p,s)}(\kappa\hat{r}) \tag{17.15}$$

That is, the components of $\vec{E}_\nu(\vec{r})$ are essentially spherical waves whose amplitudes are modulated by the 4D spatiotemporal Fourier transform of the transverse polarization density distribution. Reiterating some results from Section 9.2.1, this modulation depends on the propagation direction $\hat{r} = \frac{\vec{r}}{r}$ of the scattered field (assuming κ is fixed). Moreover, a one-to-one correspondence is established between the scattered light propagation direction and the spatial frequency of the polarization density, which, as we will see below, provides a useful basis for an intuitive understanding of scattered field radiation patterns.

Finally, as before from Eq. 9.64, we have

$$I_\nu(\vec{r}) = \left| E_\nu^{(p)}(\vec{r}) \right|^2 + \left| E_\nu^{(s)}(\vec{r}) \right|^2 \tag{17.16}$$

Plane-Wave Illumination

As a first example, let us consider an induced polarization density distribution $P_\nu(\vec{r}_0)$ driven by a nonlinear susceptibility $\chi^{(n)}$ of order n, where the driving fields $E_{\nu_1}, \ldots, E_{\nu_n}$ are all plane waves that propagate along the z axis. That is,

$$E_{\nu_m}(\vec{r}) = E_{\nu_m} e^{i2\pi\kappa_m z_0} \tag{17.17}$$

(the vectorial notation has been dropped for simplicity, and $m = \{1, \ldots, n\}$). An insertion of Eqs. 17.17 and 17.8 into Eq. 17.15 then leads to

$$E_\nu(\vec{r}) = \pi \frac{e^{i2\pi\kappa r}}{\lambda^2 r} \left(\prod_{m=1}^{n} E_{\nu_m} \right) \int \mathrm{d}^3\vec{r}_0\, e^{-i2\pi\kappa\hat{r}\cdot\vec{r}_0} e^{i2\pi\kappa_\Sigma z_0} \chi^{(n)}(-\nu; \nu_1, \ldots, \nu_n) \tag{17.18}$$

where we have taken into account the fact that the speed of light in vacuum is given by $(\mu_0 \epsilon_0)^{-1/2}$.

Before proceeding, we recall that $\chi^{(n)}(-\nu; \nu_1, \ldots, \nu_n)$ is not only frequency dependent but also spatially dependent, in general. Since the radiation pattern of the scattered field is critically dependent on this spatial dependence, we modify our notation to make this explicit, and substitute $\chi^{(n)}(-\nu; \nu_1, \ldots, \nu_n) \to \chi^{(n)}(\vec{r}_0)$.

Defining the spatial 3D Fourier transform of the susceptibility to be

$$\hat{\chi}^{(n)}(\vec{\kappa}) = \int \mathrm{d}^3\vec{r}_0\, e^{-i2\pi\vec{\kappa}\cdot\vec{r}_0} \chi^{(n)}(\vec{r}_0) \tag{17.19}$$

we obtain finally

$$E_\nu(\vec{r}) = \pi \frac{e^{i2\pi\kappa r}}{\lambda^2 r} \left(\prod_{m=1}^{n} E_{\nu_m} \right) \hat{\chi}^{(n)}(\kappa\sin\theta\cos\varphi,\ \kappa\sin\theta\sin\varphi,\ \kappa\cos\theta - \kappa_\Sigma) \tag{17.20}$$

where we have made use of the definition of $\hat{r}$ provided in Eq. 17.14.

Equation 17.20 is revealing. For example, let us examine the simplest of geometries where $\chi^{(n)}(\vec{r}_0)$ is spatially uniform in both axial and transverse directions. Such a geometry technically violates the Fraunhofer requirement of a finite-sized sample, but we will continue to assume that the observation distance is much larger than the sample size so that the Fraunhofer approximation may be considered valid. Since $\chi^{(n)}(\vec{r}_0)$ is spatially uniform, $\hat{\chi}^{(n)}(\vec{\kappa})$ is non-zero only when $\kappa_x = \kappa_y = 0$ and $\kappa_z = 0$. However, for Eq. 17.20 to yield significant scattered radiation, the first of these conditions prescribes $\sin\theta = 0$ (or equivalently $\theta = 0$ and $\cos\theta = 1$), meaning that the scattered field must be forward directed. The second of these conditions, in turn, prescribes $\kappa = \kappa_\Sigma$, which is the phase-matching condition, here along the z axis.

From our definitions of wavenumbers, this condition may be recast in the form

$$n(\nu)\nu = \sum_{m=1}^{n} n(\nu_m)\nu_m \tag{17.21}$$

where, again, $n(\nu)$ is the frequency dependent index of refraction (not to be confused with the nonlinear order).

We observe that if the sample exhibits no dispersion, that is, if $n(\nu)$ is independent of frequency, then the phase matching condition reduces to the frequency matching condition (see Eq. 17.7), and is automatically satisfied. On the other hand, if the sample is dispersive, then from our simple scalar description it is generally impossible to satisfy both the frequency and phase matching conditions. As a result, it is generally impossible to obtain nonlinear scattering from a spatially uniform sample when using colinear plane-wave driving fields.

A strategy to recover nonlinear scattering in this case is to impose some degree of sample non-uniformity. This can be achieved in a variety of ways. For example, the sample can be shortened to a finite length, thereby loosening the strict equality constraint imposed by Eq. 17.21 and allowing some degree of phase mismatch. Specifically, phase mismatch can become significant if the sample is shorter than its so-called coherence length [4] (not be confused with the optical coherence length associated with the temporal coherence of light). Alternatively, sample non-uniformity can also be achieved by imparting an axial periodicity to the medium structure, such that $\hat{\chi}^{(n)}(\vec{\kappa})$ contains a significant Fourier component when $\kappa_z = \kappa - \kappa_\Sigma$. Such a strategy is called quasi phase-matching.

Alternatively yet again, our discussion has been based on Eq. 17.9, which implicitly assumes that the sample medium is isotropic. If this is not the case, then simultaneous frequency and phase matching can be achieved by proper orientation of the medium such that Eq. 17.21 holds. This is the most common technique for harmonic generation when the nonlinear medium is crystalline.

Focused Illumination

We have based our discussion above on the premise that the polarization density was driven by a set of colinear plane waves. In practical microscopy applications, however, this is hardly ever the case. To begin, nonlinear interactions require high optical power densities to produce signals of adequate strength for imaging. To attain such high power densities, beam focusing is indispensable. In addition, an important benefit of beam focusing is that it provides intrinsic 3D resolution by confining the induced polarization density to a well-defined volume about the focal center. This principle of intrinsic 3D resolution was a defining characteristic of multiphoton microscopy, and is also a defining characteristic of coherent nonlinear microscopy, multiharmonic microscopy included. However, a focus-induced confinement of the polarization density turns out to have a significant impact on the resulting radiation pattern. This can be appreciated, at least qualitatively, from the simple description of nonlinear scattering provided by Eq. 17.20.

In particular, it was noted in Section 5.1.1 that a focused Gaussian beam undergoes a π phase shift relative to a plane wave, known as the Gouy phase shift. This phase shift is defined by $\eta(z) = \tan^{-1}\left(\frac{z}{z_R}\right)$, where z is the displacement from focus and z_R the Rayleigh length (see Eq. 5.8). Let us assume that n such Gaussian beams are focused to the same spot, such that they locally drive a nonlinear susceptibility of order n. As illustrated in Fig. 17.3, the

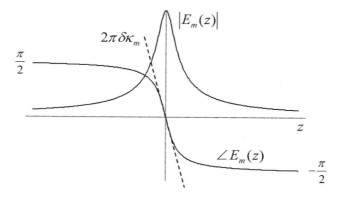

Figure 17.3. On-axis amplitude and phase (relative to a plane wave) of a Gaussian driving field $E_m(\vec{r})$.

phase shifts incurred by these beams occur mostly within the Rayleigh length, and hence in the region of largest power density where the induced polarization density is confined. In this region $\eta_m(z)$ can be roughly approximated as varying linearly with z which, in turn, can be equivalently interpreted as an effective reduction in the wavenumber κ_m of each beam about its focus. That is, for each beam we can write $\kappa_m^{\text{eff}} = \kappa_m - \delta\kappa_m$, where

$$2\pi\delta\kappa_m \approx \left.\frac{\partial\eta_m(z)}{\partial z}\right|_{z=0} = \frac{1}{z_{R_m}} = \frac{1}{\pi w_m^2 \kappa_m} \tag{17.22}$$

and z_{R_m} is the Rayleigh length of beam m focused to waist size w_m.

To appreciate the consequences of such an effective reduction in wavenumber of the driving fields, we return to our simple description of nonlinear scattering described in the previous section, noting that Eq. 17.20 must now be modified to read

$$E_\nu(\vec{r}) = \pi\frac{e^{i2\pi\kappa r}}{\lambda^2 r}\left(\prod_{m=1}^{n} E_m\right)\hat{\chi}^{(n)}\left(\kappa\sin\theta\cos\varphi,\ \kappa\sin\theta\sin\varphi,\ \kappa\cos\theta - \kappa_\Sigma + \delta\kappa_\Sigma\right) \tag{17.23}$$

where $\delta\kappa_\Sigma = \delta\kappa_1 + \cdots + \delta\kappa_n$.

This seemingly innocuous modification can cause dramatic changes in the scattered radiation pattern. As an example, let us consider a susceptibility distribution $\chi^{(n)}(\vec{r}_0)$ that is transversely uniform. In this case $\hat{\chi}^{(n)}(\kappa_x, \kappa_y, \kappa_z)$ is non-zero only when $\kappa_x = \kappa_y = 0$, which, from Eq. 17.23, prescribes the condition $\sin\theta = 0$ (or equivalently $\cos\theta = 1$), meaning that the scattered radiation is forward directed. However, for there to be significant scattered radiation, $\hat{\chi}^{(n)}(\kappa_x, \kappa_y, \kappa_z)$ must also be non-zero for $\kappa_z = \kappa - \kappa_\Sigma + \delta\kappa_\Sigma$. If the medium dispersion is negligible, this last condition reduces to $\kappa_z = \delta\kappa_\Sigma$. Inasmuch as $\delta\kappa_\Sigma > 0$ for forward-propagating focused beams, we conclude that even if the medium is dispersionless, the susceptibility distribution must contain axial structure for the focused beams to produce significant nonlinear scattering. A particularly simple way to guarantee such axial structure is with the use of a sample that features an abrupt interface, in which case $\hat{\chi}^{(n)}(\kappa_x, \kappa_y, \kappa_z)$ is non-zero over a broad range of κ_z's, including $\kappa_z = \delta\kappa_\Sigma$.

As a second example, let us consider a susceptibility distribution $\chi^{(n)}(\vec{r}_0)$ that is now axially uniform. In this case, $\hat{\chi}^{(n)}(\kappa_x, \kappa_y, \kappa_z)$ is non-zero only when $\kappa_z = 0$, and Eq. 17.23 prescribes the condition $\kappa \cos \theta - \kappa_\Sigma + \delta \kappa_\Sigma = 0$. The propagation direction of the scattered radiation is then defined by $\cos \theta = \frac{1}{\kappa}(\kappa_\Sigma - \delta \kappa_\Sigma) \approx 1 - \frac{1}{\kappa} \delta \kappa_\Sigma$, which, because $\delta \kappa_\Sigma > 0$ for forward-propagating focused beams, must be off axis. However, for there to be significant scattered radiation, $\hat{\chi}^{(n)}(\kappa_x, \kappa_y, \kappa_z)$ is also required to be non-zero for $\kappa_x = \kappa \cos \varphi \sqrt{1 - \cos^2 \theta}$ and $\kappa_y = \kappa \sin \varphi \sqrt{1 - \cos^2 \theta}$, meaning that the susceptibility distribution must contain transverse structure in either the x or y directions, or both. For example, propagation along $\varphi = 0$ results from transverse structure in the x direction, whereas propagation along $\varphi = \frac{\pi}{2}$ results from transverse structure in the y direction. Such off-axis scattering will be treated in more detail in Section 17.2 when we consider second-harmonic generation from a membrane.

The qualitative results presented above are confirmed by more rigorous calculations. In particular, an analytical derivation of the nth harmonic field generated by paraxially focused Gaussian beams is provided in [4, 15] using a slowly-varying amplitude approximation. Alternatively, derivations of the scattered field using the vectorial Green function formalism adopted here (Eq. 17.9) are provided in [8]. These calculations can be extended beyond the paraxial approximation by explicitly including the possibility of an axial polarization component at the Gaussian beam focus [7, 30].

Forward Versus Backward Scattering

Perhaps the most important application of coherent nonlinear microscopy is in biological imaging. For example, many tissue constituents possess intrinsic nonlinear susceptibilities of various orders that can be used to generate nonlinear scattering. As we have seen above, nonlinear scattering is dominantly forward directed when the susceptibility distribution is slowly varying in space, even in the case of nonlinear scattering that is driven by focused illumination. Since tissue constituents in biological tissue typically fall into this category of being only slowly varying in space, we conclude that the most appropriate detector position for coherent nonlinear microscopy is in the forward direction. Such a detector position, however, limits the applicability of coherent nonlinear microscopy to thin samples only.

Fortunately, exceptions exist to this conclusion. In particular, it is possible that the susceptibility distribution does indeed contain high enough spatial frequency components to allow nonlinear signal to be directly generated in the backward direction. Such is occasioned, for example, with abrupt interfaces or with microscopic scattering centers that are much smaller than a wavelength [9, 17].

Alternatively, if the sample is highly turbid, then nonlinear signal that is initially forward directed can be re-routed into the backward direction by random linear scattering within the sample medium. Note that such signal re-routing does not undermine the resolution of a coherent nonlinear microscope for the same reason that it does not undermine the resolution of a multiphoton microscope (see discussion in Section 16.7.1). Indeed, such re-routing of nonlinear signal is one of the rare occasions when scattering can actually be beneficial to imaging. By exploiting scattering in turbid media, the detector can thus be placed in the backward direction, allowing coherent nonlinear imaging to be possible even in thick samples.

The question as to which mechanism, direct or indirect, contributes the most to backward-propagating signal depends largely on the sample in question. Studies have suggested that the indirect mechanism is dominant for biological tissue such as brain or skin [11, 27].

17.2 SECOND-HARMONIC GENERATION (SHG) MICROSCOPY

Having developed the basic formalism common to all coherent nonlinear microscopes, we can now direct our attention specifically to multiharmonic contrast mechanisms. We begin by considering second-harmonic generation (SHG). The basic mechanism for SHG is depicted in Fig. 17.1, wherein two input fields that vary harmonically with the same frequency $\nu_1 = \nu_2 = \nu_i$ are combined by a second-order susceptibility to yield an induced polarization density at the sum frequency $2\nu_i$. That is, reverting to our tensorial notation, we have

$$\vec{P}_{2\nu_i} = \epsilon_0 \chi^{(2)}(-2\nu_i; \nu_i, \nu_i) : \vec{E}_{\nu_i}\vec{E}_{\nu_i} \tag{17.24}$$

The polarization density $\vec{P}_{2\nu_i}$, in turn, gives rise to a nonlinearly scattered field $\vec{E}_{2\nu_i}$ which represents the signal of interest in SHG microscopy [13, 14].

In practice, a SHG microscope is quite similar to a TPEF microscope. As with TPEF, the two input fields in Eq. 17.24 can be derived from the same laser beam since they are of the same frequency. Moreover, as discussed above, the laser beam is generally required to be focused, which has the effect of conferring intrinsic 3D resolution to SHG microscopy in the same manner as TPEF microscopy. In practice, then, SHG and TPEF contrasts can be simultaneously obtained from the same instrument (in which case SHG can even become resonantly enhanced).

However, there is an obvious difference between SHG and TPEF. SHG is based on scattering whereas TPEF is based on fluorescence. As such, SHG and TPEF probe very different sample properties. Moreover, as noted throughout this chapter, coherent scattering is sensitive not only to the amplitude of the illumination field but also to its phase, leading to a much more highly intricate dependence of SHG on both the excitation beam focus and the sample susceptibility distribution.

17.2.1 Symmetry Properties

There is yet another difference between SHG and TPEF whose origin lies in the spatial symmetry properties of susceptibilities. We recall that the susceptibilities $\chi^{(n)}$ are tensors whose components characterize physical properties of a medium. In cases where the medium exhibits a spatial symmetry, then this symmetry must be manifest in $\chi^{(n)}$. While a full consideration of spatial symmetry properties is quite beyond the scope of this book (see [4, 5, 24] for more details), a fundamental result can be readily derived in the electric dipole approximation for media that exhibit centrosymmetry (i.e. possess a center of inversion). This result states that if a medium is centrosymmetric then all even-ordered susceptibility tensors must vanish identically.

Inasmuch as $\chi^{(2)}$ is even ordered, the above result has major ramifications in SHG microscopy. In particular, it implies that only media that are non-centrosymmetric can produce SHG. Moreover, it implies that if the second-order susceptibility of a nonlinear medium is $\chi^{(2)}$, then upon inversion of the medium through its point of non-centrosymmetry, the second-order susceptibility must change sign and become $-\chi^{(2)}$.

As an example, let us imagine that the nonlinear medium is a non-centrosymmetric molecule that possesses a well-defined axis and orientation, and that this molecule produces a second-harmonic field $\vec{E}_{2\nu_i}$. Upon inversion of the molecule, the second-harmonic field becomes $-\vec{E}_{2\nu_i}$. That is, the phase of the second-harmonic field depends critically on the orientation of the molecule. Such an orientation dependence is absent for odd-ordered susceptibilities (e.g. linear scattering). Moreover, it is patently absent for TPEF given that the phase of fluorescence is random. As we will see below, the dependence of SHG on non-centrosymmetry can provide specific information on molecular orientation that is inaccessible to fluorescence.

As an added note, a distinction should be made between second-harmonic scattering and second-harmonic generation. In the literature, second-harmonic scattering from a single molecule is generally referred to as hyper-Rayleigh scattering [10], as is second-harmonic scattering from a collection of N randomly oriented molecules. Because random molecular orientations imply random second-harmonic phases, the total second-harmonic power scattered from N randomly oriented molecules must scale as N, on average. This is the same scaling law obeyed by incoherent emission processes such as fluorescence.

On the other hand, second-harmonic scattering from N *oriented* molecules is referred to as second-harmonic generation (SHG). When molecular orientation is well defined, so too are the phases of the second-harmonic fields, meaning that the second-harmonic fields add coherently. The total second-harmonic power generated by N molecules then depends on the spatial distribution of the molecules. For example, if the molecules are tightly confined to separation distances much less than a wavelength, then the total output power scales as N^2, whereas if they are sparsely distributed the total output power scales as N (see Section 9.4 for caveats regarding the first of these scaling laws).

17.2.2 3D-Gaussian Approximation

As emphasized above, SHG depends critically on both the excitation field profile and on the local susceptibility distribution in the sample. To highlight these dependencies, Eq. 17.24 is explicitly recast as a function of $\vec{r}_0$, obtaining

$$\vec{P}_{2\nu_i}(\vec{r}_0) = \epsilon_0 \chi^{(2)}(\vec{r}_0) : \vec{E}_{\nu_i}(\vec{r}_0)\vec{E}_{\nu_i}(\vec{r}_0) \tag{17.25}$$

The profile of the excitation field $\vec{E}_{\nu_i}(\vec{r}_0)$ depends, of course, on the illumination geometry. If the excitation beam is Gaussian and the numerical aperture of the focus is modest (NA $\lesssim 0.7$), then $\vec{E}_{\nu_i}(\vec{r}_0)$ takes on a Gaussian–Lorentzian profile, as described in Section 16.3. For higher numerical apertures, the profile becomes more complicated and a fully vectorial description is required (cf. [30]). Nevertheless, as in TPEF microscopy, this requirement can be relaxed with the adoption of a 3D-Gaussian field approximation. As argued in Section 16.3, a 3D-Gaussian approximation is reasonably accurate when considering nonlinear excitation, as we are considering here. Moreover, we will assume that the field near the focus is linearly polarized along the $\hat{x}$ direction. That is, we write $\vec{E}_{\nu_i}(\vec{r}_0) = E_{\nu_i}(\vec{r}_0)\hat{x}$, leading to the further approximation

$$E_{\nu_i}(\vec{r}_0) \approx E_{\nu_i} e^{-\rho_0^2/w_0^2 - z_0^2/w_z^2 + i2\pi(\kappa_i - \delta\kappa_i)z_0} \tag{17.26}$$

where w_0 and w_z are the field waists in the transverse and axial directions, respectively, and $\delta\kappa_i$ characterizes the focus-induced phase mismatch (see Eq. 17.22).

In general, $\vec{P}_{2\nu_i}(\vec{r}_0)$ need not be polarized in the same direction as $\vec{E}_{\nu_i}(\vec{r}_0)$ because $\chi^{(2)}(\vec{r}_0)$ is a tensor. However, for added simplicity, we assume that $\chi^{(2)}(\vec{r}_0)$ is dominated by a single diagonal component $\chi^{(2)}(\vec{r}_0)$, also in the $\hat{x}$ direction. Such is the case, for example, when $\overset{\leftrightarrow}{\chi}^{(2)}(\vec{r}_0)$ arises from a distribution of uni-axial non-centrosymmetric molecules locally aligned in $\hat{x}$, allowing us to write $\vec{P}_{2\nu_i}(\vec{r}_0) = P_{2\nu_i}(\vec{r}_0)\hat{x}$. We have then,

$$P_{2\nu_i}(\vec{r}_0) = \epsilon_0 \chi^{(2)}(\vec{r}_0) E_{\nu_i}^2(\vec{r}_0) \tag{17.27}$$

where $\chi^{(2)}(\vec{r}_0)$ scales with molecular density. The simplifications leading to Eq. 17.27 are by no means general, but they will be convenient for our discussion.

Thus, from Eq. 17.11 have

$$E_{2\nu_i}^{(p,s)}(\vec{r}) = \pi\mu_0\nu^2 \frac{e^{i2\pi\kappa r}}{r} \int d^3\vec{r}_0\, e^{-i2\pi\frac{\kappa}{r}\vec{r}\cdot\vec{r}_0} P_{2\nu_i}^{(p,s)}(\vec{r}_0) \tag{17.28}$$

From Eqs. 17.12, 17.13, and 9.58, this can be rewritten as

$$\begin{pmatrix} E_{2\nu_i}^{(p)}(\vec{r}) \\ E_{2\nu_i}^{(s)}(\vec{r}) \end{pmatrix} = \begin{pmatrix} \cos\theta\cos\varphi \\ -\sin\varphi \end{pmatrix} E_{2\nu_i}(\vec{r}) \tag{17.29}$$

where we have made use of Eqs. 17.27 and 17.26 to define

$$E_{2\nu_i}(\vec{r}) = \pi \frac{e^{i2\pi\kappa r}}{\lambda^2 r} E_{\nu_i}^2 \iint d^2\vec{\rho}_0\, dz_0\, e^{-i2\pi\frac{\kappa}{r}(\vec{\rho}_0\cdot\vec{\rho} + z_0 z)} e^{-2\rho_0^2/w_0^2 - 2z_0^2/w_z^2 + i4\pi(\kappa_i - \delta\kappa_i)z_0} \chi^{(2)}(\vec{\rho}_0, z_0) \tag{17.30}$$

Finally, from Eq. 17.16, we arrive at

$$I_{2\nu_i}(\vec{r}) = \left(1 - \sin^2\theta\cos^2\varphi\right) \left|E_{2\nu_i}(\vec{r})\right|^2 \tag{17.31}$$

Inasmuch as $E_{2\nu_i}(\vec{r})$ is proportional to $E_{\nu_i}^2$ from Eq. 17.30, we conclude that $I_{2\nu_i}(\vec{r})$ is proportional to $I_{\nu_i}^2$, similarly as TPEF.

Equation 17.30 represents the far field generated from second-harmonic sources polarized in the $\hat{x}$ direction and induced by a 3D Gaussian excitation beam. Equation 17.31 represents the corresponding SHG radiation pattern. Our description remains incomplete, however. Though

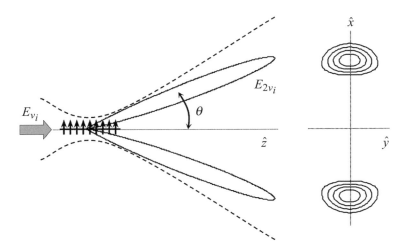

Figure 17.4. Radiation pattern of off-axis SHG produced by membrane markers.

we have specified the spatial distribution of $E_{2\nu_i}(\vec{r}_0)$, we have yet to specify the spatial distribution of $\chi^{(2)}(\vec{r}_0)$, which we turn to in the example below.

Example: SHG from a Labeled Membrane

To be specific, let us consider a standard application of SHG microscopy, which is to image labeled biological membranes, such as cellular membranes [18]. Uni-axial non-centrosymmetric molecules can be readily designed so that they attach themselves to these membranes with a preferred orientation roughly perpendicular to the membrane (see Fig. 17.4). If these molecules are illuminated with a tightly focused beam parallel to the membrane plane (i.e. side-on illumination), then the same conditions apply as those used to derive Eq. 17.30. The distribution of $\chi^{(2)}(\vec{r}_0)$ is now well defined, and we write

$$\chi^{(2)}(\vec{r}_0) = \chi^{(2)}_\rho \delta(x_0) \tag{17.32}$$

where $\chi^{(2)}_\rho$ is a surface susceptibility. Equation 17.30 then reduces to

$$E_{2\nu_i}(\vec{r}) = \pi \frac{e^{i2\pi\kappa r}}{\lambda^2 r} \chi^{(2)}_\rho E^2_{\nu_i} \iint dy_0 \, dz_0 \, e^{-i2\pi\frac{\kappa}{r}(y_0 y + z_0 z)} e^{-2y_0^2/w_0^2 - 2z_0^2/w_z^2 + i4\pi(\kappa_i - \delta\kappa_i)z_0} \tag{17.33}$$

or

$$E_{2\nu_i}(\vec{r}) = \pi \frac{e^{i2\pi\kappa r}}{\lambda^2 r} A_\rho \chi^{(2)}_\rho E^2_{\nu_i} \, e^{-\pi^2 w_0^2 \kappa^2 y^2/2r^2} e^{-\pi^2 w_z^2(\kappa z - 2(\kappa_i - \delta\kappa_i)r)^2/2r^2} \tag{17.34}$$

where $A_\rho = \left(\frac{\pi}{2}\right) w_0 w_z$ is the effective SHG active area, defined in much the same way as we defined an active volume with Eq. 13.38.

Using the associations

$$x = r \sin \theta \cos \varphi$$
$$y = r \sin \theta \sin \varphi \qquad (17.35)$$
$$z = r \cos \theta$$

and also assuming that $\kappa \simeq 2\kappa_i$, valid for a medium with negligible dispersion, then Eq. 17.34 can be recast as

$$E_{2\nu_i}(\vec{r}) = \pi \frac{e^{i4\pi\kappa_i r}}{\lambda^2 r} A_\rho \chi_\rho^{(2)} E_{\nu_i}^2 e^{-2\pi^2 w_0^2 \kappa_i^2 (\sin\theta \sin\varphi)^2} e^{-2\pi^2 w_z^2 \kappa_i^2 \left(1 - \frac{\delta\kappa_i}{\kappa_i} - \cos\theta\right)^2} \qquad (17.36)$$

A plot of the angular distribution of the SHG intensity reveals that the SHG radiation is globally forward directed; however, it is confined to two off-axis lobes at $\cos\theta \approx 1 - \frac{\delta\kappa_i}{\kappa_i}$ and $\varphi \approx \{0, \pi\}$. An interpretation of this radiation pattern is straightforward and follows directly from our qualitative arguments presented in Section 17.1.2 for focused-beam illumination. In brief, since we are dealing with a second-order nonlinearity, Eq. 17.23 reduces to

$$E_{2\nu_i}(\vec{r}) = \pi \frac{e^{i2\pi\kappa r}}{\lambda^2 r} E_{\nu_i}^2 \hat{\chi}^{(2)} (\kappa \sin\theta \cos\varphi, \ \kappa \sin\theta \sin\varphi, \ \kappa \cos\theta - 2\kappa_i + 2\delta\kappa_i) \qquad (17.37)$$

Since the susceptibility distribution $\chi^{(2)}(x, y, z)$ in question is planar along the y–z directions (Eq. 17.32), then $\hat{\chi}^{(2)}(\kappa_x, \kappa_y, \kappa_z)$ is non-zero only when $\kappa_y = 0$ and $\kappa_z = 0$. Equation 17.37 then prescribes the conditions $\sin\theta \sin\varphi = 0$ and $\cos\theta = \frac{2}{\kappa}(\kappa_i - \delta\kappa_i)$. The second of these conditions forces $\cos\theta \approx 1 - \frac{\delta\kappa_i}{\kappa_i}$ for a dispersionless medium, while the first condition forces $\sin\varphi \approx 0$, and hence $\varphi \approx \{0, \pi\}$. These are the same results as obtained from our more exact derivation of the radiation pattern given by Eq. 17.36, and illustrated in Fig. 17.4.

A word of caution should be made here. While our arguments from Section 17.1.2 are useful in providing a qualitative description of SHG radiation patterns, they remain approximate at best. In particular, these same arguments suggest that SHG cannot be generated from a focused beam in an infinite, uniform, dispersionless medium, because in such a case $\hat{\chi}^{(2)}(\kappa_x, \kappa_y, \kappa_z)$ is non-zero only when $\kappa_x = \kappa_y = \kappa_z = 0$, and it is impossible to find a radiation direction θ and φ in Eq. 17.37 matching this condition. However, a more exact calculation reveals that, in fact, SHG can be generated with a focused beam in an infinite, uniform, dispersionless medium [15], though it is generated inefficiently. The reason that our qualitative argument is somewhat inaccurate is that, while we have taken $P_{2\nu_i}(\vec{r}_0)$ to be confined about the focus, it is not sufficiently confined in SHG that our simple model of the excitation beam as a 3D-Gaussian field with an effectively reduced wavenumber along the z direction is entirely accurate. Nevertheless, in cases where $\hat{\chi}^{(2)}(\kappa_x, \kappa_y, \kappa_z)$ contains structure, this simple model is useful in providing an intuitive understanding of SHG radiation patterns.

17.2.3 Applications

As noted above, SHG and TPEF contrasts can be obtained from the same instrument. Their combination yields complementary information about non-centrosymmetric structure (SHG)

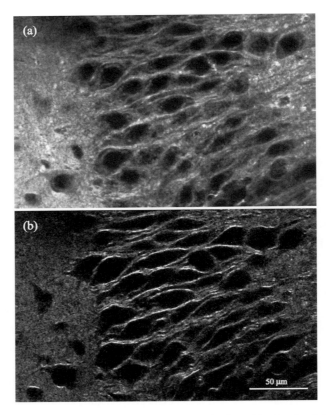

Figure 17.5. Simultaneous TPEF (a) and SHG (b) images of cultured brain slice labeled with a membrane marker. Adapted from [16].

and fluorescence (TPEF). As it happens, non-centrosymmetric structures are fairly common in biological tissue. The most prevalent of these is collagen, a structural protein, whose fibrilous geometry can provide extraordinarily large values of $\chi^{(2)}$, leading to striking SHG contrast [12, 26, 27]. Other structures that exhibit SHG contrast are protein complexes present in muscle tissue, and microtubules present in the cytoskeleton or in mitotic spindles [6]. In all cases, SHG images are obtained from intrinsic contrast that requires no tissue labeling whatsoever.

Alternatively, SHG imaging is useful for imaging cell membranes, as illustrated in Fig. 17.5. Since membranes themselves do not provide a sufficient $\chi^{(2)}$ to be imaged directly, exogenous labeling is required. Considerable efforts have been devoted to the development of non-centrosymmetric molecules that insert themselves into cell membranes in an aligned manner perpendicular to the membrane (the same geometry as illustrated in Fig. 17.4). One of the motivations for the development of such molecules comes from their possible application as membrane-potential reporters [3]. In particular, living cells exhibit large electric fields across their membranes on the order of 10^7 V/m. Electric fields of this magnitude, when applied to non-centrosymmetric molecules, can lead to changes in the molecules' effective second-order susceptibilities, leading to changes in their capacity to produce SHG power. Needless to say, any possibility of imaging of such changes in power with microscopic spatial resolution is highly attractive.

A variety of physical mechanisms can be responsible for the changes in the effective second-order susceptibility. For example, an electric field applied to polar molecules can force them to align, thereby inducing a macroscopic non-centrosymmetry that can lead to a field-dependent $\chi^{(2)}$. SHG produced by such a poling mechanism is generally referred to as electric-field induced second-harmonic generation (or EFISH). Alternatively, an electric field can alter the microscopic non-centrosymmetry already present in aligned molecules, by way of $\chi^{(3)}$. Thus, we can have

$$
\begin{aligned}
P_{2\nu_i} &= \epsilon_0 \chi^{(2)}(-2\nu_i; \nu_i, \nu_i) E_{\nu_i} E_{\nu_i} + \epsilon_0 \chi^{(3)}(-2\nu_i; \nu_i, \nu_i, 0) E_{\nu_i} E_{\nu_i} E_{dc} \\
&= \epsilon_0 \left[\chi^{(2)}(-2\nu_i; \nu_i, \nu_i) + \chi^{(3)}(-2\nu_i; \nu_i, \nu_i, 0) E_{dc} \right] E_{\nu_i} E_{\nu_i}
\end{aligned}
\tag{17.38}
$$

where E_{dc} represents the field across the membrane (essentially static compared to optical frequencies), and the terms in brackets can be thought of as an effective second-order susceptibility that is field dependent. This last mechanism has been shown to result from the Stark effect [22]. For more information on the mechanisms of membrane potential sensitivity in SHG microscopy as well as some biological applications, the reader is referred to [21].

17.3 THIRD-HARMONIC GENERATION (THG) MICROSCOPY

Third-harmonic generation (THG) microscopy is a natural extension of SHG microscopy to the third-order susceptibility. That is, we consider the induced polarization density at the sum frequency $3\nu_i$ of three driving fields of frequency ν_i, given by

$$
\vec{P}_{3\nu_i} = \epsilon_0 \chi^{(3)}(-3\nu_i; \nu_i, \nu_i, \nu_i) : \vec{E}_{\nu_i} \vec{E}_{\nu_i} \vec{E}_{\nu_i}
\tag{17.39}
$$

In turn, $\vec{P}_{3\nu_i}$ gives rise to a scattered field $\vec{E}_{3\nu_i}$ by way of Eq. 17.9. The process of THG is illustrated in Fig. 17.1. As in the case of SHG microscopy, these fields are generally derived from a single laser beam, presumably pulsed to obtain adequate power density for THG imaging.

Despite their similarities, THG and SHG microscopies differ in several fundamental aspects. The first and most obvious difference is that $P_{3\nu_i}$ is now proportional to $E_{\nu_i}^3$ rather than $E_{\nu_i}^2$ (dropping our vector notation for simplicity). Following the same arguments as led to Eq. 17.16, this implies that $I_{3\nu_i}$ is proportional to $I_{\nu_i}^3$. Thus, for the same driving field frequency, we find that THG is more tightly confined about the focal center than SHG, leading to an enhanced 3D resolution. Moreover, as a result of this tighter confinement, the simple model for focused illumination presented in Section 17.1.2 becomes more accurate than it was for SHG. Indeed the prediction derived from this model that THG cannot be generated from focused illumination in an infinite, uniform, dispersionless medium becomes confirmed by an exact calculation [4]. Part of the reason for this absence of THG from an infinite volume stems from the fact that three focused driving fields are now involved in the nonlinear scattering of light, meaning that the induced $\vec{P}_{3\nu_i}$ is subject to the cumulative effects of three Gouy phase shifts. As a result, THG generated in front of the focal center tends to be directly out of phase with THG generated behind the focal center and therefore cancels.

A second fundamental difference between THG and SHG is that the former, because it relies on an odd-ordered susceptibility, is not subject to a requirement of medium non-centrosymmetry. That is, scattering centers need not exhibit a preferred orientation to produce THG. On the one hand, such a relaxation of the non-centrosymmetry requirement considerably broadens the panoply of molecular species that can produce THG. On the other hand, a potential cause for concern might be that the environment surrounding the sample, such as the aqueous environment surrounding cells or tissue structures, can itself have the capacity to produce THG background that can overwhelm a THG signal of interest. Fortunately, this is not the case. As noted above, THG cannot be generated from a focused beam in large uniform volume and can only be generated from sample regions that contain structure. As such, THG microscopy has the advantage that it is essentially background free. The same is true of SHG microscopy, but for a different reason. While SHG from a focused beam in a large volume is possible, it only becomes significant when the volume exhibits non-centrosymmetry, which is generally not the case in practice, particularly when the large volume is amorphous. Thus SHG microscopy of cells or tissue structures in aqueous environments is also essentially background free.

THG microscopy has been demonstrated by several groups (e.g. [11, 25, 29]). The most common application of THG microscopy is the imaging of densely packed lipids, such as found as in cell vesicles or myelinated fibers. Such densely packed lipids lead to strong third-order susceptibilities. Moreover, they are often confined to size scales smaller than a wavelength, meaning they are highly structured, as is necessary for efficient THG.

17.4 PROBLEMS

Problem 17.1

Consider a pulsed laser beam whose power is written as

$$\Phi_l(t) = U_l \sum_{n=0}^{N} \delta(t - n\tau_l)$$

where U_l is the energy per pulse and τ_l is the pulse period. Now consider that each pulse is subject to a temporal jitter $\delta\tau_l$, which may be considered a random Gaussian variable. Show that the Fourier transform of $\Phi_l(t)$, averaged over large N, can be written as

$$\left\langle \hat{\Phi}_l(\nu) \right\rangle = U_l e^{-2\pi^2\nu^2\sigma_{\tau_l}^2} \frac{\sin(\pi N \bar{\tau}_l \nu)}{\sin(\pi \bar{\tau}_l \nu)}$$

where $\bar{\tau}_l$ is the mean pulse period, and $\sigma_{\tau_l}^2$ is the variance of the temporal jitter.

Hint: you may want to use a result from Appendix B.5.

Problem 17.2

The second-harmonic tensorial product $\vec{P} = \epsilon_0 \chi^{(2)} : \vec{E}\vec{E}$ (see Eq. 17.24) can be expanded as

$$P_i = \epsilon_0 \sum_{j=1}^{3} \sum_{k=1}^{3} \chi_{ijk}^{(2)} E_j E_k$$

This product depends on the coordinate system in which it is evaluated. The two relevant coordinate systems for this problem are the fixed laboratory system (denoted by L) and the molecule system (denoted by M), which may be arbitrarily oriented relative to the laboratory system.

Consider a uni-axial molecule oriented along $\hat{r}$, illuminated by a field given by $\vec{E}^{(L)}$ in the laboratory system.

(a) Defining $\mathbf{R}(\theta, \varphi)$ to be the rotation matrix linking the molecule system to the laboratory system (see Eq. 17.14), show that

$$P_l^{(L)} = \epsilon_0 \sum_{m=1}^{3} \sum_{n=1}^{3} \chi_{lmn}^{(L)} E_m^{(L)} E_n^{(L)}$$

where

$$\chi_{lmn}^{(2)(L)} = \sum_{i=1}^{3} \sum_{j=1}^{3} \sum_{k=1}^{3} R_{i,l}(\theta, \varphi) R_{j,m}(\theta, \varphi) R_{k,n}(\theta, \varphi) \chi_{ijk}^{(2)(M)}$$

Hint: recall that $\mathbf{R}(\theta, \varphi)$ is orthogonal.

(b) For simplicity, assume that all components of the molecule second-order susceptibility $\chi_{ijk}^{(2)(M)}$ are zero, except for $\chi_{111}^{(2)(M)} \equiv \chi_{rrr}^{(2)}$. Show that, in this case,

$$\vec{P}^{(L)} = \epsilon_0 \chi_{rrr}^{(2)} \left(\hat{r} \cdot \vec{E}^{(L)} \right)^2 \hat{r}$$

Problem 17.3

It can be shown that the susceptibility tensor $\chi^{(3)}$ responsible for third-harmonic generation in a homogenous isotropic medium can be written as

$$\chi_{klmn}^{(3)} = \chi_0 \left(\delta_{kl}\delta_{mn} + \delta_{km}\delta_{ln} + \delta_{kn}\delta_{lm} \right)$$

Show that no THG can be produced in such a medium if the driving field is a circularly polarized plane wave.

Problem 17.4

Consider generating SHG with a focused beam as in Fig. 17.4, but with two labeled membranes separated by a distance Δx_0. Each membrane exhibits identical, uniform second-order susceptibility $\chi_\rho^{(2)}$, but their markers are oriented in opposite directions.

(a) Use the 3D Gaussian approximation (Eq. 17.26) to derive the field $E_{2\nu_i}^{(2)}(\vec{r})$ produced by the two membranes. Express your answer in terms of $E_{2\nu_i}^{(1)}(\vec{r})$, the field produced by a single membrane (i.e. Eq. 17.36).

(b) As in Fig. 17.4, the SHG is emitted in two off-axis lobes at $\cos\theta \approx 1 - \frac{\delta\kappa_i}{\kappa_i}$ and $\varphi \approx [0, \pi]$. Plot the intensity ratio $\frac{I_{2\nu_i}^{(2)}(\vec{r})}{I_{2\nu_i}^{(1)}(\vec{r})}$ in the lobe directions, as a function of $\frac{\Delta x_0}{w_0}$ (Hint: use Eq. 17.22.) At approximately what value of $\frac{\Delta x_0}{w_0}$ is this intensity ratio peaked?

Problem 17.5

(a) Calculate the third-harmonic intensity pattern produced by a localized 3D-Gaussian susceptibility distribution given by

$$\chi^{(3)}(\vec{r}_0) = \chi^{(3)} e^{-r_0^2/w_\chi^2}$$

Assume a focused illumination beam and use the 3D-Gaussian illumination profile given by Eq. 17.26. Express your result in terms of r, θ and φ.

(b) Derive an expression for the backward/forward ratio of THG intensities emitted along the $\hat{z}$-axis. That is, derive an expression for

$$\frac{I_{backward}}{I_{forward}} = \frac{I_{3\nu_i}^{(\theta=\pi)}(\vec{r})}{I_{3\nu_i}^{(\theta=0)}(\vec{r})}$$

What does this ratio tend toward as $w_\chi \to 0$?

References

[1]   Allen, L. and Eberly, J. H. *Optical Resonance and Two-Level Atoms*, Dover (1987).

[2]   Bloembergen, N. *Nonlinear Optics*, 4th edn, World Scientific (1996).

[3]   Bouevitch, O., Lewis, A., Pinevsky, I., Wuskell, J. P. and Loew, L. M. "Probing membrane potential with nonlinear optics," *Biophys. J.* 65, 672–679 (1993).

[4]   Boyd, R. W. *Nonlinear Optics*, 3rd edn, Academic Press (2003).

[5]   Butcher, P. N. and Cotter, D. *The Elements of Nonlinear Optics*, Cambridge University Press (1990).

[6]   Campagnola, P. J. Mohler, W. and Millard, A. E. "Three-dimensional high-resolution second-harmonic imaging of endogenous structural proteins in biological tissues," *Biophys. J.* 81, 493–508 (2002).

[7]   Carrasco, S., Saleh, B. E. A., Teich, M. C. and Fourkas, J. T. "Second- and third-harmonic generation with vector Gaussian beams," *J. Opt. Soc. Am.* B 23, 2134–2141 (2006).

[8]   Cheng, J.-X. and Sunney Xie, X. "Green's function formulation of third harmonic generation microscopy," *J. Opt. Soc. Am.* B 19, 1604–1610 (2002).

[9]   Cheng, J.-X., Volkmer, A. and Sunney Xie, X. "Theoretical and experimental characterization of coherent anti-Stokes Raman scattering microscopy," *J. Opt. Soc. Am.* B 19, 1363–1375 (2002).

[10] Clays, K., Persoons, A. and De Maeyer, L. "Hyper-Rayleigh scattering in solution," in *Modern Nonlinear Optics, Vol. 85 of Advances in Chemical Physics*, Eds. I. Prigogine and S. A. Rice, Wiley (1994).

[11] Débarre, D., Olivier, N. and Beaurepaire, E. "Signal epidetection in third-harmonic generation microscopy of turbid media," *Opt. Express.* 15, 8913–8924 (2007).

[12] Freund, I. and Deutsch, M. "Second-harmonic microscopy of biological tissue," *Opt. Lett.* 11, 94–96 (1986).

[13] Gauderon, R., Lukins, P. B. and Sheppard, C. J. R. "Three-dimensional second-harmonic generation imaging with femtosecond laser pulses," *Opt. Lett.* 23, 1209–1211 (1998).

[14] Hellwarth, R. and Christensen, P. "Nonlinear optical microscope using second harmonic generation," *Appl. Opt.* 14, 247–251 (1975).

[15] Boyd, G. D. and Kleinman, D. A. "Parametric interaction of focused Gaussian light beams," *J. Appl. Phys.* 39, 3597–3639 (1968).

[16] Masters, B. R. and So, P. T. C. Eds., *Handbook of Biomedical Nonlinear Optical Microscopy*, Oxford University Press (2008).

[17] Mertz, J. and Moreaux, L. "Second-harmonic generation by focused excitation of inhomogeneously distributed scatterers," *Opt. Commun.* 196, 325–330 (2001).

[18] Moreaux, L., Sandre, O. and Mertz, J. "Membrane imaging by second-harmonic generation microscopy," *J. Opt. Soc. Am.* B 17, 1685–1694 (2000).

[19] Mukamel, S. *Principles of Nonlinear Optical Spectroscopy*, Oxford University Press (1999).

[20] Orr, B. J. and Ward, J. F. "Perturbation theory of the non-linear optical polarization of an isolated system," *Mol. Phys.* 20, 513–526 (1971).

[21] Pons, T., Moreaux, L., Mongin, O., Blanchard-Desce, M. and Mertz, J. "Mechanisms of membrane potential sensing with second-harmonic generation microscopy," *J. Biomed. Opt.* 8, 428–431 (2003).

[22] Pons, T., Mertz, J. "Membrane potential detection with second-harmonic generation and two-photon excited fluorescence: A theoretical comparison," *Opt. Commun.* 258, 203–209 (2006).

[23] Rothberg, L. "Dephasing-induced coherent phenomena," in *Progress in Optics Vol. 24*, Ed. Wolf, E. Elsevier Science (1987).

[24] Shen, Y. R. *The Principles of Nonlinear Optics*, John Wiley & Sons (1984).

[25] Squier, J. A., Müller, M., Brakenhoff, G. J. and Wilson, K. R. "Third-harmonic generation microscopy," *Opt. Express* 3, 315–324 (1998).

[26] Stoller, P., Celliers, P. M., Reiser, K. M. and Rubenchik, A. M. "Quantitative second-harmonic generation microscopy in collagen," *Appl. Opt.* 42, 5209–5219 (2003).

[27] Williams, R. M., Zipfel, W. R. and Webb, W. W. "Interpreting second-harmonic generation images of collagen I fibrils," *Biophys. J.* 88, 1377–1386 (2005).

[28] Yariv, A. *Quantum Electronics*, 3rd edn, John Wiley & Sons (1989).

[29] Yelin, D. and Silberberg, Y. "Laser scanning third-harmonic-generation microscopy in biology," *Opt. Express* 5, 169–175 (1999).

[30] Yew, E. Y. S. and Sheppard, C. J. R. "Effects of axial field components on second harmonic generation microscopy," *Opt. Express* 14, 1167–1174 (2006).

18 | Pump-Probe Microscopy

The purpose of an optical microscope is to probe matter using light. Different properties of matter can be probed. For example, the techniques described in Chapters 9 to 12 probed, in some manner or other, the local propensity of matter to scatter light. In turn, the techniques described in Chapters 13 to 15 probed, in some manner or other (typically by way of fluorescence), the propensity of matter to absorb light. These techniques share a common principle: they all rely on the direct interaction of light with electrons within the matter. That is, the properties of matter explored by these techniques are all electronic in nature. The same is true with the techniques described in Chapters 16 and 17, though generalized to nonlinear interactions.

But matter is composed of more than electrons. We turn now to microscopy techniques that involve properties of matter more nuclear in origin. In particular, most of this chapter will examine techniques that probe nuclear motion, whether vibrational or rotational. This motion can be intra-molecular, in which case the coupling with light is mediated by Raman interactions, or it can be inter-molecular, in which case it is mediated by Brillouin interactions. The coupling of light with nuclear motion, however, is not direct, and must itself be mediated electron–nuclear interactions. As such, time delays are generally involved, and the resulting interplay between light, electron and nuclear degrees of freedom can be rather complex. But this interplay can be exploited to extract invaluable information about the nature of the matter itself.

A discussion of various microscopy techniques to probe this interplay is grouped here under the arbitrary heading of pump-probe microscopies, suggesting a basic strategy of action and reaction. As can be well imagined, the variety of techniques is quite extensive, and a discussion of these techniques in a single chapter can hardly delve into any depth. Most of the techniques presented here involve nonlinear light–matter interactions, which are described in more detail in [3, 4, 5, 15, 20].

18.1 RAMAN MICROSCOPY

Vibrational states in molecules are energetically weaker than electronic states, by at least one or two orders of magnitude. Typical transition frequencies for vibration states, denoted by Ω_r, are commensurately one or two orders of magnitude smaller than typical electronic transition frequencies ν_{eg}. Various techniques are available for probing these vibrational transitions, the most obvious of which is to irradiate the sample with light of frequency Ω_r, namely mid-infrared light of wavelength on the order of 2.5–25 μm (for historical reasons, vibrational energies are generally cited in cm^{-1}, corresponding to wavenumber units $\Omega_r/c = 1/\lambda_r$, such that the above range becomes 4000–400 cm^{-1}). The excitation of a vibrational state thus becomes manifest as an absorption of mid-infrared light at the vibrational transition frequency.

While this technique works, and indeed is the basis of a discipline called infrared spectroscopy, it has several drawbacks in the context of microscopy applications. For one, absorption-based techniques where the signal is manifest as an intensity decrease within a bright background is always problematic in terms of SNR, here exacerbated by the fact that infrared detectors are notoriously noisy. A second drawback comes from poor spatial resolution, which scales as λ_r. But a third drawback is perhaps the most deleterious when performing biological imaging in aqueous environments: infrared light is severely absorbed by water, leading to penetration depths of only a few tens of microns, at best. By far, techniques employing visible or near-infrared light, for which water is nearly transparent, are preferable.

18.1.1 Incoherent Raman Microscopy

In order to address a vibrational frequency Ω_r with a much higher optical frequency, clearly a frequency-mixing interaction is required. This subsection and the next will explore a class of such interactions based on Raman scattering, respectively incoherent and coherent. Figure 18.1 provides a schematic illustration of various scattering processes. In these interactions, an incident field of frequency ν_i can be thought of as promoting a molecule to a short-lived virtual state, indicated by a thin line, which very quickly (on a time scale of femtoseconds) causes the field to scatter. In the case where the initial and final states of the molecule are the same (Fig 18.1a), the scattered field possesses the same frequency as the incident field. Such scattering is called elastic, in the sense that no energy is deposited into the molecule. We saw such scattering before in Section 9.4, where it was referred to as Rayleigh scattering. Elastic interactions, where the incident and scattered fields possess the same frequency, can give rise to complicated interferences between the fields, which of course is the basis of many of the microscope techniques we have already examined. Here we consider inelastic interactions, where energy is deposited into (Fig 18.1b) or extracted from (Fig 18.1c) the molecule. The former is called Stokes scattering; the latter is called anti-Stokes scattering. Accordingly, the former leads to a red-shifted scattered frequency given by $\nu_s = \nu_i - \Omega_r$; the latter to a blue-shifted scattered frequency given by $\nu_{as} = \nu_i + \Omega_r$. Because the molecule lands in a state uncorrelated in phase with its initial state, the phase of the scattered light is random. Raman scattering, whether Stokes or anti-Stokes, is thus an incoherent process where phase plays no role in the determination of radiation patterns or output power. In the same manner as we defined a molecular cross-section for fluorescence excitation (also incoherent), we can also define a molecular cross-section for incoherent Raman scattering (also known as spontaneous Raman scattering), such that $\Phi_R = \sigma_R I_i$. The spontaneous Raman scattered power thus scales with the number of molecules, in the same manner as fluorescence power.

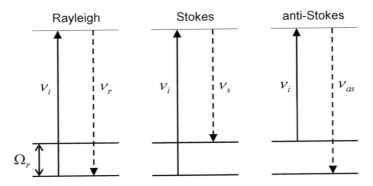

Figure 18.1. Spontaneous Rayleigh and Raman scattering.

Advantages of spontaneous Raman scattering for microscopy applications are that it provides label-free imaging with high spatial resolution, because the wavelengths involved are short, typically in the sub-micron range. Moreover, these wavelengths are typically in the range where water absorption is negligible. However, spontaneous Raman scattering suffers a serious disadvantage. Raman cross-sections are several orders of magnitude (typically 10^{-7}) smaller than fluorescence cross-sections. As such, Raman signals are extremely weak, and require very long integration times. This problem is exacerbated, in the case of Stokes scattering, by the usual presence of background auto-fluorescence from the sample, which is also red-shifted relative to the illumination wavelength and can overwhelm the Stokes signal of interest. Anti-Stokes Raman scattering is thus prescribed. But the anti-Stokes signal is even weaker than the Stokes signal, since, for standard temperatures, the initial state of the molecule is unlikely to be an excited vibrational state. In brief, spontaneous Raman microscopy is technically challenging and unavoidably quite slow.

18.1.2 Coherent Raman Microscopy

More recently, the technique of coherent Raman microscopy has gained interest. Here, the same vibrational levels are probed as in incoherent Raman microscopy, but rather than relying on a spontaneous interaction between a single field and a vibrational transition, the interaction is forced, or stimulated, by an interaction with multiple fields. Specifically, coherent Raman microscopy is based on a third-order susceptibility, similar to THG (see Section 17.3). But in contrast to THG, coherent Raman microscopy probes vibrational levels and offers a degree of molecular specificity unattainable with multiharmonic techniques. How this comes about involves a subtle interplay between fields of different frequencies. While the standard approach to treating coherent Raman scattering involves the explicit consideration of three driving fields in the frequency domain, in accord with its dependence on a third-order susceptibility, we begin instead with an alternative approach in the time domain that is perhaps more intuitive.

Time Domain

To begin, let us consider irradiating, or pumping, a sample with an ultrashort pulse of light. The sample responds in two ways. First, the electronic degrees of freedom respond almost instantaneously. Second, the nuclear degrees of freedom respond more slowly, by way of the coupling between electrons and nuclei. Our goal is to probe the resultant slower nuclear vibrations. Because we use light, which does not directly couple to nuclear motion, our only recourse is to monitor the subsequent back-action of these vibrations on the electronic degrees of freedom. Specifically, our recourse is to monitor the subsequent change in the dielectric constant of the sample resulting from the induced nuclear vibrations. Our pump-probe procedure is thus summarized as follows:

$$\delta\epsilon\,(t) = \int dt'\, R^{(3)}\,(t')\, I_p\,(t - t') \tag{18.1}$$

$$P\,(t, \tau) = \epsilon_0\, \delta\epsilon(t + \tau) E_{pr}\,(t) \tag{18.2}$$

Here, $R^{(3)}\,(t)$ comprises both a fast (electronic) and slower (nuclear) response resulting from the illumination pulse, leading to a change in dielectric constant $\delta\epsilon\,(t)$ of the sample. The reason the pump intervenes as an intensity $I_p\,(t)$ and not as a field is because we are most interested in the response involving nuclear vibrations, which are too slow to track the phase variations of light and are thus sensitive only to intensity variations. In turn, the change in dielectric constant $\delta\epsilon\,(t)$ is itself monitored by the probe field $E_{pr}\,(t)$, which near-instantaneously induces a sample polarization density $P\,(t, \tau)$. For generality, we have also included the possibility of a variable time delay τ between the pump and probe beams (see Fig. 18.2).

Finally, the induced polarization density acts as a primary source (see introduction to Section 17.1). It is precisely the radiation from this source that constitutes the Raman signal of interest. In other words, upon detection, the Raman signal is given by

$$\Phi_R\,(\tau) \propto \int dt\, |P\,(t, \tau)|^2 \tag{18.3}$$

To better understand how this pump-probe procedure yields information about vibrational transitions, we invoke a simple model for the polarizability response $R^{(3)}\,(t)$. To begin, though

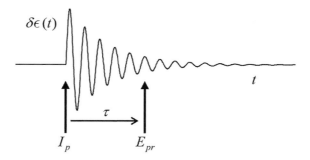

Figure 18.2. Pump intensity I_p induces a molecular vibration that is probed by E_{pr}.

we are interested only in the vibrational component of this response, we cannot, in general, neglect the faster electronic component. That is, we write

$$R^{(3)}(t) = \chi_{nr}^{(3)} \delta(t) + R_r^{(3)}(t) \tag{18.4}$$

where the first term corresponds to the near-instantaneous electronic response, and the second term to the longer-lived vibrational response. These are termed non-resonant and resonant, for reasons that will be clear below. A particularly revealing result comes from the so-called impulsive limit where both the pump and probe intensities can be modelled as ultrashort, delta-function-like pulses. A straightforward calculation then obtains

$$\Phi_R(\tau > 0) \propto I_{pr} I_p^2 \left| R_r^{(3)}(\tau) \right|^2 \tag{18.5}$$

The Raman signal thus depends quadratically on pump power and linearly on probe power. Moreover, a measurement of this signal as a function of pump-probe delay provides information about the Raman response itself, though only about its modulus. For example, a standard model for a Raman response is that of a sum of damped oscillations, written as

$$R_r^{(3)}(t) = \begin{cases} i2\pi \sum_n \mu_{r_n} \Omega_{r_n} e^{-2\pi t(\gamma_{r_n} + i\Omega_{r_n})} & \text{if } t \geq 0 \\ 0 & \text{if } t < 0 \end{cases} \tag{18.6}$$

where μ_{r_n} characterizes the relative strength of the nth vibration state, and Ω_{r_n} and γ_{r_n} characterize its frequency and inverse lifetime (see Fig. 18.2). More precisely, $R_r^{(3)}(t)$ is an analytical representation of the Raman response. The physical response is the real part of this, which vanishes as $t \to 0$.

An insertion of Eq. 18.6 into Eq. 18.5 suggests that the complexity of the Raman signal greatly increases with the number of vibrational states probed. On the other hand, if only a few vibrational states are dominant, its interpretation becomes tractable. A particular benefit of the time-domain pump-probe technique is that it provides a simple temporal separation of resonant and non-resonant Raman responses. As will be seen below, this separation becomes more problematic in the frequency domain. Another benefit is that the pump and probe beams can be extracted from the same laser, meaning that only a single pulsed laser is required, though it must produce very short pulse widths narrower than the vibration periods themselves [17]. Indeed, this technique is akin to striking a bell with a hammer: the bell only rings when the strike is sharp.

For more information on impulsive Raman scattering, the reader is referred to the seminal work by Yan and Nelson [24], as well as various applications involving coherent control [9, 21].

Frequency Domain

While a time-domain approach offers the most intuitive understanding of the pump-probe method described above, an alternative frequency domain is also quite revealing. To this end, we apply a Fourier transform to both sides of Eq. 18.2, leading eventually to

$$\mathcal{P}(\nu; \tau) = \epsilon_0 \iint d\Omega \, d\nu' \, e^{-i2\pi\Omega\tau} \chi^{(3)}(\Omega) \, \mathcal{E}_{pr}(\nu - \Omega) \, \mathcal{E}_p^*(\nu') \, \mathcal{E}_p(\nu' + \Omega) \tag{18.7}$$

where we neglect spatial coordinates here, allowing us to revert to the notation used at the beginning of Chapter 17. As before, $\chi^{(3)}(\Omega)$ is the Fourier transform of $R^{(3)}(t)$, which, according to our simple model (Eq. 18.6), is given by

$$\chi^{(3)}(\Omega) = \chi_{nr}^{(3)} + \chi_r^{(3)}(\Omega) = \chi_{nr}^{(3)} + \sum_n \frac{\mu_{r_n}\Omega_{r_n}}{\Omega_{r_n} - \Omega - i\gamma_{r_n}} \tag{18.8}$$

Finally, we note that

$$\Phi_R(\tau) \propto \int d\nu \, |\mathcal{P}(\nu;\tau)|^2 \tag{18.9}$$

from Eq. 18.3 and the Parseval theorem.

A schematic representation of Eq. 18.7 is provided in Fig. 18.3, where, for simplicity, only a single vibrational transition is depicted. This representation is different from the ones used in previous chapters, since the arrows are no longer meant to indicate energy flow but rather serve only for bookkeeping, with directions indicating whether or not the Fourier transformed fields are complex-conjugated. The dashed arrow corresponds to the complex-conjugated field resulting from $\mathcal{P}^*(\nu;\tau)$ in Eq. 18.9. This arrow returns the molecule to its initial state, thus leading to coherent Raman scattering. What is clear from Fig. 18.3 is that the pump pulse must possess a bandwidth large enough to span Ω_r in order to probe this vibrational frequency. In other words, the pump pulse must be short.

For reference, we observe that when the pump-probe time delay is set to zero, a variable change in Eq. 18.7 leads to

$$\mathcal{P}(\nu) = \epsilon_0 \iint d\nu_0 \, d\nu_1 \, \chi^{(3)}(\nu_0 - \nu_1) \, \mathcal{E}_{pr}(\nu - \nu_0 + \nu_1) \, \mathcal{E}_p^*(\nu_1) \, \mathcal{E}_p(\nu_0) \tag{18.10}$$

Two-Beam Interactions

A difficulty with pump-probe microscopy is that it is slow, in part because it requires a scanning of τ. Moreover, the interpretation of Eq. 18.5 is not always straightforward. For these reasons,

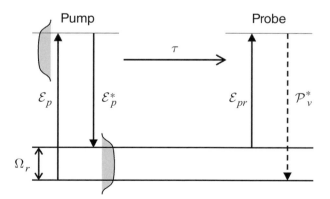

Figure 18.3. Pump-probe Raman scattering.

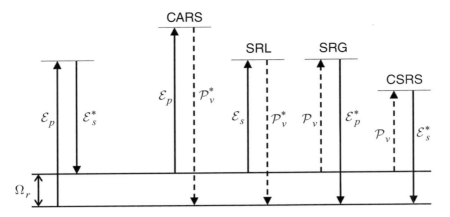

Figure 18.4. Mechanisms of coherent Raman scattering.

pump-probe coherent Raman microscopy is not very common. On the other hand, alternative implementations of coherent Raman microscopy have proven to be very successful. The most common of these are based on the use of two separate laser beams, generally of different frequencies, called pump ($\mathcal{E}_p$) and Stokes ($\mathcal{E}_s$) beams. These are directed onto the same sample location at the same time, meaning that the time delay between the two remains fixed at $\tau = 0$. A schematic of some resulting interactions that come from probing a single molecular vibrational transition is shown in Fig. 18.4.

Again, these interactions can be conceptually divided into two parts, but this time the roles of the beams are different. First, instead of a single beam exciting the molecular vibration, as was the case in Fig. 18.3, it is the combination of both the pump and Stokes beam that excites the vibration, by way of the beat frequency between the two. Once the vibration is excited, a variety of mechanisms can bring the molecule back to its initial state. We consider these one by one.

Coherent Anti-Stokes Scattering (CARS)
In the first mechanism depicted in Fig. 18.4, the pump beam itself plays the role of the probe beam. The resulting induced polarization density becomes (compare with Eq. 18.10)

$$\mathcal{P}_{\text{CARS}}(\nu) = \epsilon_0 \iint d\nu_0 \, d\nu_1 \, \chi^{(3)}(\nu_0 - \nu_1) \, \mathcal{E}_p(\nu - \nu_0 + \nu_1) \, \mathcal{E}_s^*(\nu_1) \, \mathcal{E}_p(\nu_0) \quad (18.11)$$

Let us examine this more closely. The excitation of the vibration is appreciable only near the resonance of $\chi^{(3)}(\nu_0 - \nu_1)$. From Eq. 18.8, and taking into account only a single vibrational state, this resonance condition stipulates $\nu_0 - \nu_1 \approx \Omega_r$, confirming that the vibration is excited by the beat frequency between the pump and Stokes beams. But let us now consider a pump beam that, despite being pulsed, is quasi-monochromatic. The bandwidth of $\mathcal{E}_p$ is thus relatively narrow, meaning that $\mathcal{P}(\nu)$ is only appreciable when $\nu - \nu_0 + \nu_1 \approx \nu_0$. We arrive at the conclusion that the induced polarization density must have frequency $\nu \approx 2\nu_0 - \nu_1 \approx \nu_0 + \Omega_r$.

In other words, the induced polarization density radiates at the anti-Stokes frequency of the pump beam, or, said differently, the pump beam becomes blue-shifted when scattered by the induced molecular vibration. This process is called coherent anti-Stokes Raman scattering (CARS).

Stimulated Raman Loss (SRL)

The second mechanism depicted in Fig. 18.4 is similar to the first, except that it is now the Stokes beam that plays the role of the probe beam, leading to the induced polarization density

$$\mathcal{P}_{\text{SRL}}(\nu) = \epsilon_0 \iint d\nu_0 \, d\nu_1 \, \chi^{(3)}(\nu_0 - \nu_1) \, \mathcal{E}_s(\nu - \nu_0 + \nu_1) \, \mathcal{E}_s^*(\nu_1) \, \mathcal{E}_p(\nu_0) \qquad (18.12)$$

Following the same arguments as above, we conclude that, for a quasi-monochromatic Stokes beam, the induced polarization density must have frequency $\nu \approx \nu_0$. That is, the induced polarization density radiates at the same frequency as the pump beam. This has important ramifications, as it leads to the possibility of interference between the two. In effect, this mechanism is similar to stimulated emission discussed in Section 13.1.1, where the presence of light in a particular mode (i.e. frequency and direction) stimulates, by way of a light–matter interaction, the amplification of this light. However, here, as we will see below, the phase of the stimulated emission is opposed to that of the pump beam, and the interference between the two is destructive. As a result, the pump beam experiences a de-amplification rather than an amplification – hence the name stimulated Raman loss (SRL).

Stimulated Raman Gain (SRG)

Continuing through our sequence depicted in Fig. 18.4, the third mechanism is also similar to the first, but with the difference that the pump beam, which plays the role of the probe beam, is now complex conjugated, as is the induced polarization density. The latter is given by

$$\mathcal{P}_{\text{SRG}}^*(\nu) = \epsilon_0 \iint d\nu_0 \, d\nu_1 \, \chi^{(3)}(\nu_0 - \nu_1) \, \mathcal{E}_p^*(\nu + \nu_0 - \nu_1) \, \mathcal{E}_s^*(\nu_1) \, \mathcal{E}_p(\nu_0) \qquad (18.13)$$

Following again the same arguments as above, we conclude that, for a quasi-monochromatic pump beam, the induced polarization density must have frequency $\nu \approx \nu_1$. Note that the arrows in Fig. 18.4 should not be construed as obeying any particular time sequence. Here the induced polarization density radiates at the same frequency as the Stokes beam; however, in this case the interference between the two is constructive, and the Stokes beam thus experiences amplification – hence the name stimulated Raman gain. Indeed, a variety of mid-infrared lasers are based precisely on this gain mechanism.

Coherent Stokes Raman Scattering (CSRS)

Finally, the polarization density associated with the last mechanism depicted in Fig. 18.4 is given by

$$\mathcal{P}_{\text{CSRS}}^*(\nu) = \epsilon_0 \iint d\nu_0 \, d\nu_1 \, \chi^{(3)}(\nu_0 - \nu_1) \, \mathcal{E}_s^*(\nu + \nu_0 - \nu_1) \, \mathcal{E}_s^*(\nu_1) \, \mathcal{E}_p(\nu_0) \qquad (18.14)$$

Following the same arguments as above, we conclude that the induced polarization density must have frequency $\nu \approx 2\nu_1 - \nu_0 \approx \nu_1 - \Omega_r$. That is, the induced polarization density radiates at the Stokes frequency of the Stokes beam, or, said differently, the Stokes beam becomes red-shifted when scattered by the induced molecular vibration. This process is called coherent Stokes Raman scattering (CSRS).

Microscopy Implementations

It should be emphasized that all the mechanisms depicted in Fig. 18.4 occur simultaneously, and they all serve to probe the same vibrational transition. The question remains how to exploit these mechanisms to create a microscope image. A variety of implementations can be considered. Again, for simplicity here, we assume that both the pump and Stokes beams are quasi-monochromatic, of frequencies centered about ν_p and ν_s.

Darkfield Contrast

To begin, we note that CARS and CSRS can be easily distinguished from the other stimulated Raman scattering mechanisms (collectively referred to as SRS) by virtue of their frequencies. Specifically, both CARS and CSRS produce scattered light whose frequencies differ from the incident pump and Stokes beams. As such, CARS and CSRS produce signals at frequencies that are initially absent, making the signals "darkfield" in nature. The isolation of CARS or CSRS is thus readily achieved with simple spectral separation techniques, such as with dichromatic filters or such. As in the case of spontaneous Raman scattering, CARS is preferable because it is blue-shifted relative to all other frequencies, and hence does not run the risk of being confused with background fluorescence from the sample (at least, one-photon fluorescence). In contrast to spontaneous Raman scattering where anti-Stokes tends to be much weaker than Stokes scattering, here the CARS and CSRS scattering amplitudes are of the same strength.

The technique of CARS was first demonstrated by Maker and Terhune [14], and subsequently established by Xie [8, 31] as the first coherent Raman mechanism to be implemented in microscopy applications. But despite its enormous success, some difficulties quickly became apparent. To understand these difficulties, we derive the signal power for CARS (and for CSRS, for completeness). From Eq. 18.9, we obtain in the quasi-monochromatic limit

$$\Phi_{\text{CARS}} \propto I_s I_p^2 \left| \chi^{(3)} \left(\nu_p - \nu_s \right) \right|^2 \tag{18.15}$$

$$\Phi_{\text{CSRS}} \propto I_s^2 I_p \left| \chi^{(3)} \left(\nu_p - \nu_s \right) \right|^2 \tag{18.16}$$

A first difficulty comes from the dependence of signal on the modulus squared of the third-order susceptibility. Bearing in mind that the simplest model for this susceptibility is given by Eq. 18.8, we observe that with increasing number of vibrational states, the signals very quickly become difficult to disentangle (this same difficulty was occasioned in Eq. 18.5).

An appreciation of this difficulty comes from the explicit expansion of this modulus squared, obtaining

$$\left| \chi^{(3)} \left(\nu_p - \nu_s \right) \right|^2 = \left(\chi_{nr}^{(3)} \right)^2 + 2\chi_{nr}^{(3)} \mathrm{Re} \left[\chi_r^{(3)} \left(\nu_p - \nu_s \right) \right] + \left| \chi_r^{(3)} \left(\nu_p - \nu_s \right) \right|^2 \qquad (18.17)$$

where χ_{nr} is dominantly real, by virtue of its being quasi-instantaneous.

If the resonant susceptibility is dominant, the Raman signal scales quadratically with the number of resonant molecules probed. On the other hand, if the non-resonant susceptibility is dominant (much more common), it scales linearly with the number of resonant molecules probed. This uncertainly between quadratic and linear dependence often makes quantitative measurements problematic. Even in the limiting case of where the non-resonant susceptibility is dominant, the linear dependence is only by way of the real component of $\chi_r^{(3)}$. The correspondence with conventional signals obtained from spontaneous Raman scattering thus continues to remain elusive, since the latter depends on the imaginary part of $\chi_r^{(3)}$. As a side note, this last statement may appear puzzling since spontaneous Raman scattering does not appear to involve three driving fields. However, a fully quantum mechanical treatment suggests that one of these fields may be interpreted as a fluctuation arising from the vacuum itself! The reader is referred to the more specialized nonlinear optics textbooks listed above for such arcane matters. Suffice it to say that the "darkfield" signals described above are often difficult to interpret.

Brightfield Contrast

In contrast to the CARS and CSRS techniques described above, the SRS techniques (mechanisms two and three in Fig. 18.4) produce scattered light at the same frequencies as the driving fields. As such, SRS signals can be thought of as "brightfield" in nature. Though SRS was demonstrated decades ago by Woodbury [10], for a long time it was dismissed as an impractical candidate for microscopy applications, since it was thought that the isolation of a weak SRS signal from a bright laser background would be too difficult or time consuming. This isolation, though difficult, has been made much more feasible with modern electronics. Moreover, the advantages afforded by SRS microscopy generally outweigh its technical difficulty. To understand some of these advantages, let us return to our quasi-monochromatic field approximation. In this case, we can write (neglecting units)

$$\mathcal{P}_{\mathrm{SRL}} \left(\nu \right) \approx \epsilon_0 E_p I_s \chi^{(3)} \left(\nu_p - \nu_s \right) \delta \left(\nu - \nu_p \right) \qquad (18.18)$$

$$\mathcal{P}_{\mathrm{SRG}} \left(\nu \right) \approx \epsilon_0 E_s I_p \chi^{(3)} \left(\nu_p - \nu_s \right)^* \delta \left(\nu - \nu_s \right) \qquad (18.19)$$

As emphasized throughout this and the preceding chapter, these induced polarization densities act as primary sources (see specifically the introduction to Section 17.1). For example, let us consider SRL. To determine the influence of the source $\mathcal{P}_{\mathrm{SRL}} \left(\nu_p \right)$ on the pump field E_p of the same frequency, we invoke the Helmholtz equation (Eq. 2.11). This becomes

$$\left(\nabla^2 + 4\pi^2 \kappa_p^2 \right) E_p = -4\pi^2 \frac{\nu_p^2}{c^2} I_s \chi^{(3)} \left(\nu_p - \nu_s \right) E_p \qquad (18.20)$$

where $c = (\mu_0 \epsilon_0)^{-1/2}$ is the speed of light in vacuum, and the spatial dependencies of E_p, I_s, and $\chi^{(3)}$ are implicit.

We can directly compare this with Eq. 9.21. A correspondence can thus be made with the sample scattering strength $\delta \varepsilon (\vec{r})$ utilized throughout most of this book, given by

$$\delta \varepsilon \leftrightarrow \frac{1}{n^2(\nu_p)} I_s \chi^{(3)} (\nu_p - \nu_s) \tag{18.21}$$

where $n(\nu_p)$ is the index of refraction at the pump frequency.

Once made, this correspondence provides a link to the formalism established in Section 9.3. In particular, we can derive a molecular absorption cross-section associated with SRL, defined by

$$\Phi_p^{(SRL)} = \sigma_{SRL} I_p \tag{18.22}$$

where, from Eq. 9.42 and upon integration over the molecule volume, we have (neglecting dispersion)

$$\sigma_{SRL} \propto I_s \text{Im} \left[\chi_r^{(3)} (\nu_p - \nu_s) \right] \tag{18.23}$$

We find that, in contrast to the "darkfield" techniques described above, SRL leads to an absorption cross-section that depends on the imaginary part of $\chi^{(3)}$, which, since $\chi_{nr}^{(3)}$ is dominantly real, corresponds to the imaginary part of $\chi_r^{(3)}$. The interpretation of SRL signals is thus much more straightforward. For one, SRL signals resemble well-known spontaneous Raman signals. Moreover, they depend linearly on the number of resonant molecules probed.

The SRL signal itself is manifested as a reduction in pump power. One might wonder where this power goes, as the molecule seems to return to its initial state. It turns out that part of the power is transferred from the pump beam to the Stokes beam. Indeed, a similar calculation applied to SRG yields a molecular absorption cross-section defined by

$$\Phi_s^{(SRG)} = \sigma_{SRG} I_s \tag{18.24}$$

$$\sigma_{SRG} \propto -I_p \text{Im} \left[\chi_r^{(3)} (\nu_p - \nu_s) \right] \tag{18.25}$$

Note the minus sign in Eq. 18.25, which comes from the complex-conjugation of $\chi^{(3)}$ in Eq. 18.19. On resonance, the absorption cross-section σ_{SRG} is negative. In contrast to the pump beam, the Stokes beam thus experiences an increase in power.

To summarize, the net amount of power transferred from pump to Stokes beam is given by

$$\Delta \Phi_{p \to s} \propto I_p I_s \text{Im} \left[\chi_r^{(3)} \right] \tag{18.26}$$

The question remains how to detect this transferred power, which in general is quite small (typically seven to eight orders of magnitude smaller than the laser powers themselves). A strategy is illustrated in Fig. 18.5. By modulating the power of the pump (or Stokes) beam, the modulation is transferred by SRS to the Stokes (or pump) beam. This transferred modulation can be detected using a highly sensitive lock-in amplifier. The dominant noise sources are laser and shot noise. The former decreases with frequency, prescribing modulation frequencies typically upward of 100 MHz. The latter, on the other hand, is independent of frequency, and

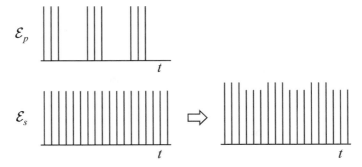

Figure 18.5. Pump modulation is transfered to Stokes beam in SRG scattering. Both pump and Stokes beams are pulsed.

can only be reduced (relative to the signal) by increasing lock-in integration time (for a given signal power).

Finally, there remains a fundamental difference we have not yet discussed between SRS on the one hand, and CARS and CSRS on the other. CARS and CSRS are parametric processes, meaning that they lead to no deposition of power into the sample. Power is only re-distributed between fields of different frequencies. In contrast, SRS is a dissipative process. While power is transferred from pump to Stokes beam, an excess power still remains. This excess power is deposited into the sample itself by way of the molecular vibration, ultimately leading to heat. The dependence of SRS on the imaginary component of $\chi_r^{(3)}$ is, indeed, a signature of a dissipative process.

Phase Matching

We close our discussion on coherent Raman scattering with a brief word about phase matching, which we have blithely neglected. Because coherent Raman scattering is, as its name indicates, a coherent process, it too must obey the condition of phase matching in the same manner as multiharmonic scattering in Chapter 17, with similar consequences when the incident laser beams are focused. But there are also differences. For example, it was found in Section 17.3 that a focused beam produces essentially no THG in a uniform volumetric sample. The reason for this is that the effective phase mismatch $\delta\kappa_\Sigma = 3\delta\kappa_i$ resulting from the cumulative effect of three incident-field Gouy shifts is so large as to essentially annul the efficacy of THG. Following the same arguments we used to arrive at Eq. 17.23, we find that for CARS the effective phase mismatch resulting from Gouy shifts is given by $\delta\kappa_\Sigma = 2\delta\kappa_p - \delta\kappa_s$. Here the Stokes Gouy-shift intervenes with an opposite sign, partially canceling the two pump Gouy shifts, resulting in a smaller overall phase mismatch than with THG. Thus, even though THG and CARS are both third-order scattering processes, while focused THG is very inefficient in volume, this cannot be said of CARS. It is for this reason that the non-resonant background in CARS is generally so much stronger than the resonant signal of interest. For example, when performing biological imaging, the water environment itself produces significant non-resonant background that is often overwhelming.

Historically, this problem of non-resonant volume background had been a serious impediment to the development of CARS microscopy, motivating a variety of countermeasures. One such countermeasure involved recording the CARS signal in the backward, or epi, direction, called E-CARS. Because volume background is produced in the forward direction, E-CARS is essentially background free [8]. However, like all reflection-based microscopes, E-CARS can only reveal sample structure that possesses very high axial spatial frequencies, such as interfaces or point-like objects. Another countermeasure involved a clever use of coherent control of the pump and Stokes fields to introduce additional phase mismatch to suppress volume background [18]. Still another involved introducing the Stokes beam in the opposite direction to the pump [11].

But a key breakthrough came from the realization that the interference of the Raman signal with a reference beam could assure its dependence on $\text{Im}\left[\chi^{(3)}\right]$, thus eliminating non-resonant background altogether and rendering the signal linearly proportional to the number of probed molecules (allowing, for example, a PSF to be well defined). The reference beam itself can be supplied externally (e.g. [1]), or internally by self-interference with either the pump or Stokes beam, as in SRS microscopy. An advantage of self-interference is that phase-matching with the reference beam becomes automatic.

To date, coherent Raman microscopy has been highly successful in a variety of biological imaging applications, including the discrimination of proteins from lipids based on their

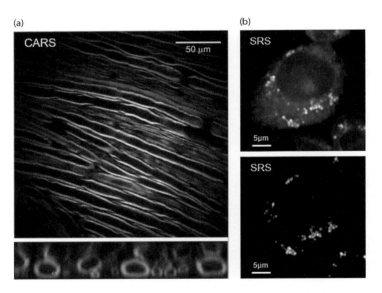

Figure 18.6. (a) En-face and cross-sectional CARS images of myelin sheaths in *ex vivo* unstained rat spinal cord tissue. Courtesy of E. Bélanger and D. Côté, Centre de Recherche Université Laval Robert Giffard, Québec. (b) SRS images of lipid uptake in cells revealing C-H (top) and C-D (bottom) bonds. Adapted from [28].

characteristic vibrational modes. For example, proteins contain more C-H$_3$ stretching modes (2950 cm^{-1}) while saturated lipids contain more C-H$_2$ stretching modes (2850 cm^{-1}). Example CARS and SRS images are shown in Fig. 18.6. Excellent reviews of coherent Raman microscopy and its applications are provided in [7].

18.2 BRILLOUIN MICROSCOPY

Vibrations in matter can exist in many forms. While Raman scattering provides access to molecular vibrations in matter, an alternative type of scattering, called Brillouin scattering, provides access to mesoscopic vibrations, called acoustic vibrations. These tend to be much slower than molecular vibrations by several orders of magnitude. That is, while molecular vibrations have frequencies typically in the range 10^{13}–10^{14}Hz, acoustic vibrations that can be probed by light are typically in the range 10^9–10^{10}Hz (for liquids). The frequency shifts resulting from Brillouin scattering are thus much smaller, and impose the use of specialized high-resolution spectrometers (e.g. [27]). Nevertheless, despite the technical hurdles involved, Brillouin microscopy is emerging as a promising new technique to probe mesoscopic properties of matter at micron spatial scales.

Just as Raman scattering can be spontaneous or stimulated, so too with Brillouin scattering. The basic mechanism leading to Brillouin scattering is a phenomenon called electrostriction, whereby density variations $\delta\rho$ within a sample are transformed into dielectric constant variations $\delta\epsilon$ by way of the electrostrictive constant

$$\gamma_e = \bar{\rho}\left(\frac{\partial\epsilon}{\partial\rho}\right)_{\bar{\rho}} \tag{18.27}$$

where $\bar{\rho}$ denotes the mean sample density. We thus have dielectric constant variations given by $\delta\epsilon = \frac{\gamma_e}{\bar{\rho}}\delta\rho$, which, in turn, can be optically probed by way of Eq. 18.2.

Whether Brillouin scattering is spontaneous or stimulated depends on the origin of the density variations. When these are thermally driven, the scattering is spontaneous. The local density fluctuations in this case are described by Einstein–Smoluchowski statistics (e.g. [12]), and the scattering becomes incoherent, obeying the usual rules that the scattered power scales with the number of fluctuation centers. Alternatively, the density variations can be themselves driven by light. As in stimulated Raman scattering, these are too slow to respond to optical phase variations, and can respond only to optical intensity variations, leading to responses in the form of Eq. 18.1, albeit slower still. As an aside, this same mechanism that couples light intensity to density variations has enabled the possibility of "optical trapping" of particles, which has been exploited to much success by the biophysics community [16].

To date, the few demonstrations of Brillouin microscopy have mostly involved spontaneous Brillouin scattering (e.g. [19]). These are preferably operated in a reflection geometry, for reasons that are made clear in Fig. 18.7. Here, $\vec{\kappa}_i$ and $\vec{\kappa}$ denote the incident and scattered optical wavevectors, and $\vec{q}$ the wavevector of the acoustic vibration. By momentum conservation, we must have $\vec{\kappa} = \vec{\kappa}_i + \vec{q}$. Moreover, because the frequency shifts caused by acoustic vibrations

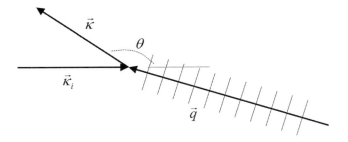

Figure 18.7. Incident wavevector $\vec{\kappa}_i$ is scattered by acoustical wavevector $\vec{q}$.

are very small compared to optical frequencies, to an excellent approximation we may treat Brillouin scattering as elastic, meaning that by energy conservation $\kappa \approx \kappa_i$. We thus arrive at the well-known Bragg scattering condition

$$q = 2\kappa \sin\frac{\theta}{2} \tag{18.28}$$

where θ is the scattering angle. This condition defines the acoustic wavenumber q required to cause light to scatter in the direction θ. This wavenumber can be quite high. For example, in a reflection geometry ($\theta = \pi$) it is twice the optical wavenumber, corresponding to roughly half the optical wavelength. But we have seen this requirement before, since it is a property of all reflection-based microscopes where the sample must cause light to effect a u-turn (see discussion in Section 11.4).

But in fact the scattering is not quite elastic. The acoustic wave oscillates in time with frequency Ω_b and can transfer energy to the scattered light, which manifests itself as red and blue-shifted frequency sidebands. That is, the scattered light, in addition to oscillating at frequency ν_i (Rayleigh scattering), also oscillates at frequencies $\nu_i \pm \Omega_b$ (Brillouin scattering). As in spontaneous Raman scattering, these are referred to as Stokes (red-shifted) and anti-Stokes (blue shifted), with the anti-Stokes scattering being much weaker than the Stokes.

Finally, it remains to determine Ω_b. This is constrained by the value of q allowed by the detection geometry (i.e. by the selection of θ), but also by the sample itself by way of the dispersion relation $\Omega_b = qv_s$, where v_s is the speed of the acoustic wave in the sample. We thus have, from Eq. 18.28

$$\Omega_b = 2\kappa v_s \sin\frac{\theta}{2} \tag{18.29}$$

As the reader may have surmised, this is exactly the Doppler shift occasioned by a moving reflector of velocity v_s in the direction $\vec{q}$, and indeed may be interpreted as such. This Doppler shift is maximized in a reflection geometry and nullified in a transmission geometry.

Ultimately, a Brillouin microscope provides local measurements of Ω_b, which in turn provides local information on v_s. The questions of how this acoustic speed is to be interpreted at such high frequencies and what it reveals about the mechanical properties of the sample itself, however, remain a subject of some speculation, particularly when imaging soft matter such as biological tissue.

18.3 PHOTOTHERMAL MICROSCOPY

As it happens, the dielectric constant is not only sensitive to density fluctuations, but also to other thermodynamic variables such as entropy, temperature, etc. [12]. The dependence on the last of these, temperature T, is the basis of yet another pump-probe technique called photothermal microscopy.

As the name suggests, photothermal microscopy relies of the transfer of light energy into heat. For example, when a molecule or particle absorbs light energy this energy must go somewhere. Throughout this chapter we have considered energy release in the form of vibrations, either molecular or mesoscopic. But these vibrations are systematically damped, in turn dissipating their energy into the sample bulk in the form of heat. The time scales involved in this dissipation are much slower, making their measurements readily accessible to conventional electronics. Actual photothermal measurements proceed in a familiar manner, using a time-varying pump intensity I_p and a fixed probe field E_{pr} (both assumed monochromatic),

$$H(\vec{r}, t) = \mu_a(\vec{r}) I_p(\vec{r}, t) \tag{18.30}$$

$$\delta T(\vec{r}_0, t) = \iint d^3\vec{r}' \, dt' \, R_{PT}(\vec{r}_0, t') H(\vec{r}_0 - \vec{r}', t - t') \tag{18.31}$$

$$\delta\epsilon(\vec{r}_0, t) = \left(\frac{\partial\epsilon}{\partial T}\right)_{\bar{T}} \delta T(\vec{r}_0, t) \tag{18.32}$$

$$\vec{P}(\vec{r}_0, t) = \epsilon_0 \delta\epsilon(\vec{r}_0, t) \vec{E}_{pr}(\vec{r}_0) \tag{18.33}$$

where $\mu_a(\vec{r})$ is an absorption coefficient (more will be said about this in Chapter 20) and $H(\vec{r}, t)$ is the rate of light energy converted into heat per unit volume. Ultimately, this absorbed energy is transformed into an induced polarization density $P(\vec{r}_0, t)$, which acts as a primary source and yields a scattered field $E_s(\vec{r}, t)$. Note that, though we are considering vector fields in Eq. 18.33, the induced change in dielectric constant is taken to be a scalar, for simplicity. Note also that the pump and probe wavelengths are generally chosen such that I_p is strongly absorbed by the sample whereas I_{pr} is not. Finally, as in stimulated Raman microscopy, the photothermal signal is given by the interference of the scattered field with the probe field $E_{pr}(\vec{r}, t)$ itself (indeed, one must take care not to confuse SRS with photothermal signals!).

It is instructive to consider a specific example of photothermal microscopy in detail, since this brings to bear much of the formalism we have developed so far. Specifically, let us consider a point absorber of absorption cross-section σ_a located at the origin. That is, we write $\mu_a(\vec{r}) = \sigma_a\delta^3(\vec{r})$. Upon taking the temporal Fourier transform of both sides of Eq. 18.31 we have

$$\delta T(\vec{r}_0; \Omega) = \sigma_a R_{PT}(\vec{r}_0; \Omega) I_p(0; \Omega) \tag{18.34}$$

where $R_{PT}(\vec{r}_0; \Omega)$ is the temporal Fourier transform of $R_{PT}(\vec{r}_0, t)$ (the temporal variations are so slow here compared with optical frequencies that we do not bother with calligraphic variables).

A derivation of $R_{\text{PT}}\left(\vec{r}_0; \Omega\right)$ can be found in classic textbooks on heat diffusion, such as [6]. We cite only the relevant result:

$$R_{\text{PT}}\left(\vec{r}_0; \Omega\right) = \frac{1}{4\pi r_0 DC} e^{-r_0\sqrt{-i2\pi\Omega/D}} \tag{18.35}$$

where D denotes the thermal diffusivity constant and C the heat capacity per unit volume, leading to

$$\delta\epsilon\left(\vec{r}_0; \Omega\right) = \frac{\sigma_a}{4\pi r_0 DC} \left(\frac{\partial\epsilon}{\partial T}\right)_{\bar{T}} e^{-r_0\sqrt{-i2\pi\Omega/D}} I_p\left(0; \Omega\right) \tag{18.36}$$

Photothermal microscopy is based on the same principle as SRS microscopy. The pump intensity is modulated at frequency Ω_p. This modulation is then transferred, by way of scattering caused by the photothermally induced polarization density, to the otherwise unmodulated probe, whereupon it is detected by interference. Bearing in mind that the induced polarization density responds quasi-instantaneously to a change in dielectric constant, we have

$$\vec{P}\left(\vec{r}_0; \Omega_m\right) = \epsilon_0 \delta\epsilon\left(\vec{r}_0; \Omega_p\right) \vec{E}_{pr}\left(\vec{r}_0\right) \tag{18.37}$$

Treating the polarization density as a primary source, the resultant scattered far field becomes, from Eqs. 9.66 and 9.62,

$$E_s^{(p,s)}(\vec{r}; \Omega_p) = \pi\mu_0\nu^2 \frac{e^{i2\pi\kappa r}}{r} \int d^3\vec{r}_0\, e^{-i2\pi\kappa\hat{r}\cdot\vec{r}_0} P^{(p,s)}\left(\vec{r}_0; \Omega_p\right) \tag{18.38}$$

At this point we must make assumptions about the probe beam. We assume this to be a $\hat{x}$-polarized plane wave directed along the $\hat{z}$ axis, again for simplicity, such that we can write $\vec{E}_{pr}\left(\vec{r}_0\right) = E_{pr}e^{i2\pi\kappa z_0}\hat{x}$. Making use of Eq. 9.58, we arrive at

$$E_s^{(p,s)}(\vec{r}; \Omega_p) = \begin{pmatrix} \cos\theta\cos\varphi \\ -\sin\varphi \end{pmatrix} E_s(\vec{r}; \Omega_p) \tag{18.39}$$

where

$$E_s(\vec{r}; \Omega_p) = \pi\frac{e^{i2\pi\kappa r}}{\lambda^2 r} E_{pr} \int d^3\vec{r}_0\, e^{-i2\pi\kappa(\hat{r}-\hat{z})\cdot\vec{r}_0} \delta\epsilon(\vec{r}_0; \Omega_p) \tag{18.40}$$

or, more succinctly,

$$E_s(\vec{r}; \Omega_p) = \pi\frac{e^{i2\pi\kappa r}}{\lambda^2 r} E_{pr}\, \delta\hat{\epsilon}(\kappa\hat{r} - \kappa\hat{z}; \Omega_p) \tag{18.41}$$

In other words, the scattered far field is a spherical wave whose amplitude is modulated by the Fourier transform (here spatiotemporal) of the dielectric constant variations. We have seen this result before (Eq. 9.34 – recall $\delta\epsilon = n^2\delta\varepsilon$), with the difference that we now have an explicit model for $\delta\epsilon$, given by Eq. 18.36. Upon performing the Fourier transform (see [2] for an equivalent derivation), we obtain finally

$$E_s(\vec{r}; \Omega_p) = \frac{e^{i2\pi\kappa r}}{8\pi n^2 r} \frac{\sigma_a}{DC} \left(\frac{\partial\epsilon}{\partial T}\right)_{\bar{T}} \left[\frac{E_{pr}I_p\left(0; \Omega_p\right)}{1 - \cos\theta - i\Omega_p/\Omega_d}\right] \tag{18.42}$$

where we have introduced a characteristic diffusion frequency $\Omega_d = 4\pi D\kappa^2$, which can be roughly interpreted as the inverse of the time required for heat to diffuse the distance of a wavelength. Bearing in mind that D is typically 10^{-8}–10^{-7} m²/s for biological samples in aqueous environments, this characteristic diffusion frequency is on the order of 1–10 MHz.

Equation 18.42 warrants some discussion. This is the radiation pattern of the scattered field, examples of which are depicted in Fig. 18.8. This pattern depends on the modulation frequency Ω_p of the pump beam. When the modulation frequency is much larger than the diffusion frequency Ω_d, then $E_s(\vec{r}; \Omega_p)$ is equally forward and backward directed. On the other hand, when it is much smaller than Ω_d, $E_s(\vec{r}; \Omega_p)$ becomes dominantly forward directed. The origin of this behavior is clear from Eq. 18.35, which suggests that we can assign a characteristic radius to the thermally induced dielectric fluctuations, defined by $r_d \approx \sqrt{D/\pi\Omega_p}$. The larger the pump frequency, the smaller this radius. In particular, when the pump frequency is much larger than Ω_d, r_d becomes much smaller than a wavelength and the thermally induced $\delta\epsilon$ resembles a point dipole source, as we have seen in Section 9.4. As expected for such a primary source, the scattered field is $\frac{\pi}{2}$ out of phase with the driving field E_{pr}. On the other hand, when the pump frequency is much smaller than Ω_d, r_d becomes much larger than a wavelength. Scattering from large objects is dominantly forward directed (see Eq. 9.15). Moreover, the scattered field becomes in phase with the driving field (for much the same reason as when emanating from a large aperture – see discussion in Section 2.4.1). In this low-frequency regime, the thermally induced $\delta\epsilon$ resembles more a mesoscopic lens than a point-like scatterer, though with an unusual cusp-shaped profile defined by Eq. 18.35.

Ultimately, what is measured with photothermal microscopy is the local absorption of the pump beam. Absorbers in a sample are typically endogenous, and need not be introduced externally. That is, as for all the microscopy techniques discussed in this chapter, photothermal microscopy can be termed "label-free" (e.g. see Fig. 18.9). A key advantage of a microscope based on absorption contrast, as opposed to scattering contrast, is that, for small objects, absorption cross-sections scale with volume while scattering cross-sections scale with volume squared (see Eq. 9.71). As such, when searching for very small objects, absorption-based

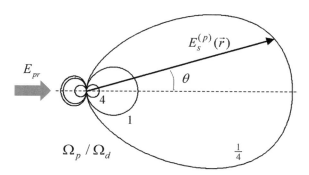

Figure 18.8. Scattered field amplitude distribution for different values of Ω_p/Ω_d.

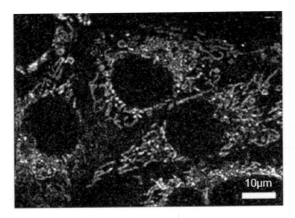

Figure 18.9. Photothermal image of mitochondria in unlabelled cells. Adapted from [13].

microscopy becomes advantageous. A particularly successful application of photothermal microscopy is in the imaging of nano-objects [2, 30]. Another application resolves some of the resolution disadvantages that come from performing Raman microscopy directly with infrared light (see introduction in Section 18.1). With photothermal microscopy, an infrared pump can be used to excite Raman vibrational levels, while a visible probe beam can be used to monitor the subsequent thermally-induced variations in the dielectric constant [29]. The spatial resolution is thus defined by the pump modulation frequency and the probe wavelength, and *not* by the pump wavelength. Moreover, because visible light is detected rather than infrared, the detector can be much less noisy.

18.4   PHOTOACOUSTIC IMAGING

We close this chapter with yet another microscopy technique based on the measurement of photo-induced vibrations. This technique is similar to photothermal microscopy in that it relies on light absorption to generate heat within a sample, except that the time scales are different. The absorbed light is delivered as a short pulse, of duration typically on the order of $\tau_p = 10^{-9}$ s. On such a short time scale, heat propagation within the sample can generally be neglected, and the delivery of heat is said to be thermally confined. In this regime the heat can directly induce acoustic pressure waves within the sample, which obey their own wave equation given by

$$\left(\nabla^2 - \frac{1}{v_s^2}\frac{\partial^2}{\partial t^2}\right)p\left(\vec{r},t\right) = -\frac{\Gamma}{v_s^2}\frac{\partial H\left(\vec{r},t\right)}{\partial t} \tag{18.43}$$

where $p\left(\vec{r},t\right)$ is acoustic pressure, v_s is the speed of sound within the medium, and Γ is the so-called Grüneisen parameter (unitless). In other words, the heat delivered by the light pulse (or rather its time-derivative) plays the role of a pressure source. The resulting acoustic wave is then readily derived using the Green function approach developed in Section 2.1, obtaining

$$p\left(\vec{r}, t\right) = \frac{\Gamma}{4\pi v_s^2} \frac{\partial}{\partial t} \int d^3 \vec{r}_0 \frac{H\left(\vec{r}_0, t - v_s^{-1}\left|\vec{r} - \vec{r}_0\right|\right)}{\left|\vec{r} - \vec{r}_0\right|} \tag{18.44}$$

We conclude that the acoustic pressure wave induced by the light absorption propagates radially outward at the speed of sound. Remarkably, this acoustic wave can be monitored directly with standard detection techniques developed for ultrasound imaging. There is no need for an optical probe beam!

The technique of photoacoustic imaging was pioneered by Wang [22] and has since garnered tremendous interest. Its key advantage comes from its ability to image very deep in thick tissue, where, as we will see in Chapter 20, optical microscopes struggle to provide any kind of resolution. For example, a rough estimate of the resolution obtainable with photoacoustic imaging can be made based on the speed of sound, which in aqueous media is about 1500 m/s. Standard ultrasound transducers operate in the range 1–50 MHz, meaning that, accordingly, the wavelengths of detected acoustic waves span 1500–30 μm. In principle, then, photoacoustic imaging can provide spatial resolutions that span roughly the same length scales. While such spatial resolutions may seem paltry compared to the optical resolutions we have become accustomed to, they very easily outcompete them at depth penetrations in tissue beyond a few millimeters. Light at such depths cannot be spatially focused because its propagation becomes diffusive. On the other hand, even though light cannot be focused, it can still largely retain its pulsed nature, even on picosecond time scales. As such, light can easily elicit photoacoustic responses from deep within the sample (see Fig. 18.10). In turn, these responses, in the form of acoustic waves, can propagate to the sample surface where they can be detected. The actual process of image formation is, of course, non-trivial and requires fairly involved

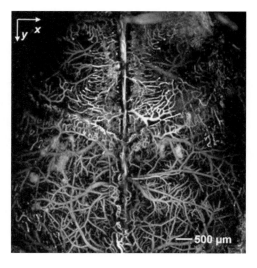

Figure 18.10. Photoacoustic images of hemoglobin absorption in unlabelled *in-vivo* mouse brain. Courtesy of Junjie Yao and Lihong Wang, and adapted from [25].

tomographic methods that go far beyond the scope of our discussion. Suffice it to say that because the illumination is pulsed, and because acoustic propagation times can be resolved with conventional electronics, time plays an all important role in the construction of photoacoustic images, in a similar manner as it does in ultrasound imaging based on echography.

The topic of photoacoustic imaging is vast, and hardly broached here. For much more information on the topic, the reader is referred to specialized references such as [23].

18.5 PROBLEMS

Problem 18.1

Derive Eqs. 18.7 and 18.10.

Problem 18.2

CARS microscopy is performed with Gaussian pump and Stokes pulses defined by $E_p(t) = E_p \exp\left(-t^2/\Delta t_p^2\right) \exp\left(-i2\pi\nu_p t\right)$ and $E_s(t) = E_s \exp\left(-t^2/\Delta t_s^2\right) \exp\left(-i2\pi\nu_s t\right)$, which overlap in time.

(a) The spectral resolution $\Delta\nu_{CARS}$ of a CARS microscope can be defined as the half-width at $1/e$-maximum of $|\mathcal{P}_{CARS}(\nu)|^2$. Show that this spectral resolution is defined by the spectral width of the pump beam alone. What is this spectral resolution?

(b) Consider adding a frequency chirp to the pump pulse, but not the Stokes pulse. The chirp rate is b. That is, the pump field and its Fourier transform are given by

$$E_p(t) = E_p e^{-t^2/\Delta t_p^2} e^{-i2\pi\left(\nu_p + bt\right)t}$$

$$\mathcal{E}_p(\nu) = \frac{1}{\sqrt{\pi\Delta\nu_p'}} E_p e^{-\left(\nu-\nu_p\right)^2/\Delta\nu_p'^2}$$

where $\Delta\nu_p' = \Delta\nu_p\sqrt{1 + i2\pi\Delta t_p^2 b}$ and $\Delta\nu_p = \frac{1}{\pi\Delta t_p}$.

What is the new spectral resolution of the CARS microscope?

(c) The pump pulse, in addition to being chirped, is also temporally broadened to a width $\Delta t_p' > \Delta t_p$. Again, calculate the CARS spectral resolution. What is the maximum chirp rate allowed for this new resolution to be better (narrower) than the original resolution calculated in part **(a)**?

Problem 18.3

Most commonly, two techniques are used to obtain a CARS spectrum. The first makes use of picosecond pump and Stokes beams (that is, both are relatively narrowband). A spectrum can then be obtained by scanning the frequency of one of the beams, and acquiring data sequentially. This has the advantage that only a single detector is required.

Alternatively, the Stokes beam can be femtosecond (i.e. broadband), and a spectrum can be obtained by recording the CARS frequencies in parallel using a grating and line camera (called a spectrograph). This has the advantage that it does not require frequency scanning.

Alternatively still, both the pump and Stokes beam can be femtosecond and chirped, with the *same* chirp rate (see Problem 18.2). For this last scenario, describe a technique to obtain a CARS spectrum without modifying the laser parameters and using only a single detector. Qualitatively, what happens if the chirp rates are not the same?

Problem 18.4

Equation 18.17 suggests that CARS microscopy cannot provide a direct measure of $\text{Im}\chi_r^{(3)}(\nu)$. But consider the case where a measurement of $\left|\chi^{(3)}(\nu)\right|^2$ is obtained over a large range of frequencies, much larger than the Raman features of interest. Assume also that $\chi_{nr}^{(3)}$ is much greater than $\chi_r^{(3)}(\nu)$, as is common in practice. Can you devise a numerical technique to estimate $\text{Im}\chi_r^{(3)}(\nu)$ from your measurement?

Hint: use Fourier transforms.

Problem 18.5

Derive Eq. 18.42. Hint: write $\vec{q} = \kappa(\hat{r} - \hat{z})$ and use a spherical coordinate system where $\vec{q} \cdot \vec{r}_0 = qr_0 \cos\theta'$ when performing the Fourier transform.

References

[1] Benalcazar, W. A., Chowdary, P. D., Jiang, Z., Marks, D. L., Chaney, E. J., Gruebele, M. and Boppart, S. A. "High-speed nonlinear interferometric vibrational imaging of biological tissue with comparison to Raman microscopy," *IEEE J. Selec. Top. Quant. Elec.* 16, 824–832 (2010).

[2] Berciaud, S., Lasne, D., Blab, G. A., Cognet, L. and Lounis, B. "Photothermal heterodyne imaging of individual metallic nanoparticles: theory versus experiment," *Phys. Rev.* B 73, 045424 (2006).

[3] Bloembergen, N. *Nonlinear Optics*, 4th edn, World Scientific (1996).

[4] Boyd, R. W. *Nonlinear Optics*, 3rd edn, Academic Press (2003).

[5] Butcher, P. N. and Cotter, D. *The Elements of Nonlinear Optics*, Cambridge University Press (1990).

[6] Carslaw, H. S. and Jaeger, J. C. *Conduction of Heat in Solids*, 2nd edn, Oxford (1986).

[7] Cheng, J.-X. and Sunney Xie, X. *Coherent Raman Scattering Microscopy*, CRC Press (2013).

[8] Cheng, J.-X. and Sunney Xie, X. Eds., "Coherent anti-Stokes Raman scattering microscopy: instrumentation, theory, and applications," *J. Phys. Chem.* B 108, 827–840 (2004).

[9] Dudovich, N., Oron, D. and Silberberg, Y. "Single-pulse coherent anti-Stokes Raman spectroscopy in the fingerprint spectral region," *J. Chem. Phys.* 118, 9208–9215 (2003).

[10] Eckhardt, G., Hellwarth, R. W., McClung, F. J., Schwartz, S. E., Weiner, D. and Woodbury, E. J. "Stimulated Raman Scattering from Organic Liquids," *Phys. Rev. Lett.* 9, 455–457 (1962).

[11] Heinrich, C., Bernet, S. and Ritsch-Marte, M. "Wide-field coherent anti-Stokes Raman scattering microscopy," *Appl. Phys. Lett.* 84, 816–818 (2004).

[12] Landau, L. D. and Lifshitz, E. M. *Statistical Physics*, 3rd edn, Butterworth-Heinemann (1980).

[13] Lasne, D., Blab, G. A., De Giorgi, F., Ichas, F., Lounis, B. and Cognet, L. "Label-free optical imaging of mitochondria in live cells," *Opt. Express* 15, 14184–14193 (2007).

[14] Maker, P. D. and Terhune, R. W. "Study of optical effects due to an induced polarization third order in the electric field strength," *Phys. Rev.* 137, A801–A818 (1965).

[15] Mukamel, S. *Principles of Nonlinear Optical Spectroscopy*, Oxford University Press (1999).

[16] Neuman, K. C. and Block, S. M. "Optical trapping," *Rev. Sci. Instrum.* 75, 2787–2809 (2004).

[17] Ogilvie, J. P., Beaurepaire, E., Alexandrou, A. and Joffre, M. "Fourier-transform coherent anti-Stokes Raman scattering microscopy," *Opt. Lett.* 31, 480–482 (2006).

[18] Oron, D., Dudovich, N., Yelin, D. and Silberberg, Y. "Quantum control of coherent anti-Stokes Raman process," *Phys. Rev.* A 65, 0434081–0434084 (2002).

[19] Scarcelli, G. and Yun, S. H. "Confocal Brillouin microscopy for three-dimensional mechanical imaging," *Nat. Phot.* 2, 39–43 (2008).

[20] Shen, Y. R. *The Principles of Nonlinear Optics*, John Wiley & Sons (1984).

[21] van Vacano, B. and Motzkus, M. "Time-resolving molecular vibration for microanalytics: single laser beam nonlinear Raman spectroscopy in simulation and experiment," *Phys. Chem. Chem. Phys.* 10, 681–691 (2008).

[22] Wang, L. V. and Hu, S. "Photoacoustic tomography: in vivo imaging from organelles to organs," *Science* 335, 1458–1462 (2012).

[23] Wang, L. V. Ed., *Photoacoustic Imaging and Spectroscopy*, CRC Press (2009).

[24] Yan, Y.-X. and Nelson, K. A. "Impulsive stimulated light scattering. I. General theory," *J. Chem. Phys.* 87, 6240 (1987).

[25] Yao, J., Wang, L., Yang, J.-M., Maslov, K. I., Wong, T. T. W., Li, L., Huang, C.-H., Zou, J. and Wang, L. V. "High-speed label-free functional photoacoustic microscopy of mouse brain in action," *Nat. Meth.* 12, 407–410 (2015).

[26] Yariv, A. *Quantum Electronics*, 3rd edn, John Wiley & Sons (1989).

[27] Xiao, S., Weiner, A. M. and Lin, C. "A dispersion law for vimaged phased-array spectral dispersers based on paraxial wave theory," *IEEE J. Quant. Elec.* 40, 420–426 (2004).

[28] Zhang, D., Slipchenko, M. N. and Cheng, J.-X. "Highly sensitive vibrational imaging by femtosecond pulse stimulated Raman loss," *J. Phys. Chem. Lett.* 2, 1248–1253 (2011).

[29] Zhang, D., Li, C., Zhang, C., Slipchenko, M. N., Eakins, G. and Cheng, J.-X. "Depth-resolved mid-infrared photothermal imaging of living cells and organisms with submicrometer spatial resolution," *Sci. Adv.* 2, e1600521 (2016).

[30] Zharov, V. P. and Lapotko, D. O. "Photothermal imaging of nanoparticles and cells," *IEEE J. Selec. Top. Quant. Elec.* 11, 733–751 (2005).

[31] Zumbusch, A., Holtom, G. R. and Sunney Xie, X. "Three dimensional vibrational imaging by coherent anti-Stokes Raman scattering," *Phys. Rev. Lett.* 82, 4142–4145 (1999).

19 | Superresolution

Needless to say, the concept of spatial resolution is essential for the characterization of any imaging device. Various definitions of resolution have been applied to microscopy, some of the more common being related to the ability of the microscope user to resolve two point objects in close proximity. Classic examples are the oft quoted Rayleigh and Sparrow resolution criteria (see [9] and references therein for a review on resolution).

Throughout this book we have favored a more general definition of spatial resolution based on the rough notion that spatial resolution and spatial-frequency bandwidth are inversely related. Accordingly, we have concentrated on characterizing the frequency bandwidth of an imaging device in terms of its 3D frequency support, from which a definition of 3D resolution naturally follows. But clearly this cannot be the full story. As a telling example, let us look at the in-focus frequency support provided by a standard incoherent widefield microscope with a circular pupil. In the transverse direction, this support is given by the frequency range over which $\mathrm{OTF}(\vec{\kappa}_\perp)$ is non-zero. From Fig. 5.6, this range is $2\Delta\kappa_\perp$. However, the spatial resolution that is generally associated with this type of microscopy is not $\frac{1}{2\Delta\kappa_\perp}$, as one might expect, but rather $\frac{1}{\Delta\kappa_\perp} \simeq \frac{\lambda}{2\mathrm{NA}}$. The reason for this is that the efficiency of the frequency transfer provided by $\mathrm{OTF}(\vec{\kappa}_\perp)$ is not uniform across its support but instead tapers to zero at high frequencies (see Fig. 5.3), which results in a blurring of the PSF and a concomitant degradation in spatial resolution compared to what might be expected from the frequency support alone. This problem of non-uniform frequency transfer becomes even more apparent when the imaging is out of focus. Certainly out-of-focus imaging provides less spatial resolution than in-focus imaging, and yet the frequency support of $\mathrm{OTF}(\vec{\kappa}_\perp; z)$ continues to span the range $2\Delta\kappa_\perp$, independently of z (again, see Fig. 5.3).

A natural question to ask is, inasmuch as $\mathrm{OTF}(\vec{\kappa}_\perp; z)$ is technically non-zero throughout the span of its frequency support (neglecting isolated points, if they occur, where it performs a zero crossing), might it be possible to numerically amplify the transfer efficiency at frequencies where it is weak? In other words, might it be possible to "equalize" the frequency response of $\mathrm{OTF}(\vec{\kappa}_\perp)$, thereby sharpening resolution? At first glance, such a procedure seems straightforward. For example, in the case of fluorescence imaging we have

$$\mathcal{I}(\vec{\kappa}_\perp) = \mathrm{OTF}(\vec{\kappa}_\perp)\mathcal{F}_0(\vec{\kappa}_\perp) \tag{19.1}$$

where $\mathcal{I}(\kappa_\perp)$ is the image spectrum obtained from an object fluorescence spectrum $\mathcal{F}_0(\kappa_\perp)$. Provided that $\mathrm{OTF}(\vec{\kappa}_\perp)$ is known, the recovery of a sharpened object spectrum from the image spectrum then seems a simple matter of inversion:

$$\mathcal{F}_0(\vec{\kappa}_\perp) = \mathrm{OTF}(\vec{\kappa}_\perp)^{-1}\mathcal{I}(\vec{\kappa}_\perp) \tag{19.2}$$

The difficulty with this procedure, of course, is that it is extremely sensitive to noise. Wherever $\mathrm{OTF}(\vec{\kappa}_\perp)$ approaches zero, its inverse correspondingly diverges, and any presence of noise in the image is amplified to the point where object recovery becomes impossible. Strategies to temper the effect of noise are the subject of an enormous field of research known as image deconvolution, or image inversion (to barely scratch the surface of this field, the reader may begin with [2, 6, 8, 10, 15, 19, 40]). We will not discuss these strategies beyond floating the general maxim that they tend to be more robust the more a priori information is available, such as information about the noise statistics or about the object itself. For example, a physical object $F_0(\vec{\rho})$ is known, a priori, to be positive-definite and of finite extent. As it happens, this last condition, called spatial boundedness, leads to the startling theoretical conclusion that if the frequency content of an object is known over a finite range of frequencies, then this range can be numerically made arbitrarily large by a process of analytic continuation. In theory, then, even though the frequency support of an imaging device is finite, it should be possible to reconstruct $F_0(\vec{\rho})$ to infinite spatial resolution! While such a possibility of superresolution was recognized early on in the development of imaging theory, it was recognized just as early on that the inevitable presence of noise would be its undoing. In practice, any strategy of superresolution based on a numerical extrapolation of the frequency support can, at best, provide only a marginal gain in resolution for only a limited class of objects, typically objects that are sparse and highly punctate or with sharp edges, and only when there is little presence of noise.

Alternatively, sparsity can be imposed by the illumination itself. For example, superresolution can be achieved, to some degree, with the use of speckled illumination (see [30]). Remarkably, though a large multiplicity of patterns is required for this, the exact illumination patterns need not be known. The deconvolution algorithms for object reconstruction in these cases are said to be "blind," though, in turn, they critically depend on additional constraints related to the illumination statistics. Again, such algorithms basically rely on the optimization of solutions to Eq. 19.2, the discussion of which would take us too far afield.

Finally, it should be noted that the theoretical possibility of infinite resolution implicitly suggests an extrapolation of the frequency support not only beyond the diffraction limit but beyond the frequency cutoff inherent to far-field radiation itself (see Section 2.2.1). Indeed, microscopy techniques based on the use of near-field light have been able to achieve superresolution, and have garnered much attention. Ultimately, these techniques rely on a conversion of near-field to far-field light by some kind of local scattering or absorption-emission process. Examples are near-field scanning optical microscopy (NSOM) or total internal reflection fluorescence (TIRF) microscopy, or even solid immersion lens microscopy. The reader is referred to textbooks such as [7] and [31] for more details on near-field techniques.

As pointed out by Simonetti [39], the conversion of near-field to far-field light is actually implicit when extending scattering theory into the multiple scattering regime beyond the Born approximation (see also [27] for an interesting application to ultrasound imaging). However, this too will not be discussed in this chapter.

Having summarized in a long-winded way what will not be discussed in this chapter, we proceed now to what will be discussed.

19.1 RESTRICTED SUPERRESOLUTION

The term superresolution has been ambiguously used in the literature, often referring to nothing more than an improvement in resolution over that obtained with an unobstructed circular pupil. An attempt will be made here to better define the term, using formalism that is largely borrowed from the concise overview by Sheppard [38].

To begin, let us consider superresolution that is restricted in the sense that it remains limited by the standard laws of far-field diffraction developed in Chapters 2 through 9. The term superresolution is a bit of a misnomer in this case, but we will use it nonetheless. Two basic strategies are considered, one where the pupil size is held fixed and another where the pupil size is effectively expanded by a technique called aperture synthesis.

19.1.1 PSF Engineering

Most of the microscopy techniques we have examined have been based on the use of a circular detection pupil (see Section 4.4.1). The property that this pupil is unobstructed serves as an important benchmark, since it can be shown that if any aberrations or phase variations $\phi(\vec{\xi})$ are introduced into such a pupil, that is, if we write $P_\phi(\vec{\xi}) = P_0(\vec{\xi})e^{i\phi(\vec{\xi})}$, then the following inequality holds:

$$\left|\mathrm{OTF}_\phi(\vec{\kappa}_\perp)\right| \leq \left|\mathrm{OTF}_0(\vec{\kappa}_\perp)\right| \tag{19.3}$$

One might infer from this inequality that the best resolution one can obtain for a given detection-pupil diameter is when the pupil is unobstructed; however, this is not always true. In particular, one must be careful how one quantifies resolution. Standard measures are based on the quality of the detection point spread function (PSF). A popular measure of this quality is based on the Strehl ratio, defined as $\mathrm{PSF}_\phi(0)/\mathrm{PSF}_0(0)$. It is clear from Eq. 19.3 that the Strehl ratio is indeed maximized only when the detection pupil is devoid of extraneous phase variations. Nevertheless, depending on the imaging application, there may be alternative measures of PSF quality that are more appropriate. For example, if a narrow PSF is required, possible measures of PSF quality might be based on its width at half maximum, or the curvature at its peak. When either of these alternative measures are adopted, then gains in PSF quality can be indeed procured with a judicious introduction of phase shifts in the pupil. Examples are discussed in [35].

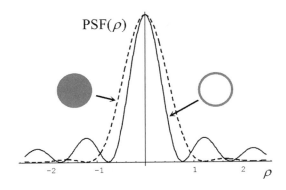

Figure 19.1. PSFs for unobstructed (●) and annular (○) pupils. The transverse coordinate ρ is normalized to $\Delta\kappa_\perp^{-1}$.

Another issue to be careful with is that we have considered only phase variations in the pupil, and not amplitude variations. As a simple though perhaps extreme example of an amplitude variation, let us consider placing a mask in a pupil such that only a thin annulus of the pupil remains unobstructed, say at its very perimeter (see [37] for more details). The PSF associated with such an annular pupil is proportional to $J_0(\pi\Delta\kappa_\perp\rho)^2$, where J_0 is the cylindrical Bessel function of order 0. Such a PSF is illustrated in Fig. 19.1, along with the standard PSF associated with an unobstructed circular pupil (Eq. 4.81). Manifestly, if one were to assess the quality of the PSFs by their widths at half maximum, then the annular-pupil PSF seems superior. However, a price is paid for this apparent superresolution when measured in terms of how effective the PSFs are at confining optical power. Specifically, the PSF envelope decays asymptotically as ρ^{-3} in the case of a standard circular-pupil PSF, whereas it decays only as ρ^{-1} in the case of an annular-pupil PSF. The latter decay is quite weak, so weak, in fact, that the annular-pupil PSF cannot even be normalized. Nevertheless, for restricted field of views, this price may be worth paying. Extensions of this idea were originally proposed by Toraldo di Francia [42], making use not of a single annulus but of many concentric annuli of well-calibrated radii and widths. While theoretically infinitely narrow PSFs can be engineered with such pupil masks, the distribution of detected power in the central peak of the PSF versus its sidelobes becomes, concomitantly, infinitely small. For more information see [3, 16], noting that for a scanning microscope, as opposed to a widefield microscope, it is the illumination PSF that should be engineered in addition to the detection PSF. More will be said about illumination-PSF engineering below.

19.1.2 Synthetic Aperture Holography

By definition, PSF engineering is equivalent to OTF engineering. While it is possible to re-adjust the balance of high frequencies relative to low frequencies in an engineered OTF, it is clear that such a strategy is fundamentally limited by the size of the pupil itself. Ultimately, the light that reaches the detector contains no spatial frequencies beyond $\Delta\kappa_\perp$. The problem then reduces to what information can be distilled from bandwidth-restricted light. If structural information is sought specifically about the light itself, then, as emphasized above, the bandwidth limit $\Delta\kappa_\perp$ should be regarded as inviolable. However, in most imaging applications it is not the light that is of interest but rather the sample. This paradigm shift, as simple as it sounds, is the key to the superresolution strategies described below. In fact, we have already seen consequences of this paradigm shift in Chapters 10 and 11 (e.g. compare Figs. 10.9 or 11.6 with Fig. 5.4). We will re-examine these consequences in the broader context of both coherent and incoherent imaging. For a prescient description of many of the ideas presented in this chapter, the reader is encouraged to consult the seminal work by Lukosz [28].

Let us consider coherent imaging, specifically holographic microscopy, as described in Chapter 11. With such imaging, a complex field $E(\vec{\rho})$ is measured at the image plane. For example, if we consider an imaging configuration where the sample of interest is characterized by a reflectance $r(\vec{\rho}_0)$, then the field at the image plane is given by the familiar equation

$$E(\vec{\rho}) = \int d^2\vec{\rho}_0\, H(\vec{\rho} - \vec{\rho}_0)r(\vec{\rho}_0)E_i(\vec{\rho}_0) \tag{19.4}$$

where, for simplicity, we have limited ourselves to 2D in-focus imaging with unit magnification.

In the frequency domain, Eq. 19.4 becomes

$$\mathcal{E}(\vec{\kappa}_\perp) = \mathcal{H}(\vec{\kappa}_\perp)\left[\hat{r}(\vec{\kappa}_\perp) * \mathcal{E}_i(\vec{\kappa}_\perp)\right]_{\kappa_\perp} \tag{19.5}$$

where $\hat{r}(\vec{\kappa}_\perp)$ is the Fourier transform of the sample reflectance.

From Eq. 19.5, we observe that the frequency support associated with $\mathcal{H}(\vec{\kappa}_\perp)$ is projected not directly onto $\hat{r}(\vec{\kappa}_\perp)$, but rather onto the convolution $\left[\hat{r}(\vec{\kappa}_\perp) * \mathcal{E}_i(\vec{\kappa}_\perp)\right]_{\kappa_\perp}$. However, since we are primarily interested in the sample reflectance, it is the frequency support projected onto $\hat{r}(\vec{\kappa}_\perp)$ that is of importance. Manifestly, from Eq. 19.5, this frequency support is dependent on $\mathcal{E}_i(\vec{\kappa}_\perp)$.

To demonstrate this, let us consider the simplest case where the illumination field is a uniform, on-axis plane wave. That is, we write $E_i(\vec{\rho}_0) = E_i$, and hence $\mathcal{E}_i(\vec{\kappa}_\perp) = E_i\delta^2(\vec{\kappa}_\perp)$. Equation 19.5 reduces to

$$\mathcal{E}(\vec{\kappa}_\perp) = E_i\mathcal{H}(\vec{\kappa}_\perp)\hat{r}(\vec{\kappa}_\perp) \tag{19.6}$$

from which we infer that the frequency support projected onto $\hat{r}(\vec{\kappa}_\perp)$ is indeed identical to that defined by $\mathcal{H}(\vec{\kappa}_\perp)$. But we have considered the simplest case only. Let us imagine a slightly more elaborate case where the incident plane wave arrives off axis, namely

$$E_i(\vec{\rho}_0) = E_i e^{-i2\pi\vec{q}_\perp \cdot \vec{\rho}_0} \tag{19.7}$$

where $\vec{q}_\perp$ is the transverse wavevector associated with the off-axis angle. We then have $\mathcal{E}_i(\vec{\kappa}_\perp) = E_i\delta^2(\vec{\kappa}_\perp + \vec{q}_\perp)$, from which Eq. 19.5 reduces to

$$\mathcal{E}(\vec{\kappa}_\perp) = E_i\mathcal{H}(\vec{\kappa}_\perp)\hat{r}(\vec{\kappa}_\perp + \vec{q}_\perp) \tag{19.8}$$

or, equivalently

$$\mathcal{E}_{\mathrm{SA}}(\vec{\kappa}_\perp) = \mathcal{E}(\vec{\kappa}_\perp - \vec{q}_\perp) = E_i\mathcal{H}(\vec{\kappa}_\perp - \vec{q}_\perp)\hat{r}(\vec{\kappa}_\perp) \tag{19.9}$$

where $\mathcal{E}_{\mathrm{SA}}(\vec{\kappa}_\perp)$ is a frequency-shifted version of $\mathcal{E}(\vec{\kappa}_\perp)$.

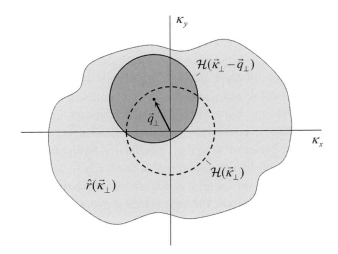

Figure 19.2. Frequency supports for $\hat{r}(\vec{\kappa}_{\perp})$, $\mathcal{H}(\vec{\kappa}_{\perp})$, and $\mathcal{H}(\vec{\kappa}_{\perp})$ displaced by a vector $\vec{q}_{\perp}$.

As illustrated in Fig. 19.2, the effective coherent transfer function associated with the imaging of the sample reflectance has now been translated by the vector $\vec{q}_{\perp}$. Accordingly, the frequency support projected onto $\hat{r}(\vec{\kappa}_{\perp})$ has also been translated. It is clear that this translated frequency support provides access to a region of sample frequencies previously inaccessible with on-axis illumination, a region that in fact extends beyond the limiting bandwidth $\Delta\kappa_{\perp}$ normally associated with $\mathcal{H}(\vec{\kappa}_{\perp})$. As a result, off-axis illumination gives rise to the possibility of superresolution. Of course, the extended region illustrated in Fig. 19.2 is confined to a limited range of spatial frequencies centered about $\vec{q}_{\perp}$. For access to a full 2π azimuthal range of spatial frequencies, several images must be acquired covering the full 2π azimuthal range of illumination angles, ultimately yielding a roughly isotropic 2D superresolution (though it should be noted that half of the 2π directions provide redundant information in the case where $r\left(\vec{\rho}_0\right)$ is real, meaning $\hat{r}(\vec{\kappa}_{\perp})$ is Hermitian). Moreover, the actual fusion of $\mathcal{E}_{\mathrm{SA}}(\vec{\kappa}_{\perp})$ must be performed with care. Generally this requires significant overlap between contiguous regions of frequency support to enable the registration of $\mathcal{E}_{\mathrm{SA}}(\vec{\kappa}_{\perp})$ from one region to another. When all is said and done, the final overall frequency support is encompassed by the term in brackets in

$$\mathcal{E}_{\mathrm{SA}}(\vec{\kappa}_{\perp}) = E_i \left[\sum_n \mathcal{H}(\vec{\kappa}_{\perp} - \vec{q}_{\perp n}) \right] \hat{r}(\vec{\kappa}_{\perp}) \qquad (19.10)$$

where the index n denotes an image acquisition with an illumination angle defined by $\vec{q}_{\perp n}$. This term represents an expanded detection aperture that has been synthesized from several displaced versions of the physical detection aperture. As described above, aperture displacements can be obtained with the use of different illumination angles, or by displacements of the aperture itself. In either case, the technique summarized by Eq. 19.10 is referred to as synthetic aperture imaging.

But it should be emphasized that superresolution by aperture synthesis does not come for free. While standard imaging requires only a single exposure, synthetic aperture imaging requires the fusion of information from several exposures. The price paid is an overall reduction in the image acquisition rate to a fraction of the standard rate. Aperture synthesis has relatively recently been applied to holographic microscopy (see [1, 5, 26, 29] for examples), though its origins date back several decades when it was first applied to radar imaging. For more general accounts of synthetic aperture optics, the reader is referred to [11] and [12]. As an added remark, synthetic aperture imaging, as described above, cannot provide unlimited resolution because of the limitations on the maximum modulus of $\vec{q}_\perp$. Fundamentally, $\left|\vec{q}_\perp\right|$ is limited by κ, the wavenumber of the illumination, and, in practice, it is even more severely limited by the illumination numerical aperture. As such, the superresolution described here remains restricted.

19.1.3 Structured Illumination Microscopy

In our example of synthetic aperture holography, the sample was characterized by its reflectance $r(\vec{\rho}_0)$. A reconstruction of this sample with imaging was then based on Eq. 19.10, which required a series of measurements of the complex field $E(\vec{\rho})$. Because standard detectors cannot directly measure complex fields, we resorted to holographic imaging to obtain these measurements.

Let us imagine a different scenario where we do not have access to the full complex field $E(\vec{\rho})$ but instead only to the intensity $I(\vec{\rho})$ at an image plane. Moreover, let us imagine that the sample is fluorescent, meaning that the quantity of interest is no longer $r(\vec{\rho}_0)$ but rather $C(\vec{\rho}_0)$, the molecular concentration. Is it possible, then, to perform the equivalent of synthetic aperture imaging? At first glance, such a prospect seems problematic since the imaging is now incoherent and all phase information is lost. Specifically, any information on the illumination direction is lost upon the generation of fluorescence, and any information on the fluorescence-light direction is lost upon the detection of $I(\vec{\rho})$. Any attempt to link aperture displacement with illumination or detection direction therefore seems unworkable. However, a technique that we have seen before can come to our aid, which is the technique of structured illumination. From Chapter 15, we found that $I(\vec{\rho})$ is rendered effectively complex by phase stepping (Eq. 15.7), allowing it to be treated in a similar manner as a field. We return to this technique of structured illumination, this time with the explicit goal of superresolution. More specifically, we return to SIM with fringe illumination, as was described in Section 15.2. The implementation of SIM remains identical here except that the numerical algorithm for image reconstruction will be slightly modified.

We recall that SIM imaging is characterized by the equation

$$\widetilde{I}(\vec{\rho}) = \sigma_f \iint \mathrm{d}^2\vec{\rho}_0 \, \mathrm{d}z_0 \, \mathrm{PSF}(\vec{\rho} - \vec{\rho}_0, z_0) C(\vec{\rho}_0, z_0) \widetilde{I}_i(\vec{\rho}_0, z_0) \tag{19.11}$$

where, both $\widetilde{I}(\vec{\rho})$ and $\widetilde{I}_i(\vec{\rho}_0, z_0)$ have been rendered effectively complex by phase stepping. We further recall that SIM largely rejects out-of-focus background, meaning that, to a good approximation, Eq. 19.11 can be simplified to the in-focus 2D imaging equation

$$\widetilde{I}(\vec{\rho}) = \sigma_f \int d^2\vec{\rho}_0 \, \mathrm{PSF}(\vec{\rho} - \vec{\rho}_0) C_\rho(\vec{\rho}_0) \widetilde{I}_i(\vec{\rho}_0) \tag{19.12}$$

where $C_\rho(\vec{\rho}_0)$ is the density of fluorescent molecules per unit area, and z_0 is implicitly taken to be equal to 0 (in-focus imaging only).

From our result in Eq. 15.8, the effective complex illumination intensity obtained from phase stepping is

$$\widetilde{I}_i(\vec{\rho}_0) = \tfrac{1}{2} I_i M(q_x) e^{-i2\pi q_x x_0} \tag{19.13}$$

where q_x is the fringe spatial frequency, taken to be aligned along x, and $M(q_x)$ is the in-focus fringe modulation contrast. Note the similarity of Eq. 19.13 and Eq. 19.7. Thus, by way of phase stepping, an effective directionality has been imparted onto $\widetilde{I}_i(\vec{\rho}_0)$, defined by the fringe frequency. This correspondence between effective directionality and fringe frequency is crucial. Indeed, with this correspondence, the strategy for obtaining superresolution becomes straightforward and follows directly from the technique of synthetic aperture holography described above.

Reformulating Eq. 19.12 in the frequency domain, we write

$$\widetilde{\mathcal{I}}(\vec{\kappa}_\perp) = \sigma_f \mathrm{OTF}(\vec{\kappa}_\perp) \left[\mathcal{C}_\rho(\vec{\kappa}_\perp) * \widetilde{\mathcal{I}}_i(\vec{\kappa}_\perp) \right]_{\vec{\kappa}_\perp} \tag{19.14}$$

where $\mathcal{C}_\rho(\vec{\kappa}_\perp)$ is the Fourier transform of $C_\rho(\vec{\rho}_0)$. Specifically,

$$\widetilde{\mathcal{I}}_i(\vec{\kappa}_\perp) = \tfrac{1}{2} I_i M(\vec{q}_\perp) \delta^2(\vec{\kappa}_\perp + \vec{q}_\perp) \tag{19.15}$$

where the fringe spatial frequency has been generalized to an arbitrary direction (in the case of Eq. 19.13, $\vec{q}_\perp = (q_x, 0)$). Insertion of Eq. 19.15 into Eq. 19.14 then leads to

$$\widetilde{\mathcal{I}}(\vec{\kappa}_\perp) = \tfrac{1}{2} \sigma_f I_i M(\vec{q}_\perp) \mathrm{OTF}(\vec{\kappa}_\perp) \mathcal{C}_\rho(\vec{\kappa}_\perp + \vec{q}_\perp) \tag{19.16}$$

or, equivalently,

$$\widetilde{\mathcal{I}}_{\mathrm{SA}}(\vec{\kappa}_\perp) = \widetilde{\mathcal{I}}(\vec{\kappa}_\perp - \vec{q}_\perp) = \tfrac{1}{2} \sigma_f I_i M(\vec{q}_\perp) \mathrm{OTF}(\vec{\kappa}_\perp - \vec{q}_\perp) \mathcal{C}_\rho(\vec{\kappa}_\perp) \tag{19.17}$$

We conclude that $\mathrm{OTF}(\vec{\kappa}_\perp)$ is displaced here in the same way that $\mathcal{H}(\vec{\kappa}_\perp)$ was displaced in Eq. 19.9, permitting access to higher frequency components of $\mathcal{C}_\rho(\vec{\kappa}_\perp)$ hitherto unavailable with standard widefield fluorescence microscopy. Schematically, Eq. 19.17 can again be illustrated by Fig. 19.2 with the modification that the circles delimit the frequency support of $\mathrm{OTF}(\vec{\kappa}_\perp)$ rather than $\mathcal{H}(\vec{\kappa}_\perp)$, and $\hat{r}(\vec{\kappa}_\perp)$ is replaced by $\mathcal{C}_\rho(\vec{\kappa}_\perp)$. As before, to distribute the frequency support over the full 2π azimuthal range, several different versions of $\widetilde{I}(\vec{\rho})$ must be acquired with different directions of the fringe frequency $\vec{q}_{\perp n}$ (each of these, in turn, requiring at least three versions of $I(\vec{\rho})$ obtained from phase stepping). The fusion of the entire set of shifted OTFs then leads to

$$\widetilde{\mathcal{I}}_{\mathrm{SA}}(\vec{\kappa}_\perp) = \left[\tfrac{1}{2}\sigma_f I_i \sum_n M(\vec{q}_{\perp n})\mathrm{OTF}(\vec{\kappa}_\perp - \vec{q}_{\perp n}) \right] \mathcal{C}_\rho(\vec{\kappa}_\perp) \qquad (19.18)$$

This is our final result. The overall span of the frequency support projected onto $\mathcal{C}_\rho(\vec{\kappa}_\perp)$ is defined here both by the span of the unshifted $\mathrm{OTF}(\vec{\kappa}_\perp)$ and by the maximum range of $\left|\vec{q}_\perp\right|$. Most SIM applications are operated in an epifluorescence mode, meaning that the illumination and detection OTFs are roughly the same (neglecting small differences in the illumination and fluorescence wavelengths due to the Stokes shift). In other words, the maximum range of $\left|\vec{q}_\perp\right|$ is roughly equivalent to the maximum range of $\mathrm{OTF}(\vec{\kappa}_\perp)$, namely $\Delta\kappa_\perp$. The overall gain in resolution based on Eq. 19.18 is therefore, at best, a factor of two compared to the resolution of a standard widefield fluorescence microscope. Nevertheless, such a resolution gain can lead to clear improvements in image quality, as illustrated in Fig. 19.3.

It might be recalled that, assuming unobstructed circular pupils, the gain in transverse frequency support of a confocal fluorescence microscope compared to a widefield microscope is also a factor of two (see Fig. 14.2), and indeed a confocal microscope may be regarded as yet another variation of SIM. However, as noted in the introduction to this chapter, the connection

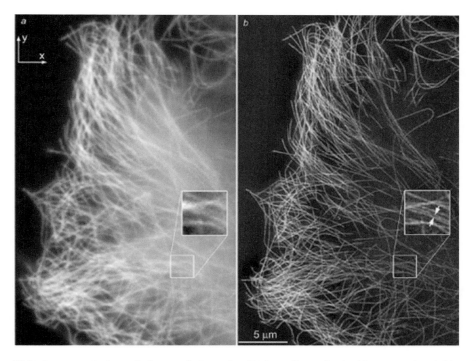

Figure 19.3. Images of microtubule cytoskeleton in a HeLa cell acquired with conventional (a) and structured-illumination microscopy (b). The latter image exhibits both higher resolution and out-of-focus background rejection. Adapted from [14].

between resolution and frequency support is only approximate. SIM with fringe illumination turns out to do a much better job at emphasizing higher spatial frequencies than confocal microscopy, and as a result provides better resolution. For a more generalized and detailed description of a SIM algorithm that extends to three dimensions, the reader is referred to [14].

19.2 UNRESTRICTED SUPERRESOLUTION

All the superresolution techniques that have been described so far, whether based on a fixed aperture size or on an aperture size that is expanded by synthesis, are fundamentally restricted by the diffraction limit. That is, ultimately they are restricted by the wavelength of light. The origin of this restriction is rooted in the fact that we are considering far-field detection only. The natural question that arises is whether far-field detection can allow true superresolution at all, where by "true" superresolution we mean a spatial frequency bandwidth that extends beyond 2κ for coherent imaging or beyond 4κ for incoherent imaging. The answer to the above question turns out again to be yes. Two very different strategies will serve to illustrate this point. The first requires some form of nonlinear interaction between the sample and the illumination light. The second, far simpler, is based on molecular localization, which will be the subject of a section in its own right.

It should be noted that, in fact, all interactions between light and matter are nonlinear to some degree. For example, if the interaction is scattering, then our linearization procedure using the Born approximation may not always be valid, particularly when dealing with strong fields or dense samples, and higher orders must be taken into account (see Chapter 17 and also [39]). If the interaction is absorption, then an even wider taxonomy of phenomena can introduce nonlinearities, including intersystem crossing, saturation, ground-state depletion, photobleaching, stimulated emission, etc. We encountered just some of these in Chapter 13. In most cases such nonlinearities are not desirable since they can lead to problems in image interpretation. However, certainly in Chapters 16–18 we saw counterexamples where nonlinear interactions were actually integral to the imaging process.

We will turn here to more such counterexamples, directed this time specifically toward the goal of superresolution imaging of fluorescent samples. In particular, we will demonstrate that nonlinear interactions, in fact, provide the key to unrestricted superresolution [18].

19.2.1 Nonlinear Structured Illumination Microscopy

As a first example, we return yet again to the ever-versatile technique of SIM with fringe illumination. It was noted above that this technique can effectively displace the detection OTF by a vector $\vec{q}_\perp$, the fringe spatial frequency, allowing access to higher spatial frequency components of the sample. However, in our discussion, the modulus of $\vec{q}_\perp$ was fundamentally restricted by the wavelength of light. As will be shown below, this limit can be surpassed.

Our starting point is a generalization of Eq. 13.2 that allows for nonlinearities in the molecular interaction between the local illumination intensity $I_i(\vec{\rho}_0)$ and the resulting emitted fluorescence power $\Phi(\vec{\rho}_0)$. That is, we write

$$\Phi(\vec{\rho}_0) = \sum_{n=0}^{N} \sigma_n I_i(\vec{\rho}_0)^n \tag{19.19}$$

where generalized fluorescence cross-sections σ_n have been introduced for each interaction order n. Clearly there can be no fluorescence without illumination intensity, and so the $n = 0$ term must vanish by default. Nevertheless, for notational convenience it is retained here with the tacit understanding that $\sigma_0 = 0$. The expansion in Eq. 19.19 is taken out to finite order N, with the assumption that higher orders are negligible.

Once again we consider illumination with a series of laterally translated fringe patterns, denoted by

$$I_i^{(k)}(\vec{\rho}_0) = I_i \left[1 + M(\vec{q}_\perp) \cos(2\pi \vec{q}_\perp \cdot \vec{\rho}_0 + \phi_k) \right] \tag{19.20}$$

where the fringe wavevector can assume arbitrary transverse direction. Note that the modulus of $\vec{q}_\perp$ remains, as before, diffraction limited. That is, $\left| \vec{q}_\perp \right|$ can be no larger than permitted by the illumination numerical aperture. So far, nothing has changed from our previous implementation of SIM. The only difference lies in the interaction itself between the illumination light and the fluorescent molecules, which, because it is nonlinear here, involves higher orders of $I_i^{(k)}(\vec{\rho}_0)$. For brevity, these higher orders are cast in the compact form

$$I_i^{(k)}(\vec{\rho}_0)^n = I_i^n \sum_{m=-n}^{n} a_{nm} e^{im\left(2\pi \vec{q}_\perp \cdot \vec{\rho}_0 + \phi_k\right)} \tag{19.21}$$

where a_{nm} are coefficients derived from expansions of Eq. 19.20 taken to the power n. These coefficients obey the properties $a_{n0} = 1$ and $a_{n(-m)}^* = a_{nm}$.

Upon insertion of Eq. 19.21 into Eq. 19.19, we obtain

$$\Phi^{(k)}(\vec{\rho}_0) = \sum_{n=0}^{N} \sigma_n I_i^n \sum_{m=-n}^{n} a_{nm} e^{im\left(2\pi \vec{q}_\perp \cdot \vec{\rho}_0 + \phi_k\right)} \tag{19.22}$$

which, upon rearrangement, takes on the more convenient form

$$\Phi^{(k)}(\vec{\rho}_0) = \sum_{m=-N}^{N} \sum_{n=|m|}^{N} \sigma_n I_i^n a_{nm} e^{im\left(2\pi \vec{q}_\perp \cdot \vec{\rho}_0 + \phi_k\right)} \tag{19.23}$$

In the frequency domain, we have, then, at the image plane

$$\mathcal{I}^{(k)}(\vec{\kappa}_\perp) = \text{OTF}(\vec{\kappa}_\perp) \left[\sum_{m=-N}^{N} \sum_{n=|m|}^{N} \sigma_n I_i^n a_{nm} e^{im\phi_k} C_\rho(\vec{\kappa}_\perp + m\vec{q}_\perp) \right] \tag{19.24}$$

where, again, we are considering imaging only of the sample plane that is in focus. Equation 19.24 represents standard 2D imaging with the added particularities that the illumination is a sinusoidal fringe pattern (of circular phase $\phi_k = \frac{2\pi k}{K}$) and the fluorescence generation depends nonlinearly on this illumination. An examination of Eq. 19.24 reveals that $\mathcal{I}^{(k)}(\vec{\kappa}_\perp)$ is composed of a superposition of several ($2N + 1$) images of $\mathcal{C}_\rho(\vec{\kappa}_\perp)$, each shifted in frequency space by an integral multiple of $\vec{q}_\perp$. Our procedure will be to isolate each of these shifted images by using our familiar trick of phase stepping. However, this time we adopt a somewhat more generalized version of the phase-stepping algorithm, defined here by

$$\widetilde{\mathcal{I}}^{(l)}(\vec{\kappa}_\perp) = \frac{1}{K} \sum_{k=0}^{K-1} e^{-il\phi_k} \mathcal{I}^{(k)}(\vec{\kappa}_\perp) \tag{19.25}$$

In comparison with the previous algorithm (Eq. 15.5), an index l has been introduced as an extra multiplicative factor in the phase. With the provision that K be at least as large as $2N+1$, the application of this generalized phase-stepping algorithm to Eq. 19.24 leads finally to

$$\widetilde{\mathcal{I}}^{(l)}(\vec{\kappa}_\perp) = \Phi_l \, \text{OTF}(\vec{\kappa}_\perp) \mathcal{C}_\rho(\vec{\kappa}_\perp + l\vec{q}_\perp) \tag{19.26}$$

where

$$\Phi_l = \sum_{n=|l|}^{N} \sigma_n I_i^n a_{nl} \tag{19.27}$$

Presumably Φ_l is a known quantity for each value of l. Equation 19.26 may be directly compared with Eq. 19.16 (the former reduces to the latter when $l = N = 1$). It is clear that nonlinear SIM provides access to a much wider range of frequency support for $\mathcal{C}_\rho(\vec{\kappa}_\perp)$, as illustrated in Fig. 19.4. The full extent of this range is no longer limited by $|\vec{q}_\perp|$ but rather by the product $|l\vec{q}_\perp|$. Since $|l|$ can assume any integer value up to N, and since N is itself

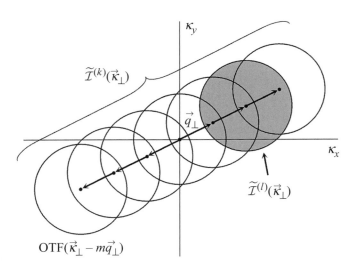

Figure 19.4. Effective frequency support range when $N = 3$. Shaded region is selected by phase stepping with $l = 2$.

prescribed only by the degree of nonlinearity of the fluorescence interaction, this range is virtually unlimited. That is, nonlinear SIM provides, in principle, unrestricted superresolution.

As it happens, we have already made use of the above connection between nonlinearity and superresolution when dealing with TPEF microscopy (Chapter 16). In this case the degree of nonlinearity was $n = 2$ (see Eq. 16.1) and so the gain in resolution was modest. Moreover, this gain was partially offset by a requirement of longer illumination wavelengths. Nevertheless, the underlying principle of resolution gain with TPEF is the same as the one described here, and can certainly be generalized to higher-order multiphoton excitation microscopy.

The technique of obtaining superresolution based on nonlinear SIM with fringe illumination was proposed by Heintzmann [17] and demonstrated by Gustafsson [13]. The nonlinearity in the latter case was the result of fluorescence saturation. More recently, nonlinearities based on molecular switching have also proved to be quite effective. More will be said about molecular switching below.

19.2.2 Stimulated Emission Depletion (STED) Microscopy

An excellent example of the connection between nonlinearity and superresolution is provided by a different type of nonlinear microscopy technique, based this time on a phenomenon called stimulated emission depletion (STED). This technique was pioneered by Hell [20] and predates the SIM techniques described above.

As we have seen in Section 13.1, an excited molecule can release the energy stored in its excited state by way of a variety of decay channels. We classified these channels as radiative or non-radiative, though we have been most interested in the radiative channel giving rise to spontaneous emission, otherwise known as fluorescence. But there is another radiative channel that we only briefly touched upon, namely that of stimulated emission. By its nature, stimulated emission requires the presence not only of an excited molecule but also of light. When the frequency of this light is matched to a downward transition of the excited state to a vibro-rotational level in the ground state, the probability of the transition is amplified. The presence of stimulating light thereby causes the release of even more light of the same frequency. This new light, aptly, is called stimulated emission.

The reason we have neglected stimulated emission is that we have operated under the assumption that the fluorescent molecules of interest in microscopy applications exhibit fairly large Stokes shifts. As such, the excitation frequency for these molecules is different from the emission frequency, meaning there is essentially no possibility of stimulated emission (see Section 13.1 for more details). However, let us imagine now that a second beam of light is directed onto the molecules whose frequency is expressly chosen to produce stimulated emission. If the intensity of this second beam is so high that the probability of stimulated emission completely overwhelms that of spontaneous emission, then the possibility of fluorescence becomes effectively extinguished by excited-state depletion. This, of course, is not the goal in fluorescence microscopy. However, if the stimulating beam is spatially structured

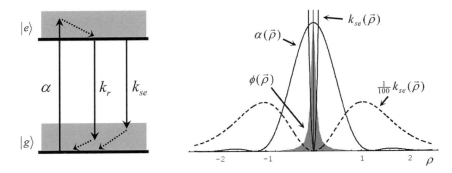

Figure 19.5. Rate constants associated with excitation (α), fluorescence (k_r), and stimulated emission (k_{se}). Resulting confined fluorescence emission profile is given by rate $\phi(\vec{\rho})$.

in a manner that it selectively turns off fluorescence everywhere *except* at a highly localized point about the focal center, then superresolution becomes possible. This is the principle of STED microscopy.

To understand how STED microscopy works in practice, let us turn to a concrete example. In its usual implementation, a STED microscope is much like a confocal fluorescence microscope except that it involves two illumination beams. The first beam is the standard excitation beam, focused in the usual way to a diffraction-limited spot typically defined by an Airy pattern at the focal plane (see Eq. 5.32). So far, nothing new. The novelty comes from the second beam, or STED beam, which is focused not to an Airy spot but rather to a beam profile that features an intensity null exactly at the focal center. Such a beam profile can be created by pupil engineering, as described in Section 19.1.1. A classic example is the donut beam generated by a spiral-phase pupil, whose profile is schematically illustrated in Fig. 19.5. We recall that the purpose of the STED beam is to quench fluorescence. It can be seen from Fig. 19.5 that such quenching occurs everywhere except at the very focal center where the STED beam intensity is identically zero. The remaining fluorescence profile at the focal center is therefore substantially sharpened.

To roughly characterize the width of the remaining fluorescence profile, we can consider a sample of uniformly distributed fluorescent molecules of, say, unit concentration. The steady-state distribution of fluorescence power generated from such molecules is given by

$$\phi(\vec{\rho}) = \frac{\Phi(\vec{\rho})}{h\nu_f} = \frac{\alpha(\vec{\rho})k_r}{\alpha(\vec{\rho}) + k_r + k_{se}(\vec{\rho})} \tag{19.28}$$

where we have made use of Eq. 13.9, and where k_{nr} in Eq. 13.8 has been effectively replaced by the stimulated-emission-induced excited-state depletion rate $k_{se}(\vec{\rho})$. For convenience, the fluorescence power has been normalized to be an emission rate in units of photons/second. Both $\alpha(\vec{\rho})$, the excitation rate, and $k_{se}(\vec{\rho})$ are spatially dependent (see Fig. 19.4). For simplicity, let us also consider the weak excitation regime. That is, we assume $\alpha(\vec{\rho}) \ll k_r$ everywhere, allowing us to neglect $\alpha(\vec{\rho})$ in the denominator of Eq. 19.28.

While there are many ways to characterize the profile of $\phi(\vec{\rho})$, we will follow the arguments in [22] by deriving the curvature at its peak. From Eq. 19.28 this curvature is given by

$$\nabla^2\phi(\vec{\rho})\big|_0 = \nabla^2\alpha(\vec{\rho})\big|_0 - \frac{\alpha(0)}{k_r}\nabla^2 k_{se}(\vec{\rho})\big|_0 \tag{19.29}$$

At this point, we can make some approximations. Inasmuch as the STED beam is generally delivered through the same optics as the excitation beam, a reasonable description of the STED profile near the focal center is provided by

$$k_{se}(\vec{\rho}) \approx \left[1 - \frac{\alpha(\vec{\rho})}{\alpha(0)}\right] k_{\text{STED}} \tag{19.30}$$

where k_{STED} is a scaling factor directly proportional to the STED beam power. Equation 19.29 then simplifies to

$$\nabla^2\phi(\vec{\rho})\big|_0 \approx \left[1 + \frac{k_{\text{STED}}}{k_r}\right]\nabla^2\alpha(\vec{\rho})\big|_0 \tag{19.31}$$

With the added assumption that the fluorescence emission peak is roughly parabolic in shape, the widths of this peak with and without the presence of the STED beam are roughly related by

$$\delta\rho_{\text{STED}} \approx \frac{\delta\rho}{\sqrt{1 + \frac{k_{\text{STED}}}{k_r}}} \tag{19.32}$$

Manifestly, the introduction of the STED beam has caused a sharpening of the fluorescence emission peak. The extent of this sharpening depends only on the maximum value allowed for k_{STED}. In practice, this should be no greater than the thermal intraband relaxation rate of the molecule, since otherwise the STED beam would cause a re-excitation of the molecule immediately following excited-state depletion. Nevertheless, the ratio $\frac{k_{\text{STED}}}{k_r}$ can be several orders of magnitude, meaning that the STED-induced enhancement in resolution can be quite spectacular, certainly well beyond the standard diffraction limit. Our derivation assumes here that both the excitation and STED beams are applied in a continuous manner. In more common implementations of STED microscopy the beams are instead pulsed, where the STED pulse is slightly delayed relative to the excitation pulse. This makes much more efficient use of the STED beam power, though leads to similar results. Finally, our evaluation here has considered resolution enhancement in the transverse direction only. The principle of STED microscopy can be extended to the third axial dimension with the use of more complicated PSF engineering, for example involving phase masks in the form of a top hat. Details on such an extension are presented in [21].

An advantage of STED is that it provides superresolution directly, without any need for image post-processing as was required with previous SIM-based techniques. Another advantage of STED is that, like two-photon microscopy, the fluorescence generation in STED microscopy is background free, at least ideally, thereby minimizing background shot noise and enabling the full dynamic range of the detector to be dedicated to signal recovery. A comparison of confocal and STED images is illustrated in Fig. 19.6.

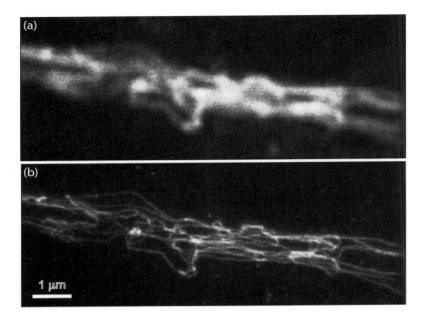

Figure 19.6. Confocal (a) and STED (b) image of vimentin in a neurite of a retinoic-acid-differentiated human neuroblastoma. Courtesy of R. Medda, D. Wildanger, L. Kastrup, and S. W. Hell, MPI Biophys. Chem., Göttingen.

A possible source of confusion should be dispelled here. The substitution of k_{nr} with $k_{se}(\vec{\rho})$ in Eq. 13.9 should not be taken literally to indicate that $k_{se}(\vec{\rho})$ is non-radiative; $k_{se}(\vec{\rho})$ is indeed radiative since it produces stimulated emission. However, because stimulated emission is coherent it is also forward directed (see Chapter 17) and hence largely undetected in an epifluorescence collection geometry. Moreover, the wavelength of the stimulated emission is on the red edge of the fluorescence of interest, and can be rejected spectrally. As such, $k_{se}(\vec{\rho})$ can indeed be interpreted as non-radiative.

19.2.3 Molecular Switching

The critical point by which superresolution is enabled with STED is Eq. 19.28, which is nonlinear in the variable $k_{se}(\vec{\rho})$. That is, a Taylor expansion of $\phi(\vec{\rho})$ about $k_{se}(\vec{\rho}) = 0$ involves $k_{se}(\vec{\rho})$s to all orders, reminiscent of Eq. 19.19. Even without the presence of $k_{se}(\vec{\rho})$, Eq. 19.28 is nonlinear in the variable $\alpha(\vec{\rho})$, involving $\alpha(\vec{\rho})$s to all orders as well (it was this second nonlinearity, in the form of saturation, that led to superresolution in [13]). In both cases, the level of superresolution ultimately depends on the degree of nonlinearity that is attained in Eq. 19.28, which in turn is characterized by the ratio $\frac{k_{STED}}{k_r}$ (see Eq. 19.32) or $\frac{\alpha}{k_r}$, depending on which nonlinearity is exploited. In both cases, this ratio should be larger than one. Since k_r is typically on the order of 10^9 Hz, the illumination intensities involved, therefore, should be rather high.

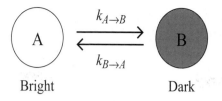

Figure 19.7. Schematic of molecular photoswitching.

However, as pointed out by Hell [22, 25], the concepts underlying Eq. 19.28 can be generalized to any two-state system in which state A can fluoresce (bright) and state B cannot fluoresce (dark). The transition rates linking these states are denoted by $k_{A \to B}$ and $k_{B \to A}$, as illustrated in Fig. 19.7. In steady state, the fluorescence emission rate then reduces to

$$\phi(\vec{\rho}) = \alpha(\vec{\rho}) \left[\frac{k_{B \to A}}{k_{A \to B}(\vec{\rho}) + k_{B \to A}} \right] \tag{19.33}$$

So far we have only treated cases where $k_{B \to A}$ is a spontaneous rate and hence spatially independent; however, a spatial dependence could be introduced here for greater generality still. Inasmuch as $k_{A \to B}(\vec{\rho})$ can be controlled by an illumination intensity $I_i(\vec{\rho})$, then superresolution becomes possible provided the ratio $\frac{k_{A \to B}(\vec{\rho})}{k_{B \to A}}$ can be brought to a high level. Notably, this need not necessarily require a large $k_{A \to B}(\vec{\rho})$, as we have considered so far. For example, if $k_{B \to A}$ is vanishingly small, then superresolution can be attained with quite modest values of $k_{A \to B}(\vec{\rho})$, and, correspondingly, modest illumination intensities.

Such is the principle behind the use of photoactivatable molecules for superresolution imaging. Photoactivatable molecules are capable of switching between two states, one bright and one dark. The switching in one direction is optically induced, typically by way of photoisomerization. The switching in the other direction is spontaneous (though this too could be optically induced at a different wavelength). If the spontaneous rate is slow, then $k_{B \to A}$ is small. For more information on different types of photoactivatable molecules used in biological imaging, the reader may consult [36] and references therein.

19.3 RESOLUTION VERSUS LOCALIZATION

Remarkably, the continued development of photoactivatable molecules has paved the way for an entirely different class of superresolution microscopy that requires neither structured illumination nor scanning. This class of microscopy is based on the imaging of fluorescent molecules one by one, in series, rather than all at once, in parallel. To understand how this technique works, we must go back to the very definition of resolution itself.

Resolution is conventionally characterized by a length scale $\delta\rho$. By definition, any object structures of smaller size than this scale cannot be resolved by imaging, whereas larger object structures can be resolved. For example, if fluorescent molecules are separated by distances less than $\delta\rho$, then standard widefield fluorescence microscopy can not identify their location to a resolution better than $\delta\rho$. That is, the collection of molecules is indistinguishable from a fluorescent blob of size $\delta\rho$.

On the other hand, let us now imagine that we have a priori information about these molecules. This information can take on a variety of forms, but let us imagine that we are able to "turn on" each of these molecules one at a time. In this case, the a priori information is that, at known times, we are imaging one and only one molecule. Such information has tremendous ramifications since it opens the door to virtually unlimited resolution. In effect, the problem of imaging reduces now from one of resolution to one of localization. If one knows with certainty that only a single molecule is being imaged (presumably fixed in space), then even though the image of this molecule appears as a blurred spot of width $\delta\rho$, one need only evaluate the baricenter of this spot to determine the location of the molecule. The precision with which one can evaluate such a baricenter depends not only on the spot size, as expected, but also on the signal-to-noise ratio. In the ideal case where the detection is strictly shot-noise limited, the localization precision is given by

$$\delta\rho_{\text{loc}} \approx \frac{\delta\rho}{\sqrt{N}} \tag{19.34}$$

where N is the number of fluorescence photons actually detected from the molecule. This expression does not take into account errors due to camera pixilation, but as shown in [32, 33, 41], may be taken as reasonably valid provided the pixel size is smaller than $\delta\rho$. The salient point here is that, with a sufficient number of detected photons, $\delta\rho_{\text{loc}}$ can be made significantly smaller than $\delta\rho$. In other words, the location of a single molecule can be determined to a much higher precision than prescribed by the diffraction limit alone.

Such a principle has been exploited for decades in the single-molecule tracking community. More recently, it has also been adopted by the imaging community. Key events that led to this adoption have been the development of fast, low-noise cameras, such as EM-CCD or scientific CMOS cameras (see Section 8.3.2), enabling true shot-noise-limited imaging, and, more importantly, the advent of photoactivatable fluorescent molecules. Specifically, it was assumed from the outset that fluorescent molecules can be "turned on" one at a time, which is indeed possible when the transition rate $k_{B\to A}$ (as opposed to $k_{A\to B}$) is controlled by light. In this manner, molecules can be promoted from dark states to bright states. The assurance that only a single molecule is promoted at a time can be based on statistics.

The technique described above is called photoactivatable light microscopy, or PALM, as was first demonstrated by Betzig [4] (though see also [34] and [24], which appeared almost simultaneously under different names). Let us assume that a sample is doubly illuminated with two uniform beams of different frequencies. The first beam drives the transition $B \to A$, while the second beam serves to excite molecules once they are in state A, causing them to fluoresce (i.e. it drives the transition rate α). By keeping $k_{B\to A}$ very weak, molecules only rarely switch into state A and the sample remains globally dark. Moreover, in the event that a molecule

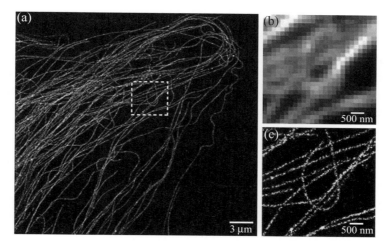

Figure 19.8. Superresolution image of microtubules in a mammalian cell acquired by localization of photoactivatable dyes (a) Comparsion of conventional (b) and superresolution (c) images are shown for zoomed inset. Adapted from [43].

does, randomly, switch into state A, it is likely to be isolated from other such molecules and separated by at least a distance $\delta\rho$. As such, at any given time the acquired images consist of well-identifiable spots each roughly of size $\delta\rho$, the diffraction limit. Each of these spots can be treated as a single molecule with a high degree of certainty. When enough photons are detected per molecule, the molecules are then localized to high precision and their locations cumulatively mapped to form an image.

The final step is to "turn off" the molecules once they have been localized. This final step occurs automatically, by virtue of the fact that molecules eventually photobleach, or it may be spurred on by yet a third beam that drives the transition $A \rightarrow B$, for example by ground-state depletion. In either case, the process is repeated until enough molecules have undergone photoactivation to generate a final image of ultra-high resolution, as illustrated in Fig. 19.8.

The resolutions achieved by PALM (or its variations) are on par with those achieved by STED microscopy, namely a few tens of nanometers. Like STED, PALM can be operated with continuous illumination, or, alternatively, by cycling through both illumination frequencies. The advantage of PALM is its extreme conceptual simplicity. Basically, any widefield microscope can perform PALM provided it is equipped with dual illumination of appropriate wavelengths and provided the sample is labeled with photoactivatable molecules. As pointed out by Lukosz [28], however, there is always a price to pay for superresolution. Since a dense image can only be assembled by PALM from a multitude of sparse images, the price to pay is time. This issue of relatively long acquisition times is common to all superresolution techniques, though it can be mitigated somewhat by parallelization strategies, for example as described in [23].

19.4 PROBLEMS

Problem 19.1

The pupil and point spread functions of a microscope are denoted by $P(\vec{\xi})$ and $\mathrm{PSF}(\vec{\rho})$ respectively. Consider introducing phase variations (or aberrations) in the pupil function, such that $P_\phi(\vec{\xi}) = e^{i\phi(\vec{\rho})}$, leading to $\mathrm{PSF}_\phi(\vec{\rho})$. A standard method for evaluating $\mathrm{PSF}_\phi(\vec{\rho})$ is with the Strehl ratio, defined by

$$S_\phi = \frac{\mathrm{PSF}_\phi(\vec{0})}{\mathrm{PSF}_0(\vec{0})}$$

where $\mathrm{PSF}_0(\vec{\rho})$ is the theoretical diffraction-limited PSF obtained when the pupil is unobstructed (i.e. $P_0(\vec{\xi}) = 0$ or 1). The larger the Strehl ratio, the better the quality of PSF_ϕ. Show that the introduction of aberrations can only lead to a degradation in the point spread function (i.e. $S_\phi \leq 1$). Proceed by first verifying Eq. 19.3.

Hint: you will find the Schwarz inequality to be useful here, which states:

$$\left| \int d^2\vec{\kappa}_\perp\, X(\vec{\kappa}_\perp) Y(\vec{\kappa}_\perp) \right|^2 \leq \left(\int d^2\vec{\kappa}_\perp\, \left| X(\vec{\kappa}_\perp) \right|^2 \right) \left(\int d^2\vec{\kappa}_\perp\, \left| Y(\vec{\kappa}_\perp) \right|^2 \right)$$

where X and Y are arbitrary complex functions.

Problem 19.2

Consider a confocal microscope whose illumination and detection PSFs are identical. The detected power from a simple two-level molecule can be written in a simplified form as

$$\phi(\vec{\rho}) = \alpha \xi^2(\vec{\rho})$$

where α is the molecule excitation rate exactly at the the focal center, and $\xi(\vec{\rho}) = \frac{\mathrm{PSF}(\vec{\rho})}{\mathrm{PSF}(0)}$. The above expression is valid in the weak excitation limit, namely $\alpha \ll k_r$ (equivalent to $\langle e \rangle \approx \frac{\alpha}{k_r}$ – see Section 13.1.1). In the strong excitation limit, then this expression must be modified to take into account saturation. In particular, we must write $\langle e \rangle = \frac{\alpha}{\alpha + k_r}$ (neglecting non-radiative decay channels – see Eq. 13.8).

(a) Derive an expression for $\phi_{\mathrm{sat}}(\vec{\rho})$ taking saturation into account (for simplicity, only keep terms to first order in $\frac{\alpha}{k_r}$). Note that $\phi_{\mathrm{sat}}(\vec{\rho})$ corresponds to an effective confocal PSF, which is now saturated.

(b) Now consider modulating the excitation rate such that $\alpha(t) = \alpha(1 + \cos(2\pi\Omega t))$. Correspondingly, $\phi_{\mathrm{mod}}(\vec{\rho}, t)$ also becomes modulated, and exhibits harmonics. Derive an expression for $\phi_{\mathrm{mod}}(\vec{\rho}, t)$.

(c) By using appropriate demodulation, assume that the components of $\phi_{\mathrm{mod}}(\vec{\rho}, t)$ oscillating at the first (Ω) and second (2Ω) harmonics can be isolated. Use the technique employed in Section 19.2.2 to compare the curvatures of $\phi_\Omega(\vec{\rho})$ and $\phi_{2\Omega}(\vec{\rho})$ to the curvature of $\phi(\vec{\rho})$ (unsaturated and unmodulated). That is, derive approximate expressions for $\delta\rho_\Omega$ and $\delta\rho_{2\Omega}$.

In particular, show that the effective first-harmonic PSF exhibits sub-resolution while the effective second-harmonic PSF exhibits superresolution.

Note: remember to normalize all $\phi(\vec{\rho})$'s to the same peak height before comparing their curvatures.

Problem 19.3

Assume that a molecule is imaged onto a unity-gain camera with unity magnification. Use maximum likelihood to estimate the error in localizing a molecule. That is, begin by defining a chi-squared error function given by

$$\chi^2(x) = \sum_i \frac{(N(x_i) - \bar{N}(x_i; x))^2}{\sigma_N^2(x_i; x)}$$

where i is a pixel index, $N(x_i)$ is the actual number of photocounts registered at pixel i, and $\bar{N}(x_i; x)$ and $\sigma_N^2(x_i; x)$ are the expected mean and variance, respectively, of the photocounts at pixel i for a molecule located at position x. Assume the photocounts obey shot-noise statistics alone. For simplicity, consider only a single dimension (the x axis).

The estimated position of the molecule $\hat{x}$ is obtained by minimizing $\chi^2(x)$. That is, $\hat{x}$ is a solution to the equation $\frac{d\chi^2(x)}{dx} = 0$.

(a) Show that the error in the estimated molecule position, defined by $\delta x = \hat{x} - x_0$, where x_0 is the actual molecule position, has a variance given by

$$\sigma_x^2 \approx \left[\sum_i \frac{1}{\bar{N}(x_i; x_0)} \left(\frac{d\bar{N}(x_i; x)}{dx} \bigg|_{x_0} \right)^2 \right]^{-1}$$

Hint: to obtain this result, it is useful to first solve for δx by writing

$$N(x_i) = \bar{N}(x_i; x_0) + \delta N(x_i; x_0)$$

$$\bar{N}(x_i; x) \approx \bar{N}(x_i; x_0) + \delta x \frac{d\bar{N}(x_i; x)}{dx} \bigg|_{x_0}$$

and keeping terms only to first order in $\delta N(x_i; x_0)$ and δx. Note that $\sigma_x^2 = \langle \delta x^2 \rangle$.

(b) Derive σ_x^2 for the specific example where the PSF at the camera plane has a normalized Gaussian profile given by

$$\bar{N}(x_i; x) = \frac{N}{\sqrt{2\pi}w_0} \int_{|x_i-x|-a/2}^{|x_i-x|+a/2} dx' \, e^{-x'^2/2w_0^2} \approx \frac{Na}{\sqrt{2\pi}w_0} e^{-(x_i-x)^2/2w_0^2}$$

where w_0 is the Gaussian waist and a is the camera pixel size (assume $a \ll w_0$).

How does your solution compare with Eq. 19.34?

Hint: approximate the summation with an integral. That is, for an arbitrary function $f(x_i)$, write $\sum_i f(x_i) \approx \frac{1}{a} \int dx_i f(x_i)$.

Problem 19.4

The purpose of this problem is to compare STED microscopy with continuous versus pulsed beams. The case of continuous-wave beams was treated in the chapter, where it was found in Eq. 19.28 that the time averaged fluorescence rate per molecule was

$$\langle \phi \rangle_{cw} = \frac{\langle \alpha \rangle k_r}{k_r + \sigma_{se} \langle I_{se} \rangle}$$

where σ_{se} and $\langle I_{se} \rangle$ are, repectively, the stimulated-emission cross-section and STED beam illumination intensity ($k_{se} = \sigma_{se} \langle I_{se} \rangle$ and assume $\langle \alpha \rangle \ll k_r$ and $k_{nr} = 0$).

Consider now using pulsed beams. The excitation beam has pulse period τ_l and infinitely narrow pulse duration. The STED beam has pulse period τ_l and pulse duration τ_p. Assume that the onsets of the STED pulses immediately follow the excitation pulses. Assume also that the molecule is excited with probability ξ at each excitation pulse, such that the average excitation rate is $\langle \alpha \rangle = \xi / \tau_l$, where $\tau_l \gg k_r^{-1}$. Finally, for fair comparison, assume that $\langle a \rangle$ and $\langle I_{se} \rangle$ are the same in both pulsed and continuous cases.

(a) Derive an expression for $\langle \phi \rangle_{pb}$ when using pulsed beams. Hint: this involves solving for the excited-state probability $e(t)$, and integrating.

(b) Show that STED is most efficient (i.e. $\langle \phi \rangle_{pb}$ is smallest) when $\tau_p \rightarrow 0$. What is $\langle \phi \rangle_{pb}$ in this case?

(c) Using your result from part **(b)**, show that pulsed-beam STED is always more efficient than continuous-wave STED (i.e. $\langle \phi \rangle_{pb} < \langle \phi \rangle_{cw}$), even for arbitrarily small values of $\langle I_{se} \rangle$, provided $\tau_l > k_r^{-1}$.

Problem 19.5

Consider a distribution of N point-like molecules located at positions $\vec{\rho}_n$, each fluorescing with time-dependent intensities $f_n(t)$. These are imaged by a standard widefield microscope. The resulting intensity at the camera is

$$I(\vec{\rho}, t) = \sum_{n=1}^{N} \text{PSF}(\vec{\rho} - \vec{\rho}_n) f_n(t)$$

(a) Show that if the time-dependent intensities $f_n(t)$ are *independent* of one another, then an image constructed by the temporal variance at each position $\vec{\rho}$ is given by

$$\sigma_I^2(\vec{\rho}) = \sum_{n=1}^{N} \text{PSF}^2(r - r_n) \sigma_n^2$$

where σ_n^2 is the temporal variance of $f_n(t)$. This is an unusual type of imaging since it provides a representation not of average fluorescence levels but rather of variances of fluorescence levels. Nevertheless, this can lead to superresolution imaging (called Superresolution Optical Fluctuation Imaging – or SOFI), by exploiting a priori knowledge of fluorescence statistics and the fact that PSF^2 is narrower than PSF.

(b) Consider molecules that produce fluorescence with Gaussian statistics (see Section 7.2.2). Rewrite the above result in terms of the average fluorescence levels $\langle f_n \rangle$.

(c) Convince yourself that you would not achieve the same result by simply squaring the initial image. In other words, compare the above result with $\langle I(\vec{\rho}) \rangle^2$.

References

[1] Alexandrov, S. A., Hillman, T. R., Gutzler, T. and Sampson, D. D. "Synthetic aperture Fourier holographic optical microscopy," *Phys. Rev. Lett.* 97, 168102 (2006).

[2] Barnes, C. W. "Object restoration in a diffraction-limited imaging system," *J. Opt. Soc. Am.* 56, 575 (1966).

[3] Bertero, M., Boccacci, P., Davies, R. E., Malfanti, F., Pike, E. R. and Walker, J. G. "Superresolution in confocal scanning microscopy: IV. Theory of data inversion by the use of optical masks," *Inverse Problems* 8, 1–23 (1992).

[4] Betzig, E., Patterson, G. H., Sougrat, R., Lindwasser, O. W., Olenych, S., Bonifacino, J. S., Davidson, M. W., Lippincott-Schwartz, J. and Hess, H. F. "Imaging intracellular fluorescent proteins at nanometer resolution," *Science* 313, 1642–1645 (2006).

[5] Binet, R., Colineau, J. and Lehureau, J.-C. "Short-range synthetic aperture imaging at 633nm by digital holography," *Appl. Opt.* 41, 4775–4782 (2002).

[6] Boccacci, P. and Bertero, M. "Image restoration methods: basics and algorithms," in *Confocal and Two-Photon Microscopy: Foundations, Applications, and Advances*, Ed. A. Diaspro, Wiley/Liss (2002).

[7] Courjon, D. *Near-field Microscopy and Near-field Optics*, Imperial College Press (2003).

[8] Cox, I. J. and Sheppard, C. J. R. "Information capacity and resolution in an optical system," *J. Opt. Soc. Am.* A 3, 1152–1158 (1986).

[9] den Dekker, A. J. and van den Bos, A. "Resolution: a survey," *J. Opt. Soc. Am.* A 14, 547–557 (1997).

[10] Gerchberg, R. W. "Super-resolution through error energy reduction," *Opt. Acta* 21, 709–720 (1974).

[11] Goodman, J. W. " Synthetic aperture optics," in *Progress in Optics VIII*, Ed. E. Wolf, North Holland (1970).

[12] Gough, P. T. and Hawkins, D. W. "Unified Framework for Modern Synthetic Aperture Imaging Algorithms," *Int. J. Imag. Syst. Tech.* 8, 343–358 (1997).

[13] Gustafsson, M. G. L. "Nonlinear structured-illumination microscopy: wide-field fluorescence imaging with theoretically unlimited resolution," *Proc. Natl. Acad. Sci. USA* 102, 13081–13086 (2005).

[14] Gustafsson, M. G. L., Shao, L., Carlton, P.M., Rachel Wang, C. J., Golubovskaya, I. N., Zacheus Cande, W., Agard, D. A. and Sedat, J. W. "Three dimensional resolution doubling in wide-field fluorescence microscopy by structured illumination," *Biophys. J.* 94, 4957–4970 (2008).

[15] Harris, J. L. "Diffraction and resolving power," *J. Opt. Soc. Am.* 54, 931–936 (1964).

[16] Hegedus, Z. S. and Sarafis, V. "Superresolving filters in confocally scanned imaging systems," *J. Opt. Soc. Am.* A 3, 1892–1896 (1986).

[17] Heintzmann, R., Jovin, T. M., Cremer, C. "Saturated patterned excitation microscopy – a concept for optical resolution improvement," *J. Opt. Soc. Am.* A 19, 1599–1609 (2002).

[18] Heintzmann, R. and Gustafsson, M. G. L. "Subdiffraction resolution in continuous samples," *Nat. Photon.* 3, 362–364 (2009).

[19] Heintzmann, R. "Estimating missing information by maximum likelihood deconvolution," *Micron* 38, 136–144 (2007).

[20] Hell, S. W. and Wichmann, J. "Breaking the diffraction resolution limit by stimulated emission: stimulated-emission depletion fluorescence microscopy," *Opt. Lett.* 19, 780–782 (1994).

[21] Dyba, M. and Hell, S. W. "Focal spots of size $\lambda/23$ open up far-field fluorescence microscopy at 33 nm axial resolution," *Phys. Rev. Lett.* 88, 163901 (2002).

[22] Hell, S. W. "Toward fluorescence nanoscopy," *Nat. Biotech.* 21, 1347–1355 (2003).

[23] Chmyrov, A., Keller, J., Grotjohann, T., Ratz, M., d'Este, E. Jakobs, S., Eggeling, C. and Hell, S. W. "Nanoscopy with more than 100,000 'doughnuts'," *Nat. Meth.* 10, 737–740 (2013).

[24] Hess, S. T., Girirajan, T. P. and Mason, M. D. "Ultra-high resolution imaging by photoactivation localization microscopy," *Biophys. J.* 91, 4258–4272, (2006).

[25] Hofmann, M., Eggeling, C., Jakobs, S. and Hell, S. W. "Breaking the diffraction barrier in fluorescence microscopy at low light intensities by using reversibly photoswitchable proteins," *Proc. Natl. Acad. Sci. USA* 102, 17565–17569 (2005).

[26] Le Clerc, F., Gross, M. and Collot, L. "Synthetic-aperture experiment in the visible with on-axis digital heterodyne holography," *Opt. Lett.* 26, 1550–1552 (2001).

[27] Lerosey, G., de Rosny, J., Tourin, A. and Fink, M. "Focusing beyond the diffraction limit with far-field time reversal," *Science* 315, 1120–1122 (2007).

[28] Lukosz, W. "Optical systems with resolving powers exceeding the classical limit," *J. Opt. Soc. Am.* 56, 1463–1472 (1966).

[29] Massig, J. H. "Digital off-axis holography with a synthetic aperture," *Opt. Lett.* 27, 2179–2181 (2002).

[30] Negash, A., Labouesse, S., Sandeau, N., Allain, M., Giovannini, H., Idier, J., Heintzmann, R., Chaumet, P. C., Belkebir, K. and Sentenac, A. "Improving the axial and lateral resolution of three-dimensional fluorescence microscopy using random speckle illuminations," *J. Opt. Soc.* A 33, 1069–1094 (2016).

[31] Novotny, L. and Hecht, B. *Principles of Nano-Optics*, Cambridge University Press (2006).

[32] Ober, R. J., Ram, S. and Ward, E. S. "Localization accuracy in single-molecule fluorescence microscopy," *Biophys. J.* 86, 1185–1200 (2004).

[33] Qu, X., Wu, D., Mets, L. and Scherer, N. F. "Nanometer-localized multiple single-molecule fluorescence microscopy," *Proc. Natl. Acad. Sci. USA* 101, 11298–11303 (2004).

[34] Rust, J. M., Bates, M. and Zhuang, X. "Sub-diffraction limit imaging by stochastic optical reconstruction microscopy (STORM)," *Nat. Meth.* 3, 793–795 (2006).

[35] Sales, T. R. M. and Morris, G. M. "Diffractive superresolution elements," *J. Opt. Soc. Am.* A 14, 1637–1646 (1997).

[36] Sauer, M. "Reversible molecular photoswitches: A key technology for nanoscience and fluorescence imaging," *Proc. Natl. Acad. Sci. USA* 102, 9433–9434 (2005).

[37] Sheppard, C. J. R. and Choudhury, A. "Annular pupils, radial polarization, and superresolution," *Appl. Opt.* 43, 4322–4327 (2004).

[38] Sheppard, C. J. R. "Fundamentals of superresolution," Micron 38, 165–169 (2007).

[39] Simonetti, F. "Multiple scattering: the key to unravel the subwavelength world from the far-field diffraction pattern of a scattered wave," *Phys. Rev. E* 73, 036619–1 (2006).

[40] Slepian, D. "Some comments on Fourier analysis, uncertainty and modeling," *SIAM review* 25, 379–393 (1983).

[41] Thompson, R. E., Larson, D. R. and Webb, W. W. "Precise nanometer localization analysis for individual fluorescent probes," *Biophys. J.* 82, 2775–2783 (2002).

[42] Toraldo di Francia, G. "Supergain antennas and optical resolving power," *Nuovo Cimento Suppl.* 9, 426–435 (1952).

[43] Zhuang, X. "Nano-imaging with STORM," *Nat. Photon.* 3, 365–367 (2009).

20 | Imaging in Scattering Media

Optical microscopy has had tremendous success over the decades. It can probe a wide variety of sample properties, and offers spatial resolution on the scale of microns or less (much less in the case of superresolution). But an enormous challenge still remains, where optical microscopy has run into an impasse. This is the challenge of imaging in scattering or turbid media. Almost all the microscopy techniques we have considered so far require some kind of focus, whether of the illumination light or of the detected signal. As we will see below, such a focus is difficult to maintain in scattering samples, and this difficulty rapidly intensifies with depth until imaging no longer becomes feasible. Some microscopies do better than others at maintaining focus, or at least prioritizing focused signal over unfocused background, such as OCT or multiphoton microscopes, but even these run into depth limitations that, for many purposes, are prohibitive. Alternative hybrid strategies such as photoacoustic imaging can bypass some of these limitations by making use of non-optical signalling modalities, but generally at the cost of key benefits, such as resolution. To date, no practical solution for high-resolution imaging at arbitrary depths has been found. On the other hand, recent advances have provided clues to promising research directions.

The purpose of this chapter is two-fold. First, the basic problem of light propagation in scattering media is outlined. Much of the formalism for this was developed decades ago from studies of light propagation through turbulent atmosphere. The reader is referred in particular to work by Tatarski [18, 19], whose impact in the field was profound. Other classic textbooks are those by Ishimaru [10] and Goodman [8], which attest to the fact that the problem of light propagation in scattering media has been understood for quite some time. Instead, it is the solution to this problem that has proved more elusive.

The final section of this chapter provides a very brief introduction to adaptive optics, the goal of which is to counter the effects of sample turbidity. This introduction will be more conceptual than practical.

20.1 RADIATIVE TRANSFER EQUATION

The study of light propagation in scattering media usually begins with the acknowledgment that a deterministic description of light fields is intractable. Instead, more tractable is a statistical description where the detailed phases and amplitudes of the fields are averaged over all their possible realizations within a sample. Light is described then not as a propagation of fields but as a flow of optical energy. We have encountered such a description before in Chapter 6, where the flow of optical energy was embodied in a single parameter called radiance. From this single parameter, all other radiometric quantities such as flux density, intensity, and radiant intensity could be derived. We return now to this concept of radiance and examine it more carefully.

To begin, radiance in a statistically homogeneous scattering sample is governed by what is known as the radiative transfer equation (RTE), given by

$$\frac{1}{\upsilon}\frac{\partial}{\partial t}\mathcal{L}\left(\vec{r};\vec{\kappa}\right) = -\hat{\kappa}\cdot\vec{\nabla}\mathcal{L}\left(\vec{r};\vec{\kappa}\right) - \mu_e\mathcal{L}\left(\vec{r};\vec{\kappa}\right) + \mu_s\int \mathrm{d}^2\Omega_{\kappa_i}\,\mathcal{L}\left(\vec{r};\vec{\kappa}_i\right)p\left(\hat{\kappa},\hat{\kappa}_i\right) + S\left(\vec{r};\vec{\kappa}\right)$$

(20.1)

Though the formal derivation of this equation is omitted here (see [22]), it becomes intuitively plausible when we consider each of its terms separately. The term on the left denotes the temporal change in $\mathcal{L}\left(\vec{r};\vec{\kappa}\right)$, where υ is the speed of light within the sample (i.e. c/n). Note that $\mathcal{L}\left(\vec{r};\vec{\kappa}\right)$ is implicitly time dependent, but we have omitted the time variable for brevity. For continuous light beams where the solution to $\mathcal{L}\left(\vec{r};\vec{\kappa}\right)$ is stationary, the left term vanishes.

The first term on the right is a divergence term. Since $\mathcal{L}\left(\vec{r};\vec{\kappa}\right)$ characterizes the distribution of propagation directions into which it flows, this divergence term simply conforms to this flow. This term does not involve an interaction of the radiance with the sample itself and occurs naturally, even in free space.

The second term on the right involves an interaction with the sample. Here, the radiance propagating in a particular direction $\vec{\kappa}$ is extinguished from this direction because of scattering or absorption, or, more generally, both, as parametrized by the sample extinction coefficient μ_e. When the sample is composed of discrete inhomogeneities of concentration C, we have $\mu_e = C\sigma_e$, where σ_e is the characteristic extinction cross-section of the inhomogeneities (see Section 9.3).

The third term on the right denotes the re-direction of radiance from different incoming directions $\vec{\kappa}_i$ into the outgoing direction $\vec{\kappa}$. This re-direction is mediated by scattering, where μ_s is the sample scattering coefficient and $p\left(\hat{\kappa},\hat{\kappa}_i\right)$ is the scattering phase function (more on this below). Similarly as above, $\mu_s = C\sigma_s$, where σ_s is the characteristic scattering cross-section of the sample inhomogeneities (see Section 9.3). Note that, in the presence of absorption, we have $\mu_e = \mu_s + \mu_a$, where μ_a is the sample absorption coefficient (these coefficients have units of inverse length).

Finally, the fourth term is a source term, which can arise from the presence of emitters inside or outside the sample. In particular, this source term can represent a light beam that is launched into the sample externally.

A word about the scattering phase function $p\left(\hat{\kappa},\hat{\kappa}_i\right)$. This is a unitless parameter characterizing the probability distribution that radiance propagating in an incoming direction $\vec{\kappa}_i$ is scattered into an outgoing direction $\vec{\kappa}$. This function is intimately related to the scattering amplitude (Eq. 9.47), by

$$p\left(\hat{\kappa},\hat{\kappa}_i\right) = \frac{1}{\sigma_s I_i}\left|f(\hat{\kappa},\hat{\kappa}_i)\right|^2$$

(20.2)

where I_i is the local intensity at the scatterer position. As can be readily inferred from Eq. 9.50 and from the Helmholtz reciprocity theorem (Section 14.2.3), the phase function is normalized so that

$$\int_{4\pi} d^2\Omega_\kappa \, p(\hat{\kappa}, \hat{\kappa}_i) = \int_{4\pi} d^2\Omega_{\kappa_i} \, p(\hat{\kappa}, \hat{\kappa}_i) = 1 \tag{20.3}$$

Very commonly, the scattering is dependent on the incoming and outgoing directions through their dot product, in which case we write

$$p(\hat{\kappa}, \hat{\kappa}_i) = p(\hat{\kappa} \cdot \hat{\kappa}_i) = p(\cos\theta) \tag{20.4}$$

where θ is the scattering tilt angle. As we have seen before, scattering can be highly anisotropic. In particular, for particles larger than a wavelength, scattering becomes mostly forward-directed. The resulting scattering anisotropy is parametrized by

$$g = \langle \cos\theta \rangle = \int_{4\pi} d^2\Omega_\kappa \, \cos\theta \, p(\cos\theta) = 2\pi \int_{-1}^{1} d\cos\theta \, \cos\theta \, p(\cos\theta) \tag{20.5}$$

As an example, a phenomenological phase function routinely used to characterize scattering media is the Henyey–Greenstein function, given by

$$p(\cos\theta) = \frac{1 - g^2}{4\pi(1 + g^2 - 2g\cos\theta)^{3/2}} \tag{20.6}$$

We observe that when $g = 0$, the scattering is isotropic since $p(\cos\theta) = 1/4\pi$ is independent of scattering angle. On the other hand, as g approaches 1, the phase function becomes increasingly forward directed. For example, for biological tissue illuminated by visible light, the phase function is generally highly forward-peaked with g in the range 0.9–0.98, suggesting that scatterers in biological tissue are typically much larger than a wavelength.

Conventionally, when scattering is anisotropic, it is convenient to define a reduced scattering coefficient, given by

$$\mu_s' = \mu_s(1 - g) \tag{20.7}$$

An interpretation of the scattering and reduced scattering coefficients is straightforward from Fig. 20.1. Light, when propagating through scattering media, travels an average distance $l_s = 1/\mu_s$ (called the scattering mean free path) before it is deflected by a scattering event. But the closer the scattering anisotropy g is to 1, the smaller the deflection angle, and the direction of the light remains largely preserved. On the other hand, if enough scattering events are accumulated, the light direction eventually becomes uncorrelated from its initial direction. This happens after the light has traveled a pathlength $l_t = 1/\mu_s'$, on average (called the transport mean free path). We note that if $g = 0$, a single scattering event is enough to lead to directional uncorrelation.

Finally, it should be noted that l_s corresponds to the average distance between scattering *events*, and not to the average distance between the scatterers themselves. The latter distance is given by $C^{-1/3}$, which is generally much smaller than l_s.

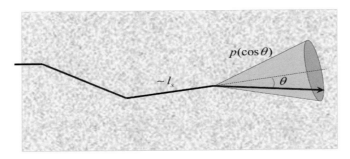

Figure 20.1. Light travels an average scattering length l_s between scattering events in a turbid medium. The deflection angle θ is governed by the phase function $p\left(\cos\theta\right)$.

20.2   DIFFUSIVE LIMIT

To investigate some implications of the RTE, we begin by integrating both sides of Eq. 20.1 over all directions. From Eqs. 6.10 and 6.19, we obtain

$$\frac{1}{v}\frac{\partial}{\partial t}I\left(\vec{r},t\right) + \mu_a I\left(\vec{r},t\right) + \vec{\nabla}\cdot\vec{F}\left(\vec{r},t\right) = \bar{S}\left(\vec{r},t\right) \tag{20.8}$$

where $\bar{S}\left(\vec{r},t\right)$ is the direction-integrated source, and we have made use of the normalization condition for $p\left(\vec{\kappa},\vec{\kappa}_i\right)$. We have also re-instated an explicit time dependence. In this equation, the intensity and flux density are coupled. A simplification comes from adopting the so-called diffusive limit where the light pathlengths in the sample become much longer than $1/\mu_s'$, meaning that light can no longer be thought of as propagating with any kind of directed motion, but rather diffuses within the sample much like heat diffuses within a solid. In this case, light can be treated in a similar manner as heat, and obeys the same laws governing heat transfer. Intensity and flux density become related then by the equivalent of Fick's first law (see Eq. 13.47), namely

$$v\vec{F}\left(\vec{r},t\right) = -D\vec{\nabla}I\left(\vec{r},t\right) \tag{20.9}$$

where we have introduced an effective optical diffusion constant, defined by

$$D = \frac{v}{3\left(\mu_a + \mu_s'\right)} \tag{20.10}$$

Equation 20.8 can then be re-written as

$$\frac{\partial}{\partial t}I\left(\vec{r},t\right) + v\mu_a I\left(\vec{r},t\right) - D\nabla^2 I\left(\vec{r},t\right) = v\bar{S}\left(\vec{r},t\right) \tag{20.11}$$

which is well-known diffusion equation, where we have made allowances for loss of intensity by way of absorption. A standard way to solve this equation is to take the temporal Fourier transform of both sides [3], leading to

$$\nabla^2 I\left(\vec{r};\Omega\right) + 4\pi^2\kappa_{\text{eff}}^2 I\left(\vec{r};\Omega\right) = -\frac{v}{D}\bar{S}\left(\vec{r};\Omega\right) \tag{20.12}$$

where we have introduced the effective complex wavenumber

$$\kappa_{\text{eff}} = \frac{1}{2\pi}\sqrt{i\frac{2\pi\Omega}{D} - \frac{\upsilon\mu_a}{D}} \tag{20.13}$$

In turn, Eq. 20.12 is the Helmholtz equation we are familiar with. The Green function for this equation has already been established (see Eq. 2.13). For example, the intensity resulting from a time-varying point source at the origin (that is, $\bar{S}(\vec{r};\Omega) = \delta^3(\vec{r})\bar{S}_0(\Omega)$) has frequency components

$$I(\vec{r};\Omega) = \frac{\upsilon\bar{S}_0(\Omega)}{4\pi Dr}e^{i2\pi\kappa_{\text{eff}}r} \tag{20.14}$$

Such an intensity distribution is called a photon density wave. It differs here from a propagating spherical wave in that it is heavily damped, since κ_{eff} is complex. In fact, even if there is no absorption (i.e. if $\mu_a = 0$) the wave is, at best, critically damped, meaning that it decays over the distance of a single oscillation. We have seen such critically damped waves before in the context of heat diffusion in Section 18.3 (see Eq. 18.35). As an aside, we have also seen the solution to the diffusion equation for a point source when it is impulsive (see Eq. 13.49).

Remarkably, it is possible to perform imaging with photon density waves at very large depths within scattering media, for example well over centimeter ranges in biological tissue. Such imaging is called diffuse optical tomography, and has generated considerable interest, particularly in the near-infrared spectrum where absorption in biological tissue is minimal (e.g. see [5] for a review). The spatial resolution obtainable with diffuse optical tomography, however, is rather poor, hardly falling under the purview of a microscopy technique. For this reason, we will not investigate this diffusive limit further.

20.3 SMALL SCATTERING ANGLES

Of more relevance to microscopy applications is the limit where light propagates over much shorter distances, on the order of a few scattering mean free pathlengths l_s. Unfortunately, in this case, full analytic solutions to the RTE are currently unavailable, and one must resort instead to more limited strategies. A common strategy is to evaluate the RTE numerically, for example using Monte-Carlo simulation (e.g. [23]). Alternatively, one can apply approximations to the RTE to make it more tractable. We will consider such an approximation here. In particular, we will assume that the phase function is highly forward peaked ($g > 0.9$), as it is in most biological tissues. This approximation is known as the small-angle approximation, and leads to an enormous simplification of the RTE, which can be recast as a function of transverse coordinates only:

$$\frac{\partial}{\partial z}\mathcal{L}_z(\vec{\rho};\vec{\kappa}_\perp) = -\hat{\kappa}_\perp\cdot\vec{\nabla}_\perp\mathcal{L}_z(\vec{\rho};\vec{\kappa}_\perp) - \mu_e\mathcal{L}_z(\vec{\rho};\vec{\kappa}_\perp) + \frac{\mu_s}{\kappa^2}\int d^2\vec{\kappa}_{\perp i}\,\mathcal{L}_z(\vec{\rho};\vec{\kappa}_{\perp i})\,p(\vec{\kappa}_\perp - \vec{\kappa}_{\perp i}) \tag{20.15}$$

($\hat{\kappa}_\perp = \vec{\kappa}_\perp/\kappa$ is not a unit vector here, but rather the transverse component of the unit vector $\hat{\kappa}$).

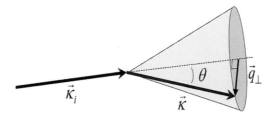

Figure 20.2. Geometry for small scattering angles, where $q_\perp = \kappa\theta$.

Again, we make the assumption that $p\left(\hat{\kappa}, \hat{\kappa}_i\right)$ depends only on the difference between incoming and outgoing wavevectors, $\vec{q} = \vec{\kappa} - \vec{\kappa}_i$, which for small angles may be approximated by the transverse 2D vector $\vec{q}_\perp$ (see Fig. 20.2). Note that we have slightly modified our notation where the z coordinate now appears as a subscript. Note also that $p\left(\vec{q}_\perp\right)$ is normalized such that

$$\frac{1}{\kappa^2}\int d^2\vec{q}_\perp\, p\left(\vec{q}_\perp\right) = 1 \tag{20.16}$$

Remarkably, Eq. 20.15 can be solved analytically (e.g. see [4], and also [1, 6, 10, 14]). To see how this is done, we introduce a new radiometric quantity, which is the double Fourier transform of the radiance $\mathcal{L}_z\left(\vec{\rho}; \vec{\kappa}_\perp\right)$, given by

$$\mathcal{F}_z(\vec{\kappa}_{\perp d}; \vec{\rho}_d) = \iint d^2\vec{\rho}\, d^2\vec{\kappa}_\perp\, \mathcal{L}_z\left(\vec{\rho}; \vec{\kappa}_\perp\right) e^{-i2\pi\vec{\rho}\cdot\vec{\kappa}_{\perp d}} e^{i2\pi\vec{\kappa}_\perp\cdot\vec{\rho}_d} \tag{20.17}$$

Though we have not seen this radiometric quantity before, it is closely analogous in behavior to radiance, so much so that we call it here a conjugate radiance. As a side note, this relation between radiance and conjugate radiance is akin to the relation between two other functions well known in the literature, namely the Wigner function and its conjugate called the ambiguity function (e.g. [16]).

In terms of the conjugate radiance, the small-angle RTE can be rewritten as

$$\left[\frac{\partial}{\partial z} + \hat{\kappa}_{\perp d}\cdot\vec{\nabla}_{\perp d}\right]\mathcal{F}_z(\vec{\kappa}_{\perp d}; \vec{\rho}_d) = \left[-\mu_e + \mu_s\hat{p}\left(\vec{\rho}_d\right)\right]\mathcal{F}_z(\vec{\kappa}_{\perp d}; \vec{\rho}_d) \tag{20.18}$$

where $\hat{p}\left(\vec{\rho}_d\right)$ is the normalized (unitless) Fourier transform of $p\left(\vec{q}_\perp\right)$, given by

$$\hat{p}\left(\vec{\rho}_d\right) = \frac{1}{\kappa^2}\int d^2\vec{\kappa}_\perp\, p\left(\vec{\kappa}_\perp\right) e^{i2\pi\vec{\kappa}_\perp\cdot\vec{\rho}_d} \tag{20.19}$$

The solution to Eq. 20.18 is

$$\mathcal{F}_z(\vec{\kappa}_{\perp d}; \vec{\rho}_d) = \mathcal{F}_0\left(\vec{\kappa}_{\perp d}; \vec{\rho}_d - \frac{z}{\kappa}\vec{\kappa}_{\perp d}\right)\exp\left[-\mu_e z + \mu_s\int_0^z dz'\,\hat{p}\left(\vec{\rho}_d - \frac{z'}{\kappa}\vec{\kappa}_{\perp d}\right)\right] \tag{20.20}$$

This result, which will be a mainstay of this chapter, is astonishing in its simplicity. Indeed, it isolates the effect of the scattering medium from that of free-space propagation. When scattering and absorption are absent, the exponent in square brackets vanishes, leading to

$$\mathcal{F}_z(\vec{\kappa}_{\perp d}; \vec{\rho}_d) = \mathcal{F}_0(\vec{\kappa}_{\perp d}; \vec{\rho}_d - \frac{z}{\kappa}\vec{\kappa}_{\perp d}) \qquad \text{(free space)} \tag{20.21}$$

(compare with the analogous free-space propagation law for radiance given by Eq. 6.18).

On the other hand, when scattering or absorption are present, their effects are confined to the exponential term in Eq. 20.20, which serves as a filter function. This becomes clearer if we recast Eq. 20.20 in terms of something more familiar, namely the mutual intensity. Just as mutual intensity can be derived from radiance, so too it can be derived from the conjugate radiance, from the relation

$$J_z\left(\vec{\rho}_c, \vec{\rho}_d\right) = \frac{1}{\kappa^2} \int d^2\vec{\kappa}_{\perp d}\, \mathcal{F}_z(\vec{\kappa}_{\perp d}; \vec{\rho}_d) e^{i2\pi \vec{\kappa}_{\perp d} \cdot \vec{\rho}_c} \tag{20.22}$$

valid for small angles (compare with the analogous Eq. 6.16).

Equation 20.20 then becomes

$$J_z\left(\vec{\rho}_c, \vec{\rho}_d\right) = \left(\frac{\kappa}{z}\right)^2 \iint d^2\vec{\rho}'_c\, d^2\vec{\rho}'_d\, J_0\left(\vec{\rho}'_c, \vec{\rho}'_d\right) e^{i2\pi \frac{\kappa}{z}\left(\vec{\rho}_d - \vec{\rho}'_d\right)\cdot\left(\vec{\rho}_c - \vec{\rho}'_c\right)} K_z\left(\vec{\rho}_d, \vec{\rho}'_d\right) \tag{20.23}$$

where we have introduced the scattering filter function defined by

$$K_z\left(\vec{\rho}_d, \vec{\rho}'_d\right) = \exp\left[-\mu_e z + \mu_s z \int_0^1 d\eta\, \hat{p}\left((1-\eta)\,\vec{\rho}_d + \eta\vec{\rho}'_d\right)\right] \tag{20.24}$$

Here again, in the absence of the scattering medium we have $K_z\left(\vec{\rho}_d, \vec{\rho}'_d\right) \to 1$ and Eq. 20.23 reduces to the free-space Zernike propagation law (see Eq. 4.26).

In the event we are interested only in evaluating the propagation of intensity, Eq. 20.23 reduces to

$$I_z\left(\vec{\rho}_c\right) = \left(\frac{\kappa}{z}\right)^2 \iint d^2\vec{\rho}'_c\, d^2\vec{\rho}'_d\, J_0\left(\vec{\rho}'_c, \vec{\rho}'_d\right) e^{-i2\pi \frac{\kappa}{z}\vec{\rho}'_d \cdot\left(\vec{\rho}_c - \vec{\rho}'_c\right)} K_z\left(0, \vec{\rho}'_d\right) \tag{20.25}$$

20.3.1 Scattering Filter Function

In all cases, it is clear that the scattering filter function plays a critical role in characterizing propagation through turbid media. To better understand this role, let us focus on two specialized cases. The first is where $\vec{\rho}_d = \vec{\rho}'_d$, in which case the scattering filter function simplifies to

$$K_z\left(\vec{\rho}_d, \vec{\rho}_d\right) = e^{-\mu_e z + \mu_s z \hat{p}\left(\vec{\rho}_d\right)} \tag{20.26}$$

The second is where $\vec{\rho}_d = 0$ (as in Eq. 20.25), or $\vec{\rho}'_d = 0$, in which case the scattering filter function simplifies instead to

$$K_z\left(\vec{\rho}_d, 0\right) = K_z\left(0, \vec{\rho}_d\right) = e^{-\mu_e z + \mu_s z \xi\left(\vec{\rho}_d\right)} \tag{20.27}$$

where

$$\xi\left(\vec{\rho}_d\right) = \int_0^1 d\eta\, \hat{p}\left(\eta\vec{\rho}_d\right) \tag{20.28}$$

In both cases, K_z varies between the extrema $K_z = \exp\left(-\mu_a z\right)$ when $\rho_d \to 0$, and $K_z = \exp\left(-\mu_e z\right)$ when $\rho_d \to \infty$.

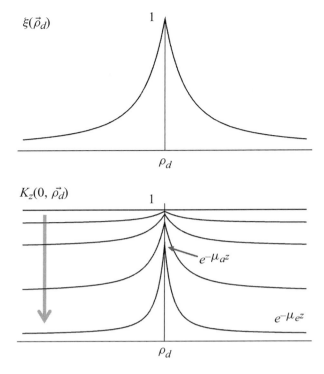

Figure 20.3. $\xi\left(\vec{\rho}_d\right)$ and $K_z\left(0,\vec{\rho}_d\right)$ for increasing values of z.

In particular, let us revisit the specific example where the phase function is characterized by the Henyey–Greenstein function (Eq. 20.6). In the limit of small scattering angles, this function becomes

$$p\left(\vec{q}_\perp\right) \approx \frac{1-g}{2\pi\left(\left(1-g\right)^2 + q_\perp^2/\kappa^2\right)^{3/2}} \qquad \text{(Henyey–Greenstein, } g > 0.9) \quad (20.29)$$

$$\hat{p}\left(\vec{\rho}_d\right) \approx e^{-2\pi(1-g)\kappa\rho_d} \qquad \text{(Henyey–Greenstein, } g > 0.9) \qquad (20.30)$$

which, in turn, leads to

$$\xi\left(\vec{\rho}_d\right) = \frac{1 - e^{-2\pi(1-g)\kappa\rho_d}}{2\pi\left(1-g\right)\kappa\rho_d} \qquad (20.31)$$

Plots of $\xi\left(\vec{\rho}_d\right)$ and the associated scattering filter function are shown in Fig. 20.3.

While the effect of K_z is not immediately apparent from Eqs. 20.26 and 20.27, a much better intuition comes from the close approximations provided by

$$K_z\left(\vec{\rho},\vec{\rho}\right) \approx e^{-\mu_e z} + e^{-\mu_a z}\left(1 - e^{-\mu_s z}\right)\hat{p}(\vec{\rho}\sqrt{1+\mu_s z}) \qquad (20.32)$$

$$K_z\left(\vec{\rho},0\right) = K_z\left(0,\vec{\rho}\right) \approx e^{-\mu_e z} + e^{-\mu_a z}\left(1 - e^{-\mu_s z}\right)\xi(\vec{\rho}\sqrt{1+\mu_s z}) \qquad (20.33)$$

In other words, the filter function can be thought of as being separated into two components. The first component, called the coherent or ballistic component, attenuates with distance according to $\exp(-\mu_e z)$ but otherwise leaves the propagation of light unchanged from free-space propagation. The second component, called the incoherent component, causes both coherence attenuation and beam spread, as we will see below.

20.4 PROPAGATION OF INTENSITY AND COHERENCE

At this point, to see how the small-angle solutions to the RTE can be put to use, it is instructive to consider a few examples by examining the propagation of different types of beams.

20.4.1 Plane-Wave Beam

We begin with the simplest type of beam, namely a plane-wave beam entering a scattering medium at $z = 0$. The mutual intensity incident on the medium is then $J_0\left(\vec{\rho}_c', \vec{\rho}_d'\right) = I_0$, and from Eqs. 20.23 and 20.26 we find

$$J_z\left(\vec{\rho}_c, \vec{\rho}_d\right) = I_0 K_z\left(\vec{\rho}_d, \vec{\rho}_d\right) \tag{20.34}$$

Several conclusions can be drawn from this simple result. For example, based on our interpretation of $K_z\left(\vec{\rho}_d, \vec{\rho}_d\right)$ in Eq. 20.32, this may be rewritten in the more revealing form

$$J_z\left(\vec{\rho}_c, \vec{\rho}_d\right) = I_0 e^{-\mu_e z} + I_0 e^{-\mu_a z}\left(1 - e^{-\mu_s z}\right)\hat{p}(\rho_d\sqrt{1 + \mu_s z}) \tag{20.35}$$

In other words, only a fraction of the beam (first term) retains its original uniform coherence associated with the incident plane wave. This fraction decreases as $\exp(-\mu_e z)$ and eventually decays to zero. The remainder of the beam (second term) breaks apart into multiple coherence areas of finite size, as illustrated in Fig. 20.4. The greater the propagation distance, the smaller the size of these coherence areas and the beam becomes increasingly spatially

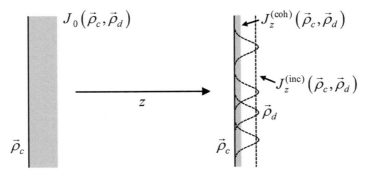

Figure 20.4. Plane-wave propagation through turbid medium. Coherence separates into coherent and incoherent components.

incoherent as this second term becomes increasingly dominant. For example, for a scattering medium characterized by a forward-peaked Henyey–Greenstein phase function (Eq. 20.30), the transverse coherence length decreases roughly as

$$l_\perp(z) \approx \frac{1}{\pi\kappa(1-g)\sqrt{1+\mu_s z}} \tag{20.36}$$

(bearing in mind, of course, that the coherence length of propagating light cannot become smaller than $\approx \kappa^{-1}$, whereupon the beam becomes fully incoherent and the small-angle approximation breaks down).

Another conclusion pertains to the beam intensity itself. In particular, from Eq. 20.34 we immediately obtain

$$I_z(\vec{\rho}_c) = I_0 K_z(0,0) = I_0 e^{-\mu_a z} \tag{20.37}$$

That is, the intensity of a plane wave decreases exponentially as it propagates through the medium, with a decay constant equal to the absorption coefficient. This intensity includes both ballistic (unscattered) and incoherent (scattered) components of the beam. If we are interested only in the intensity of the ballistic component that is undeviated from its initial on-axis propagation direction, then from Eq. 20.33 we have instead

$$I_z^{(\kappa_\perp \to 0)}(\vec{\rho}_c) = I_0 e^{-\mu_e z} \tag{20.38}$$

where the decay constant is now equal to the extinction coefficient. This is the well-known Lambert–Beer law, which we made use of in Section 16.7.1.

20.4.2 Focused Beam

We now turn to a case much more relevant to microscopy applications: a focused beam. In particular, let us consider a Gaussian beam focused a distance L within a scattering medium. To address this problem we could start from Eq. 20.23, but a simpler approach is to start from Eq. 20.20.

We begin by deriving the conjugate radiance of the Gaussian beam incident on the medium surface (defined, as before, as $z = 0$). From Eqs. 20.17 and 6.15, and also from Eq. 20.21, we find

$$\mathcal{F}_0(\vec{\kappa}_\perp; \vec{\rho}_d) = \frac{\pi}{2}\kappa^2 w_0^2 I_0 e^{-\frac{1}{2}\left(\left|\vec{\rho}_d + \frac{L}{\kappa}\vec{\kappa}_\perp\right|^2/w_0^2 + \pi^2 w_0^2 \kappa_\perp^2\right)} \tag{20.39}$$

where w_0 is the nominal waist of the beam at its focus in the absence of the scattering medium.

Our goal here is to evaluate the effect of the scattering on the spread of this beam as it propagates within the medium. That is, our goal is to evaluate $I_z(\vec{\rho})$, or, equivalently, $\mathcal{I}_z(\vec{\kappa}_\perp)$. The latter turns out to be easier to derive since it is related to the conjugate radiance simply by

$$\mathcal{I}_z(\vec{\kappa}_\perp) = \frac{1}{\kappa^2}\mathcal{F}_z(\vec{\kappa}_\perp; 0) \tag{20.40}$$

In other words, from Eq. 20.20, we immediately arrive at

$$\mathcal{I}_z\left(\vec{\kappa}_\perp\right) = \frac{\pi}{2} w_0^2 I_0 e^{-\frac{1}{2}\pi^2 w_0^2 \kappa_\perp^2 \left(1 + |L-z|^2/\pi^2\kappa^2 w_0^4\right)} K_z\left(0, \frac{z}{\kappa}\vec{\kappa}_\perp\right) \tag{20.41}$$

and our goal is achieved.

Again, we find that the effect of the turbid medium is confined to the scattering filter function $K_z\left(0, \frac{z}{\kappa}\vec{\kappa}_\perp\right)$. In other words, from Eq. 20.33, we may interpret the intensity spectrum of the beam as being separated into coherent and incoherent components

$$\mathcal{I}_z\left(\vec{\kappa}_\perp\right) \approx \mathcal{I}_z^{(\text{coh})}\left(\vec{\kappa}_\perp\right) + \mathcal{I}_z^{(\text{inc})}\left(\vec{\kappa}_\perp\right) \tag{20.42}$$

where

$$\mathcal{I}_z^{(\text{coh})}\left(\vec{\kappa}_\perp\right) = \frac{\pi}{2} w_0^2 I_0 e^{-\mu_e z} e^{-\frac{1}{2}\pi^2 w^2 (L-z)\kappa_\perp^2} \tag{20.43}$$

$$\mathcal{I}_z^{(\text{inc})}\left(\vec{\kappa}_\perp\right) = \frac{\pi}{2} w_0^2 I_0 e^{-\mu_a z} \left(1 - e^{-\mu_s z}\right) e^{-\frac{1}{2}\pi^2 w^2 (L-z)\kappa_\perp^2} \xi\left(\frac{z}{\kappa}\vec{\kappa}_\perp \sqrt{1 + \mu_s z}\right) \tag{20.44}$$

and

$$w(z) = w_0 \sqrt{1 + \frac{z^2}{\pi^2 \kappa^2 w_0^4}} \tag{20.45}$$

The first term $\mathcal{I}_z^{(\text{coh})}\left(\vec{\kappa}_\perp\right)$ is the coherent, or ballistic, term, which we are familiar with from Section 5.1.1. This is the term that arrives at the focus unaffected by the scattering medium except that it has been attenuated by a factor $\exp\left(-\mu_e L\right)$. In other words, this term, albeit increasingly small with increasing focal depth, leads to a diffraction-limited focus of waist w_0 when $z = L$.

The second term $\mathcal{I}_z^{(\text{inc})}\left(\vec{\kappa}_\perp\right)$ is more interesting. This is the incoherent term resulting from scattering. Very quickly, as $\mu_e L$ becomes greater than unity, this incoherent term becomes dominant and ultimately defines the beam spot size when the coherent term vanishes. In particular, the width of $\mathcal{I}_z^{(\text{inc})}\left(\vec{\kappa}_\perp\right)$ begins to collapse with increasing penetration into the medium, meaning that $I_z^{(\text{inc})}\left(\vec{\rho}_d\right)$ becomes increasingly broader.

At this point, we can again invoke the forward-peaked Henyey–Greenstein phase function (Eq. 20.30) to characterize the sample, which leads to

$$\xi\left(\frac{z}{\kappa}\vec{\kappa}_\perp \sqrt{1 + \mu_s z}\right) = \frac{1 - \exp\left[-2\pi\left(1 - g\right)\kappa_\perp z\sqrt{1 + \mu_s z}\right]}{2\pi\left(1 - g\right)\kappa_\perp z\sqrt{1 + \mu_s z}} \tag{20.46}$$

Manifestly, $\mathcal{I}_z^{(\text{inc})}\left(\vec{\kappa}_\perp\right)$ (and hence $I_z^{(\text{inc})}\left(\vec{\rho}\right)$) is no longer Gaussian in profile. It is thus difficult to ascribe a width to it in the conventional sense of a Gaussian beam waist. If hard-pressed, however, one could note that the width of $\xi\left(\frac{z}{\kappa}\vec{\kappa}_\perp \sqrt{1 + \mu_s z}\right)$ is very roughly given by $2\pi\left(1 - g\right)z\sqrt{1 + \mu_s z}$, allowing one to crudely ascribe an effective beam waist given by

$$w_{\text{inc}}(z) \approx \sqrt{w^2\left(z - L\right) + 4\left(1 - g\right)^2 z^2 \left(1 + \mu_s z\right)} \tag{20.47}$$

As expected, two components comprise the waist, as illustrated in Fig. 20.5. The first is a geometric component identical to the ballistic focus and independent of the scattering medium.

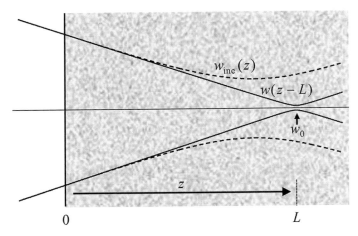

Figure 20.5. Focused-beam intensity propagation through turbid medium. Solid and dashed lines are $I_z^{(coh)}\left(\vec{\rho}\right)$ and $I_z^{(inc)}\left(\vec{\rho}\right)$, respectively.

The second arises from the scattering medium and leads to additional beam spreading that increases with propagation distance. For $z < l_s$ this additional spreading scales with z; for $z > l_s$ it scales with $z^{3/2}$. The spread of the incoherent beam at the focus itself ($z = L$) is given by roughly

$$w_{inc}\left(L\right) \approx \sqrt{w_0^2 + 4L^2 \left(1 + \mu_s L\right)\left(1 - g\right)^2} \tag{20.48}$$

It is instructive to consider here some numbers. For example, let us consider a beam that is nominally focused to a beam waist $w_0 = 1\ \mu m$ a distance $L = 100\ \mu m$ inside a medium of scattering mean free path $l_s = 100\ \mu m$ with scattering anisotropy $g = 0.95$ (we neglect here absorption). The resulting focal spot consists of two components. The coherent component is a sharply focused peak of unperturbed waist $1\ \mu m$, but attenuated by a factor 0.37. The incoherent component is a broader beam spot of waist $\approx 14\ \mu m$. If the focus is displaced an additional $100\ \mu m$ into the medium, the coherent peak is attenuated by a factor 0.14, and the surrounding incoherent component is broadened to $\approx 35\ \mu m$. As is clear from this example, the power in the coherent component rapidly becomes outweighed by that in the incoherent component, and light focusing becomes problematic.

The reader is reminded that the results in this section are based on the Henyey–Greenstein model for light scattering. This model has become quite popular and is routinely used to describe biological tissue that is relatively homogeneous, despite the fact the that it was originally designed for an altogether different purpose [9]. Though our results have relied on a fair number of small-angle approximations, they are in reasonable accord with experiment. Other models are, of course, possible. For example, models based on fractal or K-distribution statistics have also been explored (e.g. [11]). Alternatively, as we will see below, a direct link can be established between the phase function and index of refraction variations within the scattering medium.

20.5 CONNECTION WITH INDEX OF REFRACTION

An advantage of characterizing beam propagation within a scattering medium in terms of the phase function of the medium is that the latter is a readily measurable property. However, what is missing in our development is an understanding of how the phase function is connected to the medium itself. For this, we must dig deeper into the very cause of scattering. Throughout this book, we have characterized a sample by its variations in dielectric constant or index of refraction (both normalized, see Chapter 9). Here we will focus on the latter.

In Eq. 20.2 we established the connection between the phase function and the scattering amplitude. Let us examine this connection in more detail. In particular, from Eq. 9.47 we can write

$$p\left(\vec{\kappa}\cdot\vec{\kappa}_i\right) = 4\pi^2\kappa^4\sigma_s^{-1}\iint \mathrm{d}^3\vec{r}_0\,\mathrm{d}^3\vec{r}_0'\,e^{-i2\pi\vec{q}\cdot\vec{r}_0}e^{i2\pi\vec{q}\cdot\vec{r}_0'}\left\langle \delta n\left(\vec{r}_0\right)\delta n^*\left(\vec{r}_0'\right)\right\rangle \qquad (20.49)$$

where, again, $\vec{q} = \vec{\kappa} - \vec{\kappa}_i$, and the brackets here signify a statistical average.

Following the procedure outlined by Tatarski [19], we adopt our usual coordinate transformation, leading to

$$p\left(\vec{\kappa}\cdot\vec{\kappa}_i\right) = 4\pi^2\kappa^4\frac{\left\langle |\delta n|^2\right\rangle}{\sigma_s}\int_{\Delta V}\mathrm{d}^3\vec{r}_c\int_\infty \mathrm{d}^3\vec{r}_d\,e^{-i2\pi\vec{q}\cdot\vec{r}_d}\gamma_{\delta n}(r_d) \qquad (20.50)$$

where the index-of-refraction inhomogeneities are characterized by the normalized autocorrelation function

$$\gamma_{\delta n}\left(r_d\right) = \frac{\left\langle \delta n\left(\vec{r}_0\right)\delta n^*\left(\vec{r}_0'\right)\right\rangle}{\left\langle |\delta n|^2\right\rangle} \qquad (20.51)$$

taken here to be translationally invariant and spherically symmetric.

In writing Eq. 20.50 we have assumed that $\gamma_{\delta n}\left(r_d\right)$ is sufficiently localized that the integral over $\vec{r}_d$ safely converges. On the other hand, the integral over $\vec{r}_c$ is more problematic. For now, we simply extend this integral over an arbitrary volume ΔV assumed to be much larger than the volume spanned by $\gamma_{\delta n}\left(r_d\right)$.

Upon performing the integrals, we obtain

$$p\left(\cos\theta\right) = 4\pi^2\kappa^4\frac{\Delta V}{\sigma_s}\left\langle |\delta n|^2\right\rangle\hat{\gamma}_{\delta n}(\kappa\sqrt{2 - 2\cos\theta}) \qquad (20.52)$$

where $\hat{\gamma}_{\delta n}(\kappa_d)$ is the 3D Fourier transform of $\gamma_{\delta n}\left(r_d\right)$. In particular, because both are spherically symmetric (see Appendix A.4), their Fourier transform relationships are defined by

$$\hat{\gamma}_{\delta n}(q) = \frac{2}{q}\int_0^\infty \mathrm{d}r_d\,r_d\sin\left(2\pi qr_d\right)\gamma_{\delta n}\left(r_d\right) \qquad (20.53)$$

$$\gamma_{\delta n}\left(r_d\right) = \frac{2}{r_d}\int_0^\infty \mathrm{d}q\,q\sin\left(2\pi qr_d\right)\hat{\gamma}_{\delta n}(q) \qquad (20.54)$$

where Eq. 20.54, in the limit $r_d \to 0$, leads to the normalization condition

$$\gamma_{\delta n}(0) = 4\pi \int_0^\infty dq\, q^2 \hat{\gamma}_{\delta n}(q) = 1 \tag{20.55}$$

Equation 20.52 provides the formal link we have set out to establish between the angular dependence of $p(\cos\theta)$ and the index-of-refraction inhomogeneities in the medium. The overall scaling factor connecting the two remains arbitrary because of the uncertainty in ΔV, but even this difficulty can be resolved if we impose our normalization constraint on $p(\cos\theta)$.

For example, let us revert to the small-angle approximation, and assume that $p(\cos\theta)$ is highly forward-peaked. Equation 20.52 then simplifies to

$$p(\theta) = 4\pi^2 \kappa^4 \frac{\Delta V}{\sigma_s} \left\langle |\delta n|^2 \right\rangle \hat{\gamma}_{\delta n}(\kappa\theta) \tag{20.56}$$

and the normalization condition for $p(\theta)$ becomes

$$2\pi \int_0^\pi d\theta\, \sin\theta\, p(\cos\theta) \approx 2\pi \int_0^\infty d\theta\, \theta\, p(\theta) \to 1 \tag{20.57}$$

From Eqs. 20.55 and 20.57, we finally conclude

$$\hat{\gamma}_{\delta n}(\kappa\theta) \approx \frac{p(\theta)}{2\kappa^3 \bar{\theta}} \tag{20.58}$$

where $\bar{\theta}$ is the mean scattering tilt angle.

In other words, a measure of $p(\theta)$ provides a direct and simple measure of the index-of-refraction statistics in a sample medium, here within the limits of the small-angle approximation.

20.6 ADAPTIVE OPTICS

Inasmuch as microscopes invariably involve focusing of some sort, whether of the illumination or detected light, any degradation in focusing caused by sample turbidity, as illustrated for example in Fig. 20.5, can pose a significant problem. In particular, Eq. 20.48 suggests that once the ballistic focus is lost, the spot size of the remaining beam is rather large, meaning that the microscope resolution is compromised. Indeed, most of the microscope strategies outlined in this book have been based on ballistic light only, and have relied on whatever means possible to distinguish ballistic light from background. As such, it would appear that when ballistic focus is lost, micron-level resolution becomes inevitably lost as well.

Fortunately, some strategies can come to our aid. A promising strategy is called adaptive optics, which was developed decades ago by astronomers to better image starlight through turbulent atmosphere, and is only recently making its way into the field of biological microscopy. The idea is depicted in Fig. 20.6, and involves the use of an active element in the imaging optics to compensate for, in real time, the wavefront distortions caused by the sample

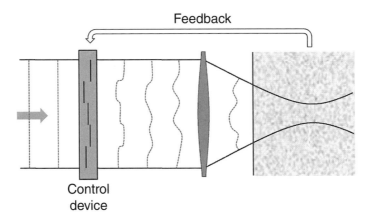

Figure 20.6. General layout for adaptive optics. The control device is usually reflective and located in the pupil plane. Here it is depicted as transmitive.

medium. Typically, this active element is a deformable mirror, which can impart user-controlled local phase shifts onto a reflecting light beam. More recently, spatial-light modulators have also been used, which have the advantage of providing a much greater number of phase-shift control elements. As we will see below, this number of control elements plays a critical role in adaptive optics.

Of course, to actively compensate for the wavefront distortions caused by the sample medium, these distortions must be measured in advance, or inferred somehow. Often, to obtain this information, a reference point is utilized in the sample, called a "guide star" (a term borrowed from astronomers). Alternatively, the wavefront distortions can be inferred by trial and error from the acquired images themselves, using some kind of iterative feedback between the images and the control device. We will not delve into the details on how wavefront distortions are measured or inferred, as this is described elsewhere (for a general review see [21]; for recent surveys pertaining to biological imaging, see [2, 15]). Instead, we will assume they are known, and limit our discussion to what happens when these wavefront distortions are corrected.

20.6.1 Focus Enhancement

To begin, let us consider the problem of focusing a light beam in a scattering medium. As described above, this beam can be thought of as separated into a coherent component, which maintains the desired focus, and an incoherent component, which becomes increasingly blurred with depth. It is this second incoherent component that is problematic, and our goal will be to actively apply wavefront variations to the incident beam to re-shape this incoherent component into a diffraction-limited focus. To do so, we must compensate for the wavefront distortions caused by the sample medium. These can be estimated, at least theoretically, by the Helmholtz reciprocity theorem (see Section 14.2.3). In other words, we can work backwards

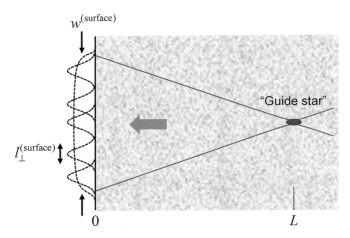

Figure 20.7. Backward propagation of beam from fictitious "guide star."

by imagining a diffraction-limited light-beam source at the focal spot (i.e. a guide star) and examining the effect of scattering on the back-propagation of this light beam to the sample surface (see Fig. 20.7). We are particularly interested in the state of coherence of this fictitious beam at the surface. Clearly, if the back propagation were through free space, the beam at the sample plane would be spatially coherent across its entire profile. But the sample is scattering, meaning that the spatial coherence becomes degraded. We could quantify this degradation with the formalism described above for Gaussian beams (Section 20.4.2), though applied this time to light expanding backwards from a Gaussian focus; but rather than becoming mired in complicated integrals, we choose instead to estimate the beam coherence using the much simpler plane-wave approximation (Section 20.4.1). Specifically, for the case of Henyey–Greenstein scattering statistics, we have already established the effect of scattering on the spatial coherence of a plane wave propagating a distance z (Eq. 20.36). Here, the propagation distance is L, corresponding to the depth of the focus in the sample. In other words, the beam from our fictitious guide-star source (more accurately, the incoherent component of this beam that arises from scattering) has spatial coherence length at the sample surface roughly given by

$$l_\perp^{(\mathrm{surface})} \approx \frac{1}{\pi \kappa \left(1 - g\right) \sqrt{1 + \mu_s L}} \tag{20.59}$$

We can compare this with the overall size of the beam itself, which is roughly given by, from Eq. 20.45,

$$w^{(\mathrm{surface})} = w_0 \sqrt{1 + \frac{L^2}{\pi^2 \kappa^2 w_0^4}} \approx \frac{L}{\pi \kappa w_0} \tag{20.60}$$

(this overall size is an underestimate, since we have neglected the effect of the additional beam spread due to scattering, which is assumed here to be small compared to the size of the beam).

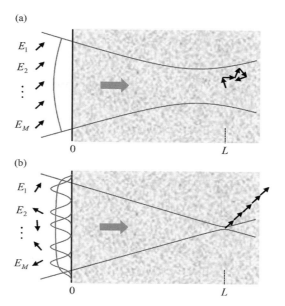

Figure 20.8. Incoherent component of focused beam without (a) and with (b) adaptive optics.

We conclude that the incoherent part of the light beam at the sample surface is decomposed into roughly M statistically independent coherence areas, where

$$M \approx \left(\frac{w^{(\text{surface})}}{l_\perp^{(\text{surface})}} \right)^2 \approx \frac{L^2}{w_0^2} (1 - g)^2 (1 + \mu_s L) \tag{20.61}$$

Making use of reciprocity again to revert back to our initial forward propagation direction, this means that M statistically independent fields are required to form a fully diffraction-limited focus of waist w_0 a distance L inside the sample, where the phases of these fields are adjusted so that they constructively interfere at the focus. As shown in Fig. 20.8, when the phases are properly adjusted, the incoherent field amplitude at the focus position is rendered coherent and increased by a factor of roughly $\sqrt{M}$ compared to when the phases are incorrectly (i.e. randomly) adjusted. The incoherent intensity is correspondingly increased by a factor M compared its initial background level, assuming the amplitudes of the fields are roughly equal.

This result is in accord with the results obtained in Section 20.4.2, Specifically, when the phases are incorrectly adjusted, the incoherent beam spot is spread over a size $w_{\text{inc}}(L)$ (see Eq. 20.48). When the phases are correctly adjusted, the spot size becomes compressed to the diffraction-limited focus size w_0. Power conservation then leads to roughly the same increase in focal-spot intensity as predicted by Eq. 20.61, namely M. We note that we are allowed to invoke power conservation here because, in the small-angle scattering limit, the light is assumed to be exclusively forward directed.

Of course, to fully recover such a diffraction-limited focus with adaptive optics, the deformable mirror or spatial light modulator must be able to supply at least M independent, controllable phase corrections, or degrees of freedom. If this is not the case, then the wavefront distortion compensation becomes incomplete. We will examine this eventuality below.

Isoplanatic Patch

Of practical importance in adaptive optics applications is the spatial range over which the wavefront distortion compensation is effective. In other words, if a diffraction-limited focus is recovered at a particular position in the sample, for example at the position of a guide star, then does the same wavefront compensation enable focus recovery in regions surrounding the guide star? The area over which this recovery remains effective is called the "isoplanatic patch" (again, a term borrowed from astronomy applications – e.g. [7]). We can estimate the size of this isoplanatic patch in a relatively straightforward manner. In effect, the question we are asking is how far can we laterally translate the phase correction pattern incident on the sample surface before it becomes ineffective, that is, before it becomes uncorrelated with the phase pattern prescribed by our reciprocity procedure. This distance is manifestly given by the span of the coherence areas at the sample surface, namely $l_\perp^{(\text{surface})}$ (Eq. 20.59).

For example, if we apply the same numbers as we did in Section 20.4.2 for light of wavelength, say, 1 μm, we conclude that for a focal depth of one scattering mean free path, the isoplanatic patch is roughly 5 μm in size; for two scattering mean free paths it is roughly 4 μm in size. These sizes are small indeed. They suggest that if we were to actively focus a beam with wavefront distortion compensation, and then scan this focus by lateral translation of the incident beam (for example, by applying phase ramps at light field in the pupil plane shown in Fig. 20.6), the focus would subsist only over a few microns. This result is borne out in practice [12].

It should be noted that this result is specific to scattering-induced wavefront distortions originating from within the sample. In many cases, the wavefront corrections resulting from adaptive optics serve more to correct instrument-induced aberrations, or aberrations produced by the abrupt change in index of refraction at the sample surface. In these cases, the isoplanatic patch can appear larger than those calculated above.

20.6.2 Deep Focusing

A difficulty with using adaptive optics to focus deeper into a sample is that the required number of degrees of freedom increases with propagation distance, to the point that it eventually surpasses the number of degrees of freedom available from the wavefront correction device. In this case, the full power of the illumination beam cannot be fully concentrated to the size of a diffraction-limited focus. Nevertheless, intensity enhancement at the focus position remains possible, for the same reasons as described above. For example, let us consider a device with $N < M$ degrees of freedom. That is, it can supply an array of N fields in parallel, with controllable phase shifts. Each of these propagates toward the focal plane, yielding N fields at the focal point. In general, the phases of these fields are uncorrelated, and the net summed field at the focal point has amplitude typically $\sqrt{N}$ times greater than each individual field at the focal point. By properly adjusting the phases of these fields so that they are aligned with one another, the summed amplitude becomes N times greater than each individual field, yielding a net summed amplitude enhancement of $\sqrt{N}$, corresponding to a net intensity enhancement of N. Remarkably, this simple argument applies to the focusing at *any* depth within the sample, even depths far beyond the transport mean free path l_t where light is best described by the diffusion equation (see Section 20.2). This may seem surprising given that all phase memory is lost between the light incident at the sample surface and at the focal plane, meaning that the output phase cannot be predicted from the input phase. But manifestly, what is required for intensity enhancement is not a control of the absolute phases of each individual field, but rather only of the relative phase shifts between these fields, which, for monochromatic light where the phases throughout the sample are stable (albeit unknown), such relative phase-shift control is straightforward. More surprising is the fact that the intensity enhancement can remain confined to extremely small ranges, even smaller than the diffraction limit range associated with the focusing optics. The reason for this is that the angular diversity of the light arriving at the focal plane is greater than that provided by the nominal focus NA, because of scattering. Indeed, in the diffusive limit where light propagates in both forward and backward directions, the angular diversity spans a full 4π steradians, meaning that, in principle, the intensity enhancement can be confined to volumes even smaller than a cubic wavelength. This is quite remarkable.

Of course, in this diffusive limit scenario, $N \ll M$, meaning that the achievable Strehl ratio for such an actively controlled focus can be no greater than N/M, which is generally quite small (see Section 19.1.1 for the definition of Strehl ratio). In other words, though the active focus can be confined to a micron-sized volume, the actual light power captured in this focus is typically only a small fraction of the diffuse background power surrounding the focus.

Nevertheless, the possibility of focusing light, even imperfectly, through arbitrarily thick scattering media has generated considerable interest, largely initiated by the seminal work of Vellekoop and Mosk [21] (see Fig. 20.9). Such focusing can even occur in the time domain for polychromatic light by way of spatiotemporal field coupling [13]. For a comprehensive review of such topics, the reader is referred to [17].

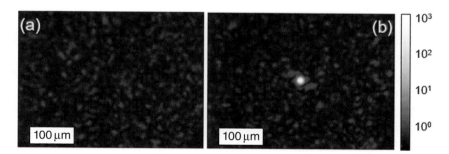

Figure 20.9. Intensity distribution of laser light after propagation through about 20 mean free paths, without (a) and with (b) adaptive optics. Intensity enhancement at focus is about 1000. Adapted from [21].

As emphasized in the introduction to this chapter, deep imaging in scattering media remains one of the greatest and most obstinate challenges in microscopy to date. The lure of this challenge remains unabated.

20.7 PROBLEMS

Problem 20.1

Consider Fig. 20.1 in the limit of small-angle scattering. A light ray enters a scattering medium from a perpendicular direction (as shown). After several scattering events, the probability distribution for the ray position is spread over a transverse area of width $w(L)$, where L is the penetration depth. Use a simple model where the transverse position of the ray undergoes a random walk as the ray propagates into the sample. That is, at each scattering event, the ray takes a step of size $l_s\theta$ in the transverse plane, with arbitrary direction. Provide rough scaling laws for $w(L)$. Specifically, how does $w(L)$ scale with L? With μ_s? With $(1-g)$?

Compare your result with the spread incurred by a Gaussian focus (Eq. 20.48).

Problem 20.2

In Section 9.2.2 we considered the Rytov solution for field propagation in an inhomogeneous medium, given by

$$E\left(\vec{r}\right) = e^{i\Psi\left(\vec{r}\right)}E_i\left(\vec{r}\right)$$

Decompose $\Psi\left(\vec{r}\right)$ into real and imaginary components, such that $\Psi\left(\vec{r}\right) = \varphi\left(\vec{r}\right)+i\chi\left(\vec{r}\right)$. Assume that both $\varphi\left(\vec{r}\right)$ and $\chi\left(\vec{r}\right)$ obey Gaussian statistics. That is, they obey the probability distributions given by

$$P_\varphi\left(\varphi\right) = \frac{1}{\sqrt{2\pi}\sigma_\varphi}e^{-\varphi^2/2\sigma_\varphi^2}$$

$$P_\chi\left(\chi\right) = \frac{1}{\sqrt{2\pi}\sigma_\chi}e^{-(\chi-\bar{\chi})^2/2\sigma_\chi^2}$$

where σ_φ and σ_χ are standard deviations (not to be confused with cross-sections), and a bias $\bar{\chi}$ is introduced to take into account a mean attenuation of the field.

Derive the corresponding probability distributions for the field amplitude $A = |E|$ and intensity $I = |E|^2$ (in terms of σ_φ, σ_χ and $\bar{\chi}$). Your results should be examples of what are known as log-normal probability distributions.

Hint: remember that probability distributions transform as $p_Y(Y) = p_X(X)|dX/dY|$.

Problem 20.3

We return here to the beam propagation method described in Section 9.1.3. Imagine that each phase screen imparts spatially random phases, such that the field just after the screen is given by $E_z'\left(\vec{\rho}\right) = e^{i\delta\psi\left(\vec{\rho}\right)}E_z\left(\vec{\rho}\right)$, where $E_z\left(\vec{\rho}\right)$ is the field just before the screen.

As illustrated in the figure, we can decompose $\delta\psi\left(\vec{\rho}\right)$ into real and imaginary components, such that $\delta\psi\left(\vec{\rho}\right) = \delta\varphi\left(\vec{\rho}\right) + i\left(\bar{\chi} + \delta\chi\left(\vec{\rho}\right)\right)$, where $\delta\varphi\left(\vec{\rho}\right)$ and $\delta\chi\left(\vec{\rho}\right)$ are (real) phase and amplitude variations, both centered on zero so that $\overline{\delta\varphi\left(\vec{\rho}\right)} = \overline{\delta\chi\left(\vec{\rho}\right)} = 0$, and a bias $\bar{\chi}$ is introduced to account for any mean amplitude reduction (note that $\delta\varphi\left(\vec{\rho}\right)$ and $\delta\chi\left(\vec{\rho}\right)$ are not the same as in Problem 20.2). Moreover, assume that the phase and amplitude fluctuations are uncorrelated, and their variances $\left\langle\delta\varphi\left(\vec{\rho}\right)^2\right\rangle$ and $\left\langle\delta\chi\left(\vec{\rho}\right)^2\right\rangle$ are independent of $\vec{\rho}$, meaning that the phase screen is statistically homogeneous.

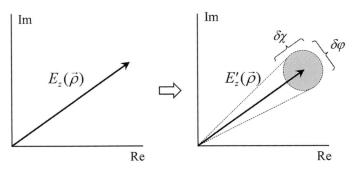

The effect of the phase screen on the mutual intensity can be written as

$$J_z'\left(\vec{\rho},\vec{\rho}'\right) = K_{\delta z}\left(\vec{\rho},\vec{\rho}'\right)J_z\left(\vec{\rho},\vec{\rho}'\right)$$

(a) Show that, with our usual coordinate transformation,

$$K_{\delta z}\left(\vec{\rho}_d\right) = \exp\left[-2\bar{\chi} + 2\left\langle\delta\chi^2\right\rangle - \left\langle\delta\chi^2\right\rangle\left(1 - \gamma_{\delta\chi}\left(\rho_d\right)\right) - \left\langle\delta\varphi^2\right\rangle\left(1 - \gamma_{\delta\varphi}\left(\rho_d\right)\right)\right]$$

where $\gamma_{\delta\chi}\left(\rho_d\right)$ and $\gamma_{\delta\varphi}\left(\rho_d\right)$ are normalized autocorrelation functions of $\delta\chi\left(\vec{\rho}\right)$ and $\delta\varphi\left(\vec{\rho}\right)$, respectively.

Hint: identities from Appendix B.5 are useful here.

(b) Assume that the phase and amplitude variations obey similar statistics, as suggested in the figure. That is, assume $\left\langle\delta\varphi\left(\vec{\rho}\right)^2\right\rangle = \left\langle\delta\chi\left(\vec{\rho}\right)^2\right\rangle \equiv \frac{1}{2}\left\langle\left|\delta\psi\left(\vec{\rho}\right)\right|^2\right\rangle$, and $\gamma_{\delta\chi}\left(\rho_d\right) = \gamma_{\delta\varphi}\left(\rho_d\right) \equiv \gamma_{\delta\psi}\left(\rho_d\right)$. Also, use the a priori knowledge that $K_{\delta z}\left(0\right) = \exp\left(-\mu_a\delta z\right)$, and $\left\langle\left|\delta\psi\left(\vec{\rho}\right)\right|^2\right\rangle = \mu_s\delta z$.

Show that $K_{\delta z}\left(\vec{\rho}_d\right)$ simplifies to

$$K_{\delta z}\left(\vec{\rho}_d\right) = \exp\left[-\mu_e\delta z + \mu_s\delta z\,\gamma_{\delta\psi}\left(\rho_d\right)\right]$$

This model for small propagation distances can be compared with the more exact Eq. 20.24 for larger propagation distances.

Problem 20.4

Problem 20.3 introduced the complex phase variations $\delta\psi\left(\vec{\rho}\right)$ imparted by a phase screen in the beam propagation method (Section 9.1.3). Here a link will be established between $\gamma_{\delta\psi}\left(\rho_d\right)$

and $\gamma_{\delta n}\left(r_d\right)$, where δn are index-of-refraction variations of the sample, assumed to be isotropic and spatially invariant, with normalized autocorrelation function given by Eq. 20.51.

(a) In accord with the beam propagation method, neglect diffraction effects when considering transmission through the phase screen. That is, write $\delta\psi\left(\vec{\rho}\right) = 2\pi\kappa \int_0^{\delta z} dz\,\delta n\left(\vec{r}\right)$

Show that

$$\gamma_{\delta\psi}\left(\rho_d\right) = \frac{\int_{-\infty}^{\infty} dz_d\,\gamma_{\delta n}\left(\rho_d, z_d\right)}{\int_{-\infty}^{\infty} dz_d\,\gamma_{\delta n}\left(0, z_d\right)}$$

Hint: make use of the following trick (similar to the trick used with Eq. 20.50)

$$\int_0^{\delta z} dz \int_0^{\delta z} dz' = \int_0^{\delta z} dz_c \int_{2|z_c-\delta z/2|-\delta z}^{\delta z-2|z_c-\delta z/2|} dz_d \approx \int_0^{\delta z} dz_c \int_{-\infty}^{\infty} dz_d$$

where the approximation makes the assumption that the characteristic size of the index-of-refraction inhomogeneities is much smaller than δz, allowing the integration limits for dz_d to be extended to infinity without significant error.

(b) Convince yourself that the first equality in the above trick is true.

Hint: make a sketch of the integration area spanned by dz_c and dz_d.

Problem 20.5

Consider a sample where the index-of-refraction variations obey Gaussian statistics given by

$$\gamma_{\delta n}(r_d) = e^{-r_d^2/l_n^2}$$

where l_n is the index-of-refraction correlation length.

(a) Show that, in the small-angle approximation (equivalent to $\kappa l_n \gg 1$), the corresponding phase function is given by

$$p\left(\theta\right) \approx \frac{1}{2\pi\left(1-g\right)} e^{-\theta^2/2(1-g)}$$

where

$$g \approx 1 - \frac{1}{2\pi^2\kappa^2 l_n^2}$$

(b) Verify that Eq. 20.58 is satisfied.

References

[1] Barabanenkov, Y. N. "On the spectral theory of radiation transport equations," *Sov. Phys. JETP* 29, 679–684 (1969).

[2] Booth, M. J. "Adaptive optical microscopy: the ongoing quest for a perfect image," *Light Sci. Appl.* 3, e165 (2014).

[3] Carslaw, H. S. and Jaeger, J. C. *Conduction of Heat in Solids*, 2nd edn, Oxford (1986).

[4] Dolin, L. S. "Light beam scattering in a turbid medium layer," *Izv. Vyssh. Uchebn. Zaved. SSSR, Radiofiz.* 7 471–478 (1964).

[5] Durduran, T., Choe, R., Baker, W. B. and Yodh, A. G. "Diffuse optics for tissue monitoring and tomography," *Rep. Prog. Phys.* 73, 076701 (2010).

[6] Fante, R. L. "Propagation of electromagnetic waves through turbulent plasma using transport theory," *IEEE Trans. Antennas Propag.* AP-21, 750–755 (1973).

[7] Fried, D. L. "Anisoplanatism in adaptive optics," *J. Opt. Soc. Am.* 72, 52–61 (1982).

[8] Goodman, J. W. *Statistical Optics*, 2nd edn, Wiley (2015).

[9] Henyey, L. C. and Greenstein, J. L. "Diffuse radiation in the galaxy," *Astrophys. J.* 93, 70–83 (1941).

[10] Ishimaru, A. *Wave Propagation and Scattering in Random Media*, IEEE and Oxford University Press (1997).

[11] Jakeman, E. and Ridley, K. D. *Modeling Fluctuations in Scattered Waves*, CRC Press (2006).

[12] Judkewitz, B., Horstmeyer, R., Vellekoop, I. M., Papadopoulos, I. N. and Yang, C. "Translation correlations in anisotropically scattering media," *Nat. Phys.* 11, 684–689 (2015).

[13] Katz, O., Small, E., Bromberg, Y. and Silberberg, Y. "Focusing and compression of ultrashort pulses through scattering media," *Nat. Phot.* 5, 372–377 (2011).

[14] Kokhanovsky, A. A. "Analytical solutions of multiple light scattering problems: a review," *Meas. Sci. Technol.* 13, 233–240 (2002).

[15] Kubby, J. A. Ed., *Adaptive Optics for Biological Imaging*, CRC Press (2013).

[16] Papoulis, A. "Ambiguity function in Fourier optics," *J. Opt. Soc. Am.* 64, 779–788 (1974).

[17] Rotter, S. and Gigan, S. "Light fields in complex media: Mesoscopic scattering meets wave control," *Rev. Mod. Phys.* 89, 015005 (2017).

[18] Rytov, S. M., Kravtsov, Y. A. and Tatarski, V. I. *Principles of Statistical Radiophysics 1–4*, Springer (2011).

[19] Tatarski, V. I. *Wave Propagation in a Turbulent Medium*, Dover (2016)

[20] Tyson, R. *Principles of Adaptive Optics*, 3rd edn, CRC Press (2010).

[21] Vellekoop, I. M. and Mosk, A. P. "Focusing coherent light through opaque strongly scattering media," *Opt. Lett.* 32, 2309–2311 (2007).

[22] Wang, L. V. and Wu, H.-i *Biomedical Optics: Principles and Imaging*, Wiley-Interscience (2007).

[23] Wang, L., Jacques, S. L. and Zheng, L. "MCML-Monte Carlo modeling of light transport in multi-layered tissues," *Comp. Meth. Prog. Biomed.* 47, 131–146 (1995).

APPENDIX A
Properties of Fourier Transforms

Spatial coordinates:

$$\vec{r} = (x, y, z) = (\vec{\rho}, z)$$

Spatial frequency coordinates (wavevectors):

$$\vec{\kappa} = (\kappa_x, \kappa_y, \kappa_z) = (\vec{\kappa}_\perp, \kappa_z)$$

Fourier transform convention $(F(\vec{\rho})$ is an arbitrary function):

$$\mathcal{F}(\vec{\kappa}_\perp) = \int d^2\vec{\rho}\, e^{-i2\pi\vec{\kappa}_\perp \cdot \vec{\rho}} F(\vec{\rho}) \quad \cdots\cdots \quad (d^2\vec{\rho} = dx\, dy) \tag{A.1}$$

$$F(\vec{\rho}) = \int d^2\vec{\kappa}_\perp\, e^{i2\pi\vec{\kappa}_\perp \cdot \vec{\rho}} \mathcal{F}(\vec{\kappa}_\perp) \quad \cdots\cdots \quad (d^2\vec{\kappa}_\perp = d\kappa_x\, d\kappa_y) \tag{A.2}$$

Corresponding 2D delta function convention:

- $\delta^2(\vec{\rho}) = \int d^2\vec{\kappa}_\perp\, e^{i2\pi\vec{\kappa}_\perp \cdot \vec{\rho}}$
- $\delta^2(\vec{\kappa}_\perp) = \int d^2\vec{\rho}\, e^{-i2\pi\vec{\rho} \cdot \vec{\kappa}_\perp}$
- $\int d^2\vec{\rho}\, \delta^2(\vec{\rho}) = \int d^2\vec{\kappa}_\perp\, \delta^2(\vec{\kappa}_\perp) = 1$

A.1 FOURIER TRANSFORM PROPERTIES

If...	then ...	
$F(x)$ real	$\mathcal{F}(-\kappa) = \mathcal{F}^*(\kappa)$	(Hermitian)
$F(x)$ imaginary	$\mathcal{F}(-\kappa) = -\mathcal{F}^*(\kappa)$	(anti-Hermitian)
$F(x)$ even	$\mathcal{F}(-\kappa) = \mathcal{F}(\kappa)$	(even)
$F(x)$ odd	$\mathcal{F}(-\kappa) = -\mathcal{F}(\kappa)$	(odd)
$F(x)$ real & even	$\mathcal{F}(\kappa)$ real & even	
$F(x)$ real & odd	$\mathcal{F}(\kappa)$ imag & odd	
$F(x)$ imag & even	$\mathcal{F}(\kappa)$ imag & even	
$F(x)$ imag & odd	$\mathcal{F}(\kappa)$ real & odd	

also

$$\text{FT}\{F^*(\vec{\rho})\} = \mathcal{F}^*(-\vec{\kappa})$$

A.2 FOURIER TRANSFORM THEOREMS

The theorems listed below are in one dimension. Generalizations to more than one dimension are straightforward.

- ## Shift theorem

$$\text{FT}\{F(x-a)\} = e^{-i2\pi\kappa_x a}\mathcal{F}(\kappa_x) \tag{A.3}$$

- ## Scaling theorem

$$\text{FT}\{F(ax)\} = \frac{1}{|a|}\mathcal{F}\left(\frac{1}{a}\kappa_x\right) \tag{A.4}$$

- ## Convolution theorem
Convolution is defined by:

$$[H*F]_x = \int_{-\infty}^{\infty} dx'\, H(x')F(x-x') = \int_{-\infty}^{\infty} dx'\, H(x-x')F(x') \tag{A.5}$$

Convolution theorem:

$$\text{FT}\{[H*F]_x\} = \mathcal{H}(\kappa_x)\mathcal{F}(\kappa_x) \tag{A.6}$$

- ## Correlation theorem
Cross-correlation is defined by:

$$[H\star F]_x = \int_{-\infty}^{\infty} dx'\, H^*(x')F(x+x') = \int_{-\infty}^{\infty} dx'\, H^*(x'-x)F(x') \tag{A.7}$$

Correlation theorem:

$$\text{FT}\{[H\star F]_x\} = \mathcal{H}^*(\kappa_x)\mathcal{F}(\kappa_x) \tag{A.8}$$

- ## Parseval theorem

$$\int_{-\infty}^{\infty} dx\, |F(x)|^2 = \int_{-\infty}^{\infty} d\kappa_x\, |\mathcal{F}(\kappa_x)|^2 \tag{A.9}$$

- ## Wiener–Khinchin theorem
For random functions of stationary mean and variance, the correlation theorem leads to the Wiener–Kninchin theorem:

$$\text{FT}\{[F\star F]_x\} = |\mathcal{F}(\kappa_x)|^2 \tag{A.10}$$

A.3 CYLINDRICAL COORDINATES

Cylindrical coordinates:

$$\vec{\rho} = (\rho, \varphi)$$
$$\vec{\kappa}_\perp = (\kappa_\perp, \phi)$$

Separability
In the case of function separability, $F(\vec{\rho}) \rightarrow F_\rho(\rho) F_\varphi(\varphi)$.

Correspondingly, the Fourier transform becomes

$$\text{FT}\{F(\vec{\rho})\} = \mathcal{F}(\kappa_\perp, \phi) = \sum_{n=-\infty}^{\infty} c_n (-i)^n e^{in\phi} \text{HT}_n\{F_\rho(\rho)\} \tag{A.11}$$

where

$$c_n = \frac{1}{2\pi} \int_0^{2\pi} d\varphi\, e^{-in\varphi} F_\varphi(\varphi) \tag{A.12}$$

and HT_n is the Hankel transform of order n, defined by

$$\text{HT}_n\{F_\rho(\rho)\} = 2\pi \int_0^{\infty} d\rho\, \rho J_n(2\pi\kappa_\perp\rho) F_\rho(\rho) \tag{A.13}$$

where J_n is the cylindrical Bessel function of order n.

Cylindrical Symmetry
In the case of cylindrical symmetry:

$$F(\vec{\rho}) \rightarrow F(\rho)$$
$$d^2\vec{\rho} \rightarrow 2\pi\rho\, d\rho$$
$$d^2\vec{\kappa}_\perp \rightarrow 2\pi\kappa_\perp\, d\kappa_\perp$$

Correspondingly, the Fourier transform becomes

$$\mathcal{F}(\kappa_\perp) = 2\pi \int_0^{\infty} d\rho\, \rho J_0(2\pi\kappa_\perp\rho) F(\rho) \tag{A.14}$$

$$F(\rho) = 2\pi \int_0^{\infty} d\kappa_\perp\, \kappa_\perp J_0(2\pi\kappa_\perp\rho) \mathcal{F}(\kappa_\perp) \tag{A.15}$$

where J_0 is the cylindrical Bessel function of order 0. These are called Fourier–Bessel transforms.

- ### jinc Function

$$\text{jinc}(2\pi\kappa_\perp\rho) = \frac{J_1(2\pi\kappa_\perp\rho)}{\pi\kappa_\perp\rho} \tag{A.16}$$

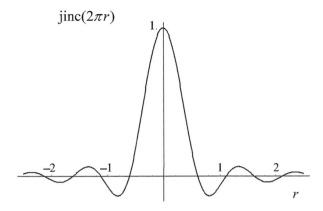

$$\text{jinc}(2\pi r)$$

where J_1 is the cylindrical Bessel function of order 1.

$$\int_0^{\Delta\rho} d^2\vec{\rho}\, J_0(2\pi\kappa_\perp\rho) = \frac{\Delta\rho}{\kappa_\perp}J_1(2\pi\kappa_\perp\Delta\rho) = \pi\Delta\rho^2\,\text{jinc}(2\pi\kappa_\perp\Delta\rho) \tag{A.17}$$

$$\int_0^{\Delta\kappa_\perp} d^2\vec{\kappa}_\perp\, J_0(2\pi\kappa_\perp\rho) = \frac{\Delta\kappa_\perp}{\rho}J_1(2\pi\Delta\kappa_\perp\rho) = \pi\Delta\kappa_\perp^2\,\text{jinc}(2\pi\Delta\kappa_\perp\rho) \tag{A.18}$$

Properties of the jinc Function
- $\text{jinc}(0) = \text{jinc}^2(0) = 1$
- $\int d^2\vec{\rho}\, \text{jinc}(2\pi\kappa_\perp\rho) = \int d^2\vec{\rho}\, \text{jinc}^2(2\pi\kappa_\perp\rho) = \frac{1}{\pi\kappa_\perp^2}$
- $\int d^2\vec{\kappa}_\perp \text{jinc}(2\pi\kappa_\perp\rho) = \int d^2\vec{\kappa}_\perp \text{jinc}^2(2\pi\kappa_\perp\rho) = \frac{1}{\pi\rho^2}$

A.4 SPHERICAL SYMMETRY

In the case of spherical symmetry:

$F(\vec{r}) \rightarrow F(r)$
$d^3\vec{r} \rightarrow 4\pi r^2\, dr$
$d^3\vec{q} \rightarrow 4\pi q^2\, dq$

Correspondingly, the Fourier transform becomes

$$\mathcal{F}(q) = \frac{2}{q}\int_0^\infty dr\, r\sin(2\pi qr)\, F(r) \tag{A.19}$$

$$F(r) = \frac{2}{r}\int_0^\infty dq\, q\sin(2\pi qr)\, \mathcal{F}(q) \tag{A.20}$$

APPENDIX B

Miscellaneous Math

B.1 GREEN THEOREM

For scalar fields $\psi(\vec{r})$ and $\phi(\vec{r})$:

$$\int_V d^3\vec{r} \left(\psi(\vec{r})\nabla^2\phi(\vec{r}) - \phi(\vec{r})\nabla^2\psi(\vec{r})\right) = \int_S d^2\vec{r} \left(\psi(\vec{r})\vec{\nabla}\phi(\vec{r}) - \phi(\vec{r})\vec{\nabla}\psi(\vec{r})\right) \cdot \hat{s}$$

(B.1)

where $\int_V$ and $\int_S$ are volume and surface integrals, respectively, and $\hat{s}$ is the outward normal to S.

B.2 A USEFUL INTEGRAL

$$\int d^2\vec{\rho}_a \, e^{i\alpha\rho_a^2 + i\beta\vec{\rho}_a \cdot \vec{\rho}_b} = \frac{i\pi}{\alpha} e^{-\frac{i\beta^2\rho_b^2}{4\alpha}}$$

(B.2)

This assumes $\text{Im}[\alpha] > 0$ or ρ_a bounded.

B.3 SYMMETRY PROPERTIES OF HERMITIAN AND ANTI-HERMITIAN FUNCTIONS

Let $F(x)$ be an arbitrary complex function. $F(x)$ may be decomposed into even and odd components

$$F_{\text{even}}(x) = \tfrac{1}{2}(F(x) + F(-x))$$
$$F_{\text{odd}}(x) = \tfrac{1}{2}(F(x) - F(-x))$$

such that $F(x) = F_{\text{even}}(x) + F_{\text{odd}}(x)$.

Or $F(x)$ may be decomposed into real and imaginary components

$$F_{\text{Re}}(x) = \text{Re}[F(x)]$$
$$F_{\text{Im}}(x) = \text{Im}[F(x)]$$

such that $F(x) = F_{\mathrm{Re}}(x) + iF_{\mathrm{Im}}(x)$.

- If $F(x)$ is Hermitian then $F(x) = F^*(-x)$ and

$$F_{\mathrm{even}}(x) = F_{\mathrm{Re}}(x)$$
$$F_{\mathrm{odd}}(x) = iF_{\mathrm{Im}}(x)$$

- If $F(x)$ is anti-Hermitian then $F(x) = -F^*(-x)$ and

$$F_{\mathrm{even}}(x) = iF_{\mathrm{Im}}(x)$$
$$F_{\mathrm{odd}}(x) = F_{\mathrm{Re}}(x)$$

B.4 PROPERTIES OF VARIANCES

The variance of a fluctuating variable X is defined to be

$$\sigma_X^2 = \left\langle (X - \langle X \rangle)^2 \right\rangle = \langle X^2 \rangle - \langle X \rangle^2 \tag{B.3}$$

where $\langle ... \rangle$ signifies a statistical average. If two variables X and Y fluctuate independently, then their sum and product also fluctuate and obey the variance rules:

- If $Z = X + Y$ then ...

$$\langle Z \rangle = \langle X \rangle + \langle Y \rangle \tag{B.4}$$

$$\sigma_Z^2 = \sigma_X^2 + \sigma_Y^2 \tag{B.5}$$

- If $Z = X \times Y$ then ...

$$\langle Z \rangle = \langle X \rangle \langle Y \rangle \tag{B.6}$$

$$\sigma_Z^2 = \langle Y \rangle^2 \sigma_X^2 + \langle X \rangle^2 \sigma_Y^2 + \sigma_X^2 \sigma_Y^2 \tag{B.7}$$

or, equivalently, if $\langle X \rangle \neq 0$ and $\langle Y \rangle \neq 0$,

$$\frac{\sigma_Z^2}{\langle Z \rangle^2} = \frac{\sigma_X^2}{\langle X \rangle^2} + \frac{\sigma_Y^2}{\langle Y \rangle^2} + \frac{\sigma_X^2}{\langle X \rangle^2} \frac{\sigma_Y^2}{\langle Y \rangle^2} \tag{B.8}$$

B.5 PROPERTY OF GAUSSIAN RANDOM VARIABLES

If α is a real-valued Gaussian random variable, then

$$\langle e^{-\alpha} \rangle = e^{-\langle \alpha \rangle + \frac{1}{2}\sigma_\alpha^2} \quad \text{and} \quad \langle e^{i\alpha} \rangle = e^{i\langle \alpha \rangle - \frac{1}{2}\sigma_\alpha^2} \tag{B.9}$$

where variance $\sigma_\alpha^2 = \langle \alpha^2 \rangle - \langle \alpha \rangle^2$.

APPENDIX C
Jones Matrix Description of Polarizers

In general, light is a vector field

$$\vec{E}(\vec{r}) = E_x(\vec{r})\hat{x} + E_y(\vec{r})\hat{y} + E_z(\vec{r})\hat{z} \tag{C.1}$$

The directions $\hat{x}$, $\hat{y}$, and $\hat{z}$ are called polarization directions, and the projections $E_x(\vec{r})$, $E_y(\vec{r})$, and $E_z(\vec{r})$ are called polarization components of the field. Because the polarization components are in orthogonal directions, they cannot interfere.

The transverse field components along a transverse plane are given by

$$\vec{E}(\vec{\rho}) = E_x(\vec{\rho})\hat{x} + E_y(\vec{\rho})\hat{y} \tag{C.2}$$

which may be written in column matrix form

$$\vec{E}(\vec{\rho}) = \begin{pmatrix} E_x(\vec{\rho}) \\ E_y(\vec{\rho}) \end{pmatrix} \tag{C.3}$$

Polarizers are optical systems that select different transverse polarization components. For example, if horizontal and vertical correspond to the directions $\hat{x}$ and $\hat{y}$ respectively, then the effect of a horizontal polarizer is to multiply the transverse field $\vec{E}(\vec{\rho})$ by the Jones matrix

$$\mathbf{M}(0°) = \begin{pmatrix} 1 & 0 \\ 0 & 0 \end{pmatrix} \tag{C.4}$$

thereby selecting only the horizontal polarization component of the field.

To evaluate the effect of an arbitrarily oriented polarizer, we introduce the rotation matrix

$$\mathbf{R}(\phi) = \begin{pmatrix} \cos\phi & -\sin\phi \\ \sin\phi & \cos\phi \end{pmatrix} \tag{C.5}$$

The Jones matrix for a polarizer oriented at an angle ϕ from vertical is then

$$\mathbf{M}(\phi) = \mathbf{R}(\phi) \begin{pmatrix} 1 & 0 \\ 0 & 0 \end{pmatrix} \mathbf{R}(-\phi) = \begin{pmatrix} \cos^2\phi & \cos\phi\sin\phi \\ \cos\phi\sin\phi & \sin^2\phi \end{pmatrix} \tag{C.6}$$

From this general result, we find for a vertical polarizer:

$$\mathbf{M}(90°) = \begin{pmatrix} 0 & 0 \\ 0 & 1 \end{pmatrix} \tag{C.7}$$

For a polarizer oriented at $\phi = 45°$:

$$\mathbf{M}(45°) = \frac{1}{2}\begin{pmatrix} 1 & 1 \\ 1 & 1 \end{pmatrix} \tag{C.8}$$

For a polarizer oriented at $\phi = -45°$:

$$\mathbf{M}(-45°) = \frac{1}{2}\begin{pmatrix} 1 & -1 \\ -1 & 1 \end{pmatrix} \tag{C.9}$$

INDEX